PLANT BIOMECHANICS

PLANT BIOMECHANICS

An Engineering Approach to Plant Form and Function

Karl J. Niklas

The University of Chicago Press *Chicago & London*

Karl J. Niklas is professor of botany at Cornell University.

Author and publisher wish to thank
the John Simon Guggenheim Memorial Foundation
for a generous grant in support of the
publication of this book.

The University of Chicago Press, Chicago 60637
The University of Chicago Press, Ltd., London

Printed in the United States of America

01 00 99 98 97 96 95 94 93 92   5 4 3 2 1

ISBN (cloth): 0–226–58630–8
ISBN (paper): 0–226–58641–6

Library of Congress Cataloging-in-Publication Data

Niklas, Karl J.
Plant biomechanics : an engineering approach to plant form and function / Karl J. Niklas.
p. cm.
Includes bibliographical references (p.) and index.
(pa : acid-free paper)
1. Plant mechanics. I. Title.
QK793.N55 1992
581.19′1—dc20 91-26312
CIP

⊗The paper used in this publication meets the minimum requirements of the American National Standard for Information Sciences—Permanence of Paper for Printed Library Materials, ANSI Z39.48–1984.

Contents

Preface

> The same thoughts sometimes put forth quite differently in the mind of another than in that of their author: unfruitful in their natural soil, abundant when transplanted.
>
> Blaise Pascal, *The Art of Persuasion*

The aim of this book is to explore how plants function, grow, reproduce, and evolve within the limits set by their physical environment. It was written in the firm belief that organisms cannot violate the laws of physics and chemistry and that knowing how these laws operate and confine the organic expression of size, form, and structure is essential to understanding biology. This perspective is shared by a number of disciplines—physiology and ecology to name just two—and traces its conceptual roots to the principal concerns of early comparative morphologists and anatomists. It differs only slightly from the bulk of biology by its emphasis on using the principles of physics and engineering to answer fundamental questions about the relation between form and function, but it clearly defines the intellectual scope of what has become known as *biomechanics*—a discipline that operates at the interface between engineering and biology.

The biological and engineering sciences have much in common. Both explore the relationships that exist between form and function. Both recognize that these relationships are contingent on local environmental conditions. Both are experimental sciences that strive for quantitative rigor by applying rich theoretical frameworks that must be constantly tested under field conditions. And both fully recognize that more often than not the phenomena they treat resist the tidy, elegant solutions so characteristic of the pure physical sciences. Despite their parallels, however, engineering and biology are not entirely compatible sciences. Typically the engineer does not deal with systems capable of altering form and substance in response to environmental changes. The engineer determines form-function relationships a priori and can find closed-form solutions to them because the geometry and the physical properties of the materials used in construction are carefully contrived and specified beforehand. By contrast, the biologist must infer function and must derive solutions for the behavior of an organism whose geometry and material properties often conform to no known fabricated structure or material and whose physical properties alter with age.

This book is not an attempt to reduce biology to a few simple equations, since those available from engineering are based on assumptions that organisms constantly violate—often with great elegance and subtlety. Rather, the equations presented here are a distillation of how mechanical and physical systems should *ideally* behave and how physical parameters *ought* to quantitatively interact—equations that reduce engineering concepts to the staccato music of mathematics but that are secondary to the harmonies and thematic variations that define the composition of the organic world.

Rather than a comprehensive review, this book must be viewed as a preliminary and admittedly idiosyncratic essay intended to illustrate representative principles rather than to delve deeply into them. The topics presented are designed to provoke readers to view plants as *structures* that grow, about which we still know comparatively little. Indeed, while I was writing this book, it often became obvious that data illustrating how some biomechanical principles relate directly to plants were unavailable in the literature. Experimentation therefore was required before some topics could be dealt with, and so the text incorporates as yet unpublished (and unconfirmed) data. For example, when I sought botanical examples of cantilevered beams other than branches, a vast, virtually unexplored research agenda (the mechanics of leaf petioles) was exposed. These and many other topics await study and detailed description.

Any book dealing with principles and organisms can be organized according to the former or the latter. Each approach has its merits, and each can irritate according to the bias of the reader. In the final analysis, neither organization is desirable, because organisms function according to principles, and principles illustrate how organisms function. But language is linear and requires that concepts be introduced according to some order, and a book like this invariably reflects the biases of its author. Accordingly, this book is organized in an ascending order of complexity tempered by a need to introduce some basic engineering and biological concepts where they first seem appropriate. The first chapter begins with a definition of plants—a dangerous enterprise in a field where even the definition of life incurs immediate debate and self-doubt—and goes on to discuss the most obvious philosophical biases that shape the fabric of this book. The next two chapters provide reference materials—to be scanned or read in detail—that I will draw on throughout the rest of the book. Chapter 2 treats the material properties and the mechanical behavior of plant substances, as well as giving a brief review of soils, and chapter 3 deals with the mechanical consequences of size and shape on mechanical performance. It is here that the behavior of tree trunks, branches, and petioles—columns, beams, and cantilevers—is discussed. Chapter 4 is devoted entirely to plant-water relations. In this chapter I discuss some fundamental concepts concerning the absorption, translocation, and conservation of water. Chapter 5 treats plant cell walls and tries to illustrate how some of the concepts introduced earlier relate to our understanding of cell wall architecture, chemistry, and that unique attribute of life we call growth. Chapter 6 treats the different plant tissues that characterize the stereotyped vascular sporophyte, and chapter 7 reviews the mechanical differences among the three types of plant organs. It is in this chapter that I take up the notion of design factors or safety margins. Chapter 8 is an overview of whole plant biomechanics and explores how organs mechanically interact with one another in the context of the individual plant. Chapter 9 discusses, all too briefly, the rich field of fluid mechanics and how it relates to plant reproduction. And finally, chapter 10 presents an overview of the major events in the evolution of land plants as seen from the idiosyncratic perspective of mechanical design, innovation, and failure. In each of these chapters I have tried to cull from the literature as many examples of plant biomechanics as are available. The literature reviewed dates back to the seminal works of Simon Schwendener (who was trained as both an engineer and a botanist) and others of similar vintage who studied plants with a quantitative inclination well before the time of D'Arcy Thompson—*nihil sub sole novum.*

Interdisciplinary is an often pompous and all too often used word whose precise definition eludes my grasp. But if such a condition truly exists, then one of its attributes is the juxtaposition of concepts that necessarily results in blending differing jargons and technical vocabularies. In terms of its collective vocabulary, no one can deny that biomechanics is interdisciplinary. Although I have tried to make the meaning of technical words clear throughout the text, a glossary of botanical and engineering terms is provided near the end of this book.

Every book is in part the product of the enthusiasm and knowledge of the author's friends. This one is no exception. Thomas D. O'Rourke (Department of Civil and Environmental Engineering, Cornell University) taught me that engineers are scientists whose theories must work. He also introduced me to the brilliant works of Stephen P. Timoshenko and guided me through a variety of experiments whose mechanical designs became clearer with time. Francis C. Moon (Department of Mechanical and Theoretical Engineering, Cornell University) introduced me to the theory of dynamic bending and taught me that one should remain skeptical of both theoretical and empirical results. Dominick J. Paolillo, Jr. (Section of Plant Biology, Cornell University), and Donald R. Kaplan (Department of Botany, University of California, Berkeley) repeatedly demonstrated the subtlety of plant growth and development and revealed the elegance of comparative morphology and anatomy. Tom L. Phillips (Department of Botany, University of Illinois, Urbana) introduced me to plant evolution and the excitement of studying fossils, while Lawrence J. Crockett (Department of Biology, City College of New York) was the first biologist who, by his brilliant lectures, turned me aside from the study of mathematics and showed me the beauty of plants. The books and papers of Stephen Wainwright, Steven Vogel (Department of Zoology, Duke University), and Julian Vincent (University of Reading, England) have inspired a whole generation of biologists to study biomechanics. Anyone interested in the subject should consult their books, which are rich with insight. My profound gratitude is extended to Thomas M. Dunn, who provided many photographs and much technical assistance, and to Edward D. Cobb, who helped beyond measure in ways too many to count.

I am particularly indebted to five individuals—two of them my editors and three "anonymous reviewers" of early draft manuscripts who have graciously revealed their identities: Susan E. Abrams (executive editor for natural sciences, University of Chicago Press) guided me through the mechanics of writing and producing this book and provided emotional support when all did not go well; Alice M. Bennett (senior manuscript editor, University of Chicago Press) patiently read and corrected the final manuscript; Lorna Gibson (Department of Civil Engineering, Massachusetts Institute of Technology) wrote a magnificently detailed review of my manuscript despite a broken hand and much pain; Bruce H. Tiffney (Department of Geological Sciences, University of California, Santa Barbara), whose friendship and subtle mind I have valued for years, provided many insights and suggestions; and finally, Steven Vogel (Department of Zoology, Duke University), whose leadership in the field of biofluid mechanics is well known and deserved, offered many helpful suggestions on improving the text. Despite the efforts of all these helpful and thoughtful people, mistakes likely remain, and responsibility for them rests entirely with me.

To these colleagues and friends who offered leadership, inspiration, hard work, and enthusiasm, I extend my sincerest thanks and acknowledgments and dedicate this book.

Finally, I take this opportunity to thank the National Science Foundation and the College of Agriculture and Life Sciences, Cornell University, as well as the John Simon Guggenheim Foundation. This book would not have been possible without the generous financial support these institutions provided my research.

One

Some Biological and Philosophical Preliminaries

When you have eliminated the impossible, whatever remains, *however improbable,* must be the truth.

Sir Arthur Conan Doyle, *The Sign of Four*

This book is about how physical laws and processes influence the growth, survival, and reproduction of plants and about how these laws and processes have routinely shaped the course of plant evolution through natural selection. Its fundamental premise is that plants, like all other types of organisms, cannot violate physical principles. This is not to say that our understanding of biology can currently be reduced to the precepts of the pure physical sciences. Rather, I wrote this book with the notion that understanding the physical sciences is requisite to understanding biology.

Much of what we currently know about how physical laws and processes influence organisms is treated by a discipline called biomechanics, so named because it exists at the interface between biology and engineering. To a certain extent its name is misleading, since bio*mechanics* implies that organisms are treated solely in the context of solid or fluid mechanics. As used here, however, *engineering* relates to the application of any physical principle, law, or process to quantify the performance of a structure or process in terms of its function. The theoretical and practical insights gained from chemical or electrical or mechanical engineering can be used to understand physiological processes of a chemical or electrical nature, just as mechanical engineering can be used to understand how organic structures operate and perform their functions. Accordingly, it is proper to say that the discipline of biomechanics includes the fields of biophysics and biochemistry as well as those associated with the traditional engineering sciences, such as solid and fluid mechanics. But the name of a discipline is never important compared with its intellectual content and pursuits. It matters little whether we call ourselves biomechanicists or biophysicists or biochemists; it matters even less whether we say we are botanists or plant biologists or simply biologists. What truly matters is that we gain an appreciation of an engineering approach to biology and that we appreciate the tremendous importance of plants.

This book was written with many philosophical biases, the most important being the conviction that an evolutionary perspective on plant biomechanics is essential. This seems the most reasonable way to approach any aspect of biology, though some may argue otherwise. The species that now compose the world's biotas are only a fraction of the species that have existed since the dawn of life billions of years ago. Neglecting this vast number of fossil organisms is unpardonable, particularly since fossils can reveal how and possibly why the organisms of today look and function as they do. Indeed, if we dwell simply on the present, then all we can say is that organisms are constructed and function in a manner that permits their survival. Logically, for them to do otherwise seems an apparent contradiction in terms. True, the way form and function relate to one another and how much they differ among differing habitats are not trivial matters. But much of our reasoning is potentially circular until we draw on the historical record of organisms, which documents the changes in form and structure that have occurred as the physical environment has changed. That differences in properties exist now and have existed in the past is really the fundamental concept, rather than the differences among the properties themselves. This fundamental concept is called evolution.

A Few Words about Evolution and the Importance of Plants

Evolutionary changes can be described in a number of ways; one is to consider each evolutionary change as a vector having *magnitude* and *direction*. The magnitude component—the degree to which the antecedent and descendant conditions differ from one another—reflects the genetic and developmental capacity of a particular type of organism to vary about the modal form within its population or species and the way these variants perform their biological functions and so have their survival differentially affected by the environment. The directional component of each vector may be either positive or negative—that is, like or contrasting to the preceding evolutionary change. It is a function of the nature, intensity, and duration of application of factors within the environment that influence the survival of the variants produced by a particular type of organism. The relations among vectors may take many forms. Regardless of the particulars, however, when the many evolutionary changes documented in the fossil record are referenced against the geological time line, their collective resultant may reveal a general overall pattern. Whether this is so depends on the capacity of many organisms to respond in like manner to their environment and on whether the environment routinely operates in the same way over relatively long periods.

If the fundamental premise of this book is true, then many of the evolutionary patterns observed in the fossil record are the result of persistent pressures brought to bear by physical laws and processes on the many potential expressions of organic form, structure, and reproduction thrown forth by genetic alteration. By virtue of their shifting genetic compositions, organisms continually propose phenotypic variants to the environment, which in turn disposes according to their various functional performance levels. Much of the way some features of the physical environment influence the survival of phenotypes can be predicted. The attenuation of light through the atmosphere, the rates at which carbon dioxide and water vapor diffuse through a small hole in the surface of a leaf, the mechanical forces generated within a stem by the wind, and the way airborne spores move in air currents are governed by physical laws and processes whose operations can be described with the aid of very precise and usually accurate equations. By the same token, the effects that the synthesis of different kinds of light-harvesting pigments and the hydraulic regulation of stomatal diameters have on metabolism, or the consequences of the location of tensile materials in stems for flexure and mechanical support, or the influence of the density and size of airborne spores on long-distance dispersal and pollination directly influence the fitness of an organism and are all quantifiable by means of these equations. Thus, in contrast to our relatively poor state of knowledge concerning the intrinsic capacity of organisms to genetically produce phenotypic variation, we know comparatively much more about how the interactions between phenotypes and their immediate environments influence fitness. When used in conjunction with the fossil record, this knowledge permits us the luxury of interpreting evolutionary patterns as adaptive or nonadaptive in nature. Yet we must always be careful, since evolutionary adaptive patterns may be artifacts. Such a view, orthogonal to the one taken here, would not deny the apparent reality of evolutionary patterns, but it would assert that they are not real in the sense that they are derived from the process of natural selection. Much as canals may be seen on the surface of Mars when the planet is viewed through a poor telescope with an imaginative eye, our interpretation of paleontological data, as well as our understanding of living organisms, is subject to our ability to resolve details and to discern relationships among them. The human mind tends to connect the dots, and all manner of observations are subject to distortion. Along these lines, the adaptive patterns in the fossil record may be statistical artifacts. Indeed, when referenced against a time line, a purely random process can manifest a general pattern. Thus, although the patterns seen in the fossil record have an undeniable existence, they may be the products of evolutionary random walks. If so, then they are canals on a statistically random landscape viewed through the distorted optics of natural selection.

One might argue that we have invented the word *random* to give proximate causality to unknown phenomena. However, such an argument advances our understanding little until we can quantify the way these phenomena operate. Given our present understanding of its complexity, we are not likely to precisely resolve the fundamental nature of evolution in my lifetime. Under any conditions, one must admit that a nonadaptationist global view of evolution is an acceptable philosophical alternative to the one taken here. Certainly we can construct a nonadaptationist scenario that accounts for all the observable facts. Driven by gradients of occupied and unoccupied habitats, evolutionary patterns would result from the passive diffusion of phenotypes throughout a vast domain of equiprobable morphological, anatomical, and reproductive expressions. With time and no perturbation in the environment, such an isotropic universe of potential organic expression would gradually become saturated. Conversely, the equilibrium condition could be indefinitely delayed by episodic or random extinction events resulting from abiotic factors. Regardless of how much we permit the environment to change, organisms would evolve over time and with apparent pattern, if for no other reason than that the first forms of life manifestly could not instantly occupy the full domain of all possible organic expression. Accordingly, neither the adaptationist nor the nonadaptationist views of evolution can be refuted based on the fact that organisms change and that patterns occur over time. To argue otherwise would be to flout observation.

However, if for the moment we at least accept the possibility that evolutionary history evinces both adaptive and nonadaptive trends, then neither the adaptationist nor the nonadaptationist view can be claimed to be right or wrong in the canonical sense. Rather, the issue becomes whether one view or the other holds in most cases: "Natural history is a domain of relative frequency, rarely of exclusivity" (Gould 1990, ix). In this sense the views of the adaptationist and the nonadaptationist establish two axes that define a surface over which the actual events of organic evolution must be plotted and analyzed. Actually, if we accept that organisms cannot directly violate the operation of physical laws and principles, then this planar surface is really a multidimensional space whose geometry can be described by the way physical laws and processes influence biological phenomena.

Biomechanics provides a superb tool whereby the outer boundaries—the impossibilities—of the universe of possible biological expressions may be mapped. Additionally, biomechanics can be used to describe the inner workings of this universe of permissible organic expression. Its power lies in its methods, which are not limited to any particular level of biological organization. The ability of the biological materials we call species to change ultimately may be distilled exclusively from the operational laws of molecular genetics and development, although we are far from this level of understanding at present. Even when completely understood and quantified, however, these laws are necessarily incomplete because they cannot predict the evolutionary fate of a population, nor can they identify the components of the external environment responsible for selecting the phenotypic variants within a population. The comprehensive evaluation of evolutionary events and patterns, whether they are real or artifacts from the point of view of natural selection, will include but must also extend beyond the molecular level of biological organization and the developmental implementation of genetic instructions. Ultimately our agenda must focus on the functional attributes of the phenotype—an entity whose properties and behavior dictate the future of the genetic materials that gave rise to it and that can never be divorced from the context of its physical environment.

If the biology of living organisms and their evolutionary histories can be dissected with the aid of biomechanical analyses—as I think they can—then we are still left with the task of identifying the organisms best suited to this research agenda. The history of science reveals that problems can result when important questions are attacked by studying organisms whose biological properties prove either recalcitrant to the techniques used or irrelevant to the questions asked. Even if we argue that history teaches us nothing, it is generally appreciated that some organisms, such as bacteria, are more or less irrelevant to some evolutionary events, such as the acquisition of flight. To parody Gilbert and Sullivan, "Let the organism fit the question." An additional consideration is that the type of organism chosen should shed as much light as possible on the panorama of evolution rather than illuminate some small, possibly trivial aspect yielding information that cannot be generally applied.

In this regard it is apropos to mention that plants constitute over 90% of the world's present (and past) biomass. Thus, simply in terms of their bulk, whatever we learn about plants has the potential to tip the balance in any debate concerning the relative frequency of occurrence of a biological phenomenon. Fortunately also, their biology makes plants well suited to biomechanical and paleontological study. From the archaic algae to the most derived multicellular terrestrial plants, from the spectral properties of light-harvesting pigments in chloroplasts to the stacking of leaves in the tree canopy, we are persistently drawn to the conclusion that the behavior of plants is in large part responsive to and intimately connected with the way the physical environment operates and is constructed. Additionally, plants tend to be exquisitely preserved in the fossil record, giving us access to the past. Many of the data necessary to study living plants are preserved by virtue of how plants grow and the materials of which they are fabricated. By scanning the cellular details along the length of a single fossil stem, we can reconstruct much of the ontogeny and development of its species. From their shape and size, we can determine the potential for long-distance dispersal and reproductive capacities of fossil spores, seeds, and fruits and so infer much about the reproductive biology of species. And finally, throughout their billion-year history, we can infer a metabolic dependency of plants' vegetative growth and survival on the availability of light, water, minerals, and space. I hope the paleontologist studying heterotrophs will excuse my ignorance, but I believe few of these benefits can be as strongly asserted for animals, whose ontogenies and mode of development can rarely be determined from a single specimen and whose manner of garnering metabolites may vary during an animal's ontogeny or differ radically among animal groups.

True, plants, like animals, have changed over millions of years. And as in animals, the operation of physical laws and processes is important to them. Neither plant or animal evolution has resulted in evolutionary stagnation or in the appearance of some ultimate optimal organism whose properties competitively exclude all other expressions of morphology, anatomy, and reproduction. Nonetheless, nowhere else in biology than in plants do we find such convincing evidence that physical laws and processes link form with function and thus have confined the scope of organic expression within boundaries that have never been breached. Our task as botanists is to identify these limitations and, at a finer level of analysis, demonstrate how physical principles have demonstrably influenced the morphological, anatomical, and reproductive directions taken by individual plant lineages during the course of their evolution.

In this chapter my objectives are much more circumspect, though equally important. We will begin by treating some fundamental issues, and definitions will play a critical role. Indeed, the very first task is to seek a preliminary definition for the organisms that will occupy our attention in subsequent chapters. This definition will be only a first-order approximation, however—indeed, in a very real sense this entire book is merely a crude attempt to define the word plant!

WHAT IS A PLANT?

Current estimates place the number of extant plant species at over 350,000, yet this is only a small fraction of all the species that have existed in the past. Collectively, the morphology and anatomy of extinct and extant species reveal a staggering diversity in form and structure, ranging from the unicellular aquatic algae to the multicellular and vascular terrestrial plants. Accordingly, any attempt to define plants must be viewed as somewhat presumptuous, particularly when (by definition) each species must either represent a unique combination of character states or possess at least one character not shared by any other species. Fortunately, however, whether they existed in a Cambrian ocean or live in our vegetable garden, all plants share features that set them apart from all other types of past and present organisms. These features constitute the foundation of an inclusive definition of *plant*. One of the most obvious and important of these features is metabolism. All organisms may be classified as either autotrophs (those capable of synthesizing their organic requirements from inorganic precursors) or heterotrophs (those that synthesize their organic requirements from organic precursors). A further distinction may be drawn, however, since not all autotrophs rely on the same energy source. Some use chemical energy (chemotrophs), while others use sunlight as their energy source. Organisms that fall into the second category, to which plants belong, are called photoautotrophs.

For simple carbohydrates, such as glucose or fructose, as well as for relatively more complex carbohydrate polymers, such as starch and cellulose, the rudiments of the process called photosynthesis can be expressed by a highly simplified formula:

$$CO_2 + H_2O \underset{\text{light}}{\overset{\text{chlorophyll}}{\rightarrow}} O_2 + (CH_2O)n \tag{1.1}$$

Equation (1.1) states that, in the presence of light and chlorophyll, the molecular structures of carbon dioxide and water are biochemically converted into those of oxygen and carbohydrates. The monomeric unit of a carbohydrate may be represented by CH_2O, where *n*-number may be linked to produce a simple or polymeric carbohydrate molecule. The metabolic process symbolized by eq. (1.1) provides the beginning of a definition for plants: except for parasitic species, which are believed to be derived from photosynthetically competent antecedents, all plants are photoautotrophs.

Aside from describing their metabolic properties, eq. (1.1) can be used to reveal the tremendous importance of phototrophic organisms. We see this when we recognize that they provide virtually all the organic carbon that heterotrophs rely on. The remaining small fraction is supplied by chemotrophic organisms. The magnitude of the organic carbon supplied by photoautotrophs can be crudely estimated because the transfer of 12 g (one gram-atom) of carbon in the form of carbon dioxide into organic matter requires 112 Kcal. Thus, roughly 9.3×10^6 Kcal is required to produce one ton of organic carbon. True, the composition and energy content of different forms of organic matter vary. But the average composition of organic matter is close to that of a carbohydrate, since well over 80% of all the organic matter on earth is in this form. We can estimate the total amount of organic carbon fixed per year if we can gauge the amount of light energy available for photosynthesis. Such an estimate of available light is clearly imprecise, because the attenuation of light energy as it passes through the earth's atmosphere is great, yet it is informative to make the attempt. To begin with, the annual solar energy flux at the outer boundary of the earth's atmosphere is roughly 1.25×10^{21} Kcal$\cdot$yr^{-1} (2 cal of solar energy$\cdot$cm$^{-2}\cdot$minute^{-1}). Only about 40% of this total reaches the earth's surface ($\approx 0.5 \times 10^{21}$ Kcal$\cdot$yr^{-1}), of which approximately 50% is in the form of photosynthetically available light. Only about 60% of this light reaching terrestrial photoautotrophs is absorbed ($\approx 0.15 \times 10^{21}$ Kcal$\cdot$yr^{-1}), while only 1% ($\approx 1.5 \times 10^{18}$ Kcal$\cdot$yr^{-1}) of the absorbed energy is photosynthetically converted into carbon. The rest is dissipated as heat or reflected back into the atmosphere. From these estimates, we can appreciate that only about 0.0012% of the annual solar energy flux at the outer boundary of the atmosphere is likely to be converted into organic carbon by photoautotrophs.

To provide a conservative estimate of the net productivity of terrestrial photoautotrophs, we will assume that only 20% of the earth's surface is fertile land area. This percentage sounds low, but remember that much of the earth's surface is covered by desert, polar ice caps, and urban sprawl. Given the annual available energy for the synthesis of carbon and the productive surface area of land (1.5×10^{18} Kcal$\cdot$yr$^{-1} \times 20\% = 3.0 \times 10^{17}$ Kcal$\cdot$yr^{-1}), our estimated gross yield would be on the order of about 32 billion tons of organic carbon per year, which translates into an annual processing of about 83 billion tons of oxygen and about 1.3 trillion tons of carbon dioxide. About 15% of the annual available energy is lost through respiration, however. Thus the estimated net productivity of terrestrial photoautotrophs is lower, on the order of 27 billion tons of organic carbon per year, which pales in comparison with the estimated 122 to 135 billion tons of organic carbon produced annually by oceanic photoautotrophs. As heterotrophs, our direct annual consumption of all the organic carbon produced by photoautotrophs is roughly 2% to 3%. This consumption of organic carbon may appear very low, but not all of the organic carbon produced by photoautotrophs is available in a digestible form—roughly 70% is cellulose and lignin, although we use some of this carbon as paper and wood products.

Returning to the definition of plants, we must recognize that photosynthesis is not a sufficient criterion for distinguishing between plants and animals. All organisms may be further categorized as either prokaryotes or eukaryotes. The former lack organelles (e.g., the bacteria and cyanobacteria formerly referred to as blue-green algae), whereas the latter possess organelles. The photosynthetic cyanobacteria share with plants the capacity for photosynthesis. To distinguish between cyanobacteria and plants, a cytological criterion must be juxtaposed with a metabolic one—that is, *plants are eukaryotic photoautotrophs*. According to this definition, all plants share more or less the same metabolic machinery for photosynthesis and the infrastructure of the chloroplast in which this machinery is stored, as well as the complement of other organelles found in eukaryotic heterotrophs. A tenet of this book is that once the metabolism and cytology of plants evolved more than a billion years ago, they set certain limits on future evolutionary expressions of plant morphology, anatomy, and reproduction. These limits have been evolutionarily explored by plants somewhat differently in the aquatic and terrestrial environments, but the expression of plant form has nonetheless always been defined by the laws of physics and chemistry whose nature and operation will occupy our attention throughout this book.

Aquatic versus Terrestrial Plants: A Simple Comparison

One of the major lessons to be learned from biomechanics is that though the same physical laws affect all organisms, the consequences of these laws differ depending on habitat. This point is well known to the ecologist and is most quickly brought home by some crude comparisons between aquatic and terrestrial plants. This broad dichotomy among otherwise very diverse habitats is justified here because the physical properties of the fluids that envelop an

TABLE 1.1 Some Physical Properties of Pure Air and Water at 20° C

Physical Property	Air	Water
D_{CO_2}	$1.47 \times 10^{-5}\ m^2 \cdot s^{-1}$	$1.80 \times 10^{-9}\ m^2 \cdot s^{-1}$
D_{H_2O}	$2.42 \times 10^{-5}\ m^2 \cdot s^{-1}$	$2.40 \times 10^{-9}\ m^2 \cdot s^{-1}$
$[CO_2]$	$0.0125\ mol \cdot m^{-3}$	$0.0117\ mol \cdot m^{-3}$
ρ	$1.205\ kg \cdot m^{-3}$	$998.2\ kg \cdot m^{-3}$
ν	$1.50 \times 10^{-5}\ m^2 \cdot s^{-1}$	$10.04 \times 10^{-5}\ m^2 \cdot s^{-1}$

organism profoundly influence every biological function (Koehl and Wainwright 1977; Vogel 1981; Koehl 1986) and because the physical properties of air and water differ dramatically (table 1.1). For example, all plants rely on mass transport to exchange oxygen and carbon dioxide with the fluids (water or air) they are submerged in; all invariably experience an exchange of energy or of momentum when the fluids surrounding them move; and all plants must sustain the force of gravity acting on their biomass in addition to the dynamic forces resulting from the movements of external fluids. Nonetheless, plants of the same size, shape, and orientation to identical ambient flow conditions will experience significantly different rates of mass transport and magnitudes of externally applied forces depending on whether they are submerged in air or in water. These differences are illustrated here by comparing two hypothetical plants (one aquatic, the other terrestrial) with the same shape and dimensions (a cylinder 0.02 m in length and 0.002 m in diameter), submitted to an ambient fluid rate of 1 $m \cdot s^{-1}$.

Considering first the matter of mass transport, the steady-state exchange of most forms of matter and energy can be expressed in terms of Fick's first law, which states that the amount of molecular species i crossing a specified area per unit time (called the flux density, symbolized by J_i) is the product of a constant (called the diffusion coefficient, symbolized by D_i) and a driving force (in this case, the negative concentration gradient along the direction x, symbolized by $-\partial C_i/\partial_x$). This law can take the following mathematical form:

$$J_i = -D_i \frac{\partial C_i}{\partial x}. \tag{1.2}$$

The negative sign in eq. (1.2) is needed because the flux density is proportional to the *negative* of the concentration gradient—net diffusion is in the direction of lower concentration. From eqs. (1.1) and (1.2), we see that carbon dioxide and water must diffuse into the plant body if carbohydrates are to be photosynthesized and the plant is to survive. Since the concentration gradient of carbon dioxide or oxygen can be expressed by taking the difference between the external (ambient) and the internal concentration of either gas, symbolized by ΔC, divided by the intervening distance, here symbolized by δ, Fick's first law can be rewritten as

$$J = D\frac{\Delta C}{\delta}. \tag{1.3}$$

To use eq. (1.3), we must know the magnitudes of δ, ΔC, and D. Fortunately, these magnitudes are easily estimated. For example, the magnitude of δ revolves around a concept (and physical reality) called the boundary layer thickness. The boundary layer is a blanket of relatively unmoving fluid (water or air) surrounding the surfaces of any object obstructing the passage of a fluid (see chap. 9). In a terrestrial habitat, the blanket of air reduces the rate at which water vapor is lost from the plant body just as the boundary layer reduces the rate at which heat is lost from our bodies. Likewise, because the fluid within the boundary layer moves little with respect to the fluid in the ambient flow, gases dissolved within the volume occupied by the boundary layer are exchanged slowly with the ambient fluid and may be depleted quickly as they diffuse into the plant body. Conversely, substances leaving the plant body can become concentrated within the volume of the boundary layer. In either case, the boundary layer thickness provides a crude measure of the resistance to exchanging a substance whose concentrations differ within and outside the plant. So the substitution of $\Delta C/\delta$ in eq. (1.3) for $\partial C_i/\partial x$ in eq. (1.2) is a reasonable approximation for our present purposes.

Intuitively, we realize that the boundary layer thickness must be influenced by the ambient flow speed and the resistance of the fluid to motion (viscosity), as well as by the size, shape, and orientation of the organism. We dissipate heat more rapidly when our bodies are exposed to high wind speeds; terrestrial plants tend to lose more water vapor as the speed of moving air increases. Since most of the parameters (ambient flow speed, size, shape, and orientation) are identical for the two hypothetical plants, the difference in the viscosities of water and air at any given temperature must play a pivotal role in determining the thickness of the boundary layer. Indeed, calculations based on a simple formula (see chap. 9) reveal that the boundary layer thicknesses for the aquatic and the terrestrial plants are on the order of 10^{-5} m and 10^{-3} m, respectively. Likewise, the values of the diffusion coefficients for carbon dioxide in air and water and the concentrations of carbon dioxide in air and water are easily found; for example, the diffusion coefficients for CO_2 are 1.47

$\times 10^{-5}$ $m^2 \cdot s^{-1}$ in air and 1.8×10^{-9} $m^2 \cdot s^{-1}$ in pure water, while the diffusion coefficients for water vapor in air and in pure water can be taken as 2.42 $\times 10^{-5}$ $m^2 \cdot s^{-1}$ and 2.40×10^{-9} $m^2 \cdot s^{-1}$ (see table 1.1). All that remains for us to specify is the difference between the concentrations of carbon dioxide within and outside the plant body. This is easy to do. For steady-state photosynthesis, a reasonable estimate of the amount of CO_2 within the plant body is 7×10^{-3} $mol \cdot m^{-3}$, whether the plant is aquatic or terrestrial. Since the concentration of carbon dioxide in air and in pure water are 0.0125 $mol \cdot m^{-3}$ and 0.0117 $mol \cdot m^{-3}$, respectively, ΔC for CO_2 in water is 0.0055 $mol \cdot m^{-3}$, while that of CO_2 in air is 0.0047 $mol \cdot m^{-3}$. Inserting the values for ΔC, D, and δ into eq. (1.3) reveals that the flux densities of carbon dioxide in pure water and in air are 8.46×10^{-7} $mol \cdot m^{-2} \cdot s^{-1}$ and 8.09×10^{-5} $mol \cdot m^{-2} \cdot s^{-1}$. From this we learn that the terrestrial plant has a distinct advantage over its aquatic counterpart in acquiring carbon dioxide, which has a much larger (two orders of magnitude) diffusion coefficient in air. (Notice the units of flux density—the amount of substance [moles, symbolized by mol] passing through a unit surface area [m^2] per unit time[s].)

Obviously, the aquatic plant is at a great advantage in terms of the rate of water loss compared with its land-dwelling counterpart, and among intertidal plants many species have evolved unique ways to conserve water upon exposure (see Vogel and Loudon 1985). For terrestrial plants the situation is much more extreme. The flux density of water (in the form of water vapor), called transpiration, is 1.18×10^{-2} $mol \cdot m^{-2} \cdot s^{-1}$ ($D_{H_2O} = 2.42 \times 10^{-5}$ $m^2 \cdot s^{-1}$ in air; $D_{H_2O} = 2.40 \times 10^{-9}$ $m^2 \cdot s^{-1}$ in pure water). From eq. (1.3) we see that the ratio of water loss to carbon dioxide gain for the land plant is 0.0118 mol $H_2O \cdot m^{-2} \cdot s^{-1}$ to 8.5×10^{-5} mol $CO_2 \cdot m^{-2} \cdot s^{-1}$; that is, 138.8 moles of water are lost per mole of CO_2 that has been photosynthetically fixed in the form of carbohydrates. Considering that most land-dwelling plants are leafy and hence have significantly greater surface area than our hypothetical cylindrical plant, the amount of water lost from real terrestrial plants during a day or an entire year should be much more impressive and thus limiting to growth. Indeed, a typical grass plant loses its own weight in water every twenty-four hours in hot, dry weather, and an acre of oats can lose hundreds of tons of water in a single growing season.

By considering the mass transport of carbon dioxide or water, we see that the physical environment has a profound influence, even though photosynthesis for both terrestrial and aquatic plants is governed by the same processes and physical laws (eqs. 1.1 to 1.3). This lesson is reinforced by considering the effects of the movement of a fluid on a plant body which can be expressed in terms of the force, symbolized by F, exerted on either of the two cylindrical plants by a fluid with density ρ_f and ambient speed U:

$$F = \rho_f S U^2, \quad (1.4)$$

where S is the area of the plant body facing the advancing fluid. For a cylindrical plant aligned perpendicular to the direction of flow, S is one-half the lateral surface area of the plant body, which, given a length of 0.02 m and a diameter of 0.002 m, equals $(1/2)(1.257 \times 10^{-4}\ \text{m}^2)$, or $0.628 \times 10^{-4}\ \text{m}^2$. At 20°C, the densities of air and pure water are $1.205\ \text{kg}\cdot\text{m}^{-3}$ and $998.2\ \text{kg}\cdot\text{m}^{-3}$, respectively (see table 1.1). Thus, from eq. (1.4), the force exerted by moving air on the terrestrial plant is found to be on the order of $7.58 \times 10^{-5}\ \text{kg}\cdot\text{m}\cdot\text{s}^{-2}$, while that of water (moving at the same speed) is $6.28 \times 10^{-2}\ \text{kg}\cdot\text{m}\cdot\text{s}^{-2}$. These values indicate that the force exerted on the aquatic plant is roughly 827 times that exerted on an identical terrestrial plant body. All things being equal, aquatic plants experience substantially greater force exerted by moving water than do their terrestrial counterparts surrounded by moving air. (Notice the units of force—mass times unit length per second squared—indicating that force is the product of mass and acceleration, which has units of length per second squared.)

As is true of many biological phenomena, what amounts to a debit in one department may appear as a dividend in another. This is seen when we compare our hypothetical plants in terms of the mechanical influence of gravity. Both plants have the same volume $(6.2 \times 10^{-8}\ \text{m}^3)$ and the same density—taken here as $1.3 \times 10^3\ \text{kg}\cdot\text{m}^{-3}$, which is slightly less than that of cellulose $(1.5 \times 10^3\ \text{kg}\cdot\text{m}^{-3})$. Naturally, both plant bodies displace the same volume of fluid, but the aquatic plant displaces a comparable volume of a much denser fluid. Therefore it is submerged in a much more buoyant medium and experiences a significantly smaller gravitational body force. The buoyancy or force (symbolized by F) that draws any body with density ρ upward or downward bathed by a fluid medium with density ρ_f is equal to the product of the difference between these two densities $(\rho - \rho_f)$, the volume of the body V, and the gravitational acceleration g (which equals $9.8\ \text{m}\cdot\text{s}^{-2}$). This verbal description can be transcribed into the language of mathematics as the formula

$$F = (\rho - \rho_f)\, V g. \quad (1.5)$$

Inserting the previously specified values for the densities of pure water and air into eq. (1.5), we find that the buoyancy of the cylindrical aquatic plant equals $(1.3 \times 10^3\ \text{kg}\cdot\text{m}^{-3} - 998.2\ \text{kg}\cdot\text{m}^{-3})\ (6.2 \times 10^{-8}\ \text{m}^3)\ (9.8\ \text{m}\cdot\text{s}^{-2})$, or $1.83 \times$

10^{-4} kg·m·s^{-2}, while the buoyancy of the cylindrical terrestrial plant body equals (1.3×10^3 kg·m^{-3} − 1.205 kg·m^{-3}) (6.2×10^{-8} m^3) (9.8 m·s^{-2}), or 7.89×10^{-4} kg·m·s^{-2}. (Once again, notice the units in which buoyancy is expressed—mass times length per second squared—indicating that buoyancy is expressed in terms of force.) Thus our calculations reveal that there is over a fourfold difference in the buoyancy of the two plant bodies even though they are identical in size, shape, and density. Clearly, our hypothetical aquatic plant is little affected by the downward (compressive) pull of gravity operating on its biomass. Indeed, many submerged aquatic plants have tissues with a significant gaseous volume fraction (intercellular lacunae, bladders, etc.) that reduces the overall density of the plant. In contrast to terrestrial plants, aquatic plants possessing flotation devices must deal mechanically with an often substantial upward (tensile) pull, even in still water. Couching this in the context of plant evolution, the transition from an aquatic to a fully terrestrial plant must have involved a biomechanical transition from a *tensile* to a *compressive* plant body. At the cellular level, regardless of the habitat, the mechanical influence of gravity continues to play a vital physical role in inducing plant growth asymmetries (Evans 1991).

THE PHYSICAL ENVIRONMENT AND CONVERGENCE IN PLANTS

Comparisons between hypothetical aquatic and terrestrial plants can only illustrate some of the interactions between the physical environment, on the one hand, and the physiological and mechanical requirements for growth, on the other. These interactions represent the Scylla and Charybdis that have molded plant morphology, anatomy, and reproduction over billions of years of evolution. An increasing awareness of the polymorphism resulting from the interplay of physical processes and physiology has fortunately shifted an older emphasis on the physiological homogeneity among diverse plant species. This older emphasis led most students to scan the tremendous diversity of biological forms with a physiological search image that viewed plants as green stuff, occasionally hoisted on a scaffold of indigestible cellulose and lignin through which gases and nutrients percolate. Such an antiquated view and its underlying doctrine flout the fact that if photosynthesis were the only biological attribute of plants upon which natural selection operated, then the world would be covered by a layer of green protoplasm roughly four centimeters thick! In point of fact, we do not live in such a world. The reasons are numerous, but certainly an important one is that the performance levels of plant

FIGURE 1.1 Examples of plant organs that can provide protection: (A) Thorns (modified branches) of *Gleditsia triacanthos* growing from a tree trunk. (B) Prickles (modified epidermal growths) produced on the stem of *Rosa vulgaris*. (C) Cluster of spines (modified leaves) produced on the stem of *Pachycereus pringlei* (a cactus).

metabolism are defined by the environmental context that dictates the availability of light, atmospheric gases, water, and mineral nutrients as well as the magnitudes of physical constants like those appearing in eqs. (1.2) to (1.5). Since the physical environment is heterogeneous, we must expect morphological heterogeneity among plant species even in the face of the relatively similar metabolic requirements of plants. From this biophysical perspective, we must anticipate that organs dissimilar in their developmental origins but fulfilling the same function will evince similar form and structure, while plants that are distantly related but growing in similar environments should evince morphological and anatomical convergence. This expectation meets considerable support from numerous studies of comparative morphology and anatomy, on the one hand, and from ecophysiology, on the other.

One of the major conclusions drawn from comparative morphology and anatomy is that organ types (stem, leaf, and root) cannot be defined based on their function (Esau 1977). When two organ types serve similar functions, they parallel one another in shape and structure (fig. 1.1), suggesting at the very least that form and function are interdependent. For example, spines derived from leaf primordia, thorns derived from branch buds, and prickles derived from the epidermis and underlying cells all tend to be nonphotosynthetic, acicular, rigid, brittle, strong, and nasty. They also convect heat (Gates 1965). When stems assume the functional role of the principal photosynthetic organ, they not infrequently take the spatulate form of foliage leaves and may preferentially orient their surfaces toward ambient light, once again much like foliage leaves (Nobel 1988). In turn, structures developmentally derived from leaf primordia can function in many capacities other than as foliage leaves. The tendrils that mechanically support the vertical stems of peas are derived from leaf primordia. Indeed, appendicular organs derived from primordia may be developmentally modified to serve as digestive organs, protective bracts, and bud scales or as reproductive structures such as petals and sepals. These and other examples illustrate the functional versatility of plant organs as well as the morphological and anatomical convergences seen among plant organs with differing developmental origins. These convergences may also illustrate Ganong's principle (1901), which states that a function will be assumed by the part of an organism that happens, at the moment, to be most available for the purpose, regardless of its developmental origin. Currently, we cannot quantify or even identify pathways of developmental least resistance. Thus we must view Ganong's principle with some skepticism. Nonetheless, sufficient evidence exists to entertain the principle as a hypothesis to be tested by future research, which may shed light on evolutionary processes at the developmen-

FIGURE 1.2 Morphological convergence between the Cactaceae and the Euphorbiaceae: (A) *Astrophytum myriostigma* (a cactus). (B) *Euphorbia valida* (a euphorb). (C) *Pachycereus pringlei* (a cactus, to the right) and *Euphorbia fruticosa* (a euphorb, to the left). (D) *Opuntia vulgaris* (a cactus, in the background) and *Euphorbia bougheyi* (a euphorb, in the foreground).

FIGURE 1.3 Examples of photosynthetic organs: (A) Scalelike leaves and photosynthetic stems of *Lycopodium* (a lycopod). (B) Photosynthetic stem and reduced leaves (borne in whorls) of *Equisetum* (a horsetail). (C) Leaf of *Pinus* (a gymnosperm). (D) Pinnately compound leaf of *Polypodium* (a fern). (E) Photosynthetic roots of *Chilochista lucifera* (an orchid), which does not produce photosynthetically competent leaves.

tal level of biological organization.

Evidence clearly exists for the dependence of both form and function on the physical environment in which a function must be performed. For example, desert species of the Old World Euphorbiaceae and the New World Cactaceae often show remarkable convergence in form despite the taxonomic distance between these two families (fig. 1.2). Species with ribbed, columnar, or globose stems bearing leaves or stems modified into spines or thorns can be found in both families, presumably in response to similar selection pressures for water storage, reduction of surface area through which water can be lost, protection against herbivores, and the physical desirability of convectively dissipating heat. Thickened cuticles, sunken stomata, and reduced leaf area—features common to many desert plant species—are also seen among many conifer species that survive in the physiological deserts produced by winter cold.

Other examples of morphological and anatomical convergence are less obvious to the nonbotanist but are just as significant. The sporophytes of vascular plants are typically constructed from flattened leaves, columnar stems, and anchoring roots. A similar structural trilogy is seen in nonvascular plants. The gametophytes of many mosses consist of numerous vertical axes, which function mechanically to support leaflike structures called phyllids. The horizontal axes of these plants may possess rhizoids, which anchor the plant body to its substrate. Similarly, many algal species have leaflike fronds, borne by stemlike stipes attached to rocks or some other substrate by means of rootlike holdfasts. Indeed, leaves appear to have evolved independently in many plant lineages (fig. 1.3), as have tissues specialized to conduct fluids. The vascular plants possess xylem and phloem that conduct water and cell sap, respectively. Functional analogues to xylem and phloem tissues and cell types, called hydroids and leptoids, are found among mosses. Among the vast number of algal species, some have evolved cells that are functional equivalents to the phloem cells conducting cell sap. The plants of the brown algal genera *Nereocystis* and *Macrocystis* can reach lengths over 60 m when mature. Within their stipes, these organisms produce trumpet cells (as much as 5 mm long) that are stacked end to end to form what have been called sieve tubes, although this terminology is more properly reserved for functionally similar but evolutionarily different devices found in vascular plants. The intervening wall between adjoining trumpet cells is perforated by numerous cytoplasmic strands (plasmodesmata) through which photosynthates can be transported; for example, in *Nereocystis* a transport rate of $0.37\ \mathrm{m \cdot hr^{-1}}$ is not uncommon, while transport rates between 0.65 and $0.78\ \mathrm{m \cdot hr^{-1}}$ have been reported for *Macrocystis*. These and other algae have evolved the capacity for secondary growth, much like that seen in many vascular plants. The plants of the brown algal genera *Laminaria* and *Pterygophora* annually deposit a layer of tissue within their stipes from an internal meristematic region of cells, much as layers of wood are deposited each year within the branches and trunks of trees. As the stipes of some brown algal plants elongate, older conducting cells are stretched and become nonfunctional, like some of the cells in the xylem of elongating stems. Secondary growth in these stipes provides new trumpet cells. Although the stipes of brown algae lack woody tissues, some genera, such as *Fucus,* produce an inner core or medullary network consisting in part of very rigid parallel aligned fibers that provide strength against the shearing effects of waves. The bulk of this inner core is increased from year to year by cellular division, much as secondary xylem is deposited in the trunks and branches of trees.

These and many other examples serve to legitimate the assertion that distantly related plant groups have converged on very similar form-function relationships, as a result of either evolution in similar, highly specific environments or because general design constraints underlie the same biological function in all manner of habitats.

Design Requirements, Compromises, and Evolution

Like all living things, plants must perform many biological functions simultaneously. In systems analysis it is generally recognized that no single function can be performed without influencing all others. Each has certain design specifications that can be incompatible with or even antagonistic to the design requirements of other functions being performed by the same system. Conflicting design specifications require compromises in overall design, implying that the performance of every function cannot be simultaneously maximized. Rather, complex systems continue to operate functionally provided their overall performance level meets some minimum specification. In turn, this minimum level is dictated by one limiting factor whose influence cascades throughout the network of operational units composing the system as a whole.

If we can identify the functions that dictate the survival of plants, as well as the design requirements for each function, then we have a much more informative definition for these organisms. Placing this definition within the context of the physical environment and its limiting factors furthers our understanding of plant biology and evolutionary history. A brief summary of the biological functions that terrestrial plants must perform is provided in figure 1.4. Significantly, the performance of each of these four functions can be evaluated in terms of underlying physical laws or principles. From these laws and principles, the maximization of each of these functions can be shown to have conflicting design requirements that must be reconciled if all four are performed simultaneously. This is not to argue that plant biology can be reduced to these physical laws and principles, but these laws and principles assist in

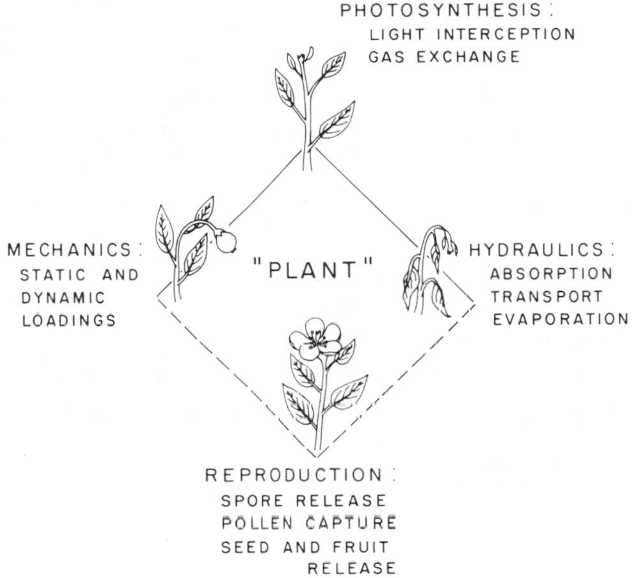

FIGURE 1.4 Principal vegetative and reproductive functions that must be performed by all plants (photoautotrophs): Vegetative growth requires light interception and gas exchange between the plant body and the external atmosphere; hydraulics involves the absorption, transport, and evaporation of water; and mechanics involves the ability of plant structures to support their own weight (and any additional weight imposed on them) against the pull of gravity. In addition to these tasks essential for vegetative growth, plants must reproduce, either sexually or asexually, to maintain the species in a habitat and through evolutionary time.

defining plant biology.

Consider the interception of light and the rate of gas exchange, which are essential to photosynthesis. Both of these functions depend on the ratio of surface area to volume. All other things being equal, a plant's capacity to intercept light and exchange gases with its external fluid medium is proportional to the ratio of surface area to volume. This is easily seen by returning to Fick's first law (eq. 1.3), which states that there is a direct dependence of the rate of diffusion (symbolized by dS/dt) on the surface area (symbolized by A) through which diffusion occurs and an inverse dependence of the rate of diffusion on distance (designated by x):

(1.6) $$\frac{dS}{dt} = -D_i\, A \left(\frac{\partial C_i}{\partial x}\right).$$

Accordingly, organisms that depend exclusively on diffusion for exchange must be small, very flat, or perforated and internally spongy (to reduce x and increase A). These stipulations are entirely compatible with the geometry of photosynthetic organisms, since very large ratios of exposed surface area to volume maximize the area that can intercept light energy. By considering these design requirements along with the fact that many plants continue to increase in absolute size as they age, we quickly see that geometric isometry during growth must be avoided by developmental subterfuge (see chap. 10)—the plant body must change shape as it gets larger in the face of an uncompromising physical law. Anisometry in the ratio of surface area to volume presents little difficulty in terms of water loss for an aquatic plant, since the tissues of these organisms have constant access to water. Indeed, some aquatic plants have the largest known ratios of surface area to volume by virtue of their monostromatic (one cell layer thick) thallus construction. Although, aquatic plants are confronted with many design trade-offs, such as between tensile forces and the display of photosynthetic tissues (Hoerner 1965; Raupach and Thom 1981; Witman and Suchanek 1984; Koehl 1986) and between mechanical sturdiness and the display of photosynthetic tissues (Gerard and Mann 1979; Mann 1982; Littler, Littler, and Taylor 1983; Koehl 1986), they do not contend with the major physiological dilemma confronting their terrestrial cousins, who must conserve water and yet have relatively large surface areas for light interception and gas exchange if they are to survive and grow. This is a serious problem, because neither nature nor modern science has invented a material that is permeable to carbon dioxide and oxygen yet impermeable to water. Accordingly, one of the first major biophysical barriers to the evolutionary transition from aquatic, essentially parenchymatous plants to fully terrestrial multicellular plants required both morphological and chemical solutions to the constraints imposed by Fick's first law. Indeed, although somewhat theatrical, it is not unfair to say that the early history of land plants was the classic laboratory experiment involving this law.

Theatrics aside, the dilemma of conflicting design specifications was fully resolved by the interplay of three evolutionary innovations: (1) an external layer of waxy material (called the cuticle) that had the capacity to reduce the rate of water loss (but through which gases could not be efficiently exchanged); (2) the internalization of exchange surfaces (in the form of internal gas-filled cavities or chambers) whose surfaces could be hydrated and whose rate of water loss could be restrained and regulated in some species by superficial pores or stomata; and (3) the development of conducting tissues through which water could flow fast enough to meet the rate of water loss by evaporation through exposed surfaces. The physiologic advantages of a cuticle and a conducting tissue are demonstrably dramatic. For example, the resistance of parenchymatous tissues to the flow of water is 10^5 to 10^6 times greater than that of the xylem tissue. This crude estimate comes from considering the volume flux density of water (symbolized by J_w), which equals the product of the difference in water potential (symbolized by ψ_w) and a proportionality factor reflecting the permeability of water flow at the cellular level, called the water conductivity coefficient (symbolized by L_w). In other words,

$$J_w = L_w\,(\Delta\psi_w). \tag{1.7}$$

The volume flux density of water has units of velocity ($m \cdot s^{-1}$). It is the volume of water flow per unit area per unit time ($m^3 \cdot m^{-2} \cdot s^{-1}$). Since water potential reflects the effective concentration of water molecules, which can be expressed in units of pressure (given in pascals, Pa; see chap. 4), a dimensional analysis of eq. (1.7) quickly shows that the water conductivity coefficient must have units of velocity per pressure ($m \cdot s^{-1} \cdot Pa^{-1}$).

If we assume that the difference in water potential ($\Delta\psi$) is the same for two tissue systems (in this case parenchyma and xylem), then the volume flux density of water for these two systems depends on the magnitude of the water conductivity coefficient. For the pathway of water through parenchyma and xylem, these two coefficients have been estimated to be on the order of 10^{-15} $m \cdot s^{-1} \cdot Pa^{-1}$ and 10^{-9} $m \cdot s^{-1} \cdot Pa^{-1}$, respectively (see Raven 1984, 116). Thus, all other things being equal, the volume flux density of water through xylem is expected to be as much as 10^6 times that of parenchyma. Some additional calculations (which need not burden us here) reveal that when the transport pathway length exceeds 2 cm, the rate of water transport through a parenchymatous hemisphere of tissue resting on pure water is insufficient to prevent desiccation even when the tissue is exposed to an ambient atmosphere at 70% relative humidity with an ambient flow rate as low as 1 $m \cdot s^{-1}$. Although 2 cm is only a very crude estimate, it seems fair to suggest that a semiaquatic, parenchymatous plant body could theoretically survive emerging above the water-air interface for only a few centimeters unless water could be rapidly transported to exposed surfaces by means of some specialized tissues or surfaces exposed to the moving air above the water were protected from desiccation by some externally applied hydrophobic material like a cuticle. Other alternatives do exist, however, such as existence in a wave-swept habitat and the retention of a large internal reservoir of water (see Vogel and Loudon 1985). But note that adding only a few micrometers of cutin and wax to the surface of a hemispherical mass of parenchyma would reduce the rate of water loss by the same amount as surrounding it by one meter of completely still air. Thus it is not surprising that land plants lacking a cuticle (like mosses and liverworts) dry quickly when exposed to moving air and either are limited to an existence within the boundary layer of their substrate or simply go dormant when they become desiccated. Nor is it surprising that most land plants (mosses, liverworts, and vascular plants) have evolved water-conducting cells that attain low resistance to the passage of water by virtue of being tubular, being stacked end to end, and lacking living protoplasts when functionally mature.

The evolution of a cuticle and conduction tissues elevated plants from a two-dimensional to a three-dimensional terrestrial existence. The subsequent exploration of the aerial realm carried many benefits in terms of shedding and transporting spores and gaining access to more light. And it is not surprising that many plant lineages evolved a vertical growth posture, perhaps as a result of a subtle arms race in which greater height and coverage by photosynthetic surfaces provided both the defensive (light gathering) and offensive (shadow casting) means of dealing with neighboring plants competing for the same resources (light, water, minerals, and space) (see Ellison 1989). But vertical growth has its own design specifications (fig. 1.4), and it can be achieved only over time and at some metabolic cost. The simple physical fact is that the orientations of photosynthetic structures, such as leaves and branches, that can maximize light interception frequently impose large mechanical forces on supporting tissues. Thus, if intercepting light has some metabolic priority, as the sun tracking of leaves suggests it has (see Zhang, Pleasants, and Jurik 1991), then mechanically supportive tissues must be deployed in the plant body at the expense of green tissues. In a buoyant fluid medium like water, we have already seen that the compressive pull of gravity is negligible, whereas in a terrestrial habitat it can be substantial. Indeed, once on land, the mechanical consequences of vertical growth on the capacity to intercept sunlight become very significant.

Consider the nearly cylindrical leaves of a pine tree deep within a stand of trees. A horizontal orientation of these foliage leaves, or needles, can be shown to maximize the capacity to intercept light, since light comes predominantly from directly above (Sprugel 1989). However, the moment of force, symbolized by M, which is the effectiveness of the mass force, for each foliage leaf is given by the formula

$$M = V \frac{l}{2} (\rho - \rho_f)\, g \sin \phi, \tag{1.8}$$

where V, l, and ρ are the volume, length, and average tissue density of the needle, ρ_f is the density of air, and ϕ is the angle of inclination of the foliage leaf (measured from the vertical). From eq. (1.8), we can easily see that for any given needle, M increases as a function of the angle ϕ (to the limiting condition where $\phi = 90°$). Thus a pine needle with a volume of 6.2×10^{-8} m^3, a length of 0.02 m, a density of $1.3 \times 10^3\ kg \cdot m^{-3}$, and a horizontal orientation ($\phi = 90°$) will experience a moment of force equal to $0.79 \times 10^{-5}\ kg \cdot m^2 \cdot s^{-2}$. If the inclination angle is reduced to 45°, then M equals $0.56 \times 10^{-5}\ kg \cdot m^2 \cdot s^{-2}$, while for a vertical orientation (the poorest in terms of light interception), $\sin \phi = 0$ and $M = 0$. From these simple calculations, we see that the orientation of the leaf that maximizes the capacity to intercept direct incident solar radiation also maximizes the moment force on the leaf. The same may be said for most foliage leaves and branches bearing leaves, particularly the latter, since branches are generally capable of indeterminate growth, increasing their length and mass with consequent effects on the moment of force. (Incidentally, if our pine needle is an appendage on an aquatic plant, then for a horizontal orientation $M = 0.18 \times 10^{-5}\ kg \cdot m^2 \cdot s^{-2}$, which is about 23% that of its terrestrial counterpart.)

Clearly, the design specifications that minimize the moment of force are antagonistic to those that can maximize a structure's capacity to intercept sunlight on land. A reversed mechanical problem exists in the aquatic habitat, where plant organs tend to be pulled upward in still water, thereby tending to collapse their foliar organs into a vertical orientation, which is the least effective one when light comes predominantly from above. Solutions are available, however; for example, the leaves of many aquatic plants float on the surface of the water, assuming a horizontal orientation. Indeed, the photosynthetic structures of some aquatic plants may be spread apart on the water by surface tension.

In terms of mechanical support, one of the truly elegant features of plant evolution was the transferal of function achieved when chemical polymers, perhaps functioning initially to ward off microbial attack, were incorporated within cell walls, primarily to resist compressive mechanical forces and permit the elevation of the plant body to higher altitudes. Lignin has been isolated from the cell walls of unicellular and multicellular algae, where it is believed to function as a chemical defense against the microbial hydrolysis of cell walls (Delwiche, Graham, and Thompson 1989). Lignin is also a major constituent in the walls of cells that provide mechanical support (sclerenchyma and vascular fibers) and transport water (tracheids and vessel members). In the case of mechanical support, lignin operates both as a bulking agent that can resist compression and as a hydrophobic chemical constituent that lessens how much the tensile strength of cellulose is reduced on hydration. In conducting tissues such as xylem, lignin, because of its hydrophobic properties, further strengthens cell walls and helps them resist implosion due to the rapid flow of water. Given the likelihood that the biosynthetic capacity to produce lignin predates its mechanical role in cell walls, lignin cannot be viewed as a chemical adaptation per se to the mechanical requirements for growth on land. It is more likely that the functional roles of lignin have diversified over the course of plant evolution. Early in the history of vascular land plants, conducting tissues constituted a very limited volume fraction of vertical organs and were consolidated into a more or less solid rod of tissue running the length of each vertical axis. From the perspective of vertical support (see chap. 3) this was a poor design, since the resistance to bending is maximized by locating structural support members as far as possible from the geometric center of cross sections. Indeed, the vascular tissues of the earliest land plants most likely provided little mechanical support, except perhaps indirectly by keeping thin-walled tissues inflated with water so they could operate as hydrostatic support tissues. However, lignification of epidermal cell walls and those of hypodermal tissues would have afforded dramatic mechanical benefits as well as a first line of defense against microbial attack or attempts by herbivores to bite into the plant body. During subsequent vascular land plant evolution the allocation of vascular tissue in the vertical stems and leaves of many taxa underwent significant modifications, in that mechanical support cell types derived from primary vascular tissues (e.g., phloem fibers) developed and matured farther from the geometric center of cross sections. The juxtaposition of nonvascular support tissues (lignified epidermis, sclerenchyma, etc.) with mechanically functioning cell types derived from the vascular tissues (primary phloem and xylem fibers, etc.) collectively provided the vertical organs of evolutionarily more recent land plants with an external rind of an extremely rigid material (which retained its biochemical legacy of lignin), as well as a core of parenchymatous, hydrostatic pith that could place this rind in hoop tension when inflated with water (and could also store metabolites). The progressive allocation of support cells toward the perimeter of cylindrical plant organs reveals that plants evolved in accordance with engineering principles long before these principles were known to the authors of textbooks.

With the advent of lignified vascular cell types that provide mechanical support and avenues for liquid transport, many terrestrial plants became capable of increasing the proportion of support tissues in their organs by the yearly accretion of secondary xylem, which when fully mature is essentially dead tissue. As new growth layers of secondary xylem are added, older layers function exclusively as mechanical support tissues, whereas newly deposited layers primarily transport water and to a limited degree store nutrients. The advent of secondary growth internalized mechanical support tissues so that the external rind of mechanical support tissues derived from primary growth could be dispensed with. This was advantageous, since the mechanical dividends accrued by the annual investment of ultimately dead tissue to the mechanical support of the plant organ are amortized over the lifetime of the plant organ. Indeed, this strategy is evident even in the leaves of some taxa, such as pine, where secondary vascular tissues are produced in foliage leaves that can remain attached to the plant for years, much like branches. Experiments indicate that the stiffness of pine needles increases as a direct function of the quantity of secondary tissues within them, while the proportion of xylem produced per leaf in the first year of growth is beautifully scaled to the length, and hence the moment force, of the leaf. Thus, regardless of their initial length and weight or subsequent changes in their size, pine needles can maintain a relatively uniform orientation with respect to the horizontal year after year.

Nonetheless, at some point in their growth the organs of many terrestrial plants reach their developmental limit for scaling mechanical strength to increasing biomass. When this limit is reached, mechanical failure can ensue, either directly from the self-imposed weight of branches and leaves that must be sustained or indirectly from an external force like wind or rain, whose exchange of momentum is more easily dealt with by smaller, more flexible organs. In this regard the capacity of trees for indeterminate growth may be defensible on the grounds of their developmental ability for continued growth from meristems, but from a mechanical perspective indeterminate growth may be a myth—at some point in the growth of every tree, the mechanical limits to stability are reached and eventually exceeded. True, over millions of years of evolution how much some plants grow before their existence is terminated by mechanical failure is impressive. For example, one of the largest species of trees is the redwood, *Sequoia sempervirens*. One specimen, called the General Sherman tree, measured 275 feet (83.9 m) in height and 82.3 feet (25.1 m) in basal girth in 1989. If we conservatively estimate the tissue density of wood as 390 $kg \cdot m^{-3}$, then the more or less conical trunk of this specimen must have had a minimum weight of 5.46×10^5 kg (more than 600 short tons). If the weight of branches, leaves, and roots is added to that of the trunk, then a total weight of approximately 8.5×10^5 kg (or 930 short tons) is not an unreasonable minimum estimate; that is, if we assumed that the trunk was

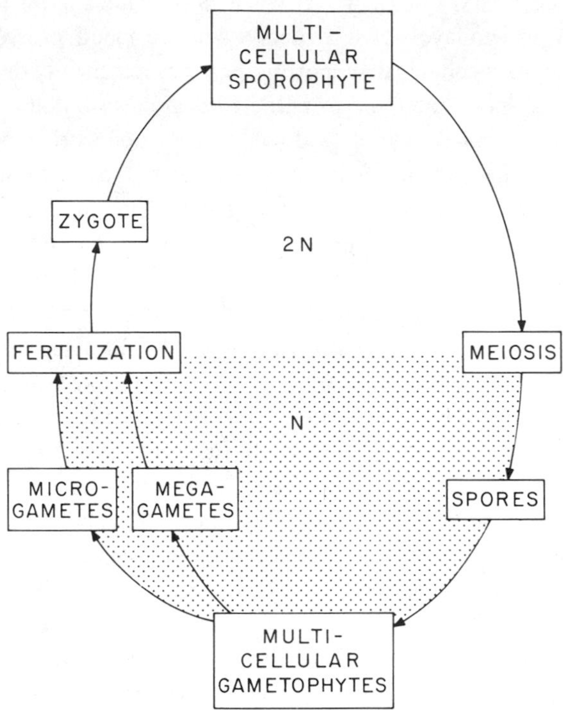

FIGURE 1.5 Diplobiontic life cycle (alternation of generations) of a vascular land plant. The multicellular sporophyte gives rise to meiospores by meiotic cell division. Meiospores develop into gametophytes that produce gametes by mitotic division. Although polyploid plants exist, the ploidy levels shown here are representative. In some plant life cycles, gametophytes have become specialized to produce either male gametes (sperm) or female gametes (eggs). These gametophytes are referred to as microgametophytes and megagametophytes, respectively. The life cycles of some algae differ significantly from the one shown here.

more cylindrical than conical, then the estimated maximum weight of just the trunk would be roughly 8.18×10^5 kg, while the total weight of the tree would be on the order of 12.7×10^5 kg. Regardless of which estimate we use (8.5×10^5 kg or 12.7×10^5 kg), the weight of this tree compares favorably with that recorded for the largest animal species that has ever existed, the great blue whale, which can reach a weight of 16×10^5 kg. (A comparison between the General Sherman tree and the great blue whale is particularly impressive when we recall that a whale is submerged in a fluid roughly 1,000 times as dense as air.) The longevity of plants is also impressive. The oldest known bristlecone pine (*Pinus longaeva*) is reported to have been 5,100 years old when it was cut down, while the vegetative growth of *Sphagnum* (a moss) in some peat bogs may have been a continuous process lasting as long as 10,000 years. Unfortunately our zoocentric view of the world neglects the fact that plants, more often than not, win in the Olympic Games of biology.

Aside from the biological functions essential for vegetative growth, for plants the game of survival also involves reproduction, and just as in the case of vegetative growth, the completion of the life cycle of plants is tightly linked to the environment and to the operation of physical laws. The life cycle of most terrestrial plant species, called the alternation of generations (also known as the diplobiontic life cycle), involves two multicellular organisms (fig. 1.5). One generation, the sporophyte, produces spores by meiosis. The other multicellular generation, the gametophyte, produces gametes by mitosis. The sporophytes of the earliest land plants shed their meiospores and most likely relied on airflow to transport their spores from one location to another, much as many extant free-sporing plants (collectively called pteridophytes) do even today. Accordingly, the reproduction of these plants was in part dictated by aerodynamic design specifications. The survival and fertilization of the gametophytes of free-sporing plants depends on access to wet microenvironments, since these small plants die relatively quickly when deprived of water, and since the movement of sperm cells across the surfaces of a gametophyte or from one gametophyte to another typically requires water. Thus the life cycle of many terrestrial plants still relies on the presence and physical properties of fluids. Some independence from external liquid water was eventually gained during the course of terrestrial plant evolution; the sporophytic generation became increasingly aggrandized, while the gametophytic generation became reduced in size and specialized in terms of the type of gametes produced. Among the seed plants (gymnosperms and angiosperms), unisexual gametophytes are produced. Those that produce only sperm are called microgametophytes, which are shed in the form of pollen grains. Those gametophytes that produce only eggs are called megagametophytes, which are retained within the sporophytic tissues and are metabolically dependent on them to some degree. Hence the megagametophytes of seed plants were released from a dependency on external bodies of water for their survival. Still, fluids continue to play a vital role in the reproductive biology of many seed plants—the pollen grains of many gymnosperms and angiosperms, even some of the most recently evolved, such as the grasses, are transported by wind.

The aerodynamic requirements for wind pollination impose stringent design specifications. For example, large and high-density pollen grains quickly settle out of the air and, all other things being equal, are transported over shorter horizontal distances than their smaller and less dense counterparts. For airborne particulates like pollen grains, the two most important parameters dictating the horizontal distance of transport are the horizontal speed of ambient airflow and how long the pollen grains remain aloft. The length of time any small particle stays airborne is inversely proportional to its rate of descent (for any given release height and ignoring convection). This is because any small airborne particulate always descends within the small volume of air enveloping it, even though the direction of the mass airflow carrying this smaller volume of air may move upward. The rate of descent, symbolized by U_t, of small spherical particles through a volume of unmoving fluid is given by the formula

$$U_t = \frac{mg}{6\pi\rho_f \upsilon r} \approx \frac{4.19\, r^2 \rho g}{6\pi\rho_f \upsilon}, \tag{1.9}$$

where m and r are the mass and unit radius of the particle and υ is known as kinematic viscosity, which is the ratio of the viscosity to the density of a fluid (see chap. 9). (The approximate solution for U_t, shown to the far right in eq. [1.9], comes from the fact that the volume of a sphere equals $4.189r^3$. Since mass is the product of density and volume, we get the approximate term shown here.) Consider two spherical pollen grains differing in size and density but not in total mass. Given the kinematic viscosity and density of air at 20°C (see table 1.1), for a spherical pollen grain with mass 1.37×10^{-10} kg and unit radius 32 μm (much like the pollen grain of the Canadian hemlock, *Tsuga canadensis*), U_t equals 0.123 $m \cdot s^{-1}$. If the radius of the pollen grain is increased by 10% (to a unit radius of 35.6 μm), while at the same time the original biomass (1.37×10^{-10} kg) is held constant, then the density of the grain would be decreased by roughly 25% (to 750 $kg \cdot m^{-3}$), and the U_t would be decreased by roughly 9% (to 0.112 $m \cdot s^{-1}$). The benefit of a 9% reduction in U_t translates into a 9% theoretical increase in the maximum horizontal dispersal distance; for example, assuming an ambient horizontal wind speed of 10 $m \cdot s^{-2}$ and a release height of 5 m, the maximum horizontal dispersal distances of our two hypothetical pollen grains would theoretically equal 407 m and 446 m.

Indeed, small and very light windborne pollen grains and seeds can be transported remarkable distances. *Senecio congestus* pollen has been reported 200 km away from the nearest parent plants, and the dust seeds of *Nepenthes ampullaria* can be carried as far as 1,100 km.

A problem ensues, however, when we consider that pollen grains must be captured from the air by the reproductive organs containing conspecific megagametophytes. The efficiency of particle capture C and its relationship to the radius R of a cylindrical obstruction to flow (such as a stigmatic filament), and the ambient speed of airflow U (carrying a spherical particle directly upstream from the obstruction) are given by the formula (see Spielman 1977)

$$C = \frac{mU}{6\pi r \rho_f \upsilon R} \approx \frac{4.19\ r^2 \rho U}{6\pi \rho_f \upsilon R}. \tag{1.10}$$

From inspection, we see that the equation for C contains the term U_t, such that

$$C = \left(\frac{U_t}{g}\right)\left(\frac{U}{R}\right). \tag{1.11}$$

Thus, for any ambient airflow speed U and for any collector radius R the efficiency of particle capture increases as U_t increases. (Recall that g is a parameter beyond the control of plants or, for that matter, the understanding of Einstein.) Thus the way the efficiency of particle capture can be maximized, in terms of pollen grain morphology, is antagonistic to the way the potential for horizontal transport of pollen grains can be maximized. Wind-pollinated plants have resolved this biophysical dilemma in a variety of ways: (1) by increasing the effective radius and decreasing the bulk density of their pollen grains, achieving a trade-off between long-distance dispersal and settling velocity; (2) by decreasing the collector radius of the reproductive structure to which grains must adhere to achieve pollination; and (3) by adapting to windswept habitats or by elevating their reproductive organs above leafy canopies into windswept microhabitats. In turn, each of these solutions has been effected in different ways among wind-pollinated species. For example, there are a variety of ways the biomass of a pollen grain (or seed) can be kept constant while at the same time decreasing ρ and increasing r. One of the most elegant is seen among species of pine and other conifers, where portions of the inner and outer pollen grain walls separate during early development (increasing the effective radius of the grain) and become filled with air (decreasing the bulk density of the grain). These flotation devices, called sacci, may also cause grains to float upward into the inner recesses of the ovule through a pool of liquid—the pollination droplet—toward megagametophytes. On the receiving end of the process called pollination, wind-pollinated plants manifest some truly elegant modifications that enhance particle capture efficiency. Most wind-pollinated flowering plant species possess featherlike stigmas (each bristle is very narrow in diameter and hence has a small R) held well above or to the side of individual flowers, which in turn typically lack petals that could interfere with the navigation of airborne pollen to stigmas and inadvertently collect pollen by inertial collision, reducing the number of grains reaching stigmas. Additionally, the inflorescences of many species are extended well above the vegetative canopy of the parent plant to maximize long-distance transport and minimize capture by their vegetative, reproductively inert surfaces. These and many other phenological and morphological adjustments to wind pollination reflect plant modifications whose benefits are entirely explicable in terms of the aerodynamic constraints imposed by physical laws.

A more global perspective for plant adaptation in terms of design requirements and compromises is seen among plants differing in phyletic affiliation. True, plants are not all alike in shape, size, and internal structure. They range from microscopic unicells to multicellular organisms and from essentially homogeneous colonial and mobile aggregates of cells, such as the algal genus *Volvox,* to large sessile organisms with discretely compartmentalized but highly integrated functional units (cells, tissues, and organs), like the plants growing in any garden or forest. This diversity in plant form and structure in part results from superimposing evolutionary relic species on those with a more recent evolutionary derivation. Once the limits imposed on metabolism and reproduction by absolute size are juxtaposed with the limits different environments place on various plant functions, however, much of the morphological and anatomical diversity seen in living and fossil plants can be explained.

THE IMPORTANCE OF MULTICELLULARITY

Like a building that must provide ventilation, plumbing, living space, and mechanical support, the plant body of most species is structurally compartmentalized by internal walls that serve as struts, beams, and columns. In multicellular species, the living protoplast of the plant body is incompletely dissected by an infrastructure of these cell walls that provides mechanical support and an avenue for transporting some nutrients. The geometry of the cell walls in a plant can be very complex, differing among tissues, or it can be simpler if the plant body evinces little or no tissue differentiation, as in many algae. The way tissues form and mature can influence many biological functions, not the least of which is mechanical support. For example, some plant tissues, such as aerenchyma, are composed of interlacing cells whose wall-to-wall contact area is low, creating many intercellular spaces filled with gas.

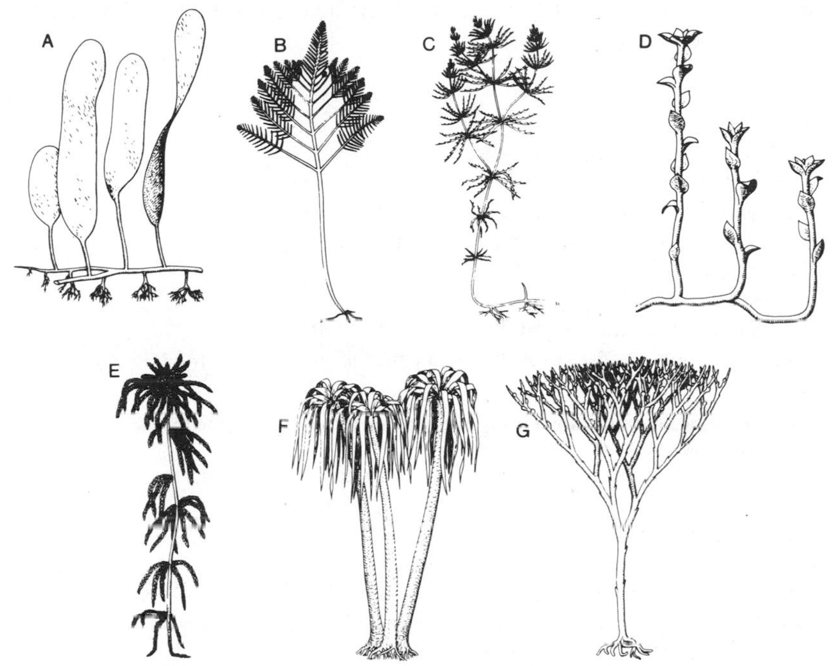

FIGURE 1.6 Examples of unicellular (A–B) and multicellular (C–G) plants: (A) *Caulerpa prolifera*, a marine green alga whose siphonaceous thallus produces vertical fronds and horizontal rhizomatous axes. (B) *Bryopsis plumosa*, a siphonaceous marine green alga whose frondlike thallus is attached to a substrate by a holdfast. (C) *Chara*, a freshwater green alga consisting of horizontal and vertical axes. (D) *Calobryum blumei*, a terrestrial moss whose gametophyte bears leaflike (phyllid) appendages. (E) *Sphagnum recurvum*, a freshwater/semiterrestrial moss whose gametophytes produce phyllid-bearing branchlike axes. (F) *Postelsia palmaeformis*, a marine brown alga with a palmlike appearance (a tubular stipe, a tough holdfast, and numerous leaflike blades). (G) *Psilotum nudum*, a terrestrial free-sporing, vascular plant with vertical and horizontal dichotomizing axes.

Similarly, the stems and leaves of some plants are hollow, reducing overall weight and producing internal surfaces that can absorb water vapor and other gases stored in cylindrical chambers when the stomata on the outer surfaces are closed. These examples illustrate the importance of multicellularity—the capacity to compartmentalize the plant body and permit physiological specialization as well as to selectively apportion cell walls in organs where mechanical forces must be dealt with. Multicellularity is not a prerequisite for the expression of complex morphology, however, nor does it intrinsically limit the absolute size of a plant (Kaplan 1987a, b). Indeed, some aquatic unicellular

TABLE 1.2 Developmental and Phylogenetic Corollaries of the Cell and Organismal Theories

Cell Theory

Developmental corollaries
- All living things are made up of cells
- Each cell is an individual of equal morphological rank
- Each multicellular organism is an aggregate of cells
- The properties of the organism are the sum of the many cells
- Ontogeny is the cooperative effort of many cells

Phylogenetic corollaries
- Unicellular organisms are primitive and "elementary"
- Elementary units formed colonial organisms through an acquired failure to separate after multiplication
- Cells within colonial organisms became specialized and interdependent, eventually producing the multicellular organisms

Organismal Theory

Developmental corollaries
- Ontogenesis is the property of the organism as a whole
- Growth and differentiation are the properties of the protoplasm
- Cell division may or may not involve septation of the protoplasm
- If septation occurs, then cells are subordinate parts of the whole
- Ontogenesis is the resolution of the whole into parts

Phylogenetic corollaries
- Unicellular and multicellular organisms are nonseptate and septate individuals, respectively
- Unicellular and multicellular organisms are homologous
- Colonial organisms are derived, not primitive organisms
- Division of labor and mechanical benefits were effected by cellularization

plants rival many bona fide terrestrial plants in size and shape (fig. 1.6). What then is the significance of multicellularity, and in what ways is it significant to the study of plant biomechanics?

To answer this question, we must first divest ourselves of some stereotypical notions, among which the logical independence of the cell from the organism is paramount. Much of current theory regarding development and evolution depends on the idea that the cell is the fundamental biological unit. This concept is part of the intellectual legacy of the cell theory that implicitly considers every organism a republic composed of essentially independent cells (see Buss 1987, for example). True, many growth processes are best studied at the level of the cell, which may be a convenient surrogate when the behavior of the protoplast is hard to see, but the cell theory has led to the idea that the concept of the cell and the concept of the organism are logically interchangeable. Nothing can be further from the truth.

The principal propositions of the cell theory, as developed by the botanist Matthias Schleiden and the zoologist Theodor Schwann and summarized in the proclamation "Omis cellula e cellula" (All is cells and cells), are that the cell is morphologically and physiologically the elementary biological unit; all cells are initially of equivalent morphological rank; and each organism is an aggregate of cells. These propositions have developmental and phylogenetic corollaries: the organism results from the collaborative efforts of cells; morphogenesis (the process by which shape is achieved) is the result of the collective actions of many specialized cells; the organism evolved from unicellular cells that, through time, formed loose aggregates by failing to disaggregate after a period of multiplication, thereby resulting in colonial organisms; and true multicellularity resulted when the cells within colonial aggregates became functionally interdependent.

The propositions and corollaries of the cell theory are not legitimized by studies of plant development or by our current understanding of plant systematics and evolution. Indeed, shortly after its publication, the cell theory had detractors who formulated the organismal theory, whose intellectual content was articulated in the nineteenth century by Heinrich de Bary's famous (but often ignored) dictum, "Die Pflanze bildet Zellen, nicht die Zelle bildet Pflanzen" (Plants make cells; cells do not make plants). The organismal theory essentially views the organism as a continuous mass of protoplasm (the symplast) that may or may not be incompletely partitioned into cells during the plant's ontogeny. Plant and animal cells are viewed as the result of ontogeny, not as its cause, and unicellular and multicellular organisms are placed in parity with one another—the latter is the septated equivalent of the former.

The differences between the cell and organismal theories, summarized in table 1.2, are fundamental to our views on development, morphogenesis, evolution, and biomechanics. Multicellularity must be viewed as a highly specialized expression of development in which the division of the protoplasm and the division of the nucleus are highly correlated. When the protoplasm partitions itself differently from the way nuclei divide, the organism may become a multinucleated but unicellular individual (e.g., *Caulerpa;* see fig. 1.6A). From this starting point we can explore the notion of multicellularity as it pertains to plant biomechanics.

Most plant cells have a cell wall external to the plasma membrane. The geometry and chemical composition of the cell wall dictate much of the mechanical behavior of single cells and the texture of plant tissues. The cell wall

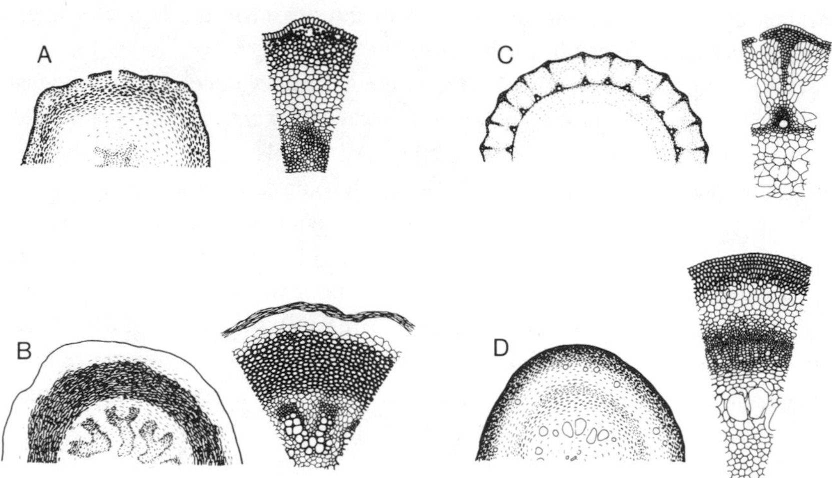

FIGURE 1.7 Examples of mechanical convergence among four phyletically distant plant groups in the distribution of thick-walled and thin-walled cells or tissues within cross sections (only half of which are shown) through stems. Regions with thick-walled cells or tissues are indicated by dense stippling; regions with thin-walled cells or tissues are indicated by less dense stippling. (Details of cellular structure are shown for a small section of each stem's cross section to the right of each stem's cross section.) (A) The fernlike plant *Psilotum*. (B) The lycopod *Lycopodium*. (C) The horsetail *Equisetum*. (D) A "typical" dicot, *Tilia*.

is functionally analogous to the skeleton in animals, but unlike metazoan evolution, in which the acquisition of a skeleton followed or closely paralleled the specialization of cells, the evolution of the plant cell wall predates the appearance of multicellularity and cellular specialization. During the course of plant evolution, the cell wall was variously apportioned by the protoplast to provide mechanical as well as physiological advantages. In some lineages cell walls form every time nuclei divide, leading to multicellularity, whereas in other lineages a network of cell wall struts and beams is formed throughout a single protoplast (e.g., the trabeculae of the coenocytic alga *Caulerpa*). This strutted infrastructure is mechanically analogous to the cell wall infrastructure found in the tissues of multicellular plants.

The extent to which the plant protoplast elaborates its cell walls appears to depend on both the size of the organism and its habitat—small aquatic plants tend to be unicellular; larger aquatic plants tend to be either unicells with internal trabeculae or multicellular, whereas all bona fide land plants, even very small ones, are multicellular. True, there are exceptions to these gross generalizations, but these exceptions do not detract from the general principle that the larger the organism the more extensively gradients of mechanical stresses develop within its body. These gradients can be dealt with in a number of ways, but at some point specialized support mechanisms become advantageous whether the organism is aquatic or terrestrial. Clearly size is not the single deciding factor, because the environment dictates the magnitude, direction, and duration of the application of mechanical forces. In a buoyant medium like water, substantially larger body sizes can be achieved without an internal mechanical support system. Besides, tensile forces can be dealt with by materials like protoplasm encased in an outer wall reinforced with cellulosic guy wires. On land or in the intertidal zone, however, where compressive forces and tensile forces operate with much larger magnitudes, a fairly specialized mechanical system becomes important for much smaller organisms (see Holbrook, Denny, and Koehl 1991). Thus the apparent abandonment of unicellularity as plants adapted to a terrestrial habitat can be rationalized based on the principles of mechanics. Physiological processes were of great importance, but the mechanical benefits of multicellularity should not be overlooked.

During the course of terrestrial plant evolution, the cell wall infrastructure of organs was modified in part because of the mechanical requirements of vertical growth. Tissues with thick cell walls become more and more evident in geologically younger plant fossils and are increasingly found toward the periphery of stems, where the internal mechanical forces induced by bending and torsion are greatest. The mechanical strategy of localizing the thickest cell walls toward the outside of stems was eventually abandoned by plants with the developmental capacity for secondary growth, which provides an internal sarcophagus of dead tissue on which living tissues are draped, but it is still very much in evidence in extant species, such as the horsetails and the common dandelion, that lack the capacity for secondary growth. Indeed, hollow stems and leaves are the extreme expression of a mechanical design that apportions support tissue where it will maximize stiffness and at the same time reduce overall weight (fig. 1.7).

One of the important consequences of the evolution of multicellularity is that, just as with a building whose external shape has only a marginal relation to its numerous floor plans, the shape of plant organs cannot be used with any great confidence to infer anatomy. Indeed, the compartmentalization of the protoplast, manifested in the shapes and relative sizes of cells, can achieve many patterns within the external boundaries of an organ. The potential for the independence of the external shape and the internal structure of plants provides two avenues (one morphological, the other anatomical) by which natural selection can operate. As a consequence, the number of possible mechanical and physiological permutations is dramatically greater than could be achieved if shape and structure were invariably correlated.

Adaptation, Reductionism, and the Engineering of Plants

The processes of convergence and adaptation raised early in this chapter highlight some additional concerns when functional analyses are performed. Among these are the objective of the analysis, the logical basis for inferring whether a structure evinces adaptation, and the more general issue of reductionism in biology. These issues are best dealt with early on in a book like this, since they are elements of every chapter.

Functional analyses have one or two primary aims. At the most fundamental level of inquiry, each analysis attempts to describe the way an organic structure or process contributes to the maintenance and survival of a particular type of organism. Additionally, it may extend its function-ascribing statement to make historical claims at the level of the species or at a higher taxonomic level. For example, based on physiological evidence, one could assert that the structure of a chloroplast contributes to the survival of eukaryotic photoautotrophs by converting light into metabolic energy. Thus the function-ascribing statement "the chloroplast contributes to the metabolic maintenance of a plant" makes no historical claims concerning the origin of this type of organelle or how it has been modified since its first evolutionary appearance. However, one might claim that free-living, prokaryotic photoautotrophs were subsequently incorporated and maintained within ancient heterotrophs, resulting in eukaryotic photoautotrophs during early evolutionary history (Margulis 1992). These protochloroplasts would have gained protection and access to the supplementary metabolic machinery of the host heterotrophic cell, while the host cell would have been released from the metabolic necessity of finding and ingesting organic carbon. Molecular and ultrastructural analyses of the chloroplasts found in a variety of algal lineages, combined with physiological studies of marine invertebrates that capitalize on the metabolic machinery of ingested plant cells, have provided critical data with which to test this endosymbiotic theory. Clearly, the intellectual agenda reflected in historical function-ascribing statements interpreting the origin of eukaryotic photoautotrophs as a consequence of the evolution of an endosymbiotic relationship differs substantially from one that simply addresses the current functional roles of chloroplasts within plants. The former approach complies with the definition of adaptation as a process, while the latter complies with its definition as a state. Adaptation as a state can be demonstrated simply when a biological structure or process is shown to contribute to the survival of an organism; adaptation as a process is demonstrated only when the functional role of a structure or process is shown to have been amplified through natural selection, thereby conferring an advantage on a species or higher taxon. The distinction between the aims of these two kinds of functional analyses is important because those that treat states of being are devoid of historical content—they look only at the immediate organism and the context in which it lives. As such, they are relatively immune to paleontological discoveries. By contrast, functional analyses that treat adaptation as a process are profoundly dependent on historical information and may vaporize in the heat of new data.

Typically, convergence provides the most robust line of evidence for adaptation as a process. The appearance of similar form-function relations in phyletically distant taxa provides a strong case for the inference that they confer advantages in survival and evolutionary success on these higher taxa, as well as on the individuals that belong to them, and that they are the products of natural selection. Demonstrable vectors in the efficiency of performance of a form-function relation within the individual lineages being considered provide additional support for function ascribing statements that make historical claims based on convergence. Additionally, the vigor of claims based on convergent form-function relations increases in proportion to the phyletic divergence that can be shown among the taxa studied. For example, it may be claimed that the presence of hydroids and leptoids in some mosses and of xylem and phloem in vascular plants reflects a convergence, from which it can be argued that the evolutionary innovation of conducting tissues is advantageous to terrestrial plant life. However, the mosses and the vascular land plants have a common ancestry. Both are embryophytes and apparently derive from the same ancestral algal plexus—the charophytes. If conducting tissues were a shared primitive character of the embryophytic plexus from which the mosses and the tracheophytes diverged, then the argument for convergence and adaptive evolution for conducting tissues in these terrestrial plant groups would be significantly depressed in vigor. By contrast, the appearance in some very large marine algae of conducting tissues having the same function and similar cellular structure as those of cells in the embryophytes provides a robust basis for asserting that conducting tissues are advantageous when the plant body reaches lengths requiring the axial transport of nutrients. However, the convergence in form and function of tissues in these two phyletically very distant plant groups (the Phaeophyta and the Bryophyta) neither supports nor refutes the claim that conducting tissues are advantageous to terrestrial plants, though this claim seems reasonable based on other lines of evidence. Accordingly, evidence for adaptive evolution based on convergence depends on current perceptions of systematic relationships, which historically have been shown to be mutable and sometimes highly unstable.

The extent to which form-function relations are couched in terms of adaptation as a process often changes as we ascend the hierarchy of biological organization. Claims concerning adaptation as a state are possible at all levels, but claims concerning adaptation as a process typically require an organismic context and therefore become more accessible, albeit difficult, as we leave the molecular level of organization behind us. In large part this is because living and nonliving processes are essentially indistinguishable at the molecular level but become increasingly disparate at the cellular level or higher. At higher levels of organization, biological processes and structures resist explication in terms of the principles of the physical sciences alone, if for no other reason than that growth, reproduction, and evolution either are not in the lexicon of the physical sciences or have vastly different meanings. Thus a chemist may speak of the growth of a crystal, or a physicist may speak of the evolution of a star, but it should be clear that the terms growth and evolution are used in contexts that deviate significantly from those of the biologist. Related to this issue is the relative ability to reduce biological phenomena to the operation of physical laws and processes. Virtually every scientist recognizes the importance of reductionism and uses this approach either implicitly or explicitly, in one form or another. However, a philosophical dichotomy exists among workers concerning our ultimate ability to reduce all biological phenomena to the principles uncovered by the pure physical sciences. Yet form-function analyses are by their nature reductionist attempts to explain biology. They are offered in the hope of providing insight into the quantitative differences among the capacities of organisms to survive, grow, and reproduce. In this manner they focus attention on the venue of natural selection.

The importance of functional analyses is seen when the image of the adaptive landscape created by Sewall Wright is rekindled (Wright 1932). The three-dimensional landscape of adaptive crests and less adaptive valleys requires an understanding of the molecular/genetic mechanisms that potentially allow organisms to move (evolve) over this landscape in their quest for greater fitness. (I hope readers will excuse this teleology, which is intended to be abrasive, if for no other reason than to point out that adaptation need not occur.) Although the capacity to alter the phenotype may ultimately be reduced to the principles of chemistry and physics, the topography of this adaptive landscape cannot be mapped by these principles. Its contours are defined by both the biotic and abiotic components of the environment, which change along the geologic time line. The adaptive landscape, if it truly exists, must be mapped with the aid of form-function analyses. Thus biomechanical analyses provide a counterpoint to those of modern genetics—both are essential to our understanding of biology in general and evolution in particular.

Although form-function analyses have grand objectives, they have many attendant practical difficulties, the most significant being the almost unavoidable necessity of viewing a process or structure in terms of a human artifact designed for a specific function or role. Aside from the philosophical fact that purposefulness has no legitimate place in evolutionary theory, there is another intrinsic difference between how the relation between form and function is dealt with in the context of biology and engineering. When an engineer designs a machine, its function and the work environment are specified and typically prescribed to remain constant over time. Other than technology and cost, there are no conceptual limits imposed on the shape, size, and material composition of a machine. The process of design and fabrication used by engineers differs substantially from the biological process of evolution and ontogeny. Machines have little or no historical legacy, they are not constructed as they function, and they are not self-assembling, as organisms are. In contrast with the engineer, the biologist must deal with an organism's evolutionary history and thus understands that biological structures and functions change in response to a changing environment and as a result of ontogeny. The biologist can never abandon the view that the organism is integrated spatially (through form and structure) and temporally (through its ontogeny and evolution).

Further, by correlative analyses, engineers can experimentally change the form of a machine and measure the effect of these changes on performance levels. By so doing, they can empirically identify and eventually construct an optimal machine—one that maximizes overall performance and safety and minimizes cost. Biologists do not have this luxury, either in theory or in practice. Our experiments involve organisms whose forms and functions are developmentally prefigured and that can be experimentally modified only with difficulty and usually trauma. Cost analyses of organic forms are likewise difficult if not impossible to make. Our conceptual models—the mechanical device or chemical process against whose performance we measure that of an organic structure or process—reflect isolated portions of an organism that has integrated many functions with many external stimuli. Although it is tempting to argue that the way an organism has reconciled its conflicting design requirements provides the basis for defining optimality, we must acknowledge that we have yet to achieve this depth of understanding for the behavior of any organism. Should this level of understanding finally be realized, I have no doubt it will lead us to realize that there are many optimal solutions nested in a domain of less satisfactory possibilities.

MODELING

The complexity of most biological phenomena may be reduced to a more manageable system by modeling. Typically, a model is a highly stylized and simplified conceptual representation of a limited aspect of reality. It is proposed as a theoretical construct that permits us to test certain of our assumptions concerning a recognizably much more complex situation. In this sense a model tests our perception of a phenomenon more than it tests the phenomenon itself.

There are two extreme mind-sets concerning modeling: one is skeptical, while the other is naive; one distrusts any attempt to reduce biological complexity to a more manageable form, while the other embraces reductionism to the extreme. Regardless of where we may fall between these two extremes, it is important to recognize that modeling is an unavoidable and fundamental aspect of any experimental design because it is a requisite to examining the interrelations among a large number of variables that would otherwise be unmanageable. Additionally, modeling is an intrinsic part of how we analyze our experimental results. All forms of statistical inference are based on a mathematical model of some sort. Thus experimental variables may be regressed against one another and correlated by means of some equation that presumes a model for the frequency distribution of our data. Accordingly, even the most skeptical cannot avoid using models in one form or another.

We should also not lose sight of the fact that the only good model is one that fails. When a model yields results that conflict with reality, as when our data points deviate from a simple linear regression line, we are shown that our assumptions about how reality operates are either inadequate or incorrect. By contrast, when a model's predictions conform to reality, the similitude may be simply fortuitous and therefore irrelevant. Good models allow us to reject our preconceptions; poor models delude us into believing we have identified ultimate causalities.

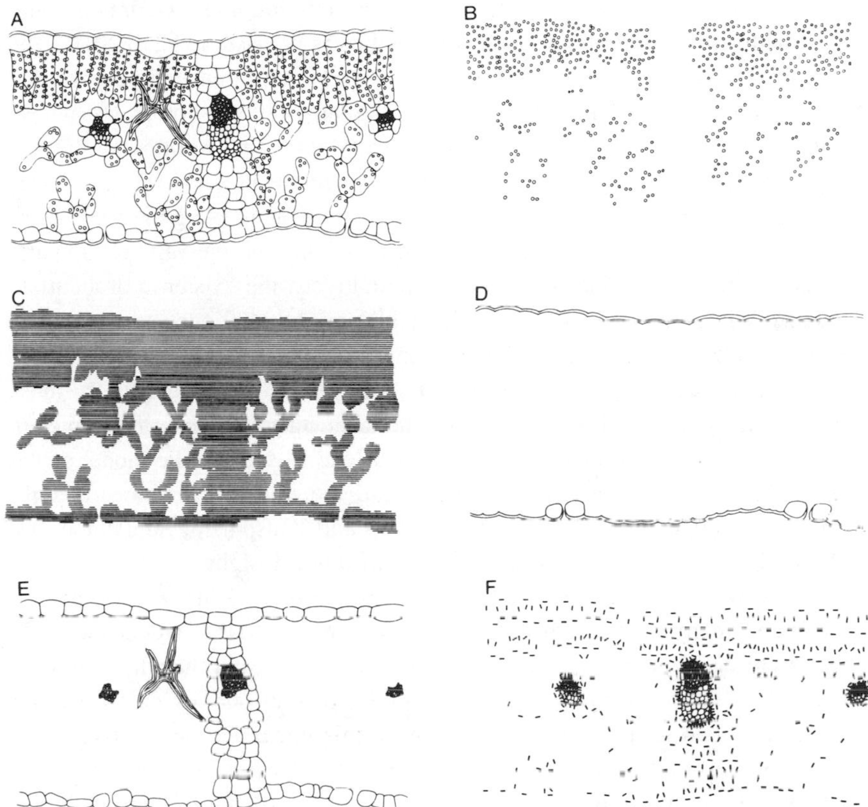

FIGURE 1.8 Different ways of viewing a single transection through a highly stereotyped dicot leaf: (A) Drawing of a transverse section through a dicot leaf revealing diversity in cell shape and size. (B–F) Tracings of various features drawn in A: (B) distribution of chloroplasts within the mesophyll; (C) contours of the air-filled cavities in the mesophyll; (D) outlines of the cuticles on the upper and lower epidermal layers (two pairs of guard cells are illustrated for the lower epidermal layer); (E) upper and lower epidermal cells, bundles of sheaths connecting the centrally positioned vascular strand and surrounding two flanking strands, and xylem that collectively compose the tissues that mechanically support the leaf lamina; (F) idealized distribution of plasmodesmata (illustrated as small dashed lines) and the location of xylem in three vascular bundles.

How we model a system often depends on our perspective, and each model reflects tunnel vision. Consider the transverse section through a typical dicot leaf shown in figure 1.8A. Our first impression might be of the diversity of cell shapes and sizes. With the appropriate qualifications, this perspective could generate a taxonomic model to distinguish one dicot species from another. If our interest is light interception, our attention might be drawn to the distribution of chloroplasts in the various layers of tissues (fig. 1.8B). This distribution could lead us to model the leaf as a multilaminated photovoltaic cell with a gradient of light-harvesting particulates called chloroplasts, whose concentration per unit volume decreases from the top to the bottom of the leaf along a gradient of light attenuation. Or we might notice large air spaces within the section, leading us to develop a model addressing the various resistances of carbon dioxide, oxygen, and water through the various compartments of the leaf lamina (fig. 1.8C). Alternatively, the existence of a cuticle studded with pairs of guard cells (fig. 1.8D) might draw our attention to the ecological distribution of species differing in these features. The location of mechanical support tissues (fig. 1.8E) could be used to model the stiffness of the leaf, while the distribution of plasmodesmata and the internal diameters of cells conducting fluids (fig. 1.8F) might lead to a hydraulic model of the leaf. Finally, we might be truly holistic and integrate the functions of light interception, gas diffusion, the conservation and transport of fluids, and mechanical support and return to figure 1.8A with new insight.

The leaf transection illustrates how the same biological structure can be viewed in many ways, each with the potential to add to our understanding as well as to deflect our ability to see that the whole is greater than the sum of its parts. It is tempting to argue that any initial bias in our perspective would eventually lead to a holistic view, but even this integrated perspective is the result of yet another model—the physiological process of photosynthesis, which tells us that any photosynthetic structure must have chlorophyll and access to light, atmospheric gases, and water. Although we have had this physiological model for many decades, we are still adding to it and cannot exclude the possibility that future modifications will alter our global view of leaf form-function relations. Indeed, we have only recently become aware of the anatomical and physiological distinctions between C_3 and C_4 plants.

PLANTS ARE NOT ANIMALS

For those who approach plant biomechanics from a strictly or predominantly zoological perspective, a note of caution is in order—plants differ in many fundamental respects from animals, particularly from the prevalent zoological paradigm, the mammal. Perhaps after this much reading such a cautionary statement is redundant, but there are many aspects of plant biology that must be constantly reemphasized to the zoocentrically minded. The fundamental metabolic machinery of plants differs radically from that of animals. The absorption of water from a substratum by roots or functional analogues to roots and the role of water gradients within the plant body for the translocation of nutrients are aspects unique to plants (chap. 4). Although some animal species possess cell walls, the roles played by the plant cell wall in growth and development, as well as in the transport of nutrients and maintaining the stiffness of the plant body, are much more diverse and often more complex than those encountered in animal systems (chaps. 5–6). Indeed, the presence of cell walls has led to a fundamental dichotomy between plant and animal development—cells and tissues must be added to the plant body to effect changes in shape and size, whereas the development of animals is characterized by the capacity of cells to migrate and of tissues to fold, contract, expand, or otherwise reconfigure in the mature organism and as ontogeny proceeds. The modularity of plants—the construction of the plant body by developmentally reiterated organs that are often indefinite in number, that frequently show signs of physiological independence, and that are shed as a normal consequence of growth (White 1979)—finds few analogues in the animal kingdom but is an intrinsic feature of virtually all metaphytes (chap. 7). Certainly another important distinction that can be drawn between most plants and animals is that plants display an alternation of generations in the life cycle. Among the embryophytes, as in many algae, the sporophyte and gametophyte generations differ significantly in appearance, size, structure, and habitat requirements and therefore represent very different organisms on which natural selection has operated. Perhaps the closest analogue between the life cycles of plants and animals is the life cycle of the Holometabola—insects whose successive developmental stages (larvae, pupae, and adults) differ radically in shape, size, and food requirements. One of the selective advantages of these polymorphic life cycles may be their capacity to introduce many representatives of the same species into the general environment, since the various phases of the life cycle occupy different microhabitats and are not in direct competition with one another. Nonetheless, during the course of terrestrial plant evolution, the gametophytic generations of evolutionarily derived lineages were smaller and less complex than those of their antecedents, while many become more dependent on the sporophytic generation for their nutrition and survival. By contrast, the sporophyte ultimately became the dominant ecological generation, thereby elevating it as the traditional, albeit biased, model for the plant body (chap. 8). Modifications in the life cycle of terrestrial plants and in the structure of reproductive organs afford some of the main features whereby the various plant groups are distinguished (fig. 1.9). Once again some parallels can

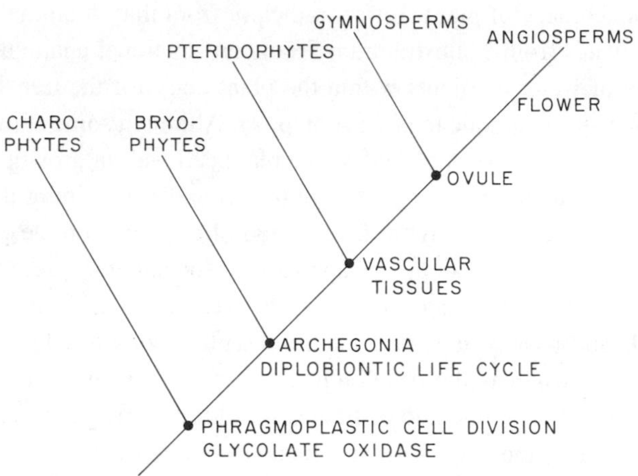

FIGURE 1.9 Evolution of major plant groups as reflected by their vegetative (cuticle, stomata, and tracheids) and reproductive features (archegonium, seed, and flower).

be drawn between the life cycles of plants and animals, but often at the expense of missing salient features of great significance. For example, the life cycle of the seed plants has evolved to a degree paralleling in some respects that of mammals—the gamete functioning as the egg and the developing embryo are retained within parental tissue and are provided nutrients as well as protection from the external environment. Nonetheless, the sporophytes of seed plants are sedentary organisms that rely on some vector to disperse their microgametophytes (chap. 9). These and other features of plant biology have played significant roles in dictating the course of plant evolution (chap. 10).

Although metabolically, structurally, and reproductively different in many significant respects, plants and animals have nonetheless evolved under the same biophysical limitations, in ways and directions particular to their own biologies. Thus biologists have at their disposal two evolutionary experiments—two biological histories that can (and must) be juxtaposed and reconciled if a grand evolutionary synthesis is to be achieved.

WHY STUDY PLANTS?

At the beginning of this chapter it was asserted that plants are the ideal organisms in which to study form-function relations. Most are sedentary, and all lack the neurologically complex behavior that often makes such relations difficult to assess in animal systems. The relatively simple shapes and structures of plant organs and cells are ideally suited to modeling, since they often conform geometrically to hollow or solid spheres and cylinders, tapered or untapered beams, or the vertical columns or horizontal cantilevers for which engineering theory provides closed-form solutions for mechanical behavior. By virtue of their physiology, all plant species are highly attuned to their physical environment. The survival of the individual and of the species clearly depends on coping with the laws of physics and chemistry. Additionally, all plants share essentially the same metabolic machinery and compete for the same resources. Thus structural solutions to deal with the physical and biotic factors in their environments have become paramount in their ecology and evolution. Because of their mechanically strong tissues, most plants resist decomposition and have left a superb fossil record from which the process of adaptation can be unraveled or refuted. Also, by virtue of their development and growth from meristematic regions, living and fossil plant organs can provide an almost continuous ontogenetic record. Finally, as I mentioned previously, plants constitute over 90% of the world's biomass. Clearly plants are as important and intriguing as they are manifestly beautiful, as I hope the following chapters will illustrate.

A Word about the Next Two Chapters

The study of plant biomechanics is far from simple. Biomechanics relies on engineering theories and practices that are far from intuitively obvious to the biologist, while much of botany may seem equally strange and nonintuitive to the formally trained engineer. This is most evident in the next two chapters, which are laden with mathematics and botanical terms and treat subjects that at first glance may appear far removed from comparative plant morphology and anatomy—or for that matter engineering. These two chapters thus may deter botanists from treating plants in terms of engineering, just as they may frustrate engineers who wish to approach plants immediately without a review of self-evident engineering "first principles." For those who dislike mathematics and feel the need to submerge themselves at once in the wonders of botany, or for those who have the mathematics of engineering well in hand and want no clumsy attempts to explain them in detail, chapters 2 and 3 can be merely glanced at as "reference sources," to be read in detail or not as required by the topics treated in the remaining chapters. In any event, their contents will resurface in one way or another and are not essential to reading the rest of this book. Nonetheless, for those who need to learn some engineering or would like to see how a botanist values engineering, chapters 2 and 3 will be useful.

Two

The Mechanical Behavior of Materials

> For all particularity in natural science reduces to the discovery of definite magnitudes and relations of magnitudes.
>
> Ernst Cassirer, *The Concepts of Natural Science*

The major premise of this book is that organisms cannot violate the fundamental laws of physics and chemistry. A corollary to this premise is that organisms have evolved and adapted to mechanical forces in a manner consistent with the limits set by the mechanical properties of their materials. No better expression of these assertions can be seen than when we examine how the material properties of different plant materials influence the mechanical behavior of plants. Nor do we find any better evidence for the evolution and adaptation of plants to mechanical forces than when we compare their mechanical attributes to those of fabricated materials, particularly since many plant materials mechanically outperform some of the most common materials used by engineers and architects. For instance, the nutshell of the macadamia, an Australian evergreen (*Macadamia ternifolia*), is as hard as annealed, commercial grade aluminum, resists twice the force necessary to fracture that and some other metals and is stronger than silicate glasses, concrete, porcelain, and domestic brick.

Yet the nutshell is less than half as dense as many of these materials. This low density is an advantage in that it contributes disproportionately less to the overall weight a plant must sustain. Strength is only one of the physical attributes that must be considered when assembling a structure like a nutshell, tree trunk, or flower stalk, since vertical construction carries with it the design constraint of self-loading (see chap. 3). Thus, in addition to being strong, materials should be light. In fact, for its density, cellulose is the strongest material known. The trade-off between strength and density within a nutshell illustrates a much more general principle: the mechanical behavior of any single material is defined by a number of material properties, and not all can be maximized. Each material must be used according to its particular qualities and the types and magnitudes of the mechanical forces it must sustain.

In addition to their low density and comparatively great strength, biological materials have other advantages over their engineered counterparts. Biological materials are versatile—they can change their material properties as they age or as a function of their immediate physiological condition. Young plant cell walls are ductile, while older cell walls tend to be much more elastic and resilient. Also, the material properties of plant substances and organs can change, through their capacity for growth, in response to the magnitudes of the forces they are subjected to. The responsiveness of plants to their immediate mechanical environment was recognized by botanists as early as the mid-nineteenth century. Vöchting (1878) investigated the anatomical responses of plant stems to different conditions of traction. He found that the stems (pedicels) of squash fruits suspended from a trellis had significantly more vascular tissue than those produced by plants growing on the ground. Similarly, Hegler (1893) stretched seedlings of sunflower (*Helianthus*) and winter rose (*Helleborus*) by attaching 150 g weights. After forty-eight hours, twice the weight required to break control plants was needed to break the seedlings held in traction. The developing fruits of many plant species are naturally suspended from the branches of trees, and the weight of these fruits can increase by more than a thousandfold as they mature, while their pedicels increase by a mere fraction of their original cross sectional area. Thus pedicels either must be very strong initially or must grow in a compensatory manner so that their mechanical attributes change as the weight of the fruits they bear increases. Likewise, as seedlings grow, the forces exerted by their expanding and elongating cells cause them to stretch. Thus, as many plant organs grow, they exert mechanical forces that can operate as development cues effecting changes in anatomy. These growth forces and their consequences on development appear to provide a feedback system allowing a plant to constantly change its internal structure and the properties of its materials in response to mechanical forces caused by growth as well as those exerted by the external environment.

Subsequent attempts to repeat the work of Vöchting and Hegler yielded mixed results. Some workers confirmed Hegler's experimental findings (Newcombe 1895; Bordner 1909), but others could not (Ball 1904; Flaskämper 1910). In large part the discrepancies were due to comparisons drawn between plant tissues of different developmental ages. Hegler, Newcombe, and Bordner used young, actively growing plant organs and tissues in their experiments; Ball and Flaskämper did not. Developmentally younger cells, tissues, and organs are more responsive to growth forces as well as externally applied forces because of their capacity for further growth, whereas the mechanical attributes of older, developmentally mature cells and tissues are locked into place once cell walls are deposited, become chemically modified, and fully mature.

Another reason for differences in the experimental results reported among early workers was that many failed to appreciate the importance of normalizing the mechanical forces applied to organs in terms of the transverse areas through which they operated. Additionally, the transverse areas of plant organs or parts of organs are difficult to measure accurately even with present-day instrumentation. Clearly, the same magnitude of force applied to structures constructed of the same materials but differing in dimensions will have very different mechanical consequences. Once the effects of developmental age and the need to control for the areas through which forces operate were incorporated into the experimental design, botanical biomechanics became a reproducible science.

The capacity of biological materials to change their material properties through growth and development confers a spatial and temporal heterogeneity on the mechanical behavior of the plant body and its constituent parts. This capacity for change sets biological materials and the structures of which they are constructed apart from all engineered artifacts. Therefore one of the goals of the biomechanicist is to understand not merely how biological materials can be studied in terms of engineering practice and theory, but how engineering theory and practice can be extended and enriched by what we learn about plants.

Nonetheless, because they *can* change over time, the mechanical properties of most plant materials are very difficult to quantify. Even at a particular developmental instant, most plant materials exhibit properties that conform to those of neither ideal solids nor ideal fluids. Fortunately, with care and a full appreciation of the many limits involved, we can approximate the behavior of many plant materials as if they were elastic solids or ideal fluids. Indeed, the fabricated materials used in everyday engineering practice are anything but ideal materials, yet they are nonetheless approximated as such with considerable success. For example, almost all solids manifest to some degree the property of elasticity—that is, they deform when subjected to an applied force and restore these deformations when the force is removed. The elementary theory of elasticity, developed by engineers, assumes that materials are ideal elastic materials—that they completely and instantly restore their deformations when the magnitude of applied forces drops to zero. Under certain boundary conditions, most metals behave very nearly as ideal elastic solids and perform within the parameters set by the elementary theory of elasticity. By the same token, when an engineer treats a problem in fluid mechanics, it is not atypical to assume that the fluid is an ideal fluid. And even though its behavior may deviate somewhat from this ideal, the extent to which theory can predict the behavior of a real fluid is typically satisfactory for most practical situations.

For pedagogical reasons, we shall use the same tactic followed in everyday engineering practice and initially treat plant materials as if they were ideal solids or ideal fluids. In this way we will learn some very fundamental concepts. For example, when an external force is applied to a material, internal forces are produced. These internal forces, called stresses, result in deformations, called strains. For any given material and over certain ranges of externally applied forces, the magnitudes of stresses and strains are related to one another by material moduli that can be used to distinguish among materials under certain conditions of loading. Once treated, these fundamental concepts will be used to examine the physical attributes of most plant materials in terms of the theory of viscoelasticity. Viscoelastic materials are those that show elastic and viscous components in their behavior. Thus, in a crude sense they are hybrid materials exhibiting the properties of both solids and fluids. Nonetheless, we will see that our ability to treat viscoelastic plant materials is limited to phenomenological descriptions of their behavior. This inadequacy stems from the fact that most plant materials show nonlinear viscoelasticity, for which no adequate mechanistic theory has as yet been developed. Thus, throughout this chapter, we must retain an awareness of when and how theory and reality fail to coincide.

As mentioned in chapter 1, the topics treated here are presented by way of reference materials to aid in understanding subjects that will be treated in chapters 4 to 10. Readers already familiar with the precepts of the material sciences can glance at this chapter and use it more as an appendix than as essential text.

TYPES OF FORCES AND THEIR FORCE COMPONENTS

Regardless of its apparent complexity, any externally applied force or combination of forces can be resolved or decomposed into two fundamental force components, distinguishable in terms of the direction in which they operate regarding a surface of interest. These components are called the normal force component, which operates perpendicular to the surface of interest, and the tangential force component, which operates parallel to the surface of interest. The normal force component results in either tension or compression, while the tangential force component results in shear. Some external forces can be applied in such a fashion that one or the other of the two force components predominates. With reference to a material, a force or forces can be directed inward or outward (resulting predominantly in compression and tension, respectively), or the external force can operate tangentially to one or two surfaces of a material (resulting in shear). By the same token, forces can be applied so that the two force components have measurable magnitudes, as when a material is bent or twisted. Nonetheless, bending or torsion is merely the deformational manifestation of the simultaneous operation of normal and tangential force components. Which of the two is more important from the perspective of subsequent mechanical performance or mechanical failure depends on the functional role of a material, how much the material deforms in response to either force component or both, and the nature of the physical environment in which the material operates.

There are two general categories of externally applied forces that invariably operate on any material or structure: surface forces and body forces. A surface force is any force distributed over the external boundaries of a material, and the most pervasive is hydrostatic pressure, since all organisms exist and physiologically operate within a fluid that exerts an external pressure. By contrast, forces distributed within the volume of a material, such as gravitational forces, are called body forces. Clearly, all plants experience surface and body forces, but depending on habitat, how much these forces influence form-function relations can vary widely (see chap. 1). For example, aquatic plants typically experience lower body (gravitational) forces because their tissues are buoyed by water, thereby lessening the influence of gravity, whereas aquatic plants typically experience higher surface (hydrostatic) forces because they are submerged in a fluid much denser than air. By contrast, the body forces induced by gravity and the surface forces generated by episodically high wind pressure have dictated many of the form-structure features typical among terrestrial plants.

Different Responses to Applied Forces

The normal and tangential force components produce deformations, and the relation between the force and deformation components can be used to distinguish among different types of materials, since the way a material deforms depends on the nature of its atomic or molecular bonds. Typically a material can respond mechanically to externally applied forces primarily in one of four ways: (1) It can add the normal and tangential force components to the bonding forces that hold its constituent atoms or molecules together and use this stored energy to return instantaneously to its original shape when the force is removed. Materials of this sort exhibit elastic behavior. (2) Once the elastic range of behavior is exceeded, even an elastic material may simply break, or it may slowly dissipate the energy supplied by the external force by deforming permanently. The second response is called plastic behavior. (3) Some materials undergo large changes in shape and internal structure that increase over time and can be partially or totally recovered slowly once the external force drops to zero or diminishes. This type of behavior is called viscoelastic behavior. (4) Finally, some materials have no elastic component, so they rapidly dissipate all the energy supplied by external forces and permanently deform by irreversibly flowing in the direction of the applied force. This response is called fluid behavior.

Clearly, some materials can exhibit more than one of these responses, as when an elastic material undergoes plastic deformations. Nonetheless, when most biological materials are tested at a particular age and in a particular metabolic condition, they primarily exhibit one of these four types of behavior. When different materials are used to assemble a structure, however, the mechanical behavior of the structure transcends the mechanical behavior of its constituent materials. What makes organisms so interesting from a mechanical perspective is that even the simplest unicellular organism consists of a variety of materials whose collective mechanical attributes often differ radically from those of its constituent materials. These material composites, which we call organisms, are sophisticated in their material properties and possess mechanical versatility and complexity that permit them to survive in habitats where many engineered artifacts perish.

STRESSES

When an external force is applied to a material, internal forces develop among its different parts. Engineers define the magnitude or intensity of these internal forces in terms of the amount of force per unit area upon or through which they act. This normalization procedure defines what is called stress, which has the units of force per unit area, F/A. In the simplest case, the stresses that develop are uniformly distributed over any cross section within a material, as

TABLE 2.1 Base and Derived Système International d'Unités (SI Units) for Physical Quantities and Decimal Multiples and Submultiples

	SI Unit or Prefix	Symbol
Physical quantity		
Length	Meter	m
Mass	Kilogram	kg
Time	Second	s
Amount	Mole	mol
Energy	Joule	J
Force	Newton	N
Power	Watt	W
Pressure	Pascal	Pa
Frequency	Hertz	Hz
Multiple		
10^1	Deka	da
10^2	Hecto	h
10^3	Kilo	k
10^6	Mega	M
10^9	Giga	G
10^{12}	Tera	T
Submultiple		
10^{-1}	Deci	d
10^{-2}	Centi	c
10^{-3}	Milli	m
10^{-6}	Micro	μ
10^{-9}	Nano	n

is seen in the case of a metal bar subjected to a pair of either tensile or compressive forces uniformly distributed over its two ends. The two opposed tensile or compressive forces are called coaxial forces. These coaxial forces operate along the same dimension, but their directions of application are opposite. A single pair of coaxial forces produces uniaxial stresses. Thus, for example, we may speak of uniaxial tensile stresses and uniaxial compressive stresses. A pair of coaxial forces can be applied to produce biaxial stresses. The pair of coaxial forces operate along two orthogonal dimensions, as when a rubber balloon is inflated and the material on its surface extends in two orthogonal directions as a result of internal hydrostatic pressure. Finally, triaxial stresses can develop, as when a material is subjected to an external, uniformly applied hydrostatic pressure. Clearly, any material can be tested in the laboratory under any of these three conditions (uniaxial, biaxial, or triaxial stresses), but it is always advisable to test a material under the conditions of loading that produce a state of stress most closely reflecting what the material naturally experiences. This is particularly true for biological materials that normally experience biaxial or triaxial stresses, such as epidermis or storage parenchyma.

We will treat the concept of stress by first examining the relatively simple uniaxial stresses that result when the material within a cylindrical bar is subjected to a tensile load. The intensity of the distribution of internal forces is uniformly distributed within the principal surface of interest, which is the plane perpendicular to the axis of length; that is, the tensile stress is calculated by dividing the applied force by the transverse area of the bar. By the same token, the tensile stress in the pedicel of an apple fruit can be calculated by dividing the weight of the fruit by the cross-sectional area of the pedicel. In the case of the bar or the pedicel, the stresses operate normal (perpendicular) to each cross section through the bar. These stresses reflect the normal force component and are called normal stresses, symbolized by σ. (Stress and other mechanical parameters are expressed in the Système International d'Unités, or SI units [table 2.1]. The unit for force is called the newton, symbolized by *N*, which equals 100 g—the weight of a modest-sized apple.)

However, externally applied forces also have tangential force components that produce internal forces operating parallel to the plane of each cross section. The stresses that result from the tangential force component are called shear stresses, symbolized by τ. Normal and shear stresses operate within each of the Cartesian coordinates of a material element. Figure 2.1 illustrates the orientation of these stresses within a cubic material element whose dimensions of depth, width, and length are designated by the Cartesian coordinates *x*, *y*, and *z*, respectively. As shown in this figure, three symbols (σ_x, σ_y, and σ_z) are needed to describe the normal stresses, while six symbols (τ_{xy}, τ_{yx}, τ_{xz}, τ_{zx}, τ_{yz}, and τ_{zy}) are required to describe the shear stresses. The subscripts used in this notation identify the direction or plane in which each stress component acts. Only one letter is required for each normal stress; for example, σ_x represents the stress operating parallel to the dimension of depth (x). In the case of the shear stresses, however, two subscripts for each stress component are required: the first letter indicates the direction perpendicular to the plane considered, while the second indicates the direction in which the component stress operates; for example, τ_{xy} is the shear stress component operating in the dimension of width (y) perpendicular to the dimension of depth (x). In figure 2.1 there are three equalities among the six shear stresses (only two of these equalities are shown). The shear stresses operating normal to the line of inter-

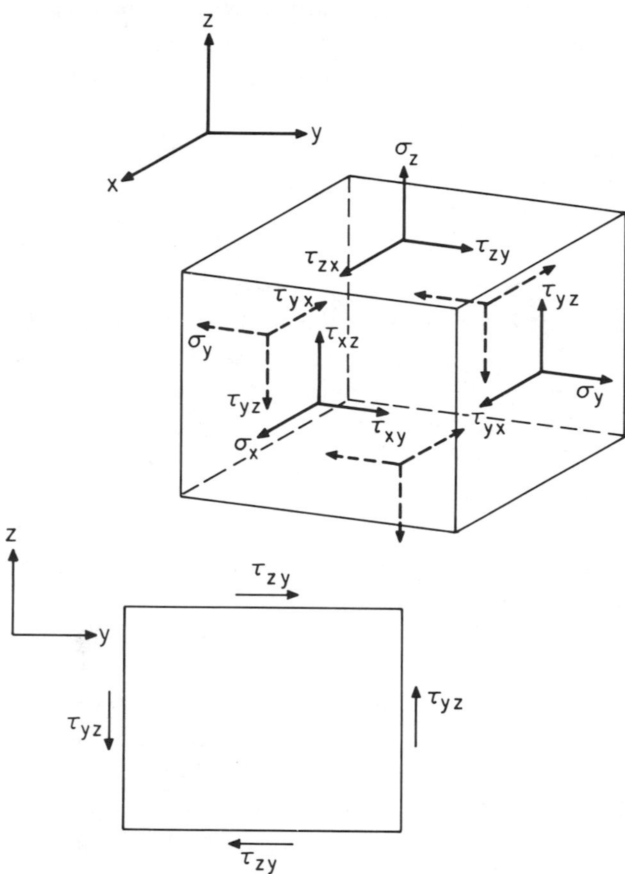

FIGURE 2.1 Normal and shear stresses within a cubical element of a material with depth x, width y, and length z. Three normal stresses, σ_x, σ_y, and σ_z, operate perpendicular to the three orthogonal axes of the element. Parallel sides of the cubical element have equivalent magnitudes of normal stress. Six shear stresses, τ_{xy}, τ_{yx}, τ_{xz}, τ_{zx}, τ_{yz}, τ_{zy}, operate parallel to the surfaces of the cubical element. The shear stresses operating perpendicular to the line of intersection of perpendicular sides of the cubical element are equivalent: $\tau_{xy} = \tau_{yx}$, $\tau_{zx} = \tau_{xz}$, $\tau_{zy} = \tau_{yz}$. Therefore only six stress components (σ_x, σ_y, σ_z, $\tau_{xy} = \tau_{yx}$, $\tau_{zx} = \tau_{xz}$, $\tau_{zy} = \tau_{yz}$) are required to specify the internal forces (= stresses) generated within the element as a result of the application of an external force. (From Timoshenko and Goodier, *Theory of Elasticity* [1970], fig. 3, reproduced with permission of McGraw-Hill, Inc.)

section of any two perpendicular sides of the cubical element are equal; that is, $\tau_{xy} = \tau_{yx}$, $\tau_{zx} = \tau_{xz}$, and $\tau_{zy} = \tau_{yz}$. Therefore only six stresses, called the components of stress, are needed to describe all the stresses operating within the

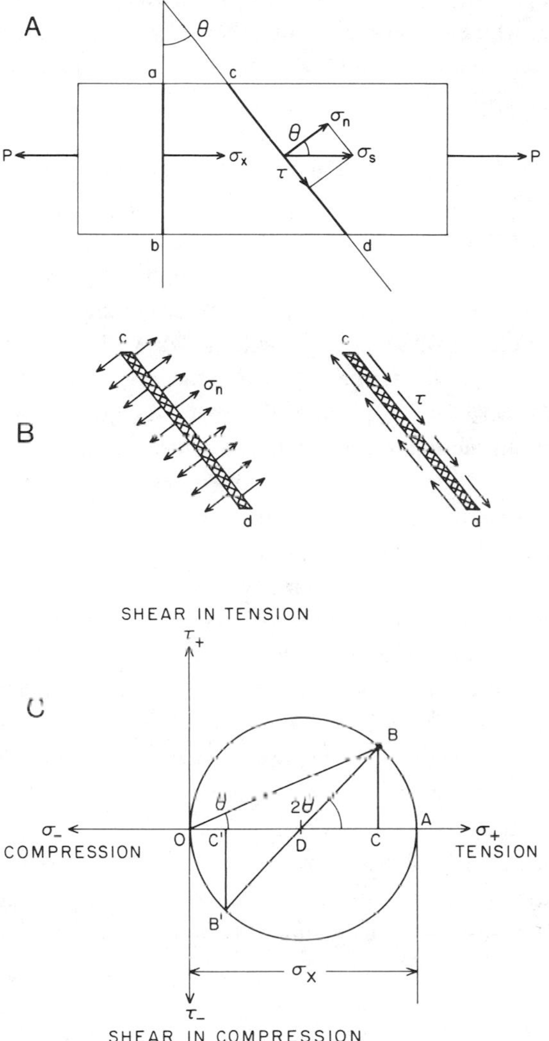

FIGURE 2.2 Normal and shear stress components within a prismatic bar submitted to a tensile force P: (A) The stress σ_s operating parallel to the axis of the bar and along an inclined plane section cd is decomposed into two stress components: the normal stress σ_n and the shear stress τ. The normal stress component reaches its largest magnitude when the inclined plane is rotated so that it is normal to the axis of the bar. The shear stress component reaches its largest magnitude when the inclination angle θ equals 45°. (B) The directions in which the two stress components operate within an inclined section. (C) Graphic technique (known as Mohr's circle) for plotting normal and shear stresses as a function of angle of inclination of the surface of interest.

material element: σ_x, σ_y, σ_z, $\tau_{xy} = \tau_{yx}$, $\tau_{zx} = \tau_{xz}$, and $\tau_{zy} = \tau_{yz}$.

(To readers who hate geometry and graphs, I sadly recommend skipping the next four paragraphs, for which no apologies are offered, though some may be richly deserved!)

The relation between normal and shear stress components within a prismatic bar submitted to uniaxial tension is shown in figure 2.2, from which we can see that the orientation of the plane through which stress components are measured and the magnitude of the applied load P play equally important roles in determining the magnitude and direction of the normal and shear stress components. Two planes of section through the bar will be considered: one section, designated a–b, is oriented normal to the axis of the bar and has an area A, while the other plane of section, designated c–d, is inclined from the normal plane by an inclination angle denoted θ and has a cross-sectional area equal to $A/\cos\theta$. Thus the resultant stress σ_s measured for the inclined plane c–d equals $(P/A)\cos\theta$. Since P/A gives the magnitude of the normal stress σ_x measured for the perpendicular plane a–b, it can be seen that the resultant stress σ_s acting on the inclined plane equals $\sigma_x \cos\theta$. This relationship indicates that the resultant stress σ_s must decrease as θ increases and that, for any inclined plane, $\sigma_s < \sigma_x$. Indeed, when $\sigma_s = \pi/2$ (when the section c–d is parallel to the section a–b), the stress σ_s vanishes, indicating that there is no stress *between* longitudinal material elements in the bar.

The resultant stress σ_s can be decomposed into two components: the normal stress component σ_n and the shear stress component τ (fig. 2.2). From inspection, we see that $\sigma_n = \sigma_s \cos\theta$, where σ_n is the normal stress component acting on the inclined plane c–d. Since $\sigma_s = \sigma_x \cos\theta$, the normal stress component is given by the formula $\sigma_n = \sigma_s \cos\theta = \sigma_x \cos^2\theta$, and the maximum normal stress component $(\sigma_n)_{\max}$ must equal σ_x (when the plane c–d is rotated so that it is normal to the axis of the bar). The tangential force component results in a shear stress τ, which equals $\sigma_s \sin\theta$. Thus the shear stress equals $\sigma_x \cos\theta \sin\theta$, or $\tau = (\sigma_x/2)\sin 2\theta$, from which we can see that the maximum shear stress occurs when $\theta = 45°$, that is, $\tau_{\max} = \sigma_x/2$. Figure 2.2 shows the directions in which the normal and the shear stress components operate within an inclined material element taken through the bar. Although the maximum shear stress is only half the maximum normal stress, the shear stress can be the controlling stress in dictating the ultimate strength of a material, since most materials are much weaker in shear than in tension. Indeed, when metals are stretched, they typically yield along planes inclined at 45° to the normal plane of section. These inclined planes reflect the plane of maximum shear stress.

The relationships among the inclination angle of a section θ and the normal and shear stress components can be represented graphically (fig. 2.2). Such a graph provides a tool whereby the normal and shear stress components acting on any inclined plane can be computed as functions of the normal stress operating in the axial direction and the inclination angle for any surface of interest within the bar. To construct the graph, a system of coordinates is selected having an origin O and orthogonal axes representing shear and normal stresses taking positive (tension) and negative (compression) values. Recall that for the transverse plane a–b ($\theta = 0$), $\sigma_n = \sigma_x$ and $\tau = 0$. This condition of stress conforms to point A on the coordinate system. For a plane parallel to the longitudinal axis of the bar ($\theta = \pi/2$), both stress components must have zero magnitude. This state of stress conforms to the origin of the coordinate system (point O in fig. 2.2). Clearly, we can construct a circle having diameter σ_x corresponding to the range $0 \leq \theta \leq \pi$, from which we can compute the stress components operating along any plane through the prismatic bar. This circle of stress is called Mohr's circle. Consider a plane of section with θ corresponding to the point B on Mohr's circle. The normal stress component acting on this plane has a coordinate value equal to OC, which equals the sum of OD and DC. From inspection of Mohr's circle, we see that $OD = \sigma_x/2$ and $DC = (\sigma_x/2) \cos 2\theta$. Thus, $\sigma_n = OC = \sigma_x \cos^2 \theta$. By the same token, the corresponding shear stress must equal CB, and on inspection we see that $\tau = CB = DB \sin 2\theta = (\sigma_x/2) \sin 2\theta$. (These values correspond to those previously computed analytically for the plane c–d inclined at an angle θ; see fig. 2.2.) It can be appreciated that as θ increases further, point B moves from A to O, with corresponding changes in the magnitudes of the shear and normal stress components expressed as functions of θ. Therefore the upper semicircle of Mohr's circle can be used to determine the magnitudes of the stress components for all values of θ within the limits $0 \leq \theta \leq \pi/2$. If the value of θ is increased beyond the upper range of these limits ($\pi/2 \leq \theta \leq \pi$, as when O moves to B′), then we can obtain a material wedge through the bar. By taking adjacent planes parallel to the planes defined by points on the upper half of Mohr's circle, we can isolate an element of material from the bar and define the normal and shear stress components operating on all surfaces of the element, as shown in figure 2.1.

Mohr's circle is a powerful device. Since we can always compute the normal stress acting on a plane perpendicular to the length of a prismatic bar of material submitted to uniaxial tension or compression ($\sigma_x = \sigma_n = P/A$), we can always compute the normal and the shear stress components acting on *any* plane of section, provided the inclination angle θ is specified. Conversely, we

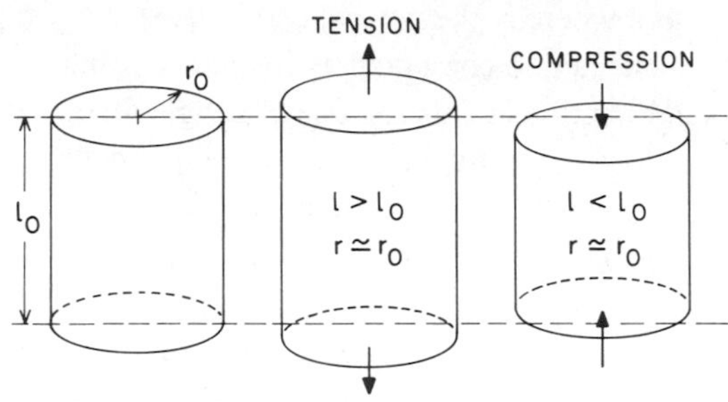

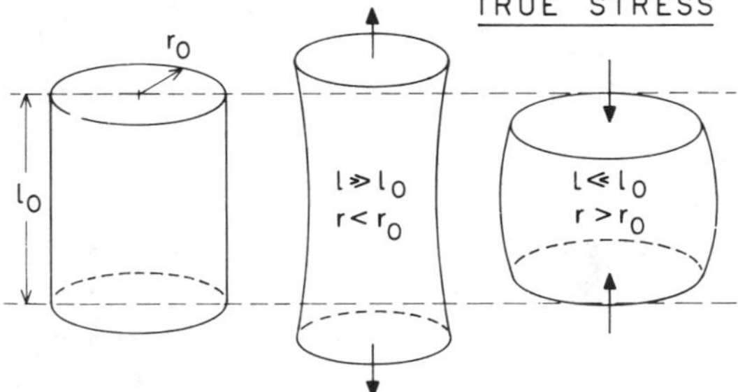

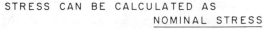

FIGURE 2.3 Two ways stress can be measured. Nominal stress can be calculated for specimens made of materials that deform little over a significant range of either tensile or compressive stress. The original (undeformed) dimension (e.g., the original radius r_o) can be used to calculate the area over which a force is applied. True stress must be calculated when a specimen undergoes significant deformation. Hence the instantaneous dimensions (r) must be used to calculate the area over which a force is applied.

can reverse the problem—if the stress components σ_n and τ are known, we can compute the normal and shear stress components in the axial direction.

The area through which normal axial stresses operate is easily measured for prepared specimens of metals, plastics, excised segments of cylindrical stems and leaves, or plugs of plant tissue. By contrast, measuring the transverse areas of intact biological structures whose cross-sectional geometries naturally vary can pose difficulties. Usually, geometric simplifications of complex or naturally irregular transverse geometries are sufficient for first-order approximations when calculating stresses. Thus we could approximate the pedicel as a circular cylinder with a uniform cross-sectional area without a significant loss of accuracy, even though pedicels are not perfectly circular cylinders. However, in many cases we need greater precision and accuracy. With the aid of computer systems and appropriate software, the transverse areas of even complex and irregular geometries can be measured empirically.

A more serious problem arises when the transverse area of a specimen deforms under an applied load (fig. 2.3). These deformations can take the form of either necking under tension or barreling under compression. Necking is regional attenuation in cross-sectional area owing to the application of tensile forces, while barreling is regional expansion in cross-sectional area owing to compression. In either case, the cross-sectional area exhibits a temporal variation in absolute dimension as the magnitude of the applied external force increases or decreases or as a specimen deforms under a constant stress level. Therefore instantaneous stresses can differ in magnitude from those that immediately precede or follow them. For some fabricated materials, like steel, the changes in transverse area may be small over a very large range of loadings. For these types of materials, stresses may be adequately calculated based on the original cross-sectional area of the unloaded material. This manner of calculating stress yields what is called engineering stress or nominal stress, symbolized by σ_n.

Nominal stresses can be calculated legitimately for materials operating under conditions of loading that do not change appreciably over time or for materials that deform little over relatively large ranges of loading. For example, nominal stress can be used to express the compressive stress at the base of a tree trunk resulting from the combined weight of the trunk and the canopy, provided this total weight changes little during a single growing season or from one year to the next. This is illustrated for the General Sherman tree (*Sequoia sempervirens*), whose trunk was estimated to have a minimum weight of 5.36 MN ($= 5.46 \times 10^5$ kg) in chapter 1. With a basal girth of roughly 25 m, the cross-sectional area of this tree is approximately 50 m^2. Based on these estimates of minimal weight and cross-sectional area, the minimal compressive nominal stress at the base of the tree is roughly 0.1 MN·m^{-2}. Likewise, the compressive stress at the base of the first Kew "flagstaff" (a trunk of a conifer measuring 67.4 m in height with a basal cross-sectional area of 0.223 m^{-2} and total estimated weight of roughly 78,122 N) is estimated to be on the order of 0.35 MN·m^{-2}. The magnitudes of these compressive stresses would be nearly doubled if the weight of branches and leaves had been considered in addition to that of the trunks or if we had taken into account the effects of dynamic wind loadings. Significantly, the ultimate strengths (see below) of conifer woods range between 45 and 114 MN·m^{-2},

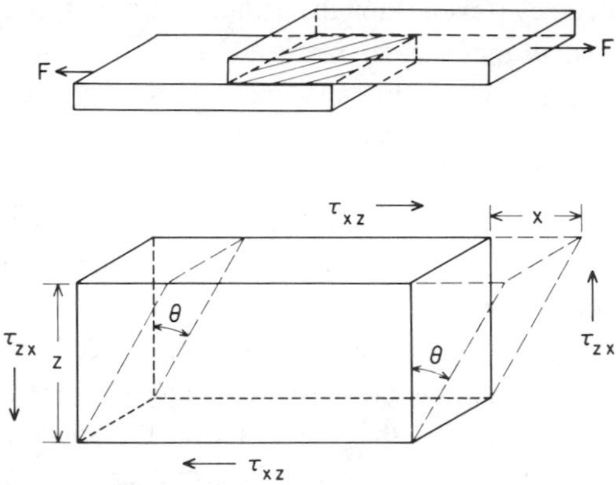

FIGURE 2.4 Shear stress τ is calculated by dividing the shear force (F) applied to a material by the area through which it operates: (A) Two rectangular elements joined by the crosshatched area, $\tau = F/A$. (B) The shear stress components τ_{xz} and τ_{zx} have a corresponding shear strain γ as illustrated by a rectangular element (solid outline) distorted into a parallelepiped (dashed outline) by a shear force, applied tangentially (in the xz plane). The figure illustrates pure shear. The magnitude of the shear strain can be calculated from the translation angle of the specimen's corners θ: $\gamma = \tan\theta = x/z$.

indicating that the compressive stresses experienced at the bases of these statically loaded trunks are well within their allowable stress limits.

Calculations of nominal stresses should be avoided when the conditions of loading change markedly over relatively short spans of time or when the deformations resulting from loadings are large ($\geq$ 5%). Since many plant materials typically deform substantially under low stress levels, and since virtually all plant materials deform more than 5% under moderate to high stress levels, stress should be calculated by dividing the applied force by the *instantaneous* transverse area. This procedure gives a stress called true stress, symbolized by σ_t (see fig. 2.3). The desirability of calculating true stress is illustrated by considering a piece of taffy suspended from a stick. As the taffy deforms under its own weight, the cross-sectional area of the portion holding it to the stick progressively decreases. Although the weight of the taffy below the attachment site does not change significantly, necking and the accompanying increase in tensile stress (resulting from the decrease in cross-sectional area) cause the taffy to deform more rapidly over time. Likewise, when a tree trunk is chipped away, its true stress changes. The magnitude of the loading (the weight of the canopy) on the trunk does not change, but the reduction of the cross-sectional area and the consequent increase in the compressive stress within the trunk leads to a very precarious mechanical situation, so that a relatively small lateral force (a slight push) may cause the tree to fall over. The safety factor of a trunk under self-loading can be approximated by comparing the minimal cross-sectional area required to maintain the tree's vertical posture with the actual cross-sectional area of the tree. Such comparisons reveal that the safety factor typically equals or exceeds four; that is, 75% or more of the cross-sectional area of a trunk can be removed before it begins to fall over under its own weight. (The "excess bulk" of a tree may operate as "ballast" to resist dynamic wind loadings.)

Like normal stresses, shear stresses are calculated by dividing the applied shear force by the area through which they operate (fig. 2.4). For example, when two planks of wood are glued together over a portion of their appressed surfaces and then pulled in directions parallel to their lengths, the shear stress is calculated by dividing the applied force by the interface surface area between the two joined planks. When dealing with shear stress, however, it is important to recognize that a solid can shear in any of three ways: direct shear, simple shear, and pure shear.

Direct shear occurs at the interface between two objects that are forced to slide past one another, as when one skids the palm of one's hand over a tabletop or when two branches rub against one another. In direct shear material moves relative to the surface of another material so that the materials deform little or not at all except as a consequence of friction and the resulting abrasion of surfaces. Simple shear happens when elements of material within a solid slide past one another and simultaneously experience tensile and compressive distortion in the direction of shearing. This is illustrated when a rectangular piece of gelatin is deformed into a parallelepiped by compressing its upper surface and pushing it parallel to the surface of a dessert plate. In a very crude fashion, simple shear is mimicked when a deck of cards is deformed into a parallelepiped by pushing it across a tabletop with one's palm. Each playing card is a crude analogue to a single rectangular material element whose surface moves parallel to the plane of shearing. Pure shear is equivalent to the state of stress produced within a material submitted to tension in one direction and equal compression in the perpendicular direction (Timoshenko 1976a, 57–58). That is, pure shear occurs when the resultant of the normal tensile and compressive stress components operating within an element of material has zero magnitude—that is, the center of Mohr's circle of the material element precisely coincides with the origin of the coordinate system and has a radius equal to $\sigma \pm$. Hence, the adjective *pure* refers to the fact that only the shear force component has magnitude. Pure shear may appear to be an abstraction, in that the normal stress components within the peripheral volume of a sheared solid typically never equal zero. However, at the very center of a large object undergoing simple shear, the net magnitudes of the normal tensile and compressive stress components approach zero, and the condition of pure shear is realized. In this sense simple shear is actually a special case of pure shear. Indeed, the behavior of a material in pure shear is often examined by placing a large specimen of the material in simple shear. Likewise, the condition of pure shear can be realized by an element of a material on the surface of a circular tube subject to torsion resulting from a very small rotation of one end of the tube with respect to the other. Thus the epidermis of a slightly twisted cylindrical stem may undergo pure shear. The boundaries of the originally undeformed element of epidermis become inclined relative to the axis of the stem as a torque is applied, and the element deforms through pure shear. We shall return to a consideration of simple and pure shear in the next chapter, where we will consider torsion in greater detail.

When a material is bent or twisted, bending stresses, torsional shear stresses, or both may develop. In bending, the maximum shear stress within any cross section develops at or near the center of the bar where material elements are forced to slide past one another. In torsion, the maximum shear stress within any cross section develops toward the perimeter. We will treat the distribution of stresses within different geometries subjected to applied forces in chapter 3. For now, the significance of bending shear stresses can be illustrated by considering a tree branch. Failure in bending generally begins as a tensile fracture, usually initiated in the most recent (outermost) annual layer of wood on the upper surface of the branch. This fracture is often quickly followed by a shear failure at the interface between xylem cells deposited in the summer and in the spring (summer and spring wood), which differ in their material properties (see Vautrin and Harris 1987). As failure progresses, shearing typically occurs between the wood and the bark above and below the attachment site of the branch to the tree trunk, causing a large ellipsoidal wound along the length of the trunk. This is particularly evident when branches fail under snow loadings, but it varies among species. The branches of willow (*Salix*) species typically shear farther from their base, whereas oak (*Quercus*) undergoes a minimum of shearing even when heavily loaded with snow and ice. The young shoots of grape (*Vitis*), when bent early in the growth season, typically snap off at the base. This failure is initially due to tension developing at the top of shoots, but it quickly develops into shearing failure at the interface between the vascular and pith tissues.

Leaf tissues also shear. The simple leaves of banana (*Musa acuminata*) and the traveler's palm (*Ravenala*) shear between the parallel vascular bundles, where wind pressure pulls the softer, nonvascular tissues apart. Young leaves or the older leaves of plants sheltered from the wind suffer little damage, but the leaves of unsheltered plants can be shredded so much that they die. By contrast, the leaves of most dicots, which have a reticulated vascular network, show little evidence of shearing, in large part because the softer tissues within these leaves are held together by an interweaving fabric of stiffer, stronger tissues.

STRAINS

The intensity of the deformations that result when a material is subjected to a force is called strain. Strain is a dimensionless quantity that reflects the ratio of the magnitude of a deformed dimension to that of the undeformed dimension. Strains can be expressed as numbers or as percentages (e.g., 0.05 or 5%). Since there are three normal stress components and three shear stress components, there are six components of strain. Three of these are the unit elongations measured along the three Cartesian axes of the material. These unit elongations are the deformations resulting from the three normal stress components. (A contraction in a dimension can be thought of as a negative elongation.) These strain components, symbolized by ε_x, ε_y, and ε_z, are referred to as the normal strain components. The remaining three components of strain are the unit shear strains, symbolized by γ_{xy}, γ_{xz}, and γ_{yz}, which are called the shear strain components. When a material is subjected solely to uniaxial tension or compression, the shear strain components have zero magnitude. But whenever forces are applied tangentially to a material's surface or when a material is bent or twisted, the resulting shear strains must be considered.

Normal strains can be calculated in one of three ways. Each method can yield a different absolute value of strain (even for the same material subjected to the same stress level), depending on the magnitude of the strains that develop. Therefore it is vital to know (and to report) precisely how strains are calculated. The first method of calculating strains gives Cauchy strains or engineering strains or conventional strains—all three terms are synonymous. These strains are calculated by dividing the difference between any deformed dimension and the original dimension by the original value of the dimension:

$$\varepsilon = \frac{l - l_o}{l_o} = \frac{l}{l_o} - 1, \tag{2.1}$$

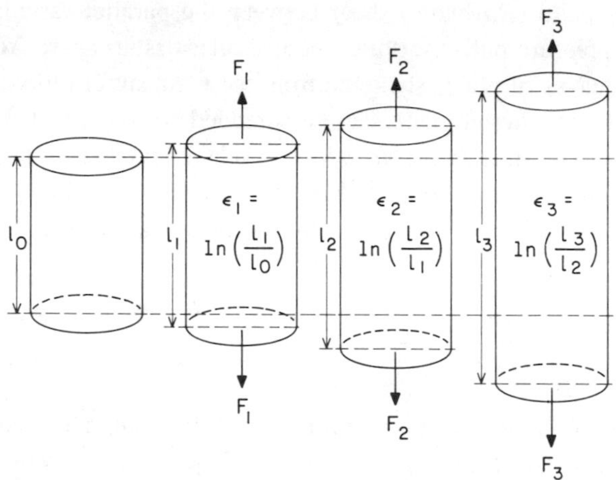

FIGURE 2.5 True strains ε_t calculated for a cylinder subjected to increasing uniaxial tension resulting from the application of tensile forces with increasing magnitudes (F_1, F_2, and F_3). As a result of the tensile forces, the cylinder extends in length and contracts in girth. The true strain is the natural logarithm (ln) of the extension ratio. The extension ratio is the ratio of the magnitude of the instantaneous dimension to that of the previous dimension (e.g., l_1/l_0 which is the extension ratio of the increase in length due to the application of the tensile force F_1 to the original length).

where l_o is the original dimension and where the fraction l/l_o is called the extension ratio. The extension ratio will be greater than unity when measured in tension and less than unity when measured in compression, reflecting the extension and contraction in the dimension of the material. Therefore conventional tensile and compressive strains have positive and negative values, respectively.

The second method of calculating normal strains gives Henchy strains or true strains or natural strains, symbolized by ε_t (fig. 2.5). The strain is calculated from the integral of the change in the reference dimension over the limits of the original to the altered dimension. This integral is the natural logarithm of the extension ratio:

$$\varepsilon_t = \int_{l_o}^{l} \frac{dl}{l} = \ln \frac{l}{l_o}. \tag{2.2}$$

The natural logarithm essentially expresses the compound interest law, where the true strain provides a reference strain for each increment of deformation relative to the preceding strain; that is, the interest (each incremental increase in the reference dimension) is added to the capital (the previous total dimension). True strains should be measured for any material exhibiting significant deformation ($\geq 5\%$). Clearly, true and conventional strains are mathematically related, $\varepsilon_t = \ln(1+\varepsilon)$, and conventional and true strains yield essentially identical values for small strains ($< 5\%$). However, it is always wise to plot a given type of strain against its comparable type of stress; we should never mix types of stresses and strains.

Typically, true strains (and stresses) provide a more realistic picture of mechanical phenomena than do conventional strains, particularly when materials undergo large elastic or plastic deformations. Consider a rubber band placed in uniaxial tension. If its original length is 10 cm and it is extended to a length of 11 cm, then the extension ratio equals 1.1 and the true strain equals 0.095, while the conventional strain equals 0.10. If the same rubber band is extended from 25 to 26 cm, then the true strain equals 0.039, whereas the conventional strain is 1.6 (where 10 cm is used as the original dimension). Indeed, some very simple calculations will reveal that a conventional extension strain of about 54 is equivalent to a true strain of about 4. Clearly, the two strains produced by equivalent extensions (1 cm increments, in the previous examples) are not equal, and deformation is best dealt with in terms of the magnitude of the dimension immediately preceding each incremental deformation.

The third way normal strains are calculated is in terms of the stretch ratio, symbolized by λ, which is simply the extension ratio of the material: $\lambda = l/l_o$. Normal strains of rubbery materials are typically (but not exclusively) reported in terms of the stretch ratio.

Unlike the normal strain components, the shear strain components in materials subjected to simple or pure shear must be calculated in terms of the substantial gradient of deformation that typically occurs within materials. (Tensile or compressive strains also have gradients of deformation but, because they tend to be very small, their gradients are largely neglected.) The gradient of shear strain is given by the formula

$$\gamma = \frac{dx}{dz}, \tag{2.3}$$

where x is the deformation in the axis that parallels the direction of the shearing force and z is the vertical dimension. This relationship is illustrated for a rectangular element of material deformed into a parallelepiped (see fig. 2.4). From inspection, $\gamma = \tan\theta = x/z$, where θ is the rotation angle. The shear strain also equals the deformation of the entire rectangular element divided by the width of the element.

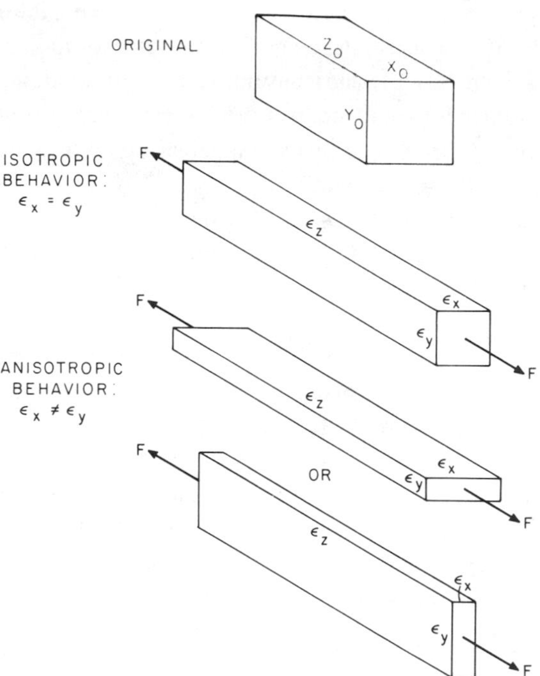

FIGURE 2.6 Poisson's ratios and the behavior of a prismatic element composed of isotropic and anisotropic materials subjected to uniaxial tension. The undeformed shape of the element, shown at the top, has depth x, width y, and length z. A tensile force applied parallel to z results in a longitudinal strain ε_z and two lateral strains ε_x and ε_y. The negative ratio of each of the two lateral strains to the longitudinal strain is called Poisson's ratio, symbolized by ν. For an isotropic material, the two Poisson's ratios are equivalent: $\varepsilon_x = \varepsilon_y$ and $\nu_{zx} = -\varepsilon_x/\varepsilon_z = \nu_{zy} = -\varepsilon_y/\varepsilon_z = \nu$. For an anisotropic material, $\varepsilon_x \neq \varepsilon_y$ and the two Poisson's ratios are not equivalent.

POISSON'S RATIOS

Virtually every common material undergoes lateral contraction when it is stretched and lateral expansion when it is compressed. The former is seen when a rubber band is placed in uniaxial tension; it simultaneously elongates in length and contracts in transverse area. The relative magnitudes of these deformations are governed by a material property known as Poisson's ratio, named in honor of the French mathematician and mechanicist Siméon-Denis Poisson (1781–1840). Poisson's ratio, symbolized by ν, is defined as the negative transverse strain divided by the axial strain in the direction of the externally applied force; that is, $\nu = -\varepsilon_t/\varepsilon_n$, where t denotes the transverse direction and n the direction of the normal axial stress component. Returning to the example of a rubber band submitted to very large tensile stresses, if the axes of depth, width, and length are designated by x, y, and z, then two Poisson's ratios describe the relation among the two transverse strains (fig. 2.6):

(2.4) $$\nu_{zx} = -\frac{\varepsilon_x}{\varepsilon_z}, \text{ and } \nu_{zy} = -\frac{\varepsilon_y}{\varepsilon_z}.$$

The first letter of each Poisson's ratio identifies the axis parallel to the direction of the applied force (in this example, the z-axis, or length), while the second letter indicates the dimension in which the transverse strain has been measured (the x- and y-axis, or depth and width). The Poisson's ratio provides an important material property because it measures the ability of a material to change in volume versus its ability to change in shape.

For most commonly occurring materials, $0 \leq \nu \leq 0.5$, although negative Poisson's ratios are thermodynamically possible (see below). As the magnitude of ν increases and approaches its maximum theoretical value of $+0.5$, the material will tend to increasingly resist a change in its volume but will tend to increasingly respond to stress by changing its shape. Accordingly, fluidlike materials have Poisson's ratios approaching 0.5, whereas crystalline solids tend to have comparably low Poisson's ratios. For example, the Poisson's ratio of cellulose, measured in the transverse direction to extension under tensile stress along its polymeric chain length, equals 0.10 (Mark 1967, 119, table 5–1), indicating that this carbohydrate changes its transverse dimensions little in response to tension. Thus cellulose has some of the material properties of a crystalline solid and provides great strength to cell walls placed in tension (see chap. 5).

For some materials, the two Poisson's ratios are equivalent in magnitude, and only one Poisson's ratio ν is required to describe the relation between the two transverse strains: $\nu_{zx} = \nu_{zy} = \nu$. Such a material is said to be isotropic. The elastic material properties of isotropic materials are the same regardless of the direction in which they are measured. For such materials, Poisson found $\nu = 0.25$, based on his analytical investigation of the molecular theory of the structure of materials. Isotropy is a consequence of an extreme molecular or infrastructural homogeneity within a material. Most metals evidence isotropic mechanical behavior and tend to have Poisson's ratios within the range 0.25–0.30. One of the symptoms of isotropic materials is that they shear in tension along planes inclined 45° to the axis submitted to tension. This property of isotropic materials is explicable in terms of the relation between the magnitudes of the maximum normal and shear stress components

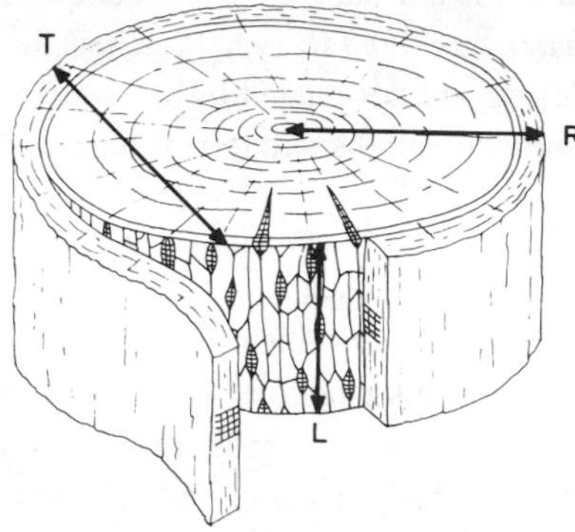

FIGURE 2.7 Three axes (designated by L, T, and R) are required to specify the anisotropic mechanical behavior of wood shown for a portion of a tree trunk (outer layer of cork has been partially peeled back to reveal wood; xylem rays and tracheary elements are not shown to scale). The longitudinal axis (L) runs along the grain of the wood; the tangential axis (T) runs along any tangent through a transverse section; the radial axis (R) runs along any radius through a transverse section. With the aid of these three axes, planes of interest can be specified for which values for Poisson's ratios and material moduli (elastic and shear moduli) may be determined. See text for further description.

resulting from uniaxial tension (see fig. 2.2).

By contrast, other materials have nonequivalent lateral strain components. These materials are called anisotropic materials. Anisotropy is illustrated by examining a rubber band that, when at rest, has a square transverse geometry but can be extended to produce a rectangular cross section, indicating that the two lateral strains are not equivalent. (At small strains, rubber behaves as an isotropic material; at large strains, it is anisotropic.) Anisotropic mechanical behavior indicates that a material has a molecular composition or microstructure evincing preferred orientations. (Rubber is a cross-linked polymer whose molecular structure reorients under tension.) Most plant tissues conferring mechanical support, such as wood and sclerenchyma, show varying degrees of anisotropy as a result of their cellular heterogeneity, the preferred orientation of their constituent cells, or both.

It is rarely feasible to treat extreme anisotropy without mathematical burden, because the principal directions of symmetry in extremely anisotropic materials can be numerous. However, anisotropic materials include axisymmetric and orthotropic materials. Axisymmetric materials have two mutually perpendicular directions of symmetry—they have equivalent material properties when measured in two of their three Cartesian dimensions. Bottle cork (the cork from *Quercus suber*) is an example. The cells in this tissue have more or less equivalent longitudinal and tangential dimensions but are prismatic in the radial direction, and when they are pulled or compressed they exhibit similar material properties in two of their three dimensions. The Poisson's ratio of cork is nearly zero when measured in compression along the prismatic axis. Thus corks can be pressed uniformly into the necks of wine bottles. (Although mechanically useful, this orientation is not desirable, since it permits wine to flow through the tubular channels within cork, called lenticels.) Orthotropic materials have three mutually perpendicular directions of symmetry and have different material properties when measured in each of their three Cartesian dimensions. Wood is an excellent example of a biomaterial that approximates an orthotropic material (fig. 2.7). Although many of the cell types in wood are axisymmetric in their individual mechanical behavior, the anatomical juxtaposition of these cell types results in three principal directions of symmetry: the longitudinal direction along the grain (denoted by L) and the radial (R) or tangential (T) direction to the grain. Although orthotropic materials require three Poisson's ratios, wood requires six because of the presence of growth rings. The six Poisson's ratios required to treat the anisotropy of wood are ν_{LR}, ν_{RL}, ν_{LT}, ν_{TL}, ν_{RT}, and ν_{TR}. (As in eq. 2.4, the first letter in each of these subscripts indicates the direction of the applied force, while the second letter indicates the direction in which the transverse strain is measured.) For balsa wood, the values for the six Poisson's ratios are 0.229, 0.488, 0.665, 0.217, 0.011, and 0.007; for yellow birch the ratios are 0.426, 0.451, 0.697, 0.447, 0.033, and 0.023. Although the absolute values of the six Poisson's ratios vary among species, for most samples of wood ν_{LT} is typically the largest of the six. More important, however, an average Poisson's ratio for a material like wood is a meaningless quantity.

By contrast, many parenchymatous plant tissues are reported to be isotropic (or nearly so). These tissues show a significant range in their Poisson's ratios; for example, 0.23 for cornstalks (Prince and Bradway 1969), 0.32 for endosperm (the storage tissue within seeds; Prince and Bradway 1969), 0.21–0.34 for apple flesh (Chappell and Hamann 1968), and 0.49–0.5 for parenchyma isolated from potato tubers (Finney and Hall 1967). The isotropic behavior of these tissues is consistent with the geometry of their constituent cells. Parenchymatous tissues are often composed of nearly isodiametric cells that have little or no preferred orientation with respect to the tissue as a whole or the organ that contains it. The Poisson's ratio of some parenchyma approaches or equals +0.5, indicating that these tissues evince a fluidlike material property; that is, they can mechanically resist compression but have little capacity to resist shearing. Such parenchymatous tissues typically shear in planes oriented 45° to the axis submitted to tension.

Before leaving the general topic of Poisson's ratio, it is worth noting that *negative* ratios are theoretically permissible for materials. That is, a material can expand orthogonally to the direction of an applied tensile load or can contract orthogonally to the direction of an applied compressive load. Based on thermodynamic considerations, the allowable range of Poisson's ratio even for an isotropic material can be -1.0 to $+0.5$. Negative Poisson's ratios have been reported for some synthetic polymers whose polymeric units separate laterally as they are extended in length. Negative Poisson's ratios have also been reported for some foams—commercially fabricated cellular solids with a beamlike or strutted infrastructure (Lakes 1987). These materials have numerous commercial applications, since they can inflate when they are pulled. They may also have botanical analogues. For example, some aerenchyma (a spongy tissue consisting of numerous strutlike interconnected cells) is anatomically similar to commercially fabricated foams. To my knowledge, no one has measured the Poisson's ratios of aerenchyma or plant organs that characteristically have a spongy infrastructure consisting of many beams and struts attached to an external wall. However, since aerenchyma is attached to the walls of otherwise hollow stems and leaves, and since these leaves and stems experience tensile and compressive stresses when they bend, negative Poisson's ratios would confer many mechanical advantages. It would be particularly interesting if the aerenchyma found in some roots had a negative Poisson's ratio, since this would produce an expansion of the root when it is placed in tension, anchoring it more firmly in its substrate.

THE MATERIAL MODULI OF ISOTROPIC MATERIALS

The relation among different types of stresses and strains, expressed in terms of ratios called material moduli, further reveal the properties of isotropic materials. Since stress has units of force per area and since strains are dimensionless, these moduli have the same units as stress. The first material modulus treated here is called the bulk modulus, symbolized by K. The bulk modulus is the ratio of a uniformly applied hydrostatic pressure, symbolized by $\mathcal{P}$, to the volumetric compressive strain, $\Delta V/V_o$ (sometimes called the cubical dilatation): $K = \mathcal{P}/(\Delta V/V_o)$. Since the magnitude of the pressure, which has units of force per area, expresses the magnitude of the resulting hydrostatic stress, and since $\Delta V/V_o$ expresses volumetric strain, which is dimensionless, K is clearly a ratio of stress to strain and has the same units as stress.

When hydrostatic pressure is applied to the exterior surface of a compressible material, the volume of the material will decrease and its density will increase. The reciprocal of K is called the compressibility of a material. Clearly, materials that have a high Poisson's ratio and resist a change in volume will have very high bulk moduli and very low compressibility. Conversely, materials with low bulk moduli have a high compressibility and densify relatively easily when subjected to modest or high hydrostatic pressure. Indeed, the bulk moduli of liquids approach infinity, whereas the bulk moduli of aerenchymatous plant tissues, which have many large air spaces within them, approach zero.

With the aid of Poisson's ratio, the bulk modulus of isotropic materials can be mathematically related to two other material moduli that reflect the ratio of normal stress to normal strain and the ratio of shear stress to shear strain. The former, symbolized by E, is called the elastic modulus, modulus of elasticity, or Young's modulus (in honor of Thomas Young, 1773–1829, who first drew attention to it). $E = \sigma/\varepsilon$. The elastic modulus of an isotropic material can be derived by superimposing the three normal strain components resulting from triaxial stresses (σ_x, σ_y, σ_z) to yield three formulas:

(2.5a) $$\varepsilon_x = \frac{1}{E_x}[\sigma_x - \nu\,(\sigma_y + \sigma_z)];$$

(2.5b) $$\varepsilon_y = \frac{1}{E_y}[\sigma_y - \nu\,(\sigma_x + \sigma_z)];$$

(2.5c) $$\varepsilon_z = \frac{1}{E_z}[\sigma_z - \nu\,(\sigma_x + \sigma_y)].$$

Since the material properties of an isotropic material are equivalent when measured in any dimension, $E_x = E_y = E_z = E$, and since the cubical dilatation ($\Delta V/V_o$) is mathematically related to the three normal strain components, eq. (2.5) can be much simplified to equate the elastic modulus with the bulk modulus:

(2.6) $$E = 3K(1 - 2\nu).$$

Since Poisson's ratio typically ranges between 0 and $+0.5$, E can range between 0 (for a fluid) and $3K$ (for a crystalline solid).

Likewise, the relationship between Poisson's ratio and the capacity of a material to resist shear can be mathematically formalized for isotropic materials. The ratio between shear stress τ and shear strain γ, symbolized by G, is called the modulus of shear or the shear modulus: $G = \tau/\gamma$. For isotropic materials, the shear modulus is related to E and ν by the formula

$$G = \frac{E}{2(1+\nu)}, \tag{2.7}$$

while G can be related to K by the formula

$$G = \frac{3K\,(1-2\nu)}{2\,(1+\nu)}. \tag{2.8}$$

Thus, for an isotropic material, the shear modulus is proportional to either the elastic modulus or the bulk modulus and inversely proportional to Poisson's ratio. Since some types of parenchyma have a Poisson's ratio equal to +0.5, we would anticipate that their shear modulus is roughly equal to one-third their elastic modulus. Indeed, experiments indicate that this prediction often holds true.

Solving eq. (2.8) explicitly for the Poisson's ratio yields $\nu = 0.50 - E/6K$. Thus, when $\nu = 0$, $E = 3K = 2G$, whereas when ν approaches +0.5, $E \approx 3G$ and K approaches infinity; that is, very large compressive hydrostatic pressures effect little change in volume. Finally, for most isotropic materials $3K >> G$, so that $E \approx 3G(1 - G/3K)$, which reveals that the ratio of normal stresses to normal strains (E) is largely controlled by the ability of a material to resist shearing (G) rather than its ability to change volume (K).

Equation (2.7) has been used to determine the shear modulus of the chemically amorphous matrix within the cell walls of secondary xylem cells (Mark 1967, 143–44). The Poisson's ratio of the matrix was taken to be 0.30, while the elastic modulus of the matrix was estimated to be 2 $GN{\cdot}m^{-2}$ based on empirical measurements. Substituting these values into eq. (2.7), the shear modulus of the cell wall matrix was calculated to be roughly 0.77 $GN{\cdot}m^{-2}$. Although the generic assumption that the cell wall matrix has a constant Poisson's ratio must be questioned, this assumption is reasonable in the case of secondary cell walls, which undergo little chemical modification once they are fully lignified. Unfortunately, eq. (2.7) cannot be used to infer G from E for anisotropic materials such as the cellulosic fibrillar infrastructure of the cell wall. And even if it could, these values would not provide us with the elastic moduli of the cell wall as a whole, since the wall is a composite structure whose mechanical behavior cannot be legitimately inferred from the behavior of any of its constituent parts.

Plants have gotten around the relation between low compressibility and high susceptibility to shearing by virtue of their composite tissue construction and the capacity to surround compressible materials like parenchyma with shear-resistant materials like sclerenchyma, which have thick-walled, lignified cells. The parenchymatous cores of stems and petioles operate as incompressible fluidlike devices in conjunction with their shear-resistant rinds. Different tissues are deployed within stems and leaves and roots so that their material properties largely conform to the nature (and sometimes the magnitude) of the types of stresses that typically develop in different portions of a single organ. In this regard, plant anatomy is strikingly different for organs that experience different types of stresses, and anatomical differences among organs can be used to infer the nature of the forces that are typically experienced once we know the material properties of different tissues. Indeed, one of the tasks of plant biomechanics is to relate anatomical differences within and among plants to the nature of the mechanical forces that organs typically experience during their functional lifetimes. This task requires us to pay attention to anatomy and to test the mechanical attributes of different tissues—topics that will be treated in chapter 6.

All of the foregoing assumes that the Poisson's ratio of an isotropic biomaterial is a constant. For many types of materials this assumption is valid. For most living plant materials, however, the Poisson's ratio can change, and great caution must be exercised in using any single Poisson's ratio to infer mechanical behavior under all loading conditions. For example, Chappell and Hamann (1968) report that the initial values of the Poisson's ratio of apple flesh (a parenchymatous tissue with a Poisson's ratio within the range 0.21–0.34) decreases over time under a constant stress level. The observed decrease in ν was adequately described by the formula $\nu = at^{-b}$, where a and b are empirically determined coefficients and t is time. Under a given loading regime the greatest decrease in ν (about 16%) was observed during the first thirty seconds of testing. But higher stress levels resulted in a more rapid decline in ν; that is, the Poisson's ratio of apple flesh depends on both the duration and the magnitude of loading. Since a decrease in ν indicates a decrease in the resistance of a material to a change in volume and an increase in the ability to resist shear, over time and under a constant stress level, apple flesh appears to have the capacity to change its state from one like that of a liquid to one tending more toward that of a solid. This change in state appears to accelerate as the level of stress is increased. From a functional perspective, the alterations in the Poisson's ratio reported for apple flesh could confer a mechanical benefit, since the tissue can resist shearing over relatively short time intervals. More important, the behavior of apple flesh highlights the desirability of considering the Poisson's ratios of organic materials as potential variables rather than as constants.

THE ELASTIC MODULUS: THE RELATION BETWEEN NORMAL STRESSES AND STRAINS

When discussing Poisson's ratio, we saw that stresses and strains are related to one another by material moduli (E, G, and K) and that these moduli are interrelated for isotropic materials in terms of the Poisson's ratio; for example, the deformations of an isotropic elastic material can be calculated from known increments of stress if only two material constants, ν and E, are specified. Some biological materials and many fabricated materials, such as metals, can be treated as isotropic elastic materials, or nearly so; therefore ν and E can be used exclusively to predict mechanical behavior.

For anisotropic materials, however, the relation between stresses and strains and the material moduli must be empirically determined. For these materials the material moduli must be reviewed in greater detail, starting with elastic materials for which stress and strain are proportionally related to one another (linearly elastic materials) and then progressing to a treatment of materials for which stresses and strains are not proportionally related (nonlinearly elastic materials). As in the previous sections, we will need to distinguish between normal and shear stresses and strains.

The relation between normal stresses and strains was first explored by Robert Hooke (1635–1702) when he was twenty-five years of age. (He himself typically spelled his family name without the *e*.) Hooke suspended long metal wires from the ceiling and measured the distance between them and the floor after attaching various weights to their free ends. At the time, this procedure was the only reliable way of measuring the very small tensile strains that typically develop in metals. Hooke was somewhat suspicious of his colleagues, and he published his findings in 1676 in the form of an anagram (all the letters were arranged in alphabetical order) that when translated read "ut tensio sic vis" (as the extension, so the force). This single phrase summarizes the behavior of all linear elastic materials. It also expresses what has come to be known as Hooke's law. Hooke finally interpreted his data in 1678, two years before his rival, Edme Mariotte, independently discovered the same relationship.

Linear elastic materials (those that obey Hooke's law) are those for which normal stress and normal strain are linearly proportional to one another: $\sigma \propto \varepsilon$. Steel and, to a limited extent, some plant fibers exhibit linear elasticity over various ranges of loadings. This can be seen by plotting normal tensile or compressive stresses against the resulting strains to produce a graph called a stress-strain diagram. The slope measured at any point along the plot is the ratio of stress to strain. For a linear elastic material, the slope measured along

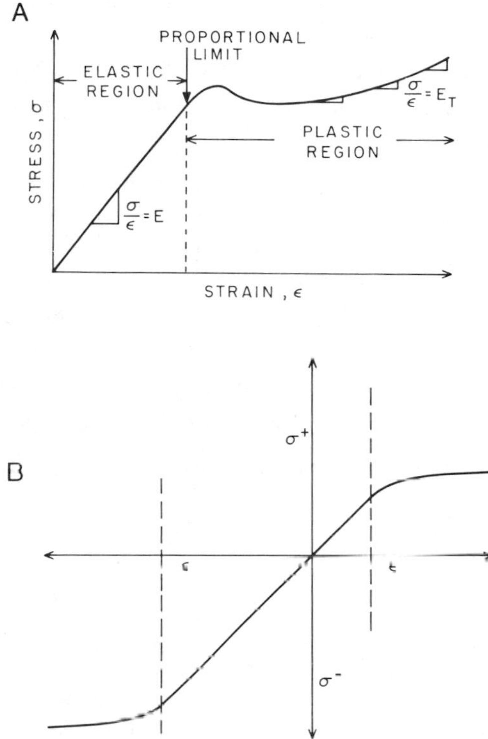

FIGURE 2.8 Linear elastic behavior and plastic behavior as revealed by a material's stress-strain diagram: (A) Linear elastic behavior is evinced by an initial linear region on the stress-strain diagram. The slope of this portion of the stress-strain diagram is called the elastic modulus (or Young's modulus), symbolized by E: that is, $E = \sigma/\varepsilon$. The portion of the stress-strain diagram where stress and strain are proportionally related to one another by the constant E is called the elastic region. The plastic region of the stress-strain diagram is nonlinear, and the ratio of stress to strain varies within this region depending on where the slope is measured. The slope measured anywhere on the stress-strain diagram is called the tangent modulus (E_T). The proportional limit is the stress level below which elastic behavior is manifest. (B) The proportional limits of the same material measured under tensile and compressive stress ($\sigma+$ and $\sigma-$, respectively) may not be equivalent even if the elastic moduli measured in tension and compression are equivalent.

the linear portion of the plot is called the elastic modulus, E (fig. 2.8A). That is, E is the proportionality factor relating normal stresses to normal strains: $\sigma = \varepsilon E$ or $E = \sigma/\varepsilon$. Since the stresses that develop within a material depend on the types of atomic bonds within the material, E provides a mechanical parameter that is frequently used to distinguish one linear elastic material from another. For isotropic linear elastic materials, the values of E measured in tension and in compression are identical.

Linear elastic materials with a high elastic modulus, such as structural steel, are desirable when assembling a stiff structure designed to resist tensile or compressive stresses. Conversely, materials with low elastic moduli may be used if large deformations are functionally desirable. Indeed, in some biological circumstances, stiff structures are not always the best from a functional perspective. For example, Lee (1981) examined ash (*Fraxinus americana*) and calculated the elastic modulus of the cell walls in phloem (the tissue that is specialized to conduct cell sap). The elastic modulus of these cell walls has a range of 5.6–7.4 $MN \cdot m^{-2}$, very low compared with that of most fabricated materials. But these low values permit the cell walls to undergo very large circumferential deformations as a function of internal pressure; for example, a pressure change of 0.3–0.4 $MN \cdot m^{-2}$ would result in a 10% change in the radius of the cell, which, under the right conditions, could increase its capacity to contain sap. This potential for large deformations has an important bearing on mathematical models (which typically rely on measurements of cell diameters made on dead, essentially unpressurized tissues) treating the flow of cell sap in trees, since the rate of flow through a phloem cell would depend in part on the radius of the cell.

When we treat axisymmetric or orthotropic elastic materials, two or three Young's moduli must be measured. Unfortunately, the literature on plant materials rarely provides all of the elastic moduli (or the Poisson's ratios from which some of the elastic moduli could be calculated). Nonetheless these elastic moduli are essential, since for axisymmetric materials $E_x \neq E_y = E_z$ or $E_x = E_y \neq E_z$, whereas for orthotropic materials $E_x \neq E_y \neq E_z$. The elastic modulus of wood submitted to uniaxial compression along the direction of the grain, symbolized by E_L, can differ by one or two orders of magnitude from the elastic moduli measured in the tangential and radial directions to the grain (denoted by E_T and E_R; see fig. 2.7). For balsa, $E_L = 3.12$ $GN \cdot m^{-2}$, while $E_R = 0.144$ $GN \cdot m^{-2}$ and $E_T = 0.0468$ $GN \cdot m^{-2}$. Thus, $E_T/E_L = 0.015$ and $E_R/E_L = 0.046$, indicating that balsa can sustain smaller lateral loadings than loadings along the grain. Unfortunately, data for E_R and E_T from other wood species are not extensive, but those that are available indicate that when a tangential elastic modulus is required but empirically unavailable, a reasonable approximation is given by $E_T \approx 0.06E_L$. Note, however, that this gives a poor estimate in the example given above.

Beyond a certain stress level, the stress-strain diagram of even a linear elas-

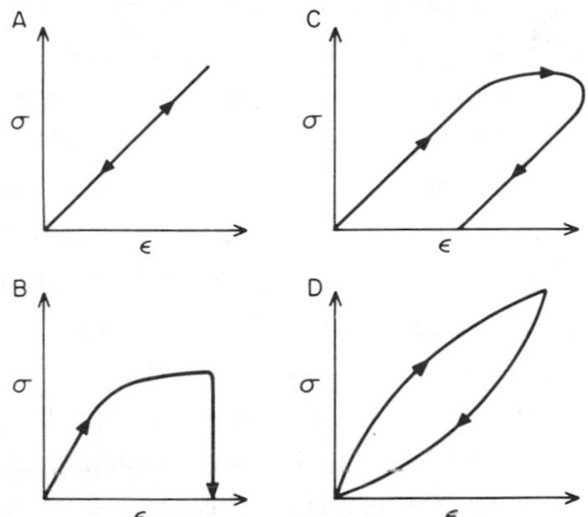

FIGURE 2.9 Comparisons among the stress-strain diagrams of an ideal elastic material (A), a hypothetical material showing no elastic recovery (B), an elastic-plastic material (C), and a viscoelastic material (D). Viscoelastic behavior is reflected by the curvilinear accumulation and recovery of strains. Viscoelastic materials may fully recover their deformations when unloaded (as shown here), or they may only partially recover them.

tic material can become nonlinear or plateau, indicating that some plastic deformations have occurred, or the stress-strain diagram can abruptly truncate, indicating that the specimen of the material has broken. In either case, the maximum stress for which the strain is directly proportional is known as the proportional limit, sometimes referred to as the elastic limit. The proportional limit defines the maximum stress level below which we can anticipate that a material will continue to behave as a linear elastic material and still be restored to its original dimensions when stresses are relieved (see fig. 2.8A). The proportional limit of a material may not be equivalent when measured in tension and in compression, however (fig. 2.8B). Thus a branch that deflects under a wind pressure that induces stresses below the proportional limit of wood measured in compression may still fail as a result of tensile stresses that exceed the proportional limit of wood measured in tension. If neither of the proportional limits is exceeded, then the branch will rebound to its original position when the wind subsides. Unfortunately, the proportional limits in tension and compression of many plant materials are not reported in the literature, and they cannot be inferred from the elastic modulus of a material. The former is the magnitude of the stress beyond which stress and strain are no longer proportionally related, while the latter is simply the proportionality factor between stress and strain. For example, the elastic modulus of structural steel

(0.15–0.25% carbon content) is about 170 GN·m^{-2}, while its proportional limit ranges between 0.17 and 0.23 GN·m^{-2}. Similarly, the elastic modulus of a typical clear-grained specimen of pine wood may be as high as 8.51 GN·m^{-2}, while the maximum stress at which elastic behavior is still retained may be as low as 0.045 GN·m^{-2}. Thus mechanical behavior cannot be inferred simply from values of E, nor can it ever be divorced from the range of loadings experienced. Indeed, when estimating the safety factor of plant tissues or organs (how closely the loads actually experienced relate to the maximum allowable loads above which failure would ensue), the allowable or working (safe) stress of plant materials (the maximum stress for which elastic behavior of material is retained) must be used in calculations (see chap. 7).

When an ideal linear elastic material is loaded and unloaded, the two portions of its stress-strain diagram will be identical provided the range of loading does not exceed the proportional limit (see fig. 2.9A). This is rarely seen for plant materials, even for those that exhibit a linear stress-strain relationship when initially loaded. Rather, the unloading portion of the stress-strain diagram is nonlinear. The nonlinear recovery of strains in an elastic material is called anelastic behavior. Two terms, the degree of elasticity and elastic hysteresis, are useful in describing anelastic behavior. The degree of elasticity is defined as the ratio of the elastic (recovered) deformation to the total deformation when a material is loaded to a given stress level and then unloaded to zero stress. The degree of elasticity of a perfectly elastic material is unity, indicating that the recovered deformation precisely equals the total deformation (fig. 2.9A). By contrast, the degree of elasticity for perfectly plastic materials equals zero, indicating that no deformation is recovered when the stress level drops to zero (fig. 2.9B). For an elastic-plastic material, the degree of elasticity can vary widely depending on how far loading has induced plastic deformations (fig. 2.9C). Between these two extremes there exists much variation. For example, the degree of elasticity for potato tuber parenchyma subjected to compression varies between 0.46 and 0.60, with an average value near 0.46, indicating that 46% of the total deformation is typically recovered when the tissue is unloaded. In general, the degree of elasticity and the maximum strain level experienced by a material are inversely related. At a total strain of 0.10, the recovered strain in potato tuber parenchyma is 0.06 and the degree of elasticity is 60%; at a total strain level of 0.28, the recovered strain is 0.13 and the degree of elasticity is 46%.

The elastic hysteresis of a material is defined as the amount of energy internally dissipated during a loading-unloading cycle. This energy can be calculated from the area spanned between the plots of the loading and unloading

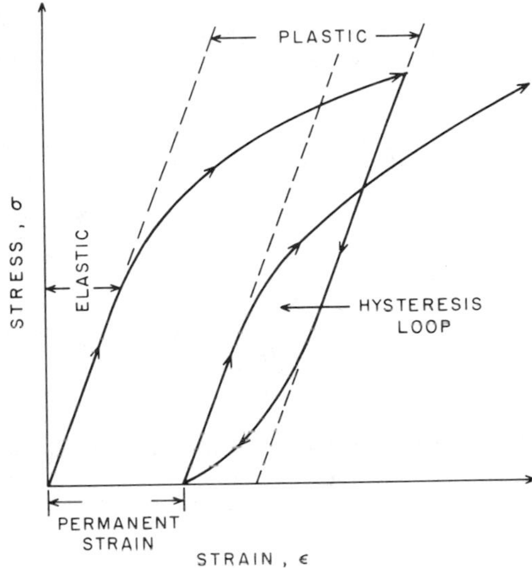

FIGURE 2.10 The stress-strain diagram of an elastic-plastic material evincing hysteresis when the stress level drops to zero and the material is reloaded. The material shows a linear elastic response during the initiation of each of the two loading episodes. The plastic deformations that result from the first loading-unloading cycle are seen as permanent strains when the stress level drops to zero. The material demonstrates a nonlinear recovery of strains when unloaded, called anelastic behavior. The area within the hysteresis loop is a measure of the energy the material consumes during its initial loading.

portions of a stress-strain diagram called the hysteresis loop; the larger the area, the more energy the material consumes during loading (fig. 2.10). For an ideal elastic material, there is no hysteresis loop, and no energy has been dissipated. Ideal elastic behavior is rarely if ever seen. Indeed, even steel has a small hysteresis loop, and those of most polymeric solids are pronounced. In the case of potato tuber parenchyma, the energy dissipated can vary between 72% and 90% of the total energy introduced during the loading cycle. Although tuber parenchyma exhibits a very nice linearity between stress and strain during its initial loading, it has a pronounced anelastic behavior when unloaded. Additionally, plastic deformations may occur, resulting in an often complex stress-strain diagram (fig. 2.10). The mechanical behavior of this tissue in particular and of most plant materials in general illustrates the need to determine stress-strain diagrams with one or more loading-unloading cycles.

Not all elastic materials are linearly elastic; some exhibit nonlinear elastic-

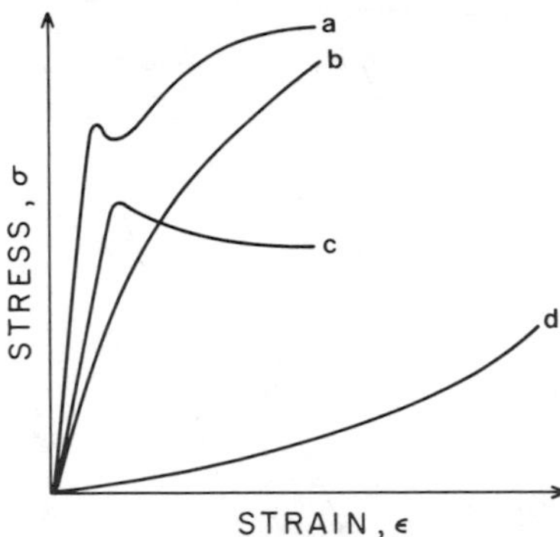

FIGURE 2.11 Representative stress-strain diagrams for four different types of materials: Linear elastic material undergoing plastic deformations and strain hardening (a); nonlinear elastic material with a decreasing tangent modulus (b); linear elastic material undergoing plastic deformations and strain softening (c); and nonlinear elastic material with an increasing tangent modulus (d).

ity. Below their proportional limit, nonlinearly elastic materials recover their original dimensions when the stress level drops to zero. Normal stresses and strains are not linearly proportional, however. Consequently these elastic materials produce curvilinear stress-strain diagrams within the limits of their elastic behavior, indicating that the ratio of stress to strain changes as a function of the magnitude of stress. The slope measured anywhere along the stress-strain diagram provides the instantaneous ratio of stress to strain, called the tangent modulus, which is sometimes referred to as the instantaneous elastic modulus and is symbolized by E_T (see fig. 2.8). Thus the tangent modulus changes within the elastic range of behavior of nonlinear elastic materials, but for any magnitude of stress $E_T = \Delta\sigma/\Delta\varepsilon$. Rubber is an excellent example of a material that exhibits nonlinear elasticity. It instantly recovers its dimensions when stresses are removed, but the strains that develop tend to decrease in proportion to the stresses. Thus the tangent modulus tends to increase as the stress level increases (fig. 2.11 d). By contrast, the stress-strain diagrams of some nonlinear elastic materials, such as silk and many of the primary plant tissues, are convex, indicating that the tangent modulus tends to decrease as the stress increases beyond a certain level (fig. 2.11 b).

Before we move on to other topics, let me discuss the way a material is tested and the influence the testing procedure has on the value of E that is empirically determined. It is typical to test materials under uniaxial tension by clamping a specimen at either end. However, these supports restrict deformations at the ends of a specimen and can increase the apparent elastic modulus of the material. End-wall effects resulting from clamps or other means of supporting a specimen are often neglected, leading to spurious values of E. Thus it is always advisable to perform uniaxial tensile tests on specimens that are at least ten times longer than they are wide. This aspect ratio is the minimum ratio for which end-wall effects can be more or less neglected. Also, it is a good idea to prepare specimens that have expanded ends (to which supports are attached) and narrower, parallel-sided midspans along which stresses and strains are actually measured. End-wall effects are particularly pronounced when anisotropic materials are stretched in the direction of their greater stiffness. Therefore it is essential to determine the preferred direction of extensibility of materials before plotting their stress-strain diagrams.

Additionally, uniaxial tests should always be avoided for biological materials that naturally experience biaxial or triaxial strains, such as the epidermis and storage parenchyma found deep within large tuberous organs. Two-dimensional straining prohibits the Poisson's effect—the ability of a material to contract in one dimension and of fibrous components within it to reorient under uniaxial tension. The Poisson's effect increases the apparent elastic modulus and decreases the ultimate strains of an anisotropic material. Thus anisotropic materials will typically yield higher elastic moduli and lower ultimate strains when tested under uniaxial tension than under biaxial tension. The values for these mechanical parameters determined under uniaxial tension will be biologically irrelevant if the material examined operates mechanically in nature so that the Poisson's effect is eliminated. For example, apple skin can sustain 7% or more uniaxial strains before rupturing, whereas when it is tested under biaxial tension, strains of 2% or less are measured before rupturing occurs.

Finally, it is important to remember that the proportional limits measured in tension and compression for the same material may have very different values. Therefore it is always advisable to test a material under loading conditions that reflect those it normally sustains.

THE SHEAR MODULUS: THE RELATION BETWEEN SHEAR STRESSES AND SHEAR STRAINS

Virtually every biological material experiences shear stresses in addition to normal stresses. Thus the shear modulus of a material is an important material modulus. From our treatment of Poisson's ratios, we know that the shear modulus G is the ratio of shear stress τ to shear strain γ. If shear stresses develop within all three orthogonal dimensions of an isotropic material, then the distortions in the angle θ subtending any two intersecting sides of the material depend exclusively on the corresponding shear stress component, and $\tau_{xy}/\gamma_{xy} = \tau_{yz}/\gamma_{yz} = \tau_{zx}/\gamma_{zx}$. Thus the three shear moduli, symbolized by G_x, G_y, and G_z, will be equivalent regardless of the direction in which external forces are applied to the material, and only one shear modulus need be measured. When treating orthotropic materials like wood, however, three shear moduli must be determined empirically from the appropriate shear stress-strain diagrams.

For wood we denote three planes—LR, LT, and RT—where L represents the direction parallel to the grain and R and T represent the directions radial and tangential to the grain (see fig. 2.7). Using this notation, the corresponding shear moduli are designated G_{LR}, G_{LT}, and G_{RT}. As in the case of the three elastic moduli of an orthotropic material, the three shear moduli can differ, often substantially, in their magnitudes; for example, for balsa $G_{LR} = 0.169$ GN·m^{-2}, $G_{LT} = 0.115$ GN·m^{-2}, and $G_{RT} = 0.0156$ GN·m^{-2}. Thus $G_{LT}/G_{LR} = 0.685$ and $G_{RT}/G_{LR} = 0.111$, indicating that balsa resists shear best in the LR plane and most poorly in the RT plane, as is generally true for most species of wood. That is, for many species of wood $G_{LT} \approx 0.98G_{LR}$ and $G_{RT} \approx 0.24G_{LR}$. The ratios of the shear to elastic moduli are always less than unity for wood, meaning that shearing failure is likely to occur in bending or torsion; for example, for balsa wood $G_{LR}/E_L = 0.054$, $G_{LT}/E_L = 0.037$, and $G_{RT}/E_L = 0.005$. However, great caution should be exercised in treating the mechanical properties of botanical materials like wood as constants, since they vary with the age and relative moisture content of a sample. The same caveat applies to treating the elastic moduli of plant tissues as constants (see chap. 6).

Finally, note that for elastic materials there exists a limit beyond which the proportionality between shear stresses and shear strains is not maintained. Beyond this proportional limit many materials yield in shear and undergo plastic deformations. Consequently the working (safe) shear stress τ_W is usually taken as some fraction of the shear yield stress τ_y. That is, $\tau_W = \tau_y/n$. For most materials the shear yield stress is significantly less than the yield stress of the material submitted to tension. Thus, for such materials, the shear yield stress typically defines the factor of safety. This is true for structural steel as well as for wood.

PLASTIC DEFORMATIONS

Plastic strains are nonrecoverable deformations. Although somewhat arbitrary, 0.2% permanent strain is taken as the standard for the onset of plastic behavior in fabricated materials, although a lower strain ($\approx 0.1\%$) can be used for biological materials. Plastic behavior occurs as a result of permanent molecular reorganizations within a simple material or microstructural deformations within a composite material. In the case of plant materials, such as potato parenchyma, plastic deformations are the product of the rupturing of cell walls and result in permanent strains when the tissue is unloaded (see fig. 2.10). Some materials exhibit plastic behavior at very low loadings; others behave plastically only after very high loadings. The initiation of plastic deformation in an elastic material is typically seen as a dramatic reduction in the slope of the stress-strain diagram just beyond the proportional limit. Once initiated, deformations continue even under a constant load. Beyond their proportional limit, many fabricated and naturally occurring polymeric materials evince one of two trends in the relation between stress and strain. The tangent modulus may increase or decrease. The former indicates strain hardening; the latter indicates strain softening (see fig. 2.11 a and c). Strain hardening is typically a physical manifestation of changes in the molecular structure of a material, as when polymeric chains (or constituent crystals within a crystalline, strain-hardening solid) increasingly align parallel to the axis of a uniaxial stress. There is an obvious advantage to strain hardening, since the instantaneous elastic modulus increases as the stress level increases. Nonetheless, the deformations that result after the onset of plastic behavior typically are greater than those that attend elastic behavior.

Strain hardening can produce some very unusual behavior and can lead to erroneous conclusions about the nature of a material. For example, there are materials with a negligible initial elastic range that will nonetheless produce an initially linear stress-strain diagram. The linear portion of the stress-strain diagram is the result of linear strain hardening in an incompressible plastic material. Thus the slope of the initially linear portion of the stress-strain diagram is *not* the elastic modulus. Rather, the slope is the ratio of the change in the generalized equivalent stress ($d\sigma^h$) to the change in the generalized plastic strain increment ($d\varepsilon^P$), which is called the strain hardening coefficient, symbolized by H'. That is, $H' = d\sigma^h/d\varepsilon^P$. Parenchyma isolated from potato tubers displays this type of behavior.

TABLE 2.2 Proportional Limits and Corresponding Strains of Plant Fibers and Some Metal Wires

Material	Proportional Limit ($MN \cdot m^{-2}$)	Strain
Secale cereale	0.147–0.196	4.4
Lilium auratum	0.186	7.6
Phormium tenax	0.196	13.0
Papyrus antiquorum	0.196	15.2
Molina coerulea	0.216	11.0
Pincenectia recurvata	0.245	14.5
Copper wire	0.119	1.0
Brass wire	0.130	1.4
Iron wire[a]	0.215	1.0
Steel wire[a]	0.241	1.2

Source: Data from Schwendener 1874.
[a]Values given for metals of Schwendener's day. Values for modern iron and steel wires are orders of magnitude higher.

To further understand the importance of the limit of elasticity and strain hardening, we can turn to the work of one of the first biomechanicists to interpret stress-strain diagrams. Simon Schwendener (1874), who was trained as both an engineer and a botanist, discovered that the fibers of monocots lying near the tissue that conducts cell sap (phloem) have limits of elasticity comparable to those of the best iron and steel of his day. For example, the phloem fibers of rye (*Secale cereale*) can withstand a stress of 0.147 to 0.196 $MN \cdot m^{-2}$ before they begin to show plastic behavior, whereas the iron and steel wires Schwendener examined had limits of elasticity of 0.215 to 0.241 $MN \cdot m^{-2}$ (modern steels begin to yield at stresses as high as 1.5 $GN \cdot m^{-2}$) (see Timoshenko 1976a, table 1). Dry phloem fibers have a modulus of elasticity of about 1 $GN \cdot m^{-2}$, so they are stiff compared with other plant tissues but not compared with steel. Additionally, the work of Schwendener (and others of a more recent vintage) reveals that the proportional limit of fibers can be very low (about 0.10 $MN \cdot m^{-2}$, which is just below that of brass wire; see table 2.2), suggesting that, in general, fibers are designed to operate elastically and that once their elastic behavior is past, they have little option but to go with the flow of things. Significantly, some plant fibers exhibit tensile strain hardening as a result of the realignment of their cellulosic molecular infrastructure. Thus, as the magnitude of externally applied forces increases and exceeds their proportional limits, the instantaneous elastic modulus of these cell walls increases, providing an increased albeit modest resistance to further deformations. Unlike steel structures, of course, plants have the capacity to add new structural components and repair many forms of mechanical damage. Growth provides a biologically unique way to deal with stresses that exceed the proportional limit. Indeed, after some cell walls plasticize under a load, they can be restiffened and resume elastic behavior after the load is removed.

Plastic deformations are not always undesirable; many are important to plant growth. Typically, each plant cell is encased in a relatively stiff cell wall, secreted by the protoplast external to its plasma membrane. Further, expansion of the protoplast is achieved by a plastification of the cell wall that allows the wall to yield under the internal pressure exerted on it by the expanding protoplast. In young or actively growing tissues, the entire cell wall infrastructure of the tissue may behave as a plastic material, deforming irreversibly as the living protoplast prevents the apoplast from becoming an elastic solid by metabolically decreasing the yield point of the cell wall.

A simple plastic theory treats the behavior of materials within their plastic range of behavior (Hill 1950), although a variety of more complex models (which cannot be treated here) have been developed to account for elastoplastic effects. There are five principal assumptions in this simple plastic theory:

1. The material is ductile; that is, it is capable of plastic deformation without fracture. The stress-strain diagram of the material is assumed to be that of an ideal elastic-plastic material. Accordingly, the consequences of strain hardening and strain history are neglected.
2. Plastic deformations are assumed to occur when any component (called a fiber—not to be confused with plant cell fibers) within a given cross section develops plastic strains. This assumption neglects the fact that within any cross section there exist structural components that will retain their elastic behavior. Thus the effects of shear between adjacent elastic and plastic fibers are neglected.
3. The loading on the material is assumed to be proportional; that is, all the loadings applied to the material are assumed to be in constant proportion to one another.
4. The deformations within a structure are assumed to be small. This assumption permits the use of the geometry of the undeformed structure to compute the equations of equilibrium.
5. Plastic deformations produce no change in the volume of the material.

For growing plant tissues, some of these assumptions are relatively reasonable (even 4, which assumes little or no change in shape; for example, the *instantaneous* shape of a tissue or organ can be used to calculate the equations of equilibrium, provided growth is evaluated in a series of small incremental steps). However, assumption 5 is probably not reasonable given the cellular nature of plant tissues, particularly porous plant tissues like wood and cork. Plastic deformations in porous plant tissues are likely to involve changes in tissue volume.

Plastic behavior in a solid can be illustrated by considering a cylindrical specimen that is flexed (Ades 1957). As the loading gradually increases and the cylinder begins to bend, four stages of behavior typically occur:

1. All the structural fibers are stressed below their proportional limits and the material behaves elastically.
2. Extreme fibers—those farthest from the centroid axis—begin to exceed their proportional limits and undergo permanent deformation.
3. As the loading increases, the number of fibers stressed beyond their proportional limits increases until all within the cross section behave plastically.
4. The deformation within the specimen increases rapidly with little or no increase in load.

If we reread these four stages in the development of plastic behavior and substitute plant fibers for the engineer's structural fibers, then we have a fairly good picture of how stems and leaves lacking woody tissues plastically deform and eventually undergo mechanical failure when subjected to extreme external surface forces like high wind pressure. Typically, in many such plant organs, primary phloem fibers or sclerenchyma (a tissue with relatively thick cell walls) are situated toward the perimeter of cross sections where tensile and compressive bending stresses reach their highest values. Fibers and sclerenchyma cells are pulled or squeezed (on the convex and concave outer surfaces of the stem or leaf) as the bending force is applied. At some stress level, however, the proportional limits of these structural fibers are exceeded, and they begin to deform plastically—but not all, and not all at once. Some cell walls, perhaps those that are thicker or drier, will have higher proportional limits than their neighbors. Provided they also have high elastic moduli, which is likely, fairly high stresses will be required to produce strains sufficient to affect neighboring but less stiff cells; that is, stiffer cells can mechanically support less stiff neighboring cells.

In some materials and plants, a plastic hinge develops—all the material in a given cross section yields plastically. However, not all the cross sections in the material may have yielded plastically. Thus a region undergoing plastic deformations within the material is spanned by two regions that have maintained their elastic characteristics. The base of young, growing stem internodes (regions of a stem spanning the attachment sites of two sequential leaves), in which cell walls are plasticized by the protoplast, can behave as a plastic hinge, particularly when wind pressure exerts a bending stress. Low wind pressures rarely exceed the proportional limits of mature grass stems, but growing stems possess regions (meristems) where new cells, with relatively thin walls, are being produced. These meristems are intercalated along the length of the stem at the base of each stem internode that is still growing, and it is within the intercalary meristem that a plastic hinge can develop, even under the relatively low bending stresses of a moderate breeze. Plastic hinges can fail mechanically, typically by the buckling of grass internodes that are hollow, particularly when the water pressure within the plant is low. The thin walls of young cells operate as hydrostats—essentially balloons inflated with a viscous liquid, the protoplast, that changes its volume as water is added or withdrawn from the cell. An interesting feature of buckling is that when a compressed protoplast is fully inflated with water, it places its cell wall in tension. Thus, even when a tissue is placed in compression, its cell walls operate as tensile elements. This is advantageous because cellulose, which is a major component of young plant cell walls, has a very high tensile strength (see below). A detailed treatment of phenomena such as plastic hinges and hydrostats, however, must be delayed until we take up the topic of beam theory (chap. 3).

Viscous and Viscoelastic Materials

It is not difficult to understand the mechanical behavior of elastic and plastic materials at the atomic level of organization. Interatomic bonding resists displacement of atoms when an external force is applied, and although the distances among neighboring atoms are altered owing to the application of an external force to an elastic solid, the energy added to the system can be used to instantly restore the solid's original shape and dimensions when the external force is removed. This ability to fully and instantly recover deformations is illustrated by the force-deformation diagram of an ideal elastic (Hookean) solid placed in tension (fig. 2.9A). By contrast, the application of an external force to a plastic material results in permanent displacement of atoms and molecules, and when the material is relieved of its stresses, strains are permanent (fig. 2.9B). Clearly, a material can exhibit an elastic-plastic response depending on the magnitude of the loadings it is subjected to (fig. 2.9C). Nonetheless, the mechanical behavior of elastic, plastic, and elastic-plastic materials can be mathematically described in terms of the parameters previously reviewed—stress, strain, and the material moduli.

Nonetheless, many biological materials do not mechanically behave as elastic, plastic, or elastic-plastic solids. Rather, they will largely return to their original shape after externally applied forces are removed (they show an elastic response), but the partial or total recovery of their deformations is not instantaneous and occurs over measurable time (they exhibit a viscous response). This type of material is called a viscoelastic material. Viscoelasticity is not the same as plasticity, although viscoelastic materials can have permanent time-dependent deformations. The magnitude of the deformations in viscoelastic materials depends on time (Bland 1960; Christensen 1971). Thus the mechanical behavior of viscoelastic materials requires mathematical descriptions involving three parameters—stress, strain, and time. A stereotyped stress-strain diagram of a viscoelastic material is presented in figure 2.9D. The time dependency of the mechanical behavior of viscoelastic materials is shown by the curvilinearity in the loading and unloading portions of the stress-strain diagram.

Two types of viscoelastic materials are distinguishable. If the stress is proportional to strain at a given time, then the material is called a linear viscoelastic material. If these two components are not linearly related, then the material is called a nonlinear viscoelastic material. Most mathematical descriptions of viscoelastic materials assume the former condition holds true, much as the theory of elastic solids assumes that a material obeys Hooke's law—that it is an ideal linear elastic material. For nonlinear viscoelastic materials, however, the assumption of linearity is legitimate only for very small strains ($\leq$ 1%). Since most biological materials are nonlinear viscoelastic materials and since most function mechanically at high or extremely high strains ($\geq$ 20–50%), the assumption of a linearity between the elastic and viscous response components is not valid. Unfortunately, no mechanistic model exists that can fully treat the behavior of nonlinear viscoelastic materials. Thus we can draw almost no generalizations about the mechanical behavior of biological viscoelastic materials experiencing large strains. Since the following deals with plant materials as if they were linear viscoelastic materials, much of it should be read with skepticism and only from a pedagogical perspective.

The elastic response component of viscoelastic materials can be largely understood in terms of our previous review of elastic solids. Therefore it is the viscous response component that must occupy our attention here. Viscos-

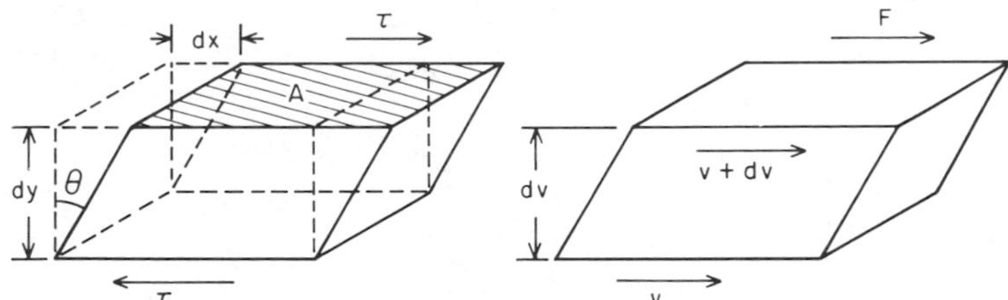

FIGURE 2.12 Physical analogy between the spatial gradient of shear strains (*dx*/*dy*) within a solid (shown to the left) and a velocity gradient (*dv*/*dy*) in a sheared fluid element (shown to the right). Viscosity (η) is defined as the ratio of the shear stress ($\tau = F/A$) to the velocity gradient (*dv*/*dy*). For the fluid element, *dv*/*dy* increases (from *v* at the base of the element to $v + dv$ at the top of the element) as a function of the vertical dimension (*dy*). Thus, $\eta = (F/A)/(dv/dy)$. For a solid, the shear modulus (*G*) is the ratio of the shear stress (τ) to the shear strain (γ). Since (*dx*/*dy*) increases as a function of the vertical dimension and since for very small rotation angles θ, $\gamma = dx/dy$, $G = (F/A)/(dx/dy) = \tau/\gamma$.

ity, symbolized by η, is defined as the ratio of stress to strain rate. It is the ratio of the shearing stress τ to the velocity gradient (dv/dy) that develops within a material when it begins to shear: $\eta = \tau/(dv/dy)$. (Viscosity is derived from the Latin word for mistletoe, *viscum*, a plant with sticky berries.) However tempting it may be to draw a comparison between viscosity and the shear modulus, therefore, these two physical parameters are not the same thing. Referring to figure 2.12, when a solid is sheared, a spatial gradient of shear (dx/dy) develops; that is, elements of material within the solid slide past one another to a degree dependent on their distance (x) from the externally applied force. For very small shearing strains, the spatial gradient equals the shear strain ($\gamma \approx dx/dy$). Thus, $\eta = \tau/(dv/dy) \approx \tau/(dx/dy)$. Nonetheless, there exists a real difference between viscosity and the shear modulus, and it is this difference that distinguishes the behavior of fluids from that of viscoelastic materials.

When a force is applied to a fluid, the force is rapidly dissipated by molecules moving past one another, and the fluid undergoes nonrecoverable deformation (fig. 2.13). By contrast, viscoelastic materials largely recover their deformations, although the rates at which strains drop to zero can be very slow. For both fluids and viscoelastic materials, the rate at which shearing strains develop ($d\gamma/dt$) is a function of the applied shear stress τ. Thus the behavior of fluids and fluidlike materials is time dependent. The rate of shearing is related to the shear stress by a proportionality factor, called dynamic viscosity, symbolized by μ:

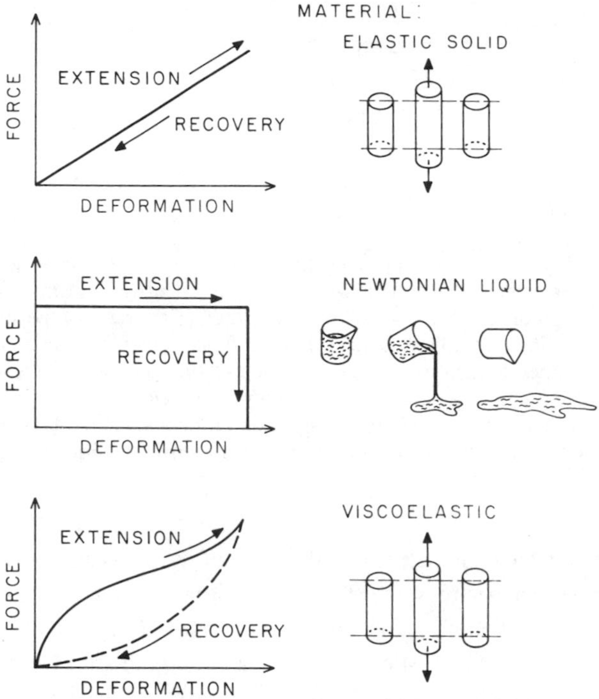

FIGURE 2.13 Force-deformation diagrams of a linear elastic solid, a Newtonian liquid, and a viscoelastic material. Linear elastic solids recover all of their deformations when applied forces are removed provided their proportional limits are not exceeded; that is, deformations are recoverable within the elastic range of behavior. Newtonian fluids do not recover any of their deformations when an applied force is removed. Viscoelastic materials recover deformations gradually when applied forces are removed.

$$\mu = \frac{\tau}{d\gamma / dt}. \tag{2.9}$$

Owing to the relative ease with which molecules can move in fluidlike materials, the dynamic viscosity of a material is a constant *only* for a given temperature, since the ability of molecules to move depends on their kinetic energy, which in turn is reflected by the material's temperature. The dynamic viscosity of a material at temperature T is calculated from the activation energy of the material H, Boltzmann's constant k:

$$\mu (T) = \mu_o \exp \frac{(H)}{kT}. \tag{2.10}$$

For liquids, μ decreases as temperature increases, but for gases the reverse is true. For example, when the temperature of pure water at 0°C is raised to 10°C, μ decreases by 27%. When pure glycerol at 20°C is heated to 25°C, μ decreases by 58.4%. Mixtures of gases, like air, tend to increase in dynamic viscosity as temperature increases. By the same token, the mechanical behavior of viscoelastic materials depends on the temperature at which they are tested; their viscosity tends to decrease as their temperature increases.

In addition to being temperature-dependent, the viscosity of organic viscoelastic materials can be physiologically modified. For example, the viscosity of cytoplasm depends on metabolic status, which can change during the cell division cycle. Viscosity generally increases in the early stage of cell division (prophase), decreases in the intermediate stage (metaphase), and increases again toward the end of the cell division cycle (telophase). When the viscosity of cytoplasm is artificially increased, the rate of mitotic division slows down (Molè-Bajer 1953). Osmotic concentration of cell sap is typically proportional to cytoplasmic viscosity and is inversely proportional to cell size and the number of chromosomes in polyploid tissues (Becker 1931).

The reciprocal of dynamic viscosity is called fluidity, symbolized by α, and the rate of shear is equal to the product of fluidity and the shear stress: $d\gamma/dt = \alpha\tau$. This linear relationship is known as Newton's law of viscous flow, which governs the behavior of ideal fluids and all linear viscoelastic materials. Newton's law is similar to Hooke's law: $\sigma = \varepsilon E$, whereas $\tau = (d\gamma/dt)\mu$. And just as not all elastic solids obey Hooke's law, so too not all fluids obey Newton's law. Indeed, there is an important distinction between those fluids that do and those that do not. In his *Principia*, published in 1687, Newton wrote, "the resistance which arises from the lack of slipperiness of the parts of a liquid, other things being equal, is proportional to the velocity with which the parts of the liquid are separated from one another." Newton's "lack of slipperiness" is clearly viscosity, which originally meant stickiness. Fluids that behave according to Newton's law are known as Newtonian fluids. Newtonian viscosity is independent of strain or the rate of shear; the shear rate will be doubled when the force applied to a Newtonian fluid is doubled. However, Newton apparently anticipated ("other things being equal") that for some fluids the shear stress and the rate of change in the shear strain are not linearly related to one another by a single proportionality constant; that is, $\tau \neq (d\gamma/dt)\mu$. These fluids are called non-Newtonian fluids. Examples of non-Newtonian fluids are silicone putty, pitch, asphalt, and some types of soil. For these fluids there exists a rate of change in the shear strain for *each* shear stress. Accordingly, the behavior of non-Newtonian fluids is stress-rate dependent. When stressed quickly, non-Newtonian fluids behave much like solids;

when stressed slowly they operate mechanically much as liquids do.

Viscoelastic materials have much in common with non-Newtonian fluids. Both materials evince time- and stress-rate dependency. Therefore, to fully characterize the mechanical behavior of viscoelastic materials, we must quantify their response under varying stress levels and their response to each stress level over time. Two fundamental types of experiments are used to quantify the mechanical properties of viscoelastic materials: dynamic experiments and transient experiments. Dynamic experiments are those in which either stress or strain is varied cyclically with time and the mechanical response of the material is measured over various frequencies of deformation (Vincent 1990a). They tend to be procedurally and mathematically complex. A material is subjected to a sinusoidally time-varying strain at a frequency ω. If the material is perfectly elastic, then stresses will always be proportional to the strains, whereas if the material has no elastic response component, then the stress will always be highest at the highest strain rate. Perhaps counterintuitively, because the strain is varying about zero in a sinusoidal fashion, the highest strain rate will occur when the strain is zero. Likewise the stress will be lowest at the lowest strain rate, which will occur when the strain is highest. Therefore the stress-strain plot (called a Lissajous figure) will be a perfect circle. Since a viscoelastic material has both elastic and viscous response components, its Lissajous figure will be elliptical. The viscous modulus, symbolized by G'', can be calculated from the Lissajous figure of a viscoelastic material. It is the ratio of the material's dynamic viscosity to the frequency at which strain is varied: $G'' = \mu/\omega$. Thus, G'' is a measure of the energy lost within the system resulting from viscous processes. Little energy is lost when the period of oscillation is dissimilar to the characteristic times describing the rates of molecular processes attending mechanical deformations; comparably greater losses of energy occur as ω approaches these characteristic times. Indeed, data from dynamic experiments can shed considerable light on the molecular processes attending the deformations that occur within viscoelastic materials.

Transient experiments are far easier to perform and are the most commonly reported in the literature. These experiments involve deforming a material (either by simple extension or in shear) and following its mechanical response over time. Since there are three variables in a transient experiment (stress, strain, and time), and since one of these variables (time) must always figure as the abscissa, there are two transient experimental formats: either a material is tested under a constant strain level and the decay in stress over time is measured, or the material is tested under a constant stress level and the

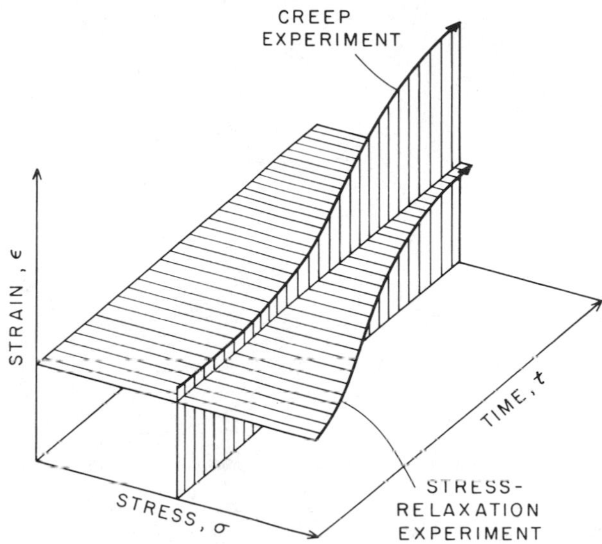

FIGURE 2.14 Comparison between two types of transient experiments used to determine the properties of a viscoelastic material. In one type, called a creep experiment, the material is subjected to a constant magnitude of stress and the resulting strains are plotted as a function of time. In another type, called a stress-relaxation experiment, the material is loaded to maintain a constant strain level and the "decay" in the resulting stresses is plotted as a function of time.

changes in strain over time are measured. The former is called a stress-relaxation experiment; the latter is called a creep experiment. These two experimental formats will be treated in the next two sections, but their characteristic features are summarized and compared in figure 2.14, from which we see that the scope of a viscoelastic material's mechanical behavior (*for a particular temperature*) conforms to a surface generated by three axes: stress, strain, and time.

STRESS-RELAXATION EXPERIMENTS AND RELAXATION TIME

The differences between a solid, a fluid, and a viscoelastic material can be expressed in terms of a quantity called the relaxation time, which is a measure of the time required to deform a material's molecular structure. The relaxation time of an ideal elastic material is infinity, whereas that of a fluid is zero. Thus the relaxation time of an ideal elastic solid or a fluid essentially cannot be measured. By contrast, a viscoelastic material has a relaxation time that lies somewhere between these two extremes—it has a measurable relaxation time.

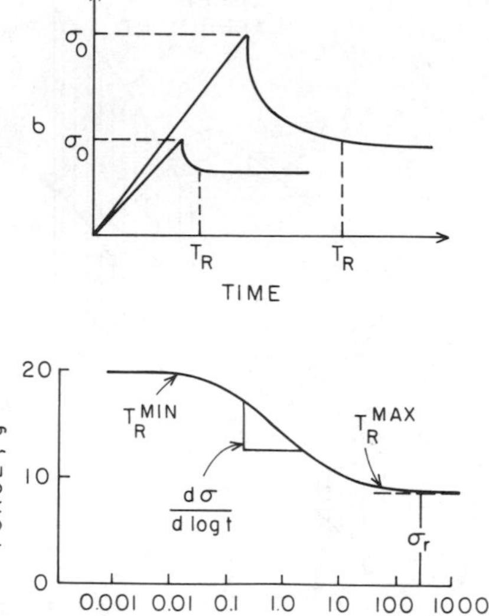

FIGURE 2.15 Stress-relaxation curves (stress versus time) for a viscoelastic material subjected to tension. The specimen is rapidly stretched to a predetermined stress level (σ_o), after which the decay in tension is measured over time (upper diagram). The decay in tension is seen as a curvilinear decrease in stress over time. The time at which the tensile stress reaches a constant level is called the relaxation time (T_R), that is, the time required to recover $1/e$ of the original stress. Data are typically plotted as stress versus log time (lower diagram). This provides information on the minimum relaxation time (T_R^{min}) and maximum relaxation time (T_R^{max}), which can be identified in a stress-relaxation curve when the applied tensile force begins to decay and when the decay of the applied tensile force levels off. The relaxation rate (the change in stress per change in log time, $d\sigma/d \log t$) is measured by the slope of the portion of the stress-relaxation curve that spans T_R^{min} and T_R^{max}. The residual stress (σ_r) is measured as the stress remaining once the stress level decays to the maximum relaxation time.

In a very real (mathematically precise) way, the relaxation time is the ratio of a material's dynamic viscosity to its modulus of elasticity. For liquids this ratio is very large; for solids it is very low; and for viscoelastic materials it lies somewhere in between.

This qualitative description can be dispensed with by recourse to some simple mathematics. For a linear elastic solid, the rates at which stress and strain change ($d\sigma/dt$ and $d\varepsilon/dt$, respectively) are related to the elastic modulus by the formula

(2.11) $$E = \left(\frac{d\sigma}{dt}\right)\left(\frac{d\varepsilon}{dt}\right)^{-1}.$$

If a material has a viscous response component, then stress will decrease at a rate dependent on the original stress σ_o (see fig. 2.13), such that the relaxation time, symbolized by T_R, can be expressed by the formula

(2.12) $$T_R = \frac{\sigma_o}{E\left(\frac{d\varepsilon}{dt}\right) - \left(\frac{d\sigma}{dt}\right)}.$$

For stress-relaxation experiments, at any time t, the rate of change in strain can be considered equal to zero and the instantaneous stress (at time t) is given by the formula

(2.13) $$\sigma(t) = \sigma_o \exp - \left(\frac{t}{T_R}\right).$$

Thus the relaxation time is the time required for the stress to decrease to $1/e$ the original stress.

The relaxation time of a viscoelastic material is measured by means of a stress-relaxation experiment in which a sample of a material is deformed to a fixed level and the decay in the stress required to maintain this strain level is recorded over time (fig. 2.15).

Relaxation time is very important to many biological phenomena, not the least of which are cell growth and the expansion of the cell wall. Recall that if the cell wall mechanically operated as an ideal elastic material, given the remarkably high elastic modulus of cellulose, tremendous internal pressures would be required to extend the cell wall, and when these forces are removed, the cell wall would return to its original dimensions and shape, provided the proportional limit of cellulose was not exceeded. Thus it is not surprising that the walls of growing cells operate as either plastic or viscoelastic materials. Indeed, relaxation curves, measured by stretching cell walls to some predetermined strain and measuring the decay in the tensile stress, indicate that both elastic and plastic deformations occur within cell walls. Usually the decrease in the applied load (force) is plotted as a function of log time, as shown in figure 2.15. (Stress relaxation for a solid subjected to shear is discussed by Nakajima and Harrell 1986.) This transient experimental procedure, much like dynamic experiments, can provide important information, even at the level of the molecular organization within the cell wall infrastructure. For example, cell wall stress-relaxation curves typically take the form of the empirically derived formula

$$\sigma(t) = \frac{d\sigma}{d\log t} \log \left[\frac{t + T_R^{\max}}{t + T_R^{\min}} \right] + \sigma_r, \tag{2.14}$$

where $T_R^{\max}$ and $T_R^{\min}$ are the maximum and minimum relaxation times and σ_r is the residual stress at the end of the relaxation period (see fig. 2.15 for an example). This equation can be interpreted in molecular terms: the minimum relaxation time depends on the average molecular weight of the viscoelastic units within the cell wall; the expression $d\sigma/d(\log t)$ depends on the number of flow units per transverse area; and the residual stress depends on the number of the cross-links among the adjacent polymers within the cell wall. As cell walls mature and cease to elongate, and therefore exhibit progressively more elastic behavior and progressively less viscoelastic behavior, the relaxation time and residual stress would be expected to increase, while the change in stress per decade of time would approach zero. Essentially, the fluidlike cell wall crystallizes into a solid.

It is important to note that some of the viscoelastic behavior evinced by growing cell walls results from the addition of new materials to the cell wall rather than from the deformation of preexisting materials. Additionally, all the available information pertaining to plant cell walls indicates that they mechanically behave as nonlinear viscoelastic materials. Nonetheless, the mechanical testing of cell walls by techniques like stress relaxation provides the opportunity to treat a complex biological phenomenon in terms of a relatively more simple physical phenomenon, permitting considerable insight.

Creep Experiments and Compliance

The extension of a viscoelastic material, such as a cell wall, is not fully reversible. An irreversible component is typically seen when a cell wall is first extended. That is, there is only a partial recovery of the wall's original dimensions when an applied force is removed. Under tension there is an instantaneous elastic deformation De_1, followed by a relatively slow, time-dependent deformation called creep. Creep occurs under a constant applied stress and consists of all the deformations that occur after the initial linear or nonlinear elastic deformation in the cell wall. The most obvious way to deal with creep is to place a material under a constant uniaxial stress and plot the resulting changes in strain over time. This is called a transient creep experiment. The rate at which deformations occur is governed by the viscosity of the material being tested. It also depends on the magnitude of the stress level and temper-

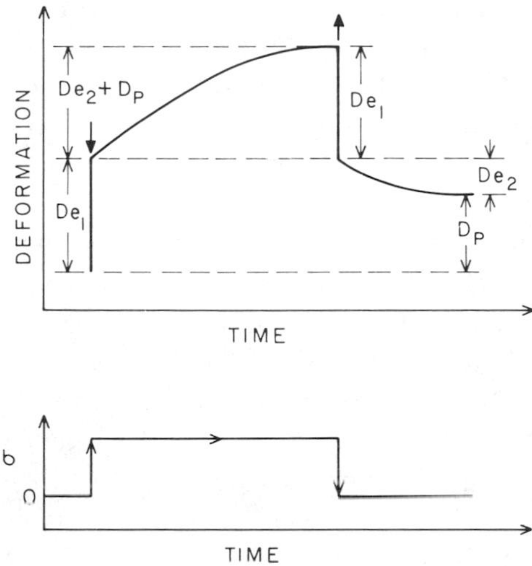

FIGURE 2.16 Cell wall deformation (under a constant tensile load) versus time. An instantaneous elastic deformation (De_1) occurs when the cell wall is loaded in tension (shown by the short downward arrow). This is followed by a retarded elastic (De_2) and plastic deformation (Dp), the sum of which is creep. When the load is removed (indicated by the short upward arrow) the cell wall undergoes a rapid elastic contraction (equal to De_1) followed by a less rapid retarded elastic contraction (equal to De_2) but maintains a plastic deformation (Dp).

ature. In growing cells, the magnitude of the stress level is governed by the extent to which water flows into a cell's protoplast, while the viscosity of the cell wall can be physiologically modified. Accordingly, it is rather naive to believe that the viscoelastic behavior of a cell wall can be circumscribed by virtue of a single creep experiment, or even many.

Nonetheless, in the case of plant cell walls, creep has been found to have two components: a retarded, linear or nonlinear elastic component De_2, and a plastic component Dp. This is illustrated in figure 2.16, from which we see that creep equals $De_1 - De_2$. When the tensile force is removed, the cell wall will undergo an instant elastic contraction, followed by a relatively slow viscoelastic contraction. The plastic component of deformation, called residual strain, is not recovered. This property of cell walls is vital to growth, since the residual strains reflect an increase in the surface area of the cell wall, which in turn reflects the increase in the volume of the enveloped protoplast. Provided the protoplast can control the material properties of its cell wall and maintain a certain level of plastic-viscoelastic behavior, it can utilize the residual strain in the cell wall to provide room for growth.

Although the emphasis has been on the viscoelastic behavior of cell walls, I should not neglect the importance of the elastic component of viscoelastic materials. The modulus of elasticity can be determined even for a complex material like a growing cell wall by preconditioning a specimen. The irreversible component, which is the plastic component of deformation, is eliminated by preextending the specimen to a small strain (between 1% and 5%). Subsequent extensions are then made fully recoverable provided they do not exceed the strain level of the initial extension (and that the direction of the externally applied force is the same, owing to the anisotropic nature of cell walls). Thus, within the range of stresses used to precondition the material, stress and strain will be related to one another by a single proportionality constant, the modulus of elasticity. Aside from being useful in studies of plant cell wall growth, this also indicates that even growing cells and tissues possess an elastic region of behavior that can mechanically sustain the weight of a tissue's or organ's biomass and that remove deformations when stresses are reduced. Note that the weight of an organ can precondition the cell walls of a growing tissue in proportion to the weight they must mechanically sustain. Thus growth provides a feedback loop in which stresses modify the mechanical properties of the materials in which they develop.

As mentioned previously, viscoelastic materials can be classified as either linear or nonlinear in behavior. For both types of material, strain increases as an applied force increases, and with time the strain decreases when the applied force is removed. However, the distinction between a linear and a nonlinear viscoelastic material is that the strains in the former show a time-additive effect. That is, for linear viscoelastic materials, if a stress σ_1 produces a strain ε_1 in time t_1 and if another stress σ_2 produces another strain ε_2 in time t_1, then when σ_1 and σ_2 are added together they will produce a strain of $\varepsilon_1 + \varepsilon_2$ in t_1. This can be mathematically expressed by the formula

$$\varepsilon\,(t) = \sum_{i=1}^{n} \sigma_i\, D\,(t - t_i), \tag{2.15}$$

where D is the compliance of the material. (Compliance is the reciprocal of the modulus of elasticity.) This formula indicates that the strain at any time t is the product of the summation of the stresses, the compliance of the material, and the time interval between measuring strain and applying stress.

Compliance under tension or compression is an important biological parameter, particularly for cell walls. As I mentioned previously, stress relaxation can be observed when cell walls are extended to a fixed strain (some predetermined length, for example) and the stress necessary to keep the cell

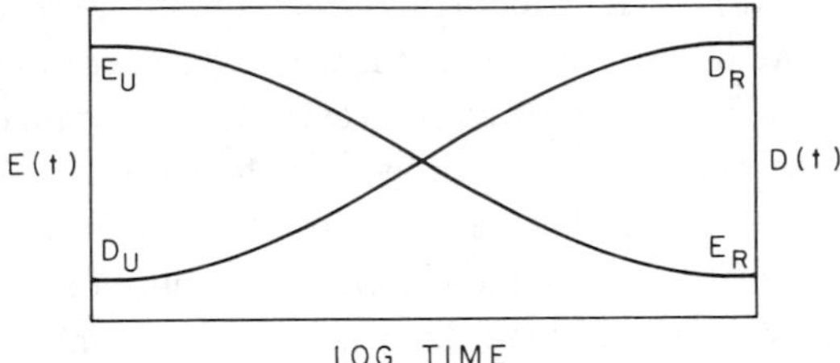

FIGURE 2.17 Relations between the tensile relaxation modulus $E(t)$ and its reciprocal, the tensile (creep) compliance $D(t)$. When measured over log time, the tensile relaxation modulus becomes asymptotic, at which point the relaxed modulus (E_R) and the relaxed compliance (D_R) can be measured.

wall at this strain is seen to gradually decline. Thus the relaxation modulus $E(t)$ is the ratio of stress to strain at time t, or

(2.16) $$E(t) = \frac{\sigma(t)}{\varepsilon},$$

where the strain ε is a constant. From eq. (2.16), we can see that the reciprocal of $E(t)$ is given by the formula

(2.17) $$D(t) = \frac{\varepsilon(t)}{\sigma},$$

where σ is a constant. $E(t)$ and $D(t)$ are referred to as the tensile relaxation modulus and the tensile compliance, respectively, since they are measured for a specimen placed in tensile stress. When measured over log time, changes in the tensile relaxation modulus and the tensile compliance of a cell wall become asymptotic, as shown in figure 2.17. The compliance, measured by taking the slope of the asymptote, is called the relaxed compliance, symbolized by D_R. Similarly, the ratio of strain to stress (when time is zero) is called the unrelaxed compliance, symbolized by D_U. From their mathematical relationship, we can derive the formula

(2.18a) $$\varepsilon(t) = \sigma\left[D_R + D_U + \frac{\varepsilon(t)}{\sigma}\right]$$

(2.18b) $$\varepsilon(t) = \sigma\left[D_R + D_U + D(t)\right].$$

In summary, the single most important parameter influencing our perception of a viscoelastic material's behavior is time. A viscoelastic material will behave as a linear elastic solid under stress when strain is measured instantly; it will behave as a viscous fluid under stress when strain is measured over a very long period (on the order of days); and it will behave as an intermediate between a solid and a fluid when strain is measured over several decades of time (on the order of hours). Since plant growth can be measured in seconds, minutes, hours, or days, the mechanical behavior of their cell walls, and in many cases their tissues, can manifest the behavior of a solid, or a viscoelastic material, or a fluid. We will return to this when dealing with plant cell walls in greater detail (chap. 5).

Composite Materials and Structures

Thus far, the term material has been used in a very cavalier manner, since in engineering it has a very precise meaning. The term refers to either a pure substance or an alloy that can be approximated as essentially homogeneous in composition. When more than one substance or material are combined, and when this combination has some internal structural heterogeneity, the term composite material is used (Bodig and Jayne 1982). Composite materials can have either a periodic or a nonperiodic ultrastructure. That is, their heterogeneity can be reiterative with the various materials distributed in a geometrically predictive manner, or not. In either case, the mechanical properties of composite materials depend on the structural relations among the various materials from which they are fabricated as well as the material properties of each constituent (Mura 1982). Plant cell walls may be viewed as periodic composite materials, since they have a highly ordered arrangement of polymers (principally cellulose) embedded within a more or less amorphous matrix (see chap. 5). Some tissues, such as parenchyma, may also be viewed as periodic composite materials consisting of a more or less geometrically ordered arrangement of cell walls holding together the fabric of the protoplast. Also, each tissue type within an organ is precisely arranged and distributed, but each tissue may differ from the others in its mechanical properties (see Kutschera 1989, for example). Accordingly, the material properties of plant cell walls and some tissues and organs can be approximated (modeled) based on the behavior of composite materials (see chaps. 6 and 7).

Both the material properties and the geometry (internal and external structure) of a plant tissue or organ contribute to mechanical behavior, however. Thus the mechanical attributes of a plant tissue or an organ are best understood in terms of the attributes of a structure rather than those of a material. That is, plant tissues and organs manifest behaviors that are best understood in terms of the arrangement and geometry of their internal parts—struts,

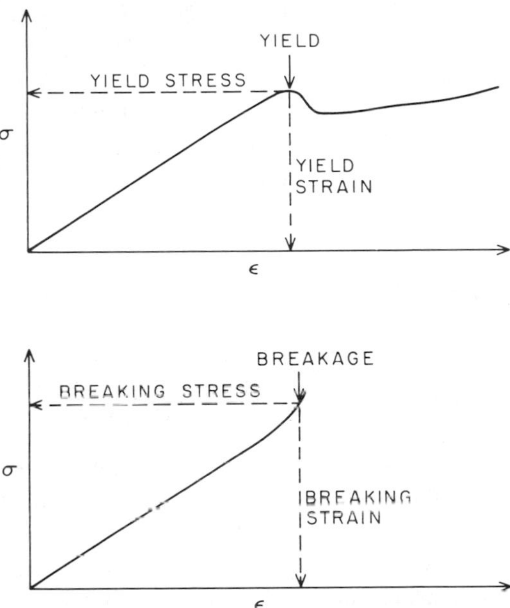

FIGURE 2.18 Stress-strain diagrams for a linear elastic ductile material (upper diagram) and a linear elastic brittle material (lower diagram). Ductile materials undergo plastic deformation beyond their proportional limits when the yield stress is achieved in their loading cycle. Brittle materials break when their proportional limits are exceeded. The stress level at which breakage occurs is called the breaking stress.

beams, and columns whose shapes and sizes are as important to the mechanical behavior of the whole as are the material properties of the substances they are made of. As we shall see (chap. 6), wood is a cellular solid. It has a highly ordered solid phase (cell walls) and a fluid phase. When compressed, the solid phase can densify the tissue until it behaves like a true solid having the material properties of the cell walls. Similarly, when plant organs lacking wood dehydrate, their material properties can dramatically alter. Tissues with thin cell walls inflated by water pressure may collapse, and their volume fraction (relative to other tissues with thicker cell walls) is reduced as the organ dehydrates. This geometric effect on structure cannot be anticipated from the material properties of either cell walls or the protoplasm alone. Additionally, the loss of water from cell walls can increase the capacity of a tissue to sustain stresses, because wet cellulose is weaker than dry cellulose. Thus dehydration results in geometric changes within a tissue or organ as well as altering the mechanical properties of the constituent materials within cell walls. Indeed, dehydrated tissues typically have higher elastic moduli than their hydrated counterparts. As I mentioned earlier, another remarkable consequence of the structure of living plant tissues is that the essentially incompressible, viscous protoplasm can place cell walls in tension when it is subjected to compression. This is particularly advantageous, since cell walls can resist remarkably high tensile stresses but are very susceptible to buckling under compression.

STRENGTH

The distinction between a material and a structure becomes very important when we discuss strength. In reference to a material, strength is the maximum stress required to cause a material to break or undergo plastic deformation. In the former case, the maximum stress is called the breaking strength, whereas the strength of a material undergoing plastic deformation beyond its proportional limit is called the plastic yield strength (fig. 2.18). Both the breaking strength and the plastic yield strength have the same units as stress (force per area), and so are referred to as the breaking stress and the plastic yield stress. Generally, the strength of a material will differ when it is measured under tension or compression or shear. The compressive strength of most species of wood is roughly 50% of the tensile strength of the same species, although there is a substantial range among species in the absolute values of both.

The strength of different biological materials tested under tension spans a few orders of magnitude: on the order of only a few $MN \cdot m^{-2}$ for marine algal tissues (e.g., *Durvillaea antarctica,* 0.7 $MN \cdot m^{-2}$; see Koehl 1979; *Fucus serratus,* 4.2 $GN \cdot m^{-2}$; see Wheeler and Neushul 1981), fresh tendon (0.08 $GN \cdot m^{-2}$), a typical wood, measured along the grain (0.1 $GN \cdot m^{-2}$), and cotton fibers (0.35 to 0.91 $GN \cdot m^{-2}$). Thus the difference in the strengths of animal tendon and plant fibers is roughly one order of magnitude, the middle range of which encompasses the tensile strength of nylon (0.45 $GN \cdot m^{-2}$). By contrast, the tensile strength of nickel-treated steel is substantially greater (1.4 to 1.6 $GN \cdot m^{-2}$). However, the tensile strength of plant materials is remarkably high in terms of their density. Venkataswamy et al. (1987) measured the tensile strength of the midrib (the main vascular bundle running the length of each leaf) of the coconut palm. They report values ranging between 0.17 and 0.30 $GN \cdot m^{-2}$, significantly greater than the tensile strength of annealed aluminum (0.059 $GN \cdot m^{-2}$) and very close to that of hard rolled aluminum bronze (0.26 $GN \cdot m^{-2}$), whose densities are more than twice that of the midrib tissues. The strongest measured plant tissue (on the order of 15 $MN \cdot m^{-2}$) for its density is the wood of the quipo tree (*Cavanillesia platanifolia,* Bombaca-

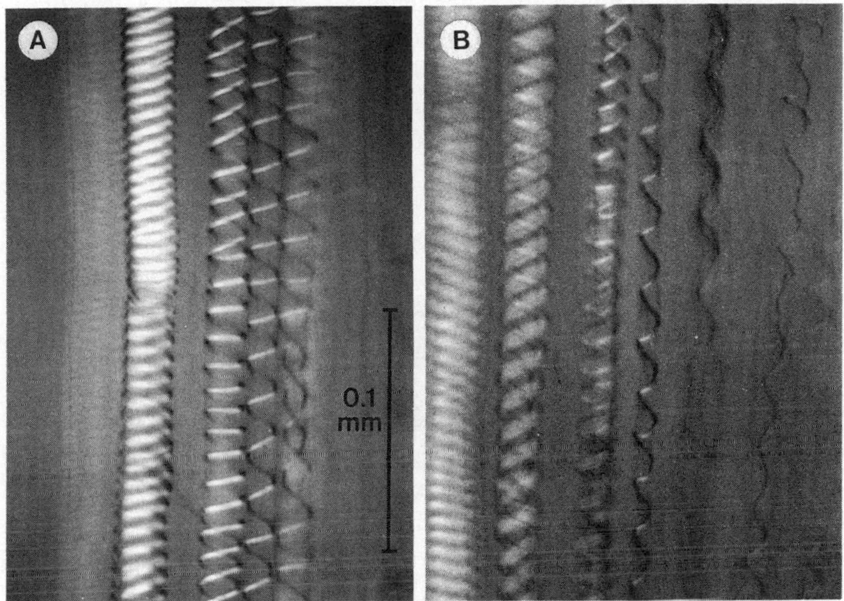

FIGURE 2.19 Two optical sections (A and B) through a cleared hypocotyl of a soybean seedling (photographed with polarized light) revealing deformations resulting from tensile stresses developed in the water-conducting cells (xylem). The relative intensity of the tensile strains developed within these cells can be gauged by the vertical displacements among the sequential gyres of the spiral secondary wall thickenings in each cell. The first cells to differentiate and mature into xylem elements (to the right of each optical section) have experienced the highest tensile strains, and the thin cell walls to which the spiral thickenings were attached have been broken. Cells that differentiate and mature subsequently have experienced progressively less tensile strain (from right to left) and have intact cell walls. These cells remain hydraulically functional in living stems. (Photographs generously provided by Dominick J. Paolillo, Jr., Section of Plant Biology, Cornell University.)

ceae), a native species of Panama, which surpasses the famous balsa wood (from the same family) in lightness.

Although the tensile strengths of many plant materials are comparable to those of some metals, most plant materials have the added advantage of considerable elastic extension before they break. Since these elastic extensions are fully recoverable, plant materials can bend and stretch when subjected to loads close to their breaking or yield stress and yet return to their original shape when stress levels drop to zero. Strands of the lichen *Usnea* are capable of elongating elastically from 60% to 100% of their original length. In general, lignification of tissues reduces their capacity for extension. Thus, woody plant stems sustain little deformation under bending stresses near their breaking stress compared with nonwoody stems.

In addition to its mechanical importance, tensile strength can play equally important roles in plant physiology and reproduction. For example, as the stems of vascular plants elongate, the first-formed cells specialized to conduct water (the primary tracheary elements) are stretched, and their cell walls may break in tension, interrupting the flow of water through the cell lumen (fig. 2.19). The tensile strains that develop from the time a cell begins to mature to the time of complete maturation have been shown to be as high as 0.13, but the primary walls of these cells may remain intact (Paolillo and Rubin 1991), indicating that their breaking stresses in tension are comparatively large (perhaps on the order of $GN{\cdot}m^{-2}$). These cells may be blocked laterally by expanding parenchyma cells flanking their sides (D. J. Paolillo, pers. comm.). Thus, in addition to their tensile properties, the compressive strength of cell walls may play an important role in maintaining avenues of water transport. As the rate of elongation of the stem diminishes, other conducting cells differentiate and mature. The walls of these cells experience lower strains than those that develop within the walls of the first cells to differentiate. Since their breaking stresses are not reached, these later-formed cells remain functionally viable, thereby providing the conduits through which water flows to the more distal, actively growing portions of the stem (fig. 2.19).

Although we have been speaking principally about solids, fluids are also typically submitted to tensile stresses. The columns of water held within the conducting tissue of plants can be placed in considerable tensile stress as a result of rapid water loss from photosynthetic tissues. When the tensile strength of these columns is exceeded, the column of water will break, sometimes with an audible click (Milburn and Johnson 1966) owing to cavitation (the formation of water vapor bubbles). (The tensile strength of water is difficult to measure because water molecules adhere to the vessels containing them [Apfel 1972].) Cavitation has been used to considerable advantage by plants to eject spores. Certain ferns have a region of specialized cells (called the annulus) that wraps around their spore-containing structures (called sporangia). The outermost cell walls of the annulus are thinner than the innermost cell walls, which gives them the capacity to mechanically operate as two adhering layers whose capacities to expand and contract differ. As spores mature, water loss causes the annulus to bend backward much like a spring held in tension. Each sporangium splits transversely as the flexing annulus causes other cells to rupture. Water continues to be lost until, quite suddenly, the tensile strength of the water in the cells composing the annulus is exceeded, water vapor is generated, and the annulus springs back to near its original position, ejecting spores much like a catapult. Haider (1954) found that the cells undergo almost simultaneous cavitation—the first cell to form a vapor bubble acts as a trigger, setting in motion a chain reaction along the entire length of the annulus. Many seed plants rely on the tensile strength of water and differences in the strengths of adjoining tissues to eject their seeds. For example, the touch-me-not and the pale snapweed (*Impatiens biflora* and *I. pallida*), as well as the wood sorrel (*Oxalis* spp.), produce fruits with longitudinally aligned strips of fairly rigid tissue laterally joined by strips of less rigid tissue. As the tissues of these fruits mature, differential tensile stresses develop, and the mechanical stability among the strips of tissues becomes increasingly precarious until even a small externally applied load results in a violent shearing failure and external flexure of the strips of tissues to which seeds are attached. The forces generated in this manner can catapult the seeds of the wood sorrel up to 3 m (vertically). The fruit of the prickly cucumber (*Echinocystis lobata*) is equally adroit at ejecting seeds. As the fruits of this plant mature, they develop very high internal hydrostatic pressures that eventually exceed the strength of a distal lid of tissue. The result is a violent ejection of seeds at a velocity that can exceed 11.5 $m \cdot s^{-1}$.

As mentioned previously, for materials with a plastic region of behavior it is usual to speak of the yield stress, rather than the breaking stress, when discussing strength, since the material does not actually break. A stress-strain diagram for such a material is illustrated in figure 2.18 and compared with that of a material that has a breaking stress. Both the yield stress of a material that exhibits plastic behavior and the breaking stress of a material that exhibits no plastic behavior share the same units (force per unit area). The yield stress is an important mechanical parameter for plant tissues that are actively growing, since it largely depends on the yield stress of the cell wall infrastructure within the tissue, which in turn relates to the physiology of cell wall growth dynamics. The yield stress of a young cell wall provides a measure of the internal hydrostatic pressure that the protoplast must exert before the cell can expand.

The strength of the structure, called the breaking load, is simply the magnitude of the loading that results in breakage. Unlike the strength of a material, which has the same units as stress, the breaking load has units of weight, and its magnitude can vary significantly even within a class of structures composed of the same material, such as tree trunks. The breaking load will not be the same for different structures made of the same material, nor will it be the same for the same structures differing in their absolute dimensions. Trees with identical shapes and sizes may have different breaking loads provided the material properties of their woods differ. Similarly, trees of the same species but differing in size may have different breaking loads, even if the strength of their wood is the same among all the individuals tested. Even though wood is a cellular structure, it can be relatively uniform in its cellular infrastructure, and within limits we can legitimately speak of its tensile or compressive strength rather than its breaking load. However, the tensile or compressive strength of wood samples possessing internal flaws may differ significantly from that of clear-grain samples. These structural defects play a significant role in how stresses are accommodated by a particular sample and when fractures will be propagated through a specimen placed in tension. The tensile strength of wood is usually specified for clear-grain samples—those lacking knots or anomalous growth patterns. This has a significant bearing on the mechanics of tree trunks and branches, which can possess many knots or structural defects in the form of fungal or bacterial or animal damage.

An obvious concern arises over the potential for variability in the tensile strength or other material properties of plant tissues. When the tensile strengths of many samples of tissue from a population of plants are measured and plotted as a function of their relative frequency (the number of individuals, expressed as a percentage, within the sampled population), the data typically take the form of a Weibull frequency distribution. This has been shown for the breaking (tensile) strength of garlic flower stalks (*Allium sativum;* Niklas 1990a) placed in bending, and the tensile strength of parenchyma from potato tubers (Lin and Pitt 1986) and wood (Woeste, Suddarth, and Galligan 1979). The Weibull frequency distribution is one of a few that has been shown to describe phenomena whose statistical behavior is responsive to environmental variability (Weibull 1939; Ang and Tang 1975). The implication is that, within a population, the strength of tissues varies in a manner that ensures the survival of some individuals over a broad range of mechanical stress levels. Some individuals are stronger than others and can perpetuate the population or the species after some unpredictable environmental event, such as a storm. The potential for a tissue's mechanical properties to vary among individuals from the same population confers a design factor or margin of safety on the population, ensuring that some will survive to reproduce and continue to occupy a site. Of less theoretical but much practical importance, variation in the strength of the same tissue among individuals highlights the importance of relying on numerous measurements of material properties when considering biomechanical models designed to predict the behavior of a tissue, an organ, or an individual plant.

When considering the strength of elastic structures capable of undergoing large-scale plastic deformations, engineers typically speak of the collapse load of the structure. For a simple prismatic bar, the collapse load is given by the product of the uniaxial yield stress and the cross-sectional area of the bar. When dealing with complex structures such as steel frameworks and plant stems, however, some parts of the structure may yield before others because of localized stresses. An example of this was seen in our previous discussion of the development of a plastic hinge. For complex structures, it is important to know the elastic load, which is the highest load the structure can tolerate before nonrecoverable deformation occurs at any point within it. In some structures the elastic load may be very small, but for plastic collapse to occur, yielding in the structure must become general and widespread.

STRAIN ENERGY

When an externally applied force causes a material to deform, energy must be introduced into the system. This energy, symbolized by U, is called strain energy. Strain energy is an important mechanical parameter for elastic solids because it can be converted into mechanical energy to elastically restore an object to its original shape and orientation when an external force is removed. Likewise, strain energy is important when considering plastic materials because it is used to produce molecular rearrangement and, in strain hardening materials, to align polymer chains in the direction of externally applied forces. The capacity of an elastic structure to convert strain energy into mechanical energy is illustrated by the mainspring of a clock. The mainspring is mechanically loaded by winding the clock, and the strain energy stored within it drives the clock's mechanism. Similarly, when a tree is bent by the wind, the strain energy stored within the tissues of its trunk and branches can be used to restore the tree to its original orientation when the wind pressure is reduced or eliminated.

The capacity of elastic materials to absorb and subsequently release strain energy is called elastic resilience. An example of the biological importance of resilience to plant reproduction is seen in the stamens of the mountain laurel, *Kalmia angustifolia*. The tips of stamens (called anther sacs—the structures in which pollen is produced) are partially enveloped within small pockets created by the foldings of the fused petals. As the flower matures and the petals expand and reflex backward, the filaments of stamens (the slender stalks of the stamen that bear the anther sacs) are bent much like a clock's mainspring. This flexure generates strain energy, which is stored within the filaments as

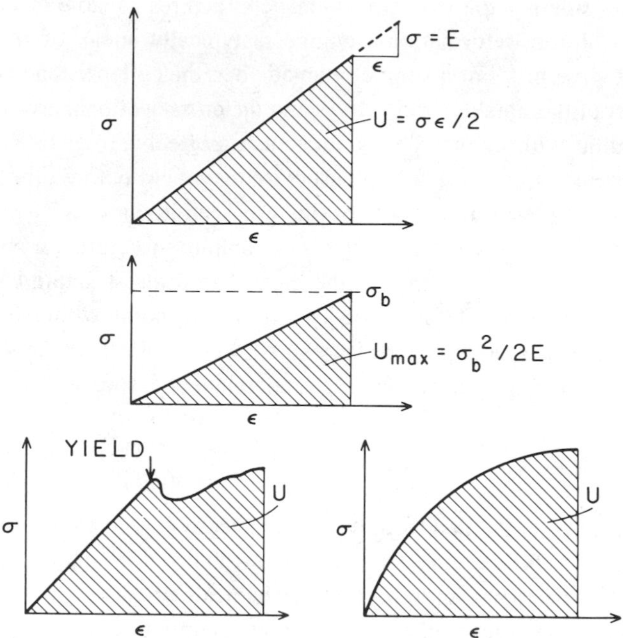

FIGURE 2.20 Relation between maximum strain energy and strength. Strain energy (U) is the energy stored within a material as it is loaded. For a linear elastic nonductile material, U is computed from the formula $U = \sigma\varepsilon/2 = \sigma^2/2E = \varepsilon^2E/2$ per unit volume of material. The maximum strain energy (U_{max}) can be computed from its elastic modulus and its breaking stress σ_b; $U_{max} = \sigma_b^2/2E$. For a ductile or plastic material the strain energy must be computed empirically from the area under the stress-strain diagram. The strength of a material is the maximum stress it can sustain before it breaks or begins to yield. For a material, strength has the same units as stress (force per area).

potential energy. As the flower matures, the stamens are placed in sufficient tension that the visit of a bee or some other pollinating insect will trigger their tips to dislodge from the petals and spring forward, striking visiting insects and dusting their bodies with pollen. Thus the resilience of the stamen aids the transport of pollen from one flower to another.

The quantity of strain energy stored within a material can be calculated from the stress-strain diagram. For linear elastic materials, the magnitude of the strain energy U is equal to the area under the material's stress-strain diagram: $U = \sigma\varepsilon/2$ per unit volume of material, as illustrated in figure 2.20. Since $E = \sigma/\varepsilon$, the strain energy for such a material is also equal to $\sigma^2/2E$ (or $\varepsilon^2E/2$) per unit volume. Thus the strain energy for a linear elastic material is proportional to the magnitude of the stress and inversely proportional to E. Since stresses and strains are not related to one another by a single proportionality constant for nonlinear elastic materials and plastic materials, the value of U for these materials must be empirically determined from measurements of the area under their stress-strain diagram.

In many cases it is important to know the maximum strain energy that a material has absorbed when stressed. We can easily calculate the maximum strain energy U_{max} for a linear elastic material that breaks from the material's elastic modulus and breaking stress σ_b; $U_{max} = \sigma_b^2/2E$. In general, the strain energy is proportional to the inverse of Young's modulus, $U \propto 1/E$. Thus materials with high Young's moduli will tend to have lower strain energies than materials with low moduli. However, linear elastic materials with high elastic moduli and high breaking stresses necessarily store large amounts of strain energy per unit volume before they break (fig. 2.20). Conversely, elastic materials with low E and low breaking stresses store relatively small amounts of strain energy per unit volume before they break. Since there is no a priori relation between a material's elastic modulus and its breaking stress, however, further generalizations about the relation between maximum strain energy and elastic moduli are not warranted. For plastic materials, the maximum strain energy must be determined by inspecting the stress-strain diagram, just as it is necessary to measure this parameter empirically for elastic materials that undergo plastic yielding beyond their proportional limit (fig. 2.20).

Since the energy required to deform a material depends on the size of a specimen, we must normalize the energy in terms of volume; that is, the energy is stored throughout the bulk of the specimen undergoing strains in all three of its principal axes—width, depth and length. When this is done, we have the strain energy density, expressed in units of energy per volume. The strain energy density required to break a material is called the breaking strain energy. The strain energy density at which a specimen yields plastically is called the yield strain energy. In this regard, the mechanical properties of cellulose are of interest. Cellulose is a very strong material with a breaking strain energy between 5×10^6 and 50×10^6 J·m^{-3}. The value for high tensile steel (20×10^6 J·m^{-3}) falls within this range. Clearly, cellulose can store amounts of energy within its molecular structure comparable to those stored by steel. Additionally, the tensile strength of cellulose is equal to or greater than that of many steels (on the order of GN·m^{-2}), though steel is roughly seven times as dense as cellulose. Indeed, for its density cellulose is the strongest and highest energy-storing material known. Nonetheless, although the properties of cellulose are remarkable, we must take care in our inferences concerning the strain energy densities occurring in plant cell walls and tissues, since these composite materials consist only partially of cellulose. Thus, for example, while the strain energy density of cellulose is on the order of 10^6 J·m^{-2}, that of the nutshells of the macadamia, which is a fibrous cellular solid, lies between 10^2 and 10^3 J·m^{-3} (Jennings and MacMillan 1986).

In contrast to materials, in structures shape plays an important role in dictating resilience. Consider a cylindrical metal bar of length l and cross-sectional area A submitted to uniaxial tension with magnitude P. The strain energy U as a function of the tensile load equals $P^2l/2AE$, or, in terms of the corresponding elongation Δl, $U = AE(\Delta l)^2/2l$, where $\Delta l = \varepsilon l$. Thus $U = A\sigma^2 l/2E$, and for a bar of given dimensions and E, the strain energy is completely determined by the magnitude of the tensile load and the elongation. But the elastic resilience of an object is maximized only if the object has a uniform cross-sectional shape (Cottrell 1964, 116–17). If the cross-sectional area of the previously considered cylinder is locally increased to A_i over a limited span of length l_i, then the strain energy absorbed in the altered span equals $[(J_bA/A_i)^2 A_i l_i]/2E$, while the strain energy absorbed in the unaltered portion of the cylinder equals $\sigma_b{}^2A(1 - l_i)/2E$. Accordingly, the total energy absorbed by the altered cylinder equals $(\sigma_b{}^2 Al/2E)[1 - (l_i/l)(1 - A/A_i)]$ and as Cottrell (1964, 117) points out, since $l_i < 1$ and $A_i > A$, the resilience of the cylinder with the regionally increased cross-sectional area is actually decreased. This interesting result is worth pondering, particularly since so many plant stems are tapered or have regional bulges and reductions in girth (see chap. 7 for a discussion of tapering).

FRACTURE MECHANICS

When an object breaks under tension, new surfaces are created in the form of a crack or fracture. To understand the mechanics of fracture, we must return to the concepts of strain energy and resilience and consider additional concepts like brittleness, ductility, and toughness.

As already noted, the energy stored within an object in the form of recoverable strains is the energy that can be used to restore an object to its original shape and dimensions after an applied load is removed. Thus, strain energy provides an elastic object with resilience. For example, when a branch is bent under an applied load and the load is removed, the branch swings back to its original position by the conversion of strain energy into kinetic energy, which is used for mechanical work. Indeed, the branch will often oscillate between positions that differ from its original equilibrium position until the strain energy stored within it is dissipated.

When an applied load produces stresses that are equal to or greater than the breaking stress, however, the strain energy stored within a biological structure can be used to fracture the structure. This typically occurs through the propagation of a crack that begins on the surface of the branch experiencing tensile stresses. Since the creation of new surfaces within a solid requires energy, breakage under tensile stress requires the expenditure of strain energy stored within the volume of an object. Indeed, as the fracture propagates, more and more strain energy must be supplied. An important feature in the expenditure of this energy is that only the strain energy stored in the immediate vicinity of a fracture can be used. Thus, as the fracture propagates, more and more strain energy must be drawn from an increasing volume within the object (see Vincent 1990b).

Fractures initiate when strain energy is released near discontinuities in the force trajectories traveling within an object. Force trajectories can be imagined by considering a small notch on a specimen placed in tension. Although the stress in the bulk of the specimen remote from the notch is uniformly distributed, the lines of tensile force operating within the specimen follow trajectories that converge and increase in concentration as they curve around a notch or imperfection. These discontinuities in the force trajectories are called stress concentrations or stress raisers. Although the applied tensile stress remote from the notch may be substantially less than is needed to cause the material to break, the tensile stress developed near the notch may exceed the tensile strength of the material. This is important to remember when comparing the tensile strengths of various materials, since strength is a property of materials while fracture strength is a measure of the structural homogeneity of a structure. The higher the structural perfection of an object, the greater its fracture strength. Since it is reasonable to predict that the probability that an imperfection, such as a notch or crack, will be found will increase as the size of an organic structure increases, it is reasonable to suppose that fracture strength is a statistical function of the size distribution of a class of organic structures. This has a bearing on the fracture mechanics of tree trunks and branches, which increase in size as they grow. In theory, as these structures get older and grow bigger, the probability of growth defects that will form the nuclei for fracture propagation should increase. Thus, older and larger branches and trees should have a higher probability of mechanical failure by means of fracturing as a result of "notch stresses." The statistical relation between an object's size and the probability of finding structural defects may in large part explain why physical parameters like bending strength and tensile

TABLE 2.3 Moduli of Rupture (M_R) and Elasticity (E) and Densities (ρ) of Thirty-three Species of Wood

Wood	M_R (GN·m^{-2})	E (GN·m^{-2})	ρ (kg·m^3)
Apple (*Pyrus malus*)	0.088	8.77	745
Ash, black (*Fraxinus nigra*)	0.088	11.04	526
Aspen (*Populus tremuloides*)	0.059	8.22	401
Basswood (*Tilia americana*)	0.060	10.10	398
Beech (*Fagus americana*)	0.101	11.57	655
Birch, yellow (*Betula lutea*)	0.117	14.53 max	668
Butternut (*Juglans cinerea*)	0.056	8.14	404
Cedar, northern white (*Chamaecyparis thyoides*)	0.047	6.42	353 min
Cherry, black (*Prunus serotina*)	0.086	10.26	534
Chestnut (*Castanea dentata*)	0.060	8.53	454
Cottonwood, eastern (*Populus deltoides*)	0.060	9.53	433
Dogwood (*Cornus florida*)	0.105	10.64	796
Elm, American (*Ulmus americana*)	0.083	9.30	554
Fir, balsam (*Abies balsamea*)	0.053	8.62	414
Gum, red (*Liquidambar styraciflua*)	0.082	10.25	530
Hemlock (*Tsuga canadensis*)	0.059	8.30	431
Hornbeam (*Ostrya virginiana*)	0.100	11.76	762
Larch (*Larix occidentalis*)	0.081	11.65	587
Locust, black (*Robinia pseudacacia*)	0.134 max	14.20	798
Maple, silver (*Acer saccharinum*)	0.062	6.22	506
Maple, sugar (*A. saccharum*)	0.108	12.65	676
Oak, black (*Quercus velutina*)	0.095	11.31	669
Oak, live (*Q. virginiana*)	0.127	13.54	977 max
Oak, white (*Q. alba*)	0.121	14.18	792
Pine, longleaf (*Pinus palustris*)	0.107	14.17	638
Pine, red (*P. resinosa*)	0.086	12.39	507
Pine, white (*P. strobus*)	0.061	8.81	373
Poplar, yellow (*Liriodendron tulipifera*)	0.064	10.38	427
Redwood (*Sequoia sempervirens*)	0.074	9.40	436
Spruce, white (*Picea glauca*)	0.063	9.82	431
Sycamore (*Platanus occidentalis*)	0.070	9.83	539
Walnut, black (*Juglans nigra*)	0.102	11.62	562
Willow, black (*Salix nigra*)	0.043 min	5.03 min	408

Note: See table 6.3 for additional values for E.

strength have Weibull frequency distributions. Indeed, these distributions may simply reflect the size distributions of structures like stems in natural populations of plants. (An interesting treatment of how tree growth can avoid notch stresses is given by Mattheck 1990.)

The fact that a structural defect decreases the fracture strength of an object is exploited by glaziers, who apply a modest bending stress to break a piece of scored glass. Even a small imperfection, such as a score, can act as a stress raiser and decrease the amount of applied energy necessary to break the glass. For brittle materials like glass, the energy required to break atomic bonds and produce a crack is supplied by the volume of the material immediately adjacent to the new surfaces that propagate near the initial imperfection. Since the energy required to break atomic bonds varies relatively little as a function of the type of atoms involved, the energy necessary to fracture brittle materials varies little from material to material, typically ranging between 1 and 10 $J \cdot m^{-2}$. Brittle solids are typically tested in bending. The highest tensile stress at fracture, which is determined from the bending moment and the cross-sectional geometry of a specimen, is called the modulus of rupture, symbolized by M_R. For highly brittle solids, $M_R \approx 2$ (tensile strength). Table 2.3 provides the moduli of rupture and elasticity for thirty-three species of wood, as well as the densities of these materials. Regression analyses of these data reveal that both E and M_R are highly correlated with ρ, as well as with each other; for example, M_R (in $GN \cdot m^{-2}$) = 0.00249 + 0.00014 ρ (in $kg \cdot m^{-3}$) ($r = .886$); $M_R \approx 0.0092\ E$ (in $GN \cdot m^{-2}$) ($r = .886$).

For ductile materials, such as many types of metal, the energy supplied by an external force is dissipated over a greater portion of the volume of the material near the crack's surfaces and can be on the order of 10^3 to 10^6 $J \cdot m^{-2}$. Necking is a physical manifestation of the dissipation of energy in ductile materials placed in tension—atoms or molecules slip past one another over a considerable volume removed from a stress raiser, so the specimen's cross-sectional area is attenuated by localized plastic straining.

The physics underlying fracture mechanics was first dealt with successfully by A. A. Griffith (1893–1963). He initially explored fracture mechanics for brittle, linearly elastic materials in terms of an energy budget—that is, the difference between the energy needed to form cracked surfaces and the strain energy released by the propagation of a crack equals net energy. Griffith determined that there exists a critical crack length below which energy is consumed by the system such that the crack will not propagate (increase in length). Put differently, even if local stresses are high at the tip of the crack, the structure will not fracture at the site provided the crack is below this critical length.

Griffith derived an equation from which the critical crack length can be calculated. This length, known as the Griffith critical crack length, is symbolized by L_G. The derivation of this equation is of some interest and is based on an estimate of the elastic energy of the crack. To illustrate this, consider a metal bar with a crack (of length l) oriented transverse to an applied tensile stress. Intuitively, we recognize that near the surfaces of the crack the local tensile stresses are relaxed to zero magnitude, but as we progress away from the crack surfaces the tensile stress σ approaches the magnitude of the applied tensile stress. Therefore a region with a radius roughly equal to $l/2$ around the crack will have been relieved of its elastic energy. This energy, which equals $\pi\sigma^2 l^2/4E$ per unit thickness of the metal bar, can be converted into the surface energy, symbolized by $\bar{\omega}$, that is used to generate the crack's new surfaces. Griffith calculated a precise integration of the true shape of the stress trajectories near a crack and found that the relationship among these parameters is given by the formula $\sigma = (2\bar{\omega}E/\pi l)^{1/2}$, where σ is the Griffith stress. Solving for l, which we now see is the Griffith critical crack length L_G (the total crack length and not the semicrack length), gives us the formula

$$L_G = \frac{2WE}{\pi\ \sigma^2}, \tag{2.19}$$

where W is the work of fracture, which is theoretically equivalent to the surface energy $\bar{\omega}$, though this is not precisely so.

Since a great many assumptions enter into the derivation of eq. (2.19), it is not unreasonable to assume $L_G \approx WE/\sigma^2$. Any brittle structure having cracks less than L_G long is essentially safe from fracturing. For any material with an elastic modulus E, the length of safe cracks depends on the ratio of the work of fracture to that of the strain energy stored in the material. Also, the Griffith critical crack length is inversely proportional to the resilience of a material. The higher the resilience, the shorter the crack that will operate as a fracture site. Accordingly, there exists a trade-off between the capacity of a material to store strain energy (and reassume its original shape when an applied load is removed) and the probability that the structure made from the material will fracture.

The work of fracture is temperature dependent, generally decreasing with decreasing temperature, though this is not so for wood. Although at any given moisture content the energy required to propagate a crack parallel to the grain decreases with temperature, the energy required to propagate a crack across the grain increases as the temperature drops (see Vincent 1990a, 158–59). The reasons for this are not clear, but the advantages to trees growing in cold climates may be significant.

The work of fracture *W* allows us to distinguish two very broad categories or classes of brittle materials: weak ($W \approx$ the true surface energy) and strong ($W >>$ true surface energy). For example, steel and glass are both brittle, and they are comparably hard. But steel is much stronger than glass. Strong brittle materials, like hard steel, retain their strength even though their surfaces are heavily scratched or slightly notched, whereas weak brittle materials, like glass and many types of stone, retain little of their strength when their surfaces are scored. Stonemasons utilize this characteristic when they score a block of granite and cleave it with a relatively small percussion. By the same token, steel girders can sustain much superficial scratching yet still support large and heavy structures. Indeed, Cottrell (1964, 363) made the observation that the advance of humanity from the Stone Age to the Iron Age was based on advancement in our understanding of fracture mechanics.

Empirical fracture mechanics indicates that, in most circumstances, unreasonably large amounts of surface energy have to be assumed if Griffith's formula is to be used. And it has been proposed that the total energy required to propagate a crack is the sum of the surface energy already discussed and the energy needed to produce plastic deformations. We see this when a very thin sheet of metal is torn—plastic rupture actually occurs well below the general yield stress. The essential condition for elastic-plastic fracture is that the plastic displacements at the tip of the propagating crack are attended by elastic displacements within regions distant from the crack. This condition is often found in ductile materials.

Bear in mind that Griffith's theory does not apply to highly deformable elastic materials. For example, rubber is a brittle material, but it strains elastically at such low stresses that notches in it can deform into mechanically harmless blunt regions well before the breaking stress is achieved. This type of behavior is called notch insensitivity, and plants have evolved several ways to achieve it. Some plant materials are ductile and can dissipate stress concentrations in the form of internal molecular slippage. Natural rubber, plant latex, and growing cell walls are only three examples. Another way is to round off naturally occurring indentations or notches. Undulating leaf margins and the rounded surfaces of stems are examples of this design. Another way to be insensitive to abrasions or scratches is to produce many parallel fibers (in a binding matrix) that will shear little. In this arrangement, if a fiber breaks the stress is distributed evenly among the remaining fibers. This design is seen in leaves possessing parallel vascular bundles running their length, as in grasses.

The work of fracture for grass leaves is relatively high for many biological structures—for example, about 40 $J \cdot m^{-2}$ for leaves of perennial English rye grass, *Lolium perenne* (Vincent 1982).

Engineers and botanists frequently refer to the fracture toughness of a material or plant tissue. Fracture toughness, symbolized by K_C, is the resistance a material offers to the propagation of a crack. Since toughness corresponds to the energy required to fracture a material, the area under a stress-strain diagram provides a measure of a material's toughness in terms of the energy absorbed per unit volume. As a general rule, toughness is maximized by an optimal combination of strength and ductility. For isotropic materials, K_C is related to the elastic modulus E, the work of fracture W, and the Poisson's ratio ν, such that $K_C = (EW)^{1/2}/(1 - \nu)$. For these materials, K_C decreases as the Poisson's ratio decreases. For nonbrittle and anisotropic materials, determining fracture toughness tends to be very complex procedurally. Thus, considerable efforts have been made to determine fracture toughness from more easily measured tensile stress-strain diagrams. Through these efforts, it has been shown that fracture toughness can be estimated from the formula

$$K_C \approx (0.25E\, S_y\, l^*\, \varepsilon_f)^{0.5}, \tag{2.20}$$

where S_y is the tensile yield strength, l^* is the plastic zone width at the onset of cracking, and ε_f is the true fracture strain in tension. Unfortunately, plastic zone widths can be analytically calculated for only a few simple cases (see Cottrell 1964, 352–56), although Hahn and Rosenfield (1968) have shown that l^* is approximately equal to the square of the material's strain hardening exponent, n. However, based on eq. (2.10) and tensile tests, Greenberg et al. (1989) calculated that the fracture toughness of leaves from perennial English rye grass ranges between 0.31 and 1.57×10^5 $N \cdot m^{-3/2}$. These values are remarkably low, suggesting that it is more efficient for an animal to nibble a leaf blade and introduce a crack than it is to pull and break it.

THE MECHANICAL PROPERTIES OF SOILS

The mechanical properties of soils are important to plant survival and growth, since they influence the ease with which roots penetrate their substrata, the establishment of seedlings (parts of which must penetrate the soil to reach the light), and the force required to dislodge roots and overturn individual plants. Therefore we shall consider soil mechanics in this chapter, particularly since many of the engineering concepts reviewed thus far apply to the behavior of soils.

For any given type of soil, its mechanical behavior depends on its water content. As the water content increases, the consistency or state of a particular soil can change from that of a solid to that of a plastic or, in extreme cases, a non-Newtonian liquid. For example, soils with a high silty clay or organic content behave like plastics or liquids when they have a moderate water content. As solids, soils can fracture, and as plastics or liquids, they will flow once a yield or threshold stress has been achieved. In large part this explains why trees dislodge from their soils. The trunks of trees act as levers when their canopies are subjected to high wind pressures. When high winds are accompanied by heavy rains, the mechanical properties of the soil trees grow in can be altered from those of a solid to those of a very plastic material or, in extreme cases, to those of a material that behaves very much like a fluid. Conversely, bending and torquing of tree trunks fracture dry soils as roots are placed alternately in tension and compression. This reduces the strength of the soil and increases the ability of rain to infiltrate deeper. As the rain continues to modify the consistency of the soil, further bending and torquing of tree trunks can produce stresses sufficient to cause the soil to yield completely, even under modest wind pressures. Soil liquefaction can have devastating consequences, particularly on steep slopes, where its consistency can change to that of a non-Newtonian fluid, resulting in rapid flow rates when yield stresses develop from a soil's own weight.

The extent to which a given soil behaves as a solid, plastic, or liquid can be experimentally determined in the field or under laboratory conditions by measuring the water content at its lower (drier) and upper (wetter) limits of plasticity. Typically, soil samples are sieved in the laboratory to remove particles larger than 2 mm in diameter; the remaining particulates are then molded into a paste by adding water (American Society for Testing and Materials 1984; British Standards Institution 1975). Needless to say, sieving a soil sample alters its material properties from what plants usually experience in the field.

Traditionally, the lower (drier) limit of a soil's plastic behavior is defined as the water content at which the soil crumbles when it is rolled into a cylinder 3 mm in diameter (Atterberg 1911). The upper (wetter) limit of plasticity is determined by a penetrometer. Water is added to the sample until the cone of the penetrometer penetrates to a depth of 30 mm. The lower and upper limits of plasticity are known as the plastic and liquid limits, while the difference between the two is called the plasticity index (Marshall and Holmes 1988). The plastic and liquid limits of most soils typically increase as the clay and organic matter contents increase, and as might be expected, they are highly correlated over a broad range of soil types. In general, soils with a low plasticity index (soils with a high silty clay or organic content) relative to their liquid limits are the least desirable for engineering purposes, since they flow easily and behave like a liquid when placed under compression.

From the perspective of root growth mechanics, however, soils that behave as a plastic with moderate water content might be considered desirable, since the growing tips of roots would encounter low resistance to the compressional stresses they generate as they advance through the soil. Since reversible elastic strains are limited in most soils, large permanent (plastic) deformations and the yield stress at which they occur dominate the mechanical environment attending root growth. These deformations can result from soil fracture or plastic flow. Since the strength of a soil decreases as the water content increases, our initial expectation that wet soils will be more mechanically desirable for root growth than dry soils appears reasonable.

This can be mathematically formalized by means of the Mohr-Coulomb equation, which provides a way to calculate the shear strength S_τ of a soil based on a constant c expressing the cohesiveness of the soil, the stress σ normal to the plane of shearing, and the tan ϕ, which is the coefficient of internal friction (see Marshall and Holmes 1988, 227):

$$S_\tau = c + \sigma \tan \phi. \tag{2.21}$$

In standard engineering practice, a soil sample is examined over a range of values for σ, and plots of S_τ versus σ are used to calculate c and tan ϕ. The soil sample's cohesiveness c is the y-intercept and tan ϕ is the slope of the linear plot of S_τ versus σ.

According to the Mohr-Coulomb equation, the shear yield stress will equal the cohesiveness of the soil when there is no stress component operating normal to the shear plane. In fact there is always some normal stress component, since the weight of the soil above the plane of shearing exerts some force, but at shallow depths this force is negligible. Not surprisingly, the Mohr-Coulomb equation indicates that for any given soil cohesiveness the shear stresses required to pull a root from the ground increase as a function of soil depth. Accordingly, if root systems are designed to operate as tensile members, then for any given soil type depth of growth is the most important factor. Further, soil cohesiveness tends to increase as the clay content increases. For example, c is zero for sand and increases to an upper limit of about 3×10^4 N·m^{-2} for clay. Therefore, all other things being equal, root systems growing in clayey soils require extremely high stresses to dislodge them. There are clear tradeoffs, however, when ease of root penetration and the requirement for roots to grip the soil are considered together. The Mohr-Coulomb equation indicates that either low normal stresses or low cohesiveness or both will aid root penetration. Shallow growth (low σ) in sandy soils (with low to zero c) aids root penetration, while deep growth (high σ) in clayey soils (high c values) ensures an efficient means of gripping the soil.

From first principles, we would expect organs to exert compressive stresses as they grow through a soil. (If roots are well supplied with oxygen, then they can exert tip pressures up to about 1 MPa.) If a soil is wetter than its plastic limit, the compressive stresses generated by root growth will cause the soil to deform plastically as the root advances, progressively reducing the pore sizes among soil aggregates in front of the root tip. Thus the loading caused by growth is progressively resisted by the larger contact area among soil particles as soil aggregates flatten against one another. How tightly aggregates in a soil are compressed can be expressed by the soil void ratio e, which is the instantaneous volume of pores within the soil divided by the volume of solid particulates. The initial void ratio e_o depends on the initial compressional stress σ_o to which the soil is subjected. It decreases as further compressional stress σ is applied. This can be mathematically expressed by the formula

$$IC = \frac{e - e_o}{\log (\sigma / \sigma_o)}, \tag{2.22}$$

where IC is the compressional index, which is empirically determined by taking a soil sample and measuring e at different values of σ. IC is the slope of the semilog plot of e versus σ. For any specified water content, IC is a constant for any sample. The data from such an experiment can be used to estimate the compressional stresses that a subterranean organ must exert to progress through a soil type as a function of the resulting changes in the void ratio of the soil. Intuitively, we can see that as growth proceeds, these stresses ought to increase. But roots typically grow around large soil aggregates and follow channels made by earthworms and former (decayed) roots, which offer comparatively little resistance to penetration by young roots. Additionally, roots expand in girth at some distance from their growing tips. This lateral expansion can propagate soil fractures, opening up low-resistance avenues for subsequent growth in length.

Measurements of soil strength by means of penetrometers invariably overestimate the actual resistance the soil offers to root growth, since the penetration tip of a penetrometer cannot avoid large soil aggregates as root tips can. Nonetheless, the rate of root elongation correlates well with penetrometer resistance measurements. Taylor and Ratliff (1969) measured the rate of root elongation of peanut plants grown in loamy sand at three water contents (0.07,

0.055, and 0.038 g $H_2O \cdot cm^{-3}$) and reported that the rate of elongation dramatically declines as soil strength increases from 0 to 6 MPa. Regression analysis of the data provided by these authors shows that, regardless of water content, the rate of elongation *e* declines exponentially with penetrometer resistance *R* according to the formula $e = 3.32 \times 10^{-0.208R}$ ($n = 18$, $r = 0.899$), where *e* is expressed in units of $mm \cdot hr^{-1}$ and *R* is in MPa. Clearly, other factors, such as the availability of oxygen to roots and leaf transpiration, influence root growth.

Even a brief review of the physical and material properties of soils demonstrates that the mechanics of underground growth is very complex. Nonetheless, a few equations and some experimentation can provide considerable insight into the obstacles that confront the delicate growing tips of underground roots and stems. Fortunately, the Poisson's ratios of young, growing tissues are high, making them essentially incompressible materials that mechanically operate as hydrostats. As such, root and stem tips can exert tremendous compressive stresses on soil. Another feature of growing roots is that they can alter the chemistry of soils by excreting compounds, and can also lubricate and slough off their surfaces, thereby reducing simple and pure shear stresses. The capacity of roots to chemically alter the soil they grow through can have surprising consequences. Palm trees that survived a hurricane that struck the Hawaiian Islands in 1979 were found to have remained anchored to their growth site by means of massive concretions of soil formed around their root systems. These cement foundations apparently provided sufficient ballast to resist the mechanical forces that washed away the foundations of many large hotels on the southern coast of the island of Kauai.

Three

The Effect of Geometry on Mechanical Behavior

Es ist dafür gesorgt, dass die Bäume nicht in den Himmel wachsen.

It is so arranged, that the trees do not grow into the heavens.

Johann Wolfgang von Goethe, *Dichtung und Wahrheit*

The distinctions drawn among elastic, viscous, and viscoelastic materials in chapter 2 provide a starting point from which to explore the mechanical behavior of plants, since a plant is not a material but rather a structure, whose shape and size also contribute to its mechanical performance. Thus far little attention has been paid to geometry, except with regard to the way stresses and strains are calculated. The object of this chapter is to redress this omission by turning attention toward solid mechanics and the fundamental problem addressed by the theory of elasticity—determining the state of stress within an object.

Our primary concern here is with how shape and size influence the relative magnitudes of stresses that develop within structural support members such as plant stems and tubular leaves. Thus our focus will be on the spatial distribution of stresses within structures such as columns and beams. A member is any structural component. There are three principal kinds of members, distinguished by how they are loaded: a beam, which is placed in bending; a column, which is placed in axial compression; and a shaft, which is placed in torsion. Support members can be attached at one end and free at the other, as with a cantilevered beam, which is anchored at its base and free at the opposite end, or they can be attached at both ends, as with a column fixed between two horizontal beams. Also, intermediary conditions between types of members are possible, as between a vertical column and a horizontal cantilever. Such a condition is called a cantilevered column. It is achieved when very slender columns deflect from the vertical as a consequence of the weight of the structure or a load at the free end, like a flagpole in a stiff breeze.

Engineers specify the material properties, geometry, size, and loading conditions of members so that they largely conform to those of ideal beams, columns, or shafts. In most practical situations, however, the support members we call plant organs rarely have the material properties and structural attributes of ideal members. Our ability to treat the more complex situations encompassing the behavior of plants, treated in chapters 5–8, requires an understanding of what happens when an ideal support member is loaded and subsequently deforms in bending, compression, or torsion. Also, with due regard to how much material and geometric properties actually vary, the mechanical behavior of many if not most plant organs can be approximated by analogy with columns or cantilevers or shafts whose behavior is treated by elementary beam theory, which is the foundation of solid mechanics.

We will deal with the mechanical behavior of beams subjected to both static and dynamic loads. Static loads are those that have long durations of application and tend not to change in magnitude or direction of application. By contrast, dynamic loadings can have very short durations of application and typically change in magnitude and direction of application. Elementary beam theory treating static loading conditions generally focuses on the practical and relatively simple situation where the weight of an object is the only body force experienced—that is, on determining the state of stress within a self-loaded object. Conversely, dynamic beam theory tends to emphasize the harmonic motion of beams resulting from the cyclical exchanges between kinetic and potential energy. An initial treatment of static loadings, particularly self-loading, is important for two reasons. First, the minimum state of stress of any object denser than its surrounding medium results from self-loading. Thus, self-loading represents the simplest state of stress in most practical circumstances. Second, the mathematics of self-loading prefigures any treatment of the states of stress that result when objects experience dynamic loadings. The botanist deals with organisms that naturally experience dynamic loadings—all plants sustain and mechanically respond to the forces generated from the motion of water or air (see chap. 1). Additional dynamic loadings occur when objects come into contact with one another, as when branches collide in storms or when the strings of a cello are sheared or plucked. Although the consequences of direct shear can be mechanically detrimental, just as in a cello, many biological structures like leaves and stamens require dynamic loadings to function.

Many of the topics treated in this chapter may appear far removed from the traditional concerns of the botanist, but most if not all are essential to understanding functional plant morphology and anatomy. Shape and size are the traditional concerns of the comparative and functional morphologist. They also intrinsically define the stresses resulting from self-loading; that is, the size of an organism dictates the weight that must be sustained, while the shape of the organism defines the cross-sectional areas through which weight operates. By the same token, the spatial distribution of materials within an object, which translates into the anatomy of a biological structure, defines the local strains resulting from a given load.

The importance of shape and size to biomechanics is self-evident when we consider engineering parameters called moments of area, which are mathematical descriptions of the spatial distribution of material within an object. The product of the appropriate moment of area and the appropriate material modulus is a quantitative measure of an object's ability to resist deformation when forces are applied. In this chapter we will consider two moments of area: the second moment of area and the polar second moment of area. The product of the second moment of area and the elastic modulus, called flexural stiffness, measures the ability of a column to resist compression or of a beam or cantilevered column to resist bending. For circular shafts the product of the polar second moment of area and the shear modulus, called torsional rigidity, measures the ability of a shaft to resist torsion. Since most self-loaded objects bend under their own weight, and since many objects naturally bend and torque when they are dynamically loaded, determining flexural stiffness and torsional rigidity is important to most if not all practical biomechanical problems. Just as important, we see that the ability to resist bending or torsion depends on shape (moments of area) and material properties (material moduli). Thus the roles of geometry and material properties constitute a syzygy—the unity of apparent opposites—in engineering theory.

As mentioned in chapter 1, the topics treated in this chapter are presented as reference materials to be drawn on in the remaining portions of this book. Readers already familiar with solid mechanics can glance at this chapter or scan it at leisure as the material covered seems appropriate.

EQUILIBRIUM

A treatment of solid mechanics begins with a discussion of static equilibrium, which occurs when the sum of all the forces operating within an object is zero. Consider a metal bar placed in uniaxial tension or compression. The two opposing tensile or compressive forces cause the bar to undergo transverse and longitudinal deformations that are increasingly resisted by the molecular and atomic forces within the metal until the resulting internal forces permit no

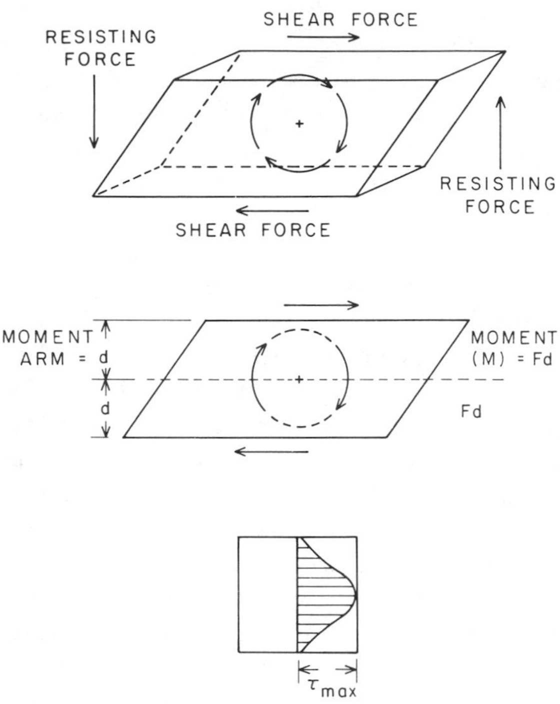

FIGURE 3.1 Moment arms within a sheared rectangular element distorted into a parallelepiped by two antiparallel shear forces (called a shear force couple). The element resists shear deformations through a pair of resisting forces (called the resisting force couple). Static equilibrium is achieved when the two opposing couples are balanced (top diagram). The moment arm is the perpendicular distance d between the line of action of the shearing forces and the center of rotation (indicated by +). The magnitude of the moment M is the product of the distance d and the applied shear force F; that is, $M = Fd$ (middle diagram). The distribution of the shear stresses (τ) within any cross section taken through a very long beam is shown in the bottom diagram. The maximum shear stress (τ_{max}) occurs just at the center of each cross section; the minimum shear stress occurs at the top and bottom of the cross section (the intensity of the shear stresses is indicated by the length of the arrows in the bottom diagram). This distribution is not true for the rectangular element shown at the top.

further deformation. At this point the bar is said to have reached static equilibrium. In principle, the same molecular process operates to achieve static equilibrium when a bar or specimen of wood is sheared, bent, or twisted. In these cases, however, a zero net body force is not sufficient to ensure the condition of static equilibrium. Consider the case of a rectangular material element subjected to shear forces, as shown in figure 3.1. Any pair of opposing forces is called a couple. A couple comprises two parallel forces acting in opposite directions separated by a moment arm. What is meant by a moment arm can be illustrated by considering the case of a sheared element (fig. 3.1). The couple operates to rotate material lines by the application of a moment, whose magnitude is the applied force times the perpendicular distance (which is the moment arm) between the lines of action of the two shear forces and the center around which rotation occurs. Thus a moment has units of force times length. In the case of a sheared element, two moments are applied, one from each of the two lines of action of the opposing pair of forces, and the magnitude of each moment arm is half the thickness of the sheared rectangular element (see fig. 3.1). For static equilibrium to exist, the applied shear couple must be balanced. That is, the shear forces shown in figure 3.1 must be counteracted by an opposing pair of forces (another couple) that resists rotation in the direction opposite to the shear couple. Equilibrium in shear is the result of a shear couple that parallels the lines of action of the two externally applied forces and another shear couple that is perpendicular to the first, generated within the sheared element in response to the externally applied forces.

By the same token, when a beamlike organ bends under its own weight, it experiences a bending force with a corresponding bending moment. (This was discussed in chap. 1 with regard to the foliage leaf of a pine tree; see eq. 1.6.) The magnitude of the bending moment depends both on the weight of the object and on the moment arm, which in turn depends on the orientation. When an object is perfectly vertical, the bending moment arm equals zero, whereas the moment arm increases as either object is increasingly tilted from the vertical. When either object is perfectly horizontal, the moment arm achieves its maximum value. Further inclination from the horizontal results in a decrease in the moment arm, hence a decrease in the bending moment, which drops to zero when an object is perfectly decumbent. Indeed, the relation among the bending moment, the moment arm, and the orientation of a beamlike plant organ give insight into the significance of tapering. Tapered stems typically droop and bend under their own weight, reducing the bending moment by virtue of a reduction in the moment arm. The mechanical strategy of tapering is elegantly discussed in terms of the panicle of the rice plant by Silk, Wang, and Cleland (1982), and sapling tree trunks are treated by Leiser and Kemper (1973).

Returning to the immediate issue of static equilibrium, recall from chapter 2 that normal and shear stress components (symbolized by σ and τ, respectively) occur whenever a material is subjected to a mechanical force (see fig. 2.1). In the example of a bar subjected to uniaxial tensile or compressive forces, only the normal stress components need be considered, whereas in the example of a sheared rectangular element, the shear stress components are largely relevant to a consideration of static equilibrium. However, in the general case of any object subjected to a force or combination of forces (as happens when a beam bends or a shaft experiences torque), static equilibrium involves all the stress components. Consider the stress components operating within a representative plane (x–y) through an object. The equations of equilibrium for the forces operating in the x and y directions are given by the formulas

(3.1a)
$$\frac{\partial \sigma_x}{\partial x} + \frac{\partial \tau_{xz}}{\partial y} + F_x = 0$$

(3.1b)
$$\frac{\partial \sigma_y}{\partial y} + \frac{\partial \tau_{xy}}{\partial x} + F_y = 0,$$

where F_x and F_y are the components of the body force per unit volume operating within the x–y plane. The remaining body force component F_z is likewise resolved and takes the same mathematical expression as that shown in eq. (3.1). And in the general case, the equilibrium state of an object must be satisfied within every element within the entire volume of the object, as well as at the external boundaries of the object. The external forces are mathematically regarded as a continuation of the internal forces (stresses); that is, the stress components must be in equilibrium with the external forces as we arrive at the boundaries of an object. Since an object's external boundaries are a physical manifestation of shape, its geometry must dictate the internal distribution of stresses.

The equations of equilibrium are used to describe the state of stress within any object subjected to any body or surface force, such as gravity or hydrostatic pressure. If the weight of an object (with mass per unit volume ρ) is the only body force experienced, and if y denotes the dimension of length, then the equations of equilibrium take the form

(3.1c)
$$\frac{\partial \sigma_x}{\partial x} + \frac{\partial \tau_{xy}}{\partial y} = 0$$

(3.1d)
$$\frac{\partial \sigma_y}{\partial y} + \frac{\partial \tau_{xy}}{\partial x} + \rho g = 0,$$

where g is gravitational acceleration. Solving these equations for the normal (σ_x, σ_y) and the shear stress components (τ_{xy}), we find that $\sigma_x = -\rho g F_y$, $\sigma_x = -\rho g F_y$, and $\tau_{xy} = 0$. These stress relationships describe the equilibrium state for the stresses produced by gravity, as well as the equilibrium state for hydrostatic pressure $\rho g F_y$ in two dimensions, with a zero stress at $F_y = 0$ (see chap. 9).

THE IMPORTANCE OF SHAPE

The importance of shape can be qualitatively illustrated by returning to a consideration of a bar placed in coaxial tension and a rectangular element that is sheared. In the former, shape is largely irrelevant unless the transverse area through which the tensile forces act varies abruptly along the length of the bar; that is, parallel force trajectories are ensured provided cross-sectional area is relatively uniform, even if the bar is slightly tapered. This is why normal stress components are relatively easy to calculate when untapered or slightly tapered bars are placed in uniaxial tension or compression (see chap. 2). By contrast, even the most geometrically uniform object subjected to shear will have an unequal distribution of stress throughout its thickness. The spatial heterogeneity in stress results from the fact that the magnitude of the opposing perpendicular couple in a sheared object depends on the amount of material above the horizontal planes of the two opposing shear forces. Since there is no material above the surface of the object along which the shear force is applied, the shear stresses at the top and bottom surfaces have zero magnitude. Since the amount of material within the bar increases toward the center, the shear stresses will increase toward the center. In fact, the shear stresses may increase nonlinearly; that is, they can increase parabolically toward the center of a very long beam placed in bending whose depth is small compared with its length (it is not, as shown in fig. 3.1 for beams with square sections, placed in pure shear or a squat rectangular section in pure shear). We shall shortly see why this is so.

Regardless of the way forces are applied to an object, the salient point is that the amount of material in relation to the direction and location of an applied external force dictates the distribution of stresses within the object's volume. This point is illustrated by considering the stress distributions that result when a bar is subjected to more than one kind of force. The simplest example (and one that has numerous applications to plants) is that of a beam that experiences two bending moments, each symbolized by M. This condition occurs whenever a beam is held at some angle from the vertical and bends under its own weight—a cantilevered beam. In bending, tensile and compressive stresses develop along the upper and lower surfaces of the beam, respectively. This distribution of stress can be readily appreciated, since the upper surface of a bent bar is convex while the lower surface is concave (as shown in fig.

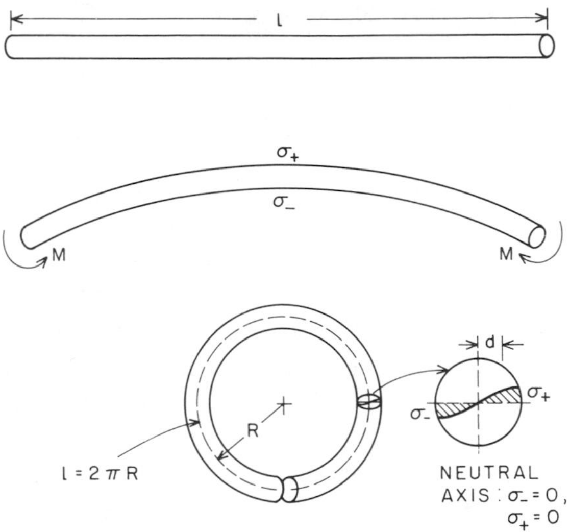

FIGURE 3.2 Tensile and compressive bending stresses generated within a cylindrical beam composed of an isotropic, linearly elastic material subjected to a bending moment. Tensile bending stresses ($\sigma+$) occur along the convex surface; compressive bending stresses ($\sigma-$) occur along the concave surface. If the isotropic beam (with length l) is bent into a perfect circle with a radius of curvature R, then the centroid axis of the beam precisely coincides with the neutral axis. The centroid axis is defined as the longitudinal axis created by connecting the center of mass of sequential cross sections through a beam. The neutral axis is the axis running through the beam along which tensile ($\sigma+$) and compressive ($\sigma-$) stresses are zero. Tensile and compressive stresses increase along the distance d from the neutral axis and reached their maximum magnitudes at the surface of the beam in the plane of bending.

3.2); that is, convexity reflects an extension in length, concavity a contraction in length. The magnitudes of the tensile and compressive stresses are not uniform throughout each cross section, however. Rather, as we shall now see, they are dictated by the distance material lies from the center of each cross section.

Our analysis of bending stress distributions is made comparatively easy provided we assume that the beam is constructed of an isotropic, linearly elastic material. This assumption ensures that the material behaves mechanically in an identical fashion both in tension and in compression. We further assume that the bar is bent into a complete circle, as shown at the bottom of figure 3.2. This assumption ensures that the length of the centroid axis of the bar equals $2\pi R$, where R is the radius of bending. The centroid axis is defined as the longitudinal axis running through the centers of mass within consecutive cross sections. (The assumed relation between the length of the centroid axis and the radius of curvature is highly impractical from a biological perspective, but it makes our treatment of bending stresses easy mathematically.) Since tensile and compressive stresses develop on each side of the bar, there must be another axis parallel to the centroid axis in which stress levels drop to zero. That is, tensile and compressive stresses must diminish from opposite sides of the bar toward the center of each cross section. This axis is called the neutral axis. Since the bar is composed of an isotropic material, the centroid axis can be precisely superimposed on the neutral axis. (Once again, this condition is rarely if ever met in real plant stems and leaves. The physical correspondence between the centroid and the neutral axes stems from our assumption that the bar is made of an isotropic material.) Since the neutral and centroid axes precisely coincide, the neutral axis will have the same length l as the undeformed bar.

The strains that will develop anywhere within a cross section through the bent beam are determined by specifying the distance d from the neutral axis at which ε is measured. Since $l = 2\pi R$, ε is given by the formula

$$\varepsilon = \frac{2\pi\,(R \pm d)}{2\pi R} - 1 = \pm\,\frac{d}{R}\,. \tag{3.2}$$

The $\pm$ in eq. (3.2) indicates that both tensile $(+)$ and compressive strains $(-)$ develop in each cross section. The magnitudes of the corresponding stresses can be related to d and R by noting that, for any linearly elastic material stressed below its proportional limit, $E = \sigma/\varepsilon$. Thus, eq. (3.2) can be rearranged to give the formula

$$\sigma = \pm\,\frac{Ed}{R}\,, \tag{3.3}$$

which indicates that the magnitude of stress at any distance d from the neutral axis is directly proportional to E and varies inversely with R. This makes intuitive sense, since the intensity of tensile and compressive stresses are expected to increase as the force of bending increases and since the radius of bending is a physical manifestation of the magnitude of the bending force.

Significantly, eqs. (3.2) and (3.3) reveal the salient design specifications for any beam that must be stiff and resist bending. Materials with the highest elastic moduli and proportional limits should be placed at the perimeter of each cross section within the beam, because they will deform comparatively little even at high normal stress levels and can elastically restore their deformations provided stress levels do not exceed their proportional limits. Indeed, plant cells and tissues (fibers or sclerenchyma) that function as tensile materials within beamlike stems and leaves are frequently very near the epidermis, precisely where the tensile and compressive stresses resulting from bending are anticipated by eq. (3.3) to nearly reach their maximum values. Accordingly, tensile stress-bearing elements in these locations would confer the greatest mechanical benefit. Stems with peripherally located tensile materials are much stiffer in bending than stems with equivalent amounts of mechanical tissues situated toward the center of their cross sections. Provided a stem has a terete (circular) cross section, eq. (3.3) would also lead us to predict that a concentric ring of tensile and compressive resistant material should be situated just beneath the epidermis, since the direction of the bending moment could vary over 360°. In fact, many plant stems have a more or less concentric ıyer of hypodermal sclerenchyma.

The apportionment of tissues with high elastic moduli and high proportional limits in regions where stresses are likely to be maximized is also seen in stems that have noncircular cross sections, such as the square or triangular stems of mints (Labiatae) and sedges (Cyperaceae). For these cross-sectional geometries, eq. (3.3) indicates that the corners of nonterete cross sections should be occupied by tissues with high elastic moduli and high proportional limits. Indeed, this expectation is not frustrated, since the bulk of the mechanical tissues in the noncircular stems of plants is concentrated in regions farthest from the center of cross sections.

As mentioned in chapter 2, with the principal exception of wood, which operates well mechanically in columns (tree trunks) under comparatively large compressive loadings, most plant materials are much stronger in tension than in compression. This generality results from the fact that the mechanical behavior of the cell wall infrastructure is dominated by the material properties of cellulose, which has a very high tensile modulus, a high tensile strength, and the capacity for considerable elastic extension in the direction of cellulose molecules (see chap. 5). Additionally, in living tissues with cell walls rich in cellulose, compressive stresses can be transmitted to cell walls as tensile stresses by the essentially incompressible turgid protoplast (see chap. 6). Thus, most stems and leaves that lack wood mechanically operate principally in tension rather than in compression, even in regions that experience compressive stresses.

Incidentally, the shear stresses that result from bending reach their maximum magnitudes toward the center of cross sections. Typically, the center of most plant stems and petioles is occupied by parenchyma, which can accommodate considerable shearing by changing its shape and which is more or less incompressible (Poisson's ratio up to 0.5). In this regard, note that some of the earliest known vascular land plants had their conducting tissues (xylem and phloem) at the centroid (neutral) axis, suggesting that the first vascular systems to evolve most likely occupied a mechanical environment dominated by bending shear stresses and not by tensile or compressive stresses. This indicates that the earliest vascular land plants did not directly rely on their vascular tissues for mechanical support (see chap. 10).

RADIUS OF CURVATURE

In the previous section, a very simple mathematical trick was used to calculate the radius of bending for a bent bar. The beam was bent into a circle so that the lengths of the centroid and neutral axes were equivalent. Since plant organs are rarely found bent to this degree or in this shape, a quantitative measure of the radius of bending for any curved organ is desirable. This goal can be realized by noting that the radius of bending is the inverse of the curvature of bending, symbolized by K. (Do not confuse the curvature of bending with the bulk modulus, which shares the same symbol.) Thus, eq. (3.3) becomes

$$\sigma = \pm K E d. \tag{3.4}$$

Provided K can be precisely defined, the assumption that the beam is bent into a perfect circle can be relaxed, and a variety of more complex bent forms can be treated. Fortunately, any curvature is mathematically defined in terms of the absolute values of the first and second derivatives of the curved line segments making up the entire geometry of curvature. The general equation for K for any two points on a curve separated by an arc length s, symbolized as $K(s)$, is given by the formula

$$K(s) = \left| \frac{\frac{dx}{dy}\left(\frac{d^2y}{ds^2}\right) - \frac{dy}{ds}\left(\frac{d^2x}{ds^2}\right)}{\left[\left(\frac{dx}{ds}\right)^2 + \left(\frac{dy}{ds}\right)^2\right]^{3/2}} \right|, \tag{3.5}$$

where x and y define the axes of the plane in which curvature occurs. Fortunately, a variety of computational methods exist to calculate K, even for very complex curves (see Silk, Wang, and Cleland 1982). Thus, eq. (3.5) is not unduly intimidating.

THE SECOND MOMENT OF AREA

In the introduction to this chapter, mention was made of flexural stiffness—the ability of an object to resist compression or bending—and flexural stiffness of an object was said to be the product of the elastic modulus and the second moment of area. We are now in a position to mathematically derive the second moment of area, symbolized by I. To do so, we return to eq. (3.4), which reveals that magnitude of stress is related to the distance a unit area of material is from the neutral axis. This relationship provides a method of calculating the magnitude of the couple required to bend a beam with curvature K. Since $\sigma = F/A$ and the bending moment M at each unit area A equals Fd, eq. (3.4) can be rewritten to reflect the local magnitude of the applied force

$$F = \sigma A = \pm E\, d\, K\, A \tag{3.6}$$

and the local magnitude of the bending moment

$$M = Fd = \pm E\, d^2\, K\, A\ . \tag{3.7}$$

From eq. (3.7) we can see that the total bending moment of the beam, symbolized by M_T, must be equal to the summation of all the infinitesimally small bending moments across each transection; that is,

$$M_T = \int_{-d_{\max}}^{d_{\max}} E\, d^2\, K\, dA = E\, K \int_{-d_{\max}}^{d_{\max}} d^2 dA. \tag{3.8}$$

This formula has an integral over the limits of $\pm d_{\max}$ because we must sum across the entire transverse section. The solution of eq. (3.8) is greatly simplified if we note that E and K are constants that can be removed from the integration process. Thus eq. (3.8) is reduced to finding the integral of $d^2\, dA$ over the limits $\pm\, d_{\max}$, that is, $I = \int d^2\, dA$ over the limits $\pm\, d_{\max}$. This integral equals the second moment of area, which is the sum of the products of each infinitesimally small area and the square of the distance each area lies from the neutral axis. Thus, I has units of length raised to the fourth power.

The second moment of area is sometimes referred to as the moment of inertia. This terminology reflects the fact that the magnitude of an object's mass is also a measure of the object's inertia, that is, force is the product of mass m and acceleration a; therefore force is also the product of inertia and acceleration. When an angular acceleration ω is imparted to an object, the moment force Fd equals the moment of inertia of a mass ($\int y^2\, dM$) times the angular acceleration ω: $Fd = \int(y^2\, dM)\omega$. By analogy, therefore, $\int(y^2\, dA)$ is the moment of inertia of an area.

Regardless of what we call I, eq. (3.8) can be used to compute I for any cross-sectional geometry. Closed-form solutions of I for a variety of geometries are given in many engineering handbooks and texts. Some of these solutions, which are relevant to calculating I for many plant organs, are given in figure 3.3, where it will be noted that I takes on different expressions depending on the plane of bending. If a closed-form solution of I for the cross-sectional geometry of an organ is not available, then I can be computed by using the simple fact that the neutral axis always coincides with the centroid axis when beams experience very small bending moments. Thus an organ's cross-sectional geometry can be drawn and its center of mass (which lies along the centroid axis) determined in the following way. The drawing should be suspended from a string by as many of its lobes or tips as is convenient. For each orientation, a plumb line should be drawn on the section. The centroid axis lies at the intersection of these plumb lines. Draw the plane of bending (a diagonal line intersecting the centroid axis), cut narrow strips of paper parallel to the plane of bending, and measure each of their surface areas (approximated as rectangles). The sum of these areas multiplied by the square of their respective distances from the plane of bending equals the second moment of area. Voids or empty spaces in the outline of the geometry (e.g., the drawing of a hollow stem) are not included in these calculations. Alternatively, a simple computer algorithm can be written in conjunction with a digitization software package that will accomplish the same thing with a minimum of fuss.

FLEXURAL STIFFNESS

Clearly, shape alone does not define the capacity of any object to resist bending. The material moduli of an object dictate the relation between stresses and strains for any level of stress. Thus the magnitude of the bending moment must be related both to shape and to a material modulus. Indeed, all the foregoing mathematical contortions reduce to the formula

$$M = E\,I\,K\,, \tag{3.9}$$

which, though deceptively simple, provides remarkable insight into the significance of shape and the elastic modulus. Equation (3.9) indicates that the magnitude of the moment M required to bend a beam to a curvature K varies directly as the product of the elastic modulus E and the second moment of area I. The product of E and I is called flexural stiffness or flexural rigidity, which has units of force times area. Thus, the curvature of bending can be reduced by increasing either E or I or both.

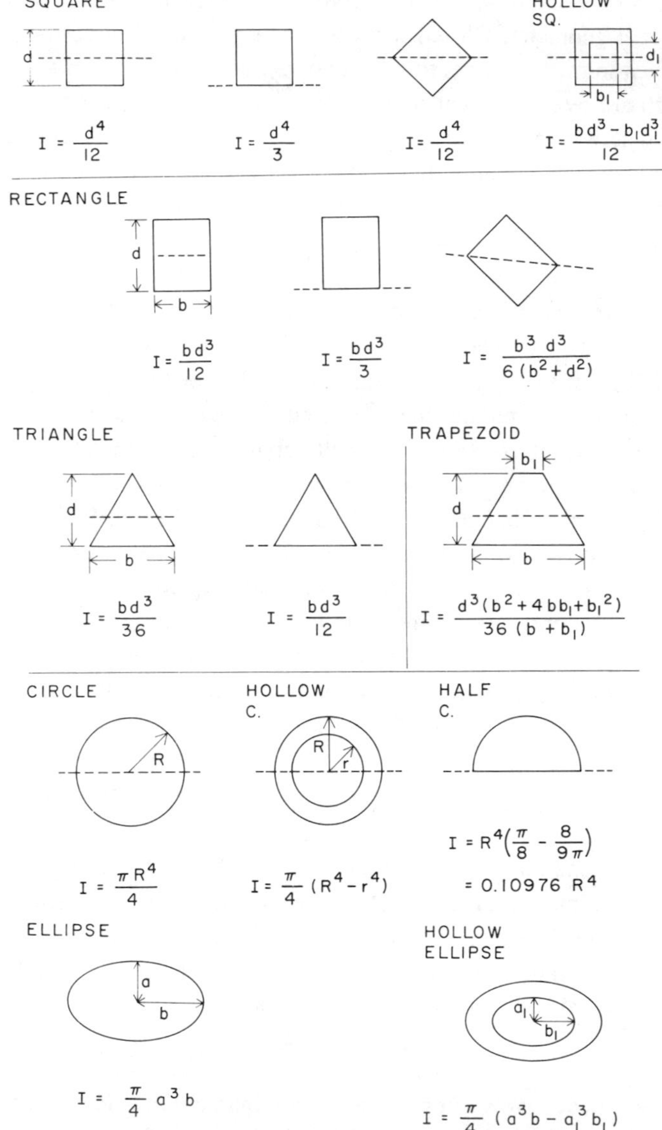

FIGURE 3.3 Formulas for the second moment of area I for members with different cross-sectional geometries. The geometry of each member is indicated by a single representative cross section. The formula for the second moment of area, even for the same geometry, depends on the plane of bending, which is shown by a dashed line in each drawing. The second moment of area always has units of length raised to the fourth power, since it is computed as the sum of the products of each infinitesimally small area A and the square of the distance d of the area from the centroid axis.

SECTORS OF CIRCLE

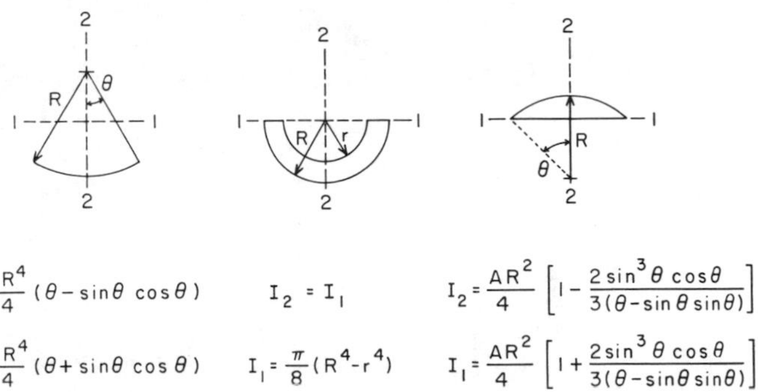

$$I_2 = \frac{R^4}{4}(\theta - \sin\theta\cos\theta) \qquad I_2 = I_1 \qquad I_2 = \frac{AR^2}{4}\left[1 - \frac{2\sin^3\theta\cos\theta}{3(\theta - \sin\theta\sin\theta)}\right]$$

$$I_1 = \frac{R^4}{4}(\theta + \sin\theta\cos\theta) \qquad I_1 = \frac{\pi}{8}(R^4 - r^4) \qquad I_1 = \frac{AR^2}{4}\left[1 + \frac{2\sin^3\theta\cos\theta}{3(\theta - \sin\theta\sin\theta)}\right]$$

SECTORS OF HOLLOWS

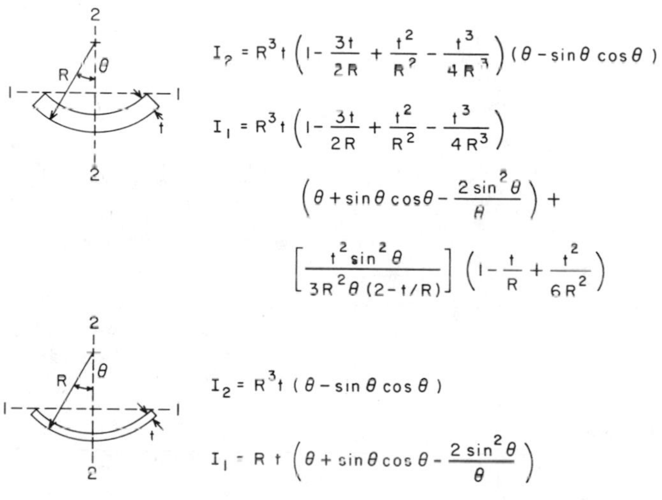

$$I_2 = R^3 t\left(1 - \frac{3t}{2R} + \frac{t^2}{R^2} - \frac{t^3}{4R^3}\right)(\theta - \sin\theta\cos\theta)$$

$$I_1 = R^3 t\left(1 - \frac{3t}{2R} + \frac{t^2}{R^2} - \frac{t^3}{4R^3}\right)\left(\theta + \sin\theta\cos\theta - \frac{2\sin^2\theta}{\theta}\right) + \left[\frac{t^2\sin^2\theta}{3R^2\theta(2 - t/R)}\right]\left(1 - \frac{t}{R} + \frac{t^2}{6R^2}\right)$$

$$I_2 = R^3 t(\theta - \sin\theta\cos\theta)$$

$$I_1 = R\,t\left(\theta + \sin\theta\cos\theta - \frac{2\sin^2\theta}{\theta}\right)$$

PARABOLA

d
1
3
b
2

$$I_1 = \frac{16}{175}a^3 b$$

$$I_2 = \frac{4}{15}ab^3$$

$$I_3 = \frac{32}{105}a^3 b$$

FIGURE 3.3 (*continued*)

REGULAR POLYGON (n = SIDES)

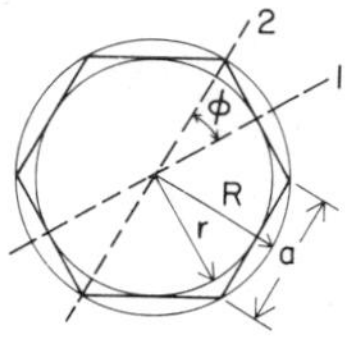

$$I_1 = I_2 = \frac{A}{24}(6R^2 - a^2) = \frac{A}{48}(12r^2 + a^2)$$

$$A = \frac{1}{4} na^2 \cot\phi = \frac{1}{2} nR^2 \sin 2\phi$$
$$= nr^2 \tan\phi$$

I-BEAM

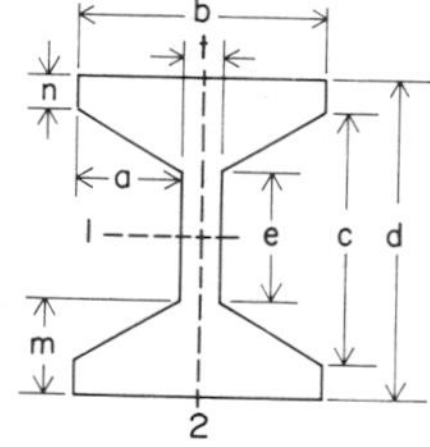

$$I_1 = \frac{1}{12}\left[bd^3 - \frac{a(c^4 - e^4)}{4(m-n)}\right]$$

$$I_2 = \frac{1}{12}\left[2nb^3 + et^3 + \frac{m-n}{4a}(b^4 - t^4)\right]$$

ANGLE

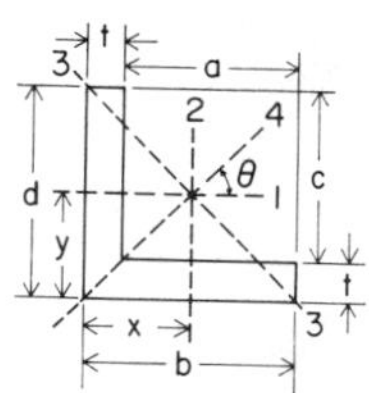

$$I_1 = \frac{1}{3}\left[t(d-y)^3 + by^3 - a(y-t)^3\right]$$

$$I_2 = \frac{1}{3}\left[t(b-x)^3 + dx^3 - c(x-t)^3\right]$$

$$I_3 = I_1 \sin^2\theta + I_2 \cos^2\theta + K \sin^2\theta$$

$$I_4 = I_1 \cos^2\theta + I_2 \sin^2\theta + K \sin^2\theta$$

K = negative when heel of angle (with respect to center of gravity) is in Ist or 3rd quadrant ; 3-3 is axis of minimum I.

CHANNEL

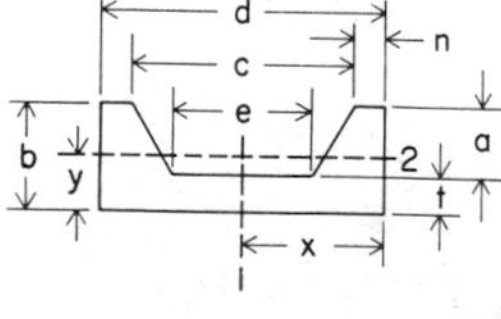

$$I_1 = \frac{1}{12}\left[bd^3 - \frac{a(c^4 - e^4)}{8(m-n)}\right]$$

$$I_2 = \frac{1}{3}\left[2nb^3 + et^3 + \frac{(m-n)(b^4 - t^4)}{2a}\right] - Ay^2$$

$$A = dt + a(m+n)$$

FIGURE 3.3 (*continued*)

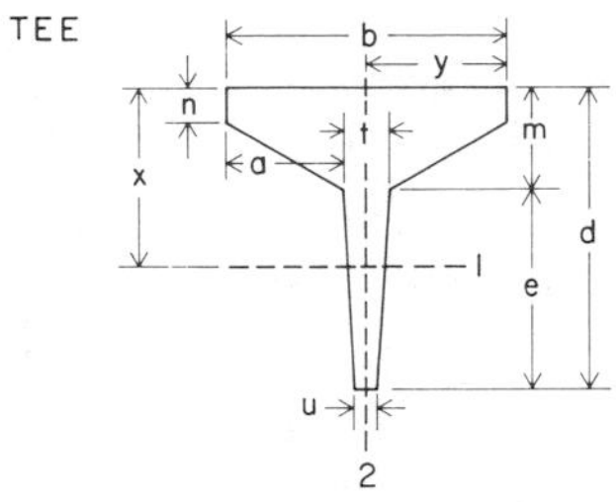

$$I_1 = \frac{1}{12}\left[e^3(3u+t) + 46m^3 - 2a(m-n)^3\right] - A(x-m)^2$$

$$I_2 = \frac{1}{12}\left[nb^3(m-n)t^3 + eu^3\right] + \frac{1}{36}\left\{a(m-n)\left[2a^2 + (2a+3t)^2\right]\right\} + \frac{1}{144}\left\{e(t-u)\left[(t-u)^2 + 2(t+2u)^2\right]\right\}$$

$$A = \frac{1}{2}\left[e(t-u)\right] + mt + a(m+n)$$

In biological contexts, I largely reflects morphology, while the value of E largely depends on anatomy. With the aid of a few simple calculations, we can assess the influence of morphology on the capacity of stems and leaves to resist bending. These calculations involve comparing organs with equivalent cross-sectional tissue areas but differing in their cross-sectional geometry (table 3.1). Since the cross section of an organ can have radial or bilateral symmetry, the second moment of area must be computed in terms of bending in the horizontal x–x and in the vertical y–y planes of cross-sectional symmetry, that is, I_{xx} and I_{yy}. For an organ with radial symmetry $I_{xx} = I_{yy}$, whereas $I_{xx} \neq I_{yy}$ for a bilaterally symmetrical cross section. This is shown in table 3.1, which further reveals that the extent of bending due to self-loading can be reduced by increasing the second moment of area measured in the vertical plane. This observation sheds light on the functional significance of why swaying plants tend to grow more in girth in the direction in which they repeatedly bend; that is, the differential growth in girth increases the second moment of area measured in the plane of bending and affords a greater geometric capacity to resist bending. This was first discovered by T. A. Knight (1811), who caused plants to sway by means of a clockwork and pendulum. Knight observed that the cross sections of swaying stems tend to become elliptical as tissue is added during the course of growth. The differential expansion in the girth of swaying trees observed by Knight is precisely the engineering solution that would be anticipated. A comparison between I_{xx} and I_{yy}

for an elliptical beam is given in table 3.1, from which we see that a fourfold difference exists between the two second moments of area; that is, $I_{xx}:I_{yy} =$ 1:4. Asymmetrical cross sections are particularly effective when the plane of bending can be anticipated, as when a plant is artificially swayed or when a leaf has a characteristic orientation with respect to the vertical. Indeed, the petioles of many dicot leaves possess asymmetrical cross sections. The petiole of *Rhus typhina* is elliptical in transection, with the major semiaxis aligned in the vertical direction. This alignment permits leaves to resist bending when self-loaded and allows comparatively larger lateral deflections when leaves are dynamically loaded from either side. By contrast, the elliptical petioles of *Carya tomentosa* are aligned with their major semiaxis in the horizontal plane. The leaves of these species bend little from side to side when they are dynamically loaded compared with the bending that results from their own weight. An interesting and as yet unanswered question is whether trees that differ in their branching architecture also differ in the direction their petioles are permitted to bend.

When the direction of dynamic loadings is more or less random or when self-loading can cause bending in any direction, an overall increase in symmetrical cross sections is desirable. Jacobs (1954) attached guy wires to the trunks of young pine trees some twenty feet from the ground. The lower parts of the trunks, held in place against the effects of wind pressure, grew less rapidly in girth than did the unrestrained portions, even though the lower parts were developmentally older than the unrestrained portions. The more or less uniform expansion in girth that Jacobs observed provides for a symmetrical increase in the second moment of area, resulting in an overall increase in flexural stiffness regardless of the direction of the applied bending force produced by wind pressure.

The differential growth in girth of swaying trunks and branches is an example of the ability of plants to respond geometrically to self-loadings and dynamic loadings. Plants can also manifest growth responses that alter their material properties. This ability has been appreciated for many years. Venning (1949) grew celery (*Apium*) seedlings in windless and windy environments. (Anemometers were difficult to get after the Second World War.) The latter were found to produce 50% (by volume) more collenchyma than the former. Collenchyma is a remarkable plant tissue with nonlinear viscoelastic cell walls that permit growth while still conferring a good measure of stiffness (see chap. 6). Many plants like celery are capable of responding developmentally to the magnitude of dynamic loadings attending growth by altering their material properties through modifications of the volume fractions of tissues.

TABLE 3.1 Comparisons (Percentage) between the Second Moments of Area, I_{xx} and I_{yy}

	I_{xx}	I_{yy}	$I_{xx}:I_{yy}$
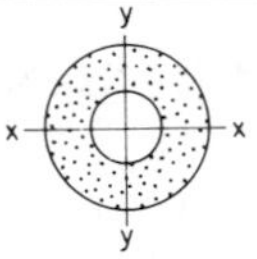 (hollow circular tube)	100%	100%	1:1
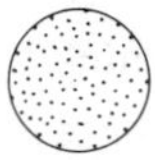 (solid circle)	60	60	1:1
 (solid ellipse)	30	120	1:4
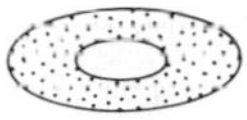 (hollow ellipse)	50	204	1:4
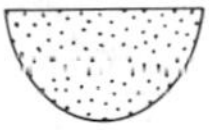 (semicircle)	77	56	1.4:1
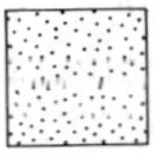 (solid square)	63	63	1:1
 (hollow square)	105	105	1:1

Note: Beams differ in cross-sectional geometry but have equivalent cross-sectional areas. For asymmetrical sections, the larger of the two dimensions is taken as twice the value of the smaller dimension. The second moment of area calculated for a hollow circular tube is used as the baseline for comparisons among the other cross sections shown.

Thigmomorphogenesis refers to any growth phenomenon evincing a responsiveness to mechanical perturbation. It is most often studied in terms of the changes in morphology or anatomy attending variations in the condition of loading. Thigmomorphogenesis has been reported for over 80% of all the species examined (Jaffe 1973). More recently, it has been studied in terms of the molecular events preceding changes in shape and size. Braam and Davis (1990) have shown that ten to twenty minutes after mechanical stimulation by handling, rain, or wind, the mRNA levels of mouse-ear cress (*Arabidopsis*) increase up to a hundredfold. Four touch-induced (TCH) genes are involved.

These genes encode for calmodulin, suggesting that calcium ions are required for the transduction of mechanical signals, thereby enabling plants to sense and respond to dynamic as well as self-imposed mechanical forces.

THE SECTION MODULUS AND MAXIMUM BENDING STRESSES

The maximum tensile and compressive stresses resulting in a beam submitted to a bending couple can be computed with the aid of eq. (3.4), provided we know the radius of curvature K and the elastic modulus E of the material. By combining eqs. (3.4) and (3.9), we can compute these maximum stresses in terms of the bending moment M and the second moment of area I: that is, from eq. (3.9), $E = M/IK$; substituting this expression for E into eq. (3.4), we find $\sigma = \pm\ KMd/IK = \pm\ Md/I$, where d is the distance from the neutral axis. The maximum tensile and compressive stresses occur in the outermost structural fibers within each cross section, when $d = D/2$, where D is the dimension of depth of the cross section. Therefore $\sigma_{max} = \pm\ MD/2I$. The quantity $2I/D$ is called the section modulus, symbolized by Z. Thus, the maximum tensile and compressive bending stresses are given by the formula $\sigma_{max} = \pm\ M/Z$. For a terete support member with diameter D, $I = \pi D^4/64$ and $Z = \pi D^3/32$; therefore $\sigma_{max} = \pm\ 32M/\pi D^3$ or, in terms of the unit radius R, $\sigma_{max} = \pm\ 4M/\pi R^3$.

For any cross-sectional geometry, the maximum stresses within a prismatic beam depend on the section modulus for any given bending moment. This leads us to suspect that any increase in cross-sectional area would decrease these stresses for a given bending moment. This is not always the case, however, since there are geometries such that an increase in cross-sectional area does not result in a decrease in the normal bending stresses. (This topic is discussed in considerable detail by Timoshenko [1976a, 100–103]. The following discussion is primarily based on his brilliant book.) Consider a beam with a square cross section having a side width a. If the cross section is oriented such that one of the two diagonals of the square is vertical, and if the beam is bent by couples acting in the vertical plane through this diagonal, then it can be shown that the maximum bending stresses are decreased by removing two small triangular sections from the opposite corners along the vertical diagonal. Why this is so can be shown analytically. The section modulus of the bent beam with a square cross section equals $a^3\sqrt{2}/12$. The section modulus corresponding to the beam for which two small triangular sections have been removed equals $(\sqrt{2}/12)a^3(1-\alpha)^2(1+3\alpha)$, where α is the fraction of the side of the square cross section removed by excising the two triangular sections. Timoshenko demonstrates that when $\alpha = 1/9$, the section modulus is maximized and the bending stresses are reduced by roughly 5%. This somewhat counterintuitive result can be understood by noting that the section modulus is the quotient of the second moment of area and half the depth of the cross section. By removing the corners from the square cross section, the second moment of area is decreased in a smaller proportion than is the depth. Therefore Z increases and σ_{max} decreases. The same result can be found for beams with circular or triangular cross sections. The section modulus of a terete cross section can be increased by 0.7% if two small segments having a depth of $0.011D$ are removed from the top and the bottom of the cross section. Likewise, removing a small triangular segment from the uppermost corner of a triangular prismatic beam increases the section modulus.

Clearly, the cross-sectional geometry conferring the highest section modulus is the more economical among cross-sectional geometries, with equivalent cross-sectional areas satisfying the same condition of strength. It is easily demonstrated that a beam with a square cross section is more economical than a beam with a circular cross section provided the cross-sectional areas of the two beams are the same, while a rectangular cross section becomes more economical as its dimension (depth in the plane of bending) increases. Likewise, the ratio of the section moduli of a hollow tube Z_h to a solid cylinder Z with equivalent cross-sectional areas equals $(D/d)[2-(4A/\pi D^2)]$, where D and d are the outer and inner diameter, from which it may be seen that Z_h approaches Z as the wall thickness of the tubular section increases, while $Z_h : Z$ approaches $2D/d$ for very thin-walled tubes; that is, $D >> d$. We shall have occasion to use the section modulus when addressing the mechanical significance of tapered plant organs and when dealing with columns and cantilevers of uniform strength (see chap. 7). For the time being, however, it is important to note that there are limits to how and how far cross-sectional geometries can be modified to increase economy. These limits are proscribed by the susceptibility of thin bilaterally symmetrical cross sections to sidewise buckling and torsion and of thin-walled tubes to lateral crimping (Brazier buckling).

THE EULER COLUMN FORMULA

Provided the flexural stiffness of a structure is known, engineering theory can be used to calculate many important mechanical relationships, such as the extent to which a stem can grow vertically before it will deflect from the vertical under an applied axial compressive load. From a mechanical perspective, the self-loaded trunk of an oak tree and the flower stalk of the common onion are the same thing, a column that sustains a compressive load (the weight of branches and leaves or of globose clusters of flowers) at its free end. True, trees and onions also experience dynamic loadings—they bend and torque when subjected to wind pressure. Nonetheless, the formulas that describe the capacity of a column to sustain a compressive load and the geometry of flexure when these loads exceed a critical level are informative and, with due care, can be used in many meaningful ways.

When the axial compressive load on a tall, slender column reaches a certain magnitude, the column will deflect from the vertical. Provided the proportional limit of the material is not exceeded, the deflection is recoverable when the axial load is removed—the deflection is elastic. The flexural stiffness and the gross geometry (length:radius) of the column will determine the magnitude of the compressive load at which deflections will occur. The elementary theory of elastic buckling of columns is credited to Leonhard Euler (1707–83), a renowned German-Swiss mathematician who spent most of his life at the Saint Petersburg court of the Empress Elizabeth. (A full analytical treatment of the theory of elastic buckling is given in Timoshenko and Gere 1961.) Euler noted that the lateral deflection of a column has a number of modes. The primary mode is a simple C-shaped geometry; secondary (S-shaped) and tertiary (double S-shaped) modes of flexure are also possible. But no matter what the mode of flexure, the smallest axial compressive load that produces any of these deflection modes is called the critical load, symbolized by P_{cr}. If the compressive load on a column is less that the critical load, then the column will remain perfectly vertical and undergo only axial compression. This loading condition is said to be stable; that is, if an additional lateral force is applied to the column and a small deflection occurs, then the deflection is recovered when the lateral force is removed. If the compressive load is increased, then at some point a condition is reached such that the column becomes unstable—a small lateral force will produce a deflection that does not disappear when the lateral force is removed. Thus the critical load is defined as the smallest axial load sufficient to keep the column in a slightly bent form. It can be calculated from the Euler column formula:

$$P_{cr} = \frac{n\pi^2 EI}{l^2}, \qquad (3.10a)$$

where n is a proportionality factor that depends on the way the column is supported at its two ends and l is the length of the column (fig. 3.4). Provided

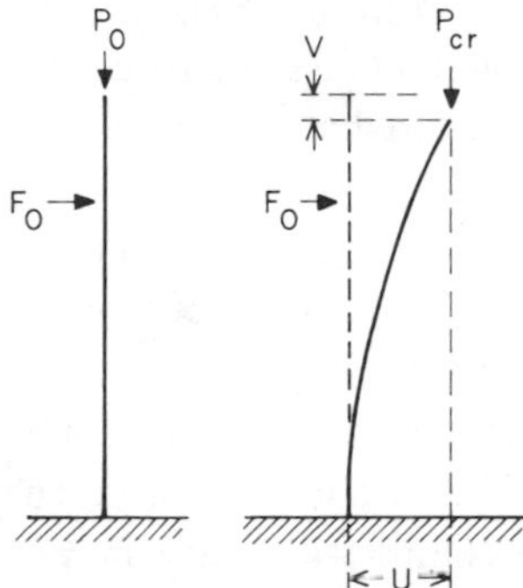

FIGURE 3.4 Euler buckling of a column and small deflections from the vertical. (The bending in this figure is magnified for convenience.) The effect of a laterally applied force F_o on the static equilibrium of a vertical column bearing an axially compressive point load P_o (applied precisely along the centroid axis of the column) at its free end depends on the assumptions that the column has a uniform cross-sectional geometry, that originally it is perfectly vertical, and that it is composed of a linearly elastic material. It is also assumed that the column is never loaded beyond its elastic (proportional) limit. When the lateral force (F_o) is removed the column elastically returns to its vertical orientation (on the left). At some critical load (P_{cr}), however, applying a lateral force causes the column to deflect from the vertical and to remain in this bent configuration. The critical load P_{cr} is the smallest point load that causes a column to maintain a deflection when a lateral force is applied. The deflection of the column from the vertical can be measured in terms of the displacements from the vertical (V) and from the horizontal (U).

the base of the column is firmly anchored and its other end is free, $n = 1/4$; that is,

$$P_{cr} = \frac{\pi^2}{4} \frac{EI}{l^2} \approx 2.4674 \frac{EI}{l^2} . \tag{3.10b}$$

Although the value of n depends on the boundary conditions of loading (the way the column is restrained at each of its two ends), eq. (3.10b) is a reasonable approximation for most plant stems, which are free to deflect under compression at the top and are anchored by roots at the base.

The Euler column formula is appropriate only when the following conditions of loading apply:

1. The column must be perfectly straight and must be uniform in its cross-sectional geometry; that is, it must have a uniform I throughout its span.
2. The column must be constructed from an isotropic material; that is, it must have a uniform E throughout its girth and span.

3. The weight of the column must be significantly less than the weight it supports; that is, the column is essentially considered to be weightless.
4. The column must be anchored at its base and must be free to move at the end that supports the load.
5. The loading cannot exceed the proportional limit of the material.
6. The column must be loaded through its centroid axis; that is, it must experience only compressive loading.

If a column conforms to all these conditions, then it is called an ideal column. Clearly, stems are rarely if ever ideal columns. They typically taper in girth and lack a uniform tissue composition; they are not uniform in I or E. Also, they are anything but weightless. They often weigh as much as the loads they must support. Accordingly, the Euler column formula must be viewed as a pedagogical tool that offers insights into the relations among variables that are much more complex in most real biological contexts.

It is worth noting that the critical load does not depend on the strength of the material used to fabricate the column, but only on the dimensions of the column and the elastic modulus of the material used. Thus two equally proportioned columns, one composed of a high-strength material and the other of a low-strength material, will elastically buckle at the same axial compressive force (Timoshenko, 1976b, 145). The immunity of elastic buckling to the strength of materials has a number of implications not explored regarding the elastic stability of trees whose woods differ in strength. For the time being, however, it is sufficient simply to note the tremendous importance of shape in dictating the way a vertical beam fails under a compressive load. When the critical load P_{cr} is reached, an ideal column either will undergo crushing failure or will elastically deform by bending from the vertical. Conversely, for any axial load, there exists a critical buckling length, L_{cr}, at which elastic buckling failure will occur; from eq. (3.10b), $L_{cr} = (\pi^2 EI/4P)^{1/2}$. Whether crushing or buckling failure occurs depends on the ratio of the column's length to its thickness, the latter clearly being a function of the second moment of area. The critical value for this ratio ranges between five and ten. Accordingly, tall and slender columns will deflect from the vertical rather than crush under their critical axial load; short and wide columns will tend to undergo crushing failure when their loading conditions exceed a critical level.

We can quickly appreciate that the Euler column formula provides a mechanical perspective on the adaptive significance of the morphological variation seen among trees of the same species growing in habitats that differ in the frequency and magnitude of wind loadings that exert lateral forces on trunks and on canopies of branches and leaves. Although there is a limit to the slenderness of tree trunks, taller and more slender, columnar trunks, which can grow relatively rapidly through dense stands of trees in their quest for light, can maintain mechanical stability for longer periods of vertical growth in habitats characterized by infrequent and low-magnitude dynamic loadings, whereas shorter and stouter trunks would be advantageous in open habitats where dynamic loadings tend to have higher magnitudes but where sunlight may not be a limiting growth factor (see chap. 9). Indeed, when forested areas are cleared from around mechanically stable trees, the specimens left behind often undergo buckling failure on exposure to even relatively low wind pressures. An interesting example of how mechanical stability depends on the capacity of neighboring trees to buffer one another is seen in the case of the first Kew "flagstaff," a trunk of a British Columbian conifer measuring 221 feet in height and having a basal diameter of 21 inches, presented as a gift to the Royal Botanical Gardens at Kew, Richmond. Owing to its great height and the fact that it was "replanted" in a relatively exposed site on the grounds of the Kew Botanical Gardens, the Kew flagstaff had to be supported by four pairs of guy wires attached at various points along its height. True, the dead tree was deprived of its root buttresses for anchorage. But equally important, when alive and growing the mechanically safe trunk of this tree, which bore the additional weight of branches and leaves, was sheltered from the wind by its cohorts.

The Euler column formula may also shed light on the importance of the rate of vertical growth, which for some species may decrease with the age of the plant. For these species, plant development can delay the time when the critical buckling height L_{cr} is reached while at the same time diminishing the annual increment of self-loading, dP/dt, perhaps in a fashion analogous to Zeno's dichotomy. (Zeno of Elea, a disciple of Parmenides, wrote forty *logoi* —loosely translated as paradoxes—against the reality of motion, one of which, called the dichotomy, argues that an object cannot pass through an unlimited number of things in a limited time. Zeno affirmed that, since distance can be divided into an unlimited number of "halfway" points, an object leaving its point of origin can never arrive at its destination. Aristotle, in his *Physics,* refuted only four of the logoi, one of which was the dichotomy.) One of the best examples of this developmental strategy is seen in the growth dynamics of the saguaro cactus (*Cereus giganteus*). Data presented by Steenbergh and Lowe (1977, 140–44, figs. 40–41) indicate that the rate at which the columnar stems of this species grow in height decreases with the age of the plant; that is, the yearly increment of growth decreases as overall stem height increases. Unfortunately these authors do not provide data for the basal girth and tapering of the specimens for which height measurements were annually tabulated. These additional data could have been used to test the hypothesis that the development of this cactus results in a mechanical procrastination—that the diminution in growth rate delays the advent of mechanical instability through elastic buckling. Photographs of plants growing in the field (Steenbergh and Lowe 1977, 136–37, fig. 38) reveal a change in growth form (a transition from club-shaped juveniles to bottle-shaped adults), suggesting that tapering also plays a critical role in the biomechanics of this species.

However, we should bear in mind that Euler buckling is not always disastrous. Plants that are composed of sufficiently flexible materials and are characterized by a slender growth habit can assume the primary elastic buckling mode as their growth posture with few if any negative consequences. Such a growth posture can be quite safe mechanically and represents a minimum weight solution for growing high even in very windy habitats. Many species of tall palms use this growth strategy, suggesting that we should not be too enamored with the popular, but often incorrect, notion that Euler buckling is always disadvantageous.

Although the strength of the material from which a column is fabricated has no bearing on when a column will buckle, the extent to which the Euler column formula can be applied to a particular case is limited in all instances by the yield or breaking stress of that material (see assumption 5). This is because the proportional limit of a material defines the range of stresses over which elastic behavior will occur. From the Euler column formula, we can see that the critical compressive load P_{cr} divided by the cross-sectional area A of the column equals the critical compressive stress: $\sigma_{cr} = P_{cr}/A = \pi^2 EI/(4l^2A)$. For a column with a circular cross section, $I = \pi R^4/4$, where R is the uniform radius of the cross section. Therefore $\sigma_{cr} = c_1 E(R/l)^2$, where $c_1 = \pi^2/16$, from which we can see that the critical compressive stress depends only on the elastic modulus and the fraction R/l; that is, the magnitude of σ_{cr} will increase as the square of the ratio of the column's radius to length increases. We can extend this derivation to encompass any point-loaded anchored column by letting $(I/A)^{1/2} = r$, where r is the radius of gyration. This leads to the general formula $\sigma_{cr} = c_2 E/(l/r)^2$, where l/r is the slenderness ratio and $c_2 = \pi^2/4$. With the aid of this general formula, the limiting case for the Euler column formula for any column is easily calculated. This calculation can be illustrated for balsa wood and structural steel. The elastic modulus and the proportional limit measured under compression along the grain of kiln-dried balsa wood are 2.16 $GN \cdot m^{-2}$ and 3.87 $MN \cdot m^{-2}$, respectively. Thus the critical load for a column made of balsa can be calculated from the Euler column formula only provided $l/r \geq 37.1$, a cylindrical column with a ratio of length to radius ≥ 18.6. The elastic modulus and proportional limit of structural steel are 18.9 $GN \cdot m^{-2}$ and 0.170 $GN \cdot m^{-2}$, respectively. Thus the limiting slenderness ratio for columns composed of structural steel is roughly 100; length:radius ≥ 50.

Provided the column has a solid cross section, the effect of shearing and the self-weight of the column on the critical load are typically ignored. Nonetheless, shear forces operate along the cross sections of any column that bends, and the total deflection of an originally vertical column will differ from that predicted by the equations that neglect the effects of these forces. Consider a beam, submitted to bending, with a rectangular cross section having depth h in the y-direction and width b in the x-direction. At each point in the cross section at distance d from the neutral axis, the two perpendicular shear stress components (τ_{yx} and τ_{xy}) operating in the plane of the cross section must have equivalent absolute magnitudes. Each of the two components of shear resulting from the shear force component F equals $(F/2I)[(h^2/4) - d^2]$, indicating that the bending shear stresses diminish as an exponential function of d and that the maximum shear stress τ_{max} occurs at the neutral axis where $d = 0$; that is, $\tau_{max} = Fh^2/8I$ or, since $I = bh^3/12$, $\tau_{max} = (3/2)(F/bh)$. (Note that the maximum shear stress in the case of a rectangular cross section is 50% greater than the average shearing stress [the total shearing force divided by the area of the cross-section].) Likewise, for a circular cylindrical beam with radius R, the shear stress measured at any point in a cross section at distance d from the neutral axis equals $RF(R^2 - d^2)^{1/2}/3I$. Since $I = \pi R^4/4$, $\tau = 4F(R^2 - d^2)^{1/2}/3\pi R^3$. Thus the maximum shear stress τ_{max} for this beam also occurs at the neutral axis (when $d = 0$), and $\tau_{max} = 4F/3A$, where A is the cross-sectional area (πR^2); that is, the maximum shear stress for a cylindrical beam is 33% larger than the average value of shear stress obtained by dividing the shearing force by the cross-sectional area. Regardless of cross-sectional geometry, when the resulting shear strains become significant ($\geq$ 5%), the ratio of the actual critical load P_a to the critical load calculated from the Euler column formula P_{cr} equals $[1 + (aP_{cr}/AG)]^{-1}$, where $P_{cr} = \pi^2 EI/4l^2$ (see eq. 3.10b), a is a numerical factor dependent on the shape of the column (e.g., $a = 1.11$ for a circular cross section, $a = 1.20$ for a rectangular cross section), A is the total cross section of the column, and G is the shear modulus of the material used to construct the column; that is, $P_a{:}P_{cr} = 1/[1 + (aP_{cr}/AG)]$. As the shear modulus decreases, the ratio of the actual to the predicted critical load will diminish. In general, however, the ratio $P_a{:}P_{cr}$ is just slightly less than unity for columns with solid cross sections, whereas it is significantly diminished for any column with a hollow cross-sectional geometry.

Clearly also, the weight of a column can be substantial and dictates the total deflection. When the weight of a column is significant compared with the axial compressive load on the column, the critical value of the uniformly distributed load intensity q per unit length of the column can be computed from the formula $(ql)_{cr} \approx 7.84\ EI/l^2$. The effect of the uniform load ql on the magnitude of the critical load P_{cr} that can be sustained is given by the approximate formula $P_{cr} \approx (\pi^2 EI/4l^2) - 0.3ql$, from which we can see that the load intensity per unit column length decreases the critical compressive load that can be sustained. (When $ql > (ql)_{cr}$, P_{cr} becomes negative, indicating that a tensile force is required to prevent the column from buckling under the applied load.)

Through artifice, we can contrive laboratory conditions for testing plant materials or structures so that most or all of the six boundary conditions are met. For example, we could select a plant organ or surgically isolate a cylindrical plug of plant tissue such that our column was uniform in I and relatively uniform in E. We could then apply loads of varying magnitude to the free end of the column and estimate its elastic modulus (since we can measure I empirically). By the same token, the Euler column formula provides a geometric criterion for testing cylindrical samples of plant material under uniaxial compression that must avoid flexure, which would otherwise give spurious strain measurements; that is, cylindrical samples should have ratios of length to radius ≤ 20 (length : radius ≤ 10 would be even better).

Finally, we must always remember that the Euler column formula applies only to columns experiencing very small deflections under compressive loads. If we are truly interested in the large deflections of flower stalks, columnar leaves, or tree trunks subjected to high wind pressures or eccentrically applied loads, then we must abandon the Euler column formula and seek other closed-form solutions. Fortunately there is a substitute formula, as we shall now see.

THE ELASTICA

The Euler column formula for bending is derived from equations using approximate expressions for very small curvatures of bending. A precise expression for curvature (given by eq. 3.5) provides the opportunity to deal with large deflections. That portion of engineering theory that deals with the large deflections of beams and columns is called the Elastica. Detailed treatments of the Elastica are mathematically complex and often difficult to follow (see, for example, Timoshenko and Gere 1961), but as is often the case, very complex mathematical proofs often condense into relatively simple final form. (The Euler column formula results from the calculus of variations. Yet the final form of the Euler equation is algebraically simple and thus sometimes fatally attractive.) Fortunately, some very useful and relatively simple algebraic equations stem from the Elastica. One of these equations treats large deflections of bending in columns and column beams:

$$P = [K\,(\sin \alpha/2)]^2 \frac{EI}{l^2}\,, \tag{3.11}$$

where P is the load at the tip of a column, α is the deflection angle measured in the horizontal plane at the tip of the column, and $[K\,(\sin \alpha/2)]$ is the complete elliptic integral of the first kind of the deflection angle (the physical meaning of some of these parameters is illustrated in fig. 3.5). Although terminologically imposing, the complete elliptic integral of the first kind is actually simple to use, since many texts provide tables that allow us to convert α into $[K\,(\sin \alpha/2)]$ (for example, table 7.1). Equation (3.11) indicates an explicit relation among a number of relatively easily measured variables. The length and second moment of area of a stem can be easily determined, as can the deflection angle. By applying weights to the tip of a vertical stem or a cylindrical, vertical leaf (such as those of chive or onion) and measuring the deflection angle, the modulus of elasticity of a stem or leaf can be estimated.

Similarly, eq. (3.11) can be manipulated to provide some very useful dimensionless expressions with which to examine the mechanical behavior of a population of plants that are similar in structure but dissimilar in size. One of these dimensionless expressions is the load parameter, which equals the square of the complete elliptic integral of the first kind:

$$[K\,(\sin \alpha/2)]^2 = \frac{Pl^2}{EI}\,. \tag{3.12}$$

The load parameter can be plotted as a function of the deflection angle or some other dimensionless parameter, such as the ratio of the horizontal deflection U to column length l: U/l. This is shown in figure 3.6, which is based on data from a study of the bending of the hollow, tubular leaves of chive (*Allium schoenoprasum*) in response to water deprivation (Niklas and O'Rourke

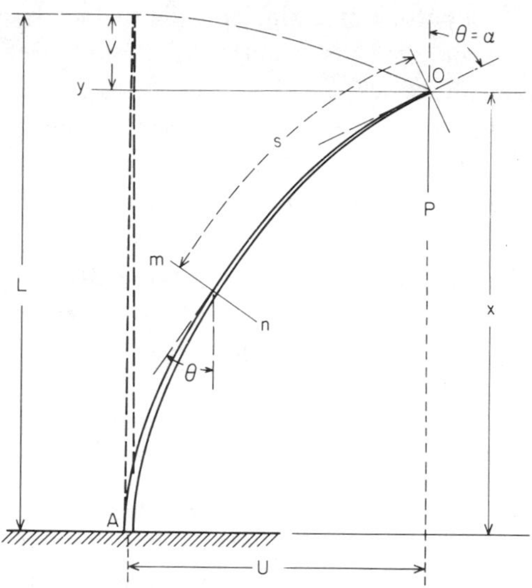

FIGURE 3.5 Large bending deflections of a beam anchored at one end (A) and subjected to load P at its free end (O). Before the application of the load, the beam is perfectly vertical (dashed outline to the left). With the application of the load, the beam undergoes vertical (V) and horizontal (U) displacements. A bending angle θ is defined at any cross section through the beam (along the plane m–n) at a distance, measured as a function of the arc length s, from the tip of the beam. The tip deflection angle ($\theta = \alpha$), in conjunction with equations from the Elastica (see text), is used to compute the relation among beam length L, flexural stiffness EI, and the magnitude of the load P.

1987). The lines of these plots represent theoretical predictions, while the actual data points are plotted to illustrate how far theory and observation coincide. In this study the load P applied to the tips of leaves and leaf length l were held constant, and EI was determined for varying tissue water content. The data indicate that the load parameter increases as a function of the decrease in tissue water content; since P and l are constant, EI must decrease for the load parameter to increase. As leaves lose water, their deflection angle increases (see insert in fig. 3.6). Notice that U/l increases by a factor of four (from about 0.2 to 0.8) when the load parameter is less than doubled (from 2.25 to 4.20). This indicates that a very small reduction in EI produces very large lateral displacements. When E and I were measured independently, it was found that I was much the more significant in determining alterations in the flexural stiffness of leaves. Chive leaves are hollow, and when their tissues

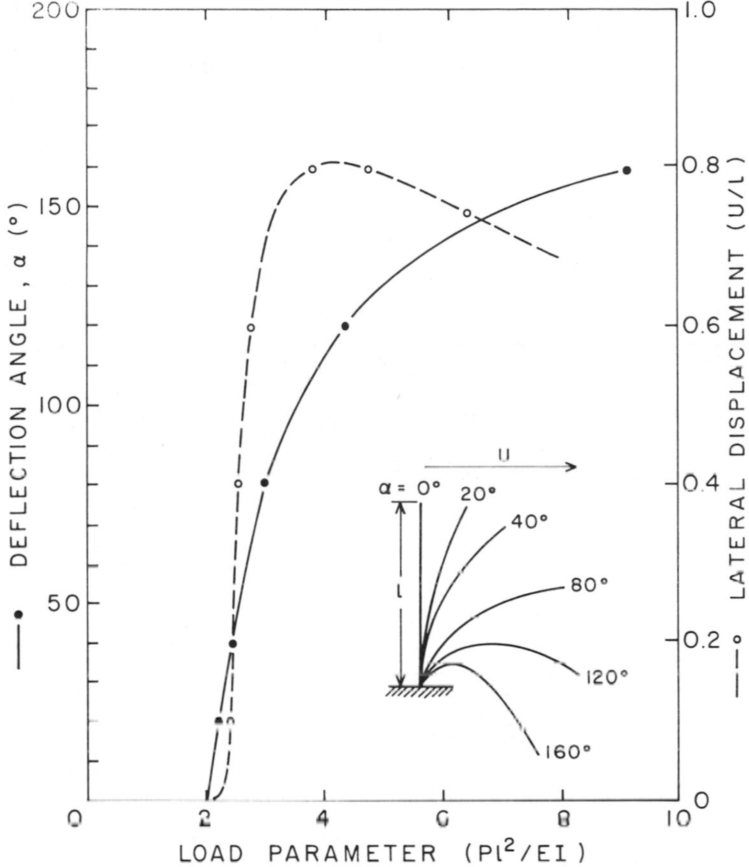

FIGURE 3.6 Predicted and observed changes in the tip deflection angle (α) and in the ratio of the lateral displacement (*U*) to leaf length (*l*) plotted as a function of the load parameter (Pl^2/EI) of a progressively bending chive leaf (see insert). The predicted changes are plotted as lines; observed values are plotted as data points (solid circles and open circles). The load parameter is a dimensionless ratio that relates the product of the load applied (*P*) at the free end of a beamlike structure (the chive leaf) and the square of the beam length (*l*) to the flexural stiffness (*EI*). As the load parameter increases, the beam experiences greater deflections from the vertical. See text for further details. (Data from Niklas and O'Rourke 1987.)

lose water and are subjected to a bending load, their cross-sectional geometry deforms into an ellipse with the minor axis aligned 90° to the plane of bending. This geometric bias dramatically reduces the second moment of area measured in the plane of bending and reduces the capacity of leaves to sustain even their own weight, let alone a weight attached to the tip. At some point, determined by the extent of water loss from tissues, the elliptical cross sections of the leaves crimp and leaves undergo irreversible mechanical failure.

HOLLOW COLUMNS AND BEAMS

Hollow stems and leaves have the advantage of having a relatively low weight per unit length while still being stiff. Additionally, hollow organs have a larger I than their solid counterparts provided the equivalent amount of tissue is used within cross sections; for example, a hollow elliptical or square cross section has an I that is $\geq 40\%$ of its solid counterpart (see table 3.1). The significance of hollow stems and leaves can be illustrated further by considering the critical buckling length that a solid versus a hollow beam with equivalent external radii can achieve. An approximate formula for the critical buckling length L_{cr} of a beam is

$$L_{cr} = \left(\frac{7.84\ EI}{W}\right)^{1/3}, \tag{3.13}$$

where W is the weight per unit length. For a solid cylindrical beam with external radius R, the weight per unit length W_s equals $\rho\ \pi R^2$, where ρ is unit density. For a hollow cylindrical beam with wall thickness t, W_h equals $\rho\ 2\pi\ R\ t$. From figure 3.3, the second moments of area for a solid (I_s) and a hollow, very thin-walled (I_h) cylinder are $\pi R^4/4$ and $\pi R^3 t$, respectively. Thus

$$\frac{(L_h)^3\ W_h}{(EI)_h} = \frac{(L_s)^3\ W_s}{(EI)_s}. \tag{3.14}$$

Since we are primarily interested in the significance of shape, not material properties, we can assume that the moduli of elasticity of the two beams are equal. By substituting the appropriate equations for the second moments of area and solving for the ratio of L_h to L_s, we derive the formula

$$\left(\frac{L_h}{L_s}\right)^3 = \left(\frac{\pi\ \rho\ R^2}{\rho\ 2\pi Rt}\right)\left(\frac{4\pi\ R^{3t}}{\pi R^4}\right) = 2\ . \tag{3.15}$$

Since $2^{1/3} = 1.26$, $L_h = 1.26\ L$, which indicates that a very thin-walled, hollow beam, such as the leaves of chive, can be extended in length by 26% the length of a solid counterpart with the same diameter before it reaches its critical length. This could be very significant for a photosynthetic organ, particularly since a hollow organ with an outer diameter equivalent to the diameter of a solid counterpart requires much less material in its construction.

The flexural stiffness of hollow and solid cylindrical stems and leaves that differ in their elastic moduli can also be considered. To do so, however, we must use a more general formula for the second moment of area of hollow organs—a formula that treats both thick- and thin-walled organs. This is important because $I = \pi R^3 t$ only for very thin-walled tubes. When t is large compared with the outer radius of the tube ($\approx t/R > 0.07$), the equation $I = \pi (R^4 - R_i^4)/4$ should be used, where R and R_i are the outer and inner radii of the tube. Thus the ratio of the flexural stiffness of any circular hollow to solid beam is given by the formula

$$\frac{(EI)_h}{(EI)_s} = \frac{E_h}{E_s}\left[\frac{R^4 - (R-t)^4}{R^4}\right] = \frac{E_h}{E_s}\left[1 - \left(1 - \frac{t}{R}\right)^4\right], \tag{3.16}$$

which indicates that, regardless of the ratio of the elastic moduli, 50% or more of the maximum possible EI is achieved *provided* $t/R \geq 0.20$. Thus a hollow stem or leaf whose wall thickness $\geq$ 20% of its external radius will be at least half (50%) as stiff as its solid counterpart regardless of the material used. This may explain why the limiting case of t/R for the hollow organs of many plant species is very nearly 0.20.

Intuitively, we recognize that the wall thickness of a hollow tube cannot be decreased indefinitely without incurring some mechanical liability. If not, then by reductio ad absurdum, a tube with zero wall thickness would be stiffer than one with a measurable wall thickness. Clearly, thin-walled tubes are susceptible to crimping and buckling when they bend under a load, as are water-stressed chive leaves (see below).

BRAZIER BUCKLING

A very long tube under axial compression can buckle in one of two ways—globally or locally. Global buckling, which is a long-wave mode of deformation, is evinced by longitudinal deformations treated by the Euler column formula or the Elastica. By contrast, local buckling, evinced by transverse deformations, is a short-wave mode of failure. Brazier (1927) recognized that when a very long, thin-walled cylindrical tube undergoes long-wave deformation, its cross-sectional geometry can become oval in the plane of bending. As a consequence, the bending moment of tubular beams is no longer a linear function of the curvature of bending, and when the applied bending moment is plotted as a function of the radius of curvature, the resulting graph has a decreasing slope. Tubes typically buckle locally by crimping at some maximum (critical) bending moment M_{cr}. This mode of failure is known as Brazier buckling, and an expression for the critical bending moment can be obtained from a treatment of the total strain energy per unit length of a beam in terms of the change in axial curvature; that is, for very long tubes (long enough that end-wall effects can be entirely neglected), the critical bending moment is given by the formula

$$M_{cr} = 0.31426 \frac{\pi E R t^2}{(1 - \nu^2)^{1/2}} . \quad (3.17)$$

As might be expected, the critical bending moment increases nonlinearly as a function of wall thickness and increases as a linear function of either the outer radius of the tube or the elastic modulus.

All of the preceding assumes that the material in a hollow tube is isotropic. That is, the classic formulas used to treat Brazier buckling neglect the fact that most plant materials have a larger elastic modulus measured in the longitudinal direction than measured in the circumferential direction ($E_L >> E_C$). For most samples of wood, E_L:E_C varies between 60:1 and 10:1; for bamboo this ratio is about 20:1, while the anisotropy ratio of delignified cell walls is about 9:1. Thus it becomes abundantly clear that most if not all plant tissues and organs are significantly stiffer longitudinally than circumferentially. Unfortunately, many workers, including myself, have used values of E_L to compute M_{cr}. Thus the maximum bending moment at which Brazier buckling will occur is typically overestimated. One solution is to use the geometric mean of E_L and E_C for the value of E in eq. (3.17). That is, $E \approx (E_L E_C)^{1/2}$. If this tactic is followed, it immediately becomes obvious that the critical bending moment is overestimated by the square root of the anisotropy ratio. Note also that the anisotropy of plant tissues influences the value of the Poisson's ratio used in eq. (3.17). A reasonable approach to this problem would be to use the product of ν_{LC} and ν_{CL}. Based on the symmetry requirement of classical engineering theory, $\nu_{CL}/\nu_{LC} = E_C/E_L$ for anisotropic materials. Thus, if the ratio of E_L to E_C is either known or inferred, eq. (3.17) can be modified to give accurate results.

It is often important to determine the critical stress level at which thin-walled tubes with finite lengths will undergo Brazier buckling. Unfortunately, analytical expressions for the critical Brazier buckling stress σ_B for thin-walled tubes having end-wall effects are extremely difficult to achieve, since a number of empirical factors must be considered. However, we can derive an approximate solution neglecting the influence of end-wall effects by noting that for any curvature of bending K and flexural stiffness EI, $M = EIK$ and $\sigma_{max} = KER$, where R is the external radius of the tube (see eqs. 3.4 and 3.9). Thus $M_{max} = I\sigma_{max}/R$. Substituting this expression for M_{cr} in eq. (3.17) and solving for the critical buckling stress ($\sigma_B \approx \sigma_{max}$) yields the formula $\sigma_B = kEt/R$, where $k = 0.31426/(1 - \nu^2)^{1/2}$. Since $0 \leq \nu \leq 0.5$, k should range between 0.314 and 0.363. For reasons that cannot be treated here, however, in fact k empirically ranges between 0.5 and 0.8 for most practical situations. Nonetheless, regardless of the numerical value of k, for any given material we see that σ_B depends simply on the dimensionless ratio of wall thickness to external radius. Once again we see the tremendous importance of shape in influencing mechanical behavior.

The role of shape is further illustrated by noting that a very long, slender tube can mechanically deform by undergoing either Brazier buckling or Euler deflections, and that one of these two modes of failure will arise when the corresponding critical stress (σ_{cr} or σ_B) is reached. The most efficient tube, therefore, would be one that maximizes both σ_{cr} and σ_B; that is, by maximizing the geometric mean, $(\sigma_{cr}\sigma_B)^{1/2} = [(\pi^2 EI/4Al^2)(kEt/R)]^{1/2} = k^{1/2}\pi Et/2l$. This expression indicates that, for any tube of length l and elastic modulus E, the single critical factor that maximizes mechanical safety is the wall thickness of the tube. This conclusion also applies when bending stresses exceed the proportional limit of a material and plastic deformations occur within a tube. Ades (1957) devised a method for determining the total work expended when a tube plastically deforms. He was also able to determine the shape of the oval cross section of the tube as a function of the radius of curvature of bending, as well as the maximum bending moment the tube sustained before failure occurred. With this information, the bending strength of the tube could be calculated from the area A under the elastic-plastic ranges of the stress-strain curve; that is, $A = (\sigma_i^2/E)\{(1/2) + [3n/7(n+1)](\sigma_i/\sigma_{0.7})^{n-1}\}$, where σ_i is the intensity of stress, $\sigma_{0.7}$ is the secant yield stress corresponding to the secant line of slope $0.7E$, and n is shape parameter in the stress-strain curve equation. Ades (1957, 608) found that the ratio of tube diameter D to thickness t dictates whether material or structural failure occurs (excessive plastic deformation of the material or ovalization of the tube's cross section). In the range of $D/t = 2$ (a solid cross section), failure is material failure, with some (but not appreciable) ovalization or other form of cross-sectional deformation. In this range, strength can be computed conservatively by standard analytical techniques. When $3 \leq D/t \leq 10$, failure is still a material failure, but ovalization can play a significant role (the amount of ovalization increases as D/t increases within this range). In the range of $30 \leq D/t \leq 50$, failure occurs primarily as a result of ovalization and Brazier buckling, although some material failure also occurs. Finally, when $D/t > 50$, structural rather than material failure predominates. One of the interesting aspects of Ades's results is that experimental data from tubes made of slightly anisotropic materials were similar to those from tubes made of isotropic materials, yet the theory he devised assumed material isotropy.

Wainwright et al. (1976) also discuss the importance of the geometry of thin-walled tubes in terms of Euler or Brazier buckling. They set the limiting case at $\sigma_{cr} = \sigma_B$; that is, $\sigma_{cr} = P_{cr}/A = n\pi^2 EI/Al^2 = kEt/D = \sigma_B$, where D is the average diameter of the tube. Since the elastic modulus can be removed from both sides of their equation, and since $D/t = (k/n)\,(Al^2/\pi I)$, Wainwright et al. (1976, 258) conclude that "the limiting dimensions for a given length are dictated *entirely* by the geometry and not the material" (italics added). Although this conclusion is not strictly correct (k is a function of Poisson's ratio, the mathematical elimination of E from both sides of an equation is something of a biological non sequitur, and material failure can occur when $D/t < 50$), their discussion of the importance of geometry (= morphology) offers valuable insights into an extremely complex problem.

Aside from maximizing the geometric mean of the critical stresses influencing Euler and Brazier buckling, there are additional geometric solutions to the general problem of local buckling. One of these is to introduce transverse septa or diaphragms into thin-walled cylindrical plant organs. These struts reduce the effective length of the tube by amplifying end-wall effects; that is, they restrict deformations at the ends of each tubular segment by regionally increasing stiffness. For example, the hollow stems of many grasses and all the species of horsetails are septated into smaller cylindrical units by relatively thin transverse diaphragms. Although these diaphragms contribute as little as 2% of a stem's total weight, they can increase stiffness by as much as 16%–20% (Niklas 1989a).

A rather interesting example of how plants might actually use Brazier buckling to advantage was pointed out by Sharon Lubkin (unpublished data). She noticed that the hollow flower stalk of the Baltic onion is regionally inflated, often dramatically, at about the midpoint of its length. We measured the wall thickness of stalks along their length and found that the region of inflation is truly an aneurysm. That is, it is much thinner than portions of the stalk above or below it. As plants mature, their stalks dry and their flower heads get heav-

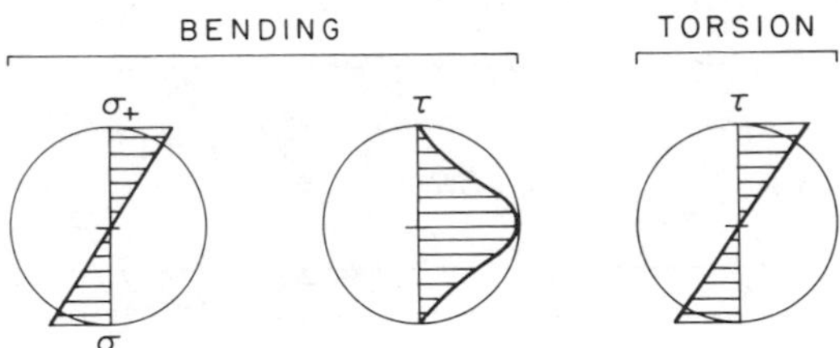

FIGURE 3.7 Normal (σ) and shear (τ) stress distributions in cross sections through a circular cylindrical member subjected to bending and torsion. Although the distributions shown are true for each transection through the member, the relative magnitude of these stresses varies as a function of the distance along the length of the member (see plate 1).

ier. Late in the growing season, the aneurysms in stalks crimp under the weight of plantlets developing at their tops, and Brazier buckling deposits plantlets some distance from their parent plants. True, the optimal design would have been to place the point of failure at the base of each stalk, maximizing the distance at which the next generation is displaced from its elders. Why this optimal design does not occur deforms with the least touch of interpretation.

TORSION

In addition to the tensile, compressive, and shear stress components that result from bending, torsional shear stresses develop when a shaft is twisted under dynamic loading or when self-loadings on a column or beam are eccentric in their application. A comparison between the distributions of torsional shear stresses and of the various types of bending stresses within a circular cross section is provided in figure 3.7, from which it is evident that torsional shear stresses have a distribution similar to that of the tensile and compressive stress distribution; that is, the magnitude of torsional shear stress increases as a function of the distance from the neutral axis. By contrast, torsional and bending shear stresses have more or less opposite distributions within the same cross section. In general, for very long members the magnitude of the maximum torsional shear stress in each transection increases toward the base and the tip of the member and reaches its minimum value at midspan (fig. 3.8).

The parameters that define the resistance of a shaft to torsion may be inferred from our previous treatment of flexural stiffness; the resistance to torsion is described by the product of a material modulus and some mathematical

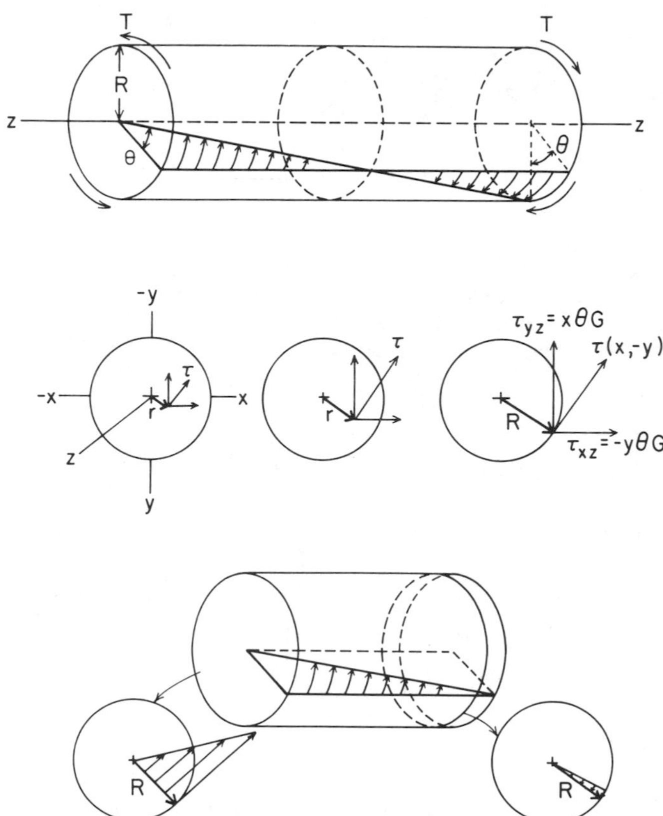

FIGURE 3.8 A cylindrical shaft with uniform radius R subjected to a torque (T) couple. The magnitude of the torsional shear strains (γ) is a function of the angle of twist θ. The intensity of the torsional shear stresses (τ) increases as a function of the distance r from the centroid axis of the shaft. The maximum torsional shear stresses are reached when $r = R$, as shown in the three cross sections drawn beneath the full view of the cylindrical shaft. For a shaft fixed at one end and subjected to a torque at the other (lowest figure), the magnitudes of the torsional shear strains increase in each cross section as a function of the distance from the centroid axis and increase from one cross section to another as a function of the cross section's distance from the fixed end.

expression of the distribution of the material in a cross section. The counterpart of flexural stiffness for shafts is called torsional rigidity, symbolized by C, which is the product of the shear modulus G and the torsional constant, symbolized by J; $C = GJ$. Like EI, torsional rigidity has the units of force times area. The torsional constant is defined as the moment of torque, symbolized by T, required to produce a torsional rotation of one radian per unit

TABLE 3.2 Torsional Constants J for Various Cross Sections

Cross Section		J
	Circle, with radius R	$\frac{\pi R^4}{2}$
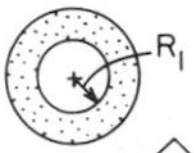	Hollow circle, with inner radius R_i	$\frac{\pi}{2}(R^4 - R_i^4)$
	Thin-walled sector of hollow circle, with outer radius R, inner radius R_i, swept by angle θ, and uniform thickness t	$\frac{\theta t^3}{2}(R + R_i)$
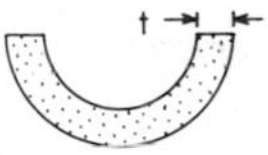	Thin-walled semiannulus, $\theta = \pi$	$\frac{\pi\theta t^3}{6}$
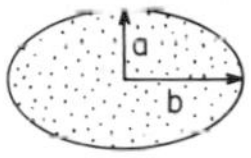	Ellipse, with semimajor and minor axes a and b	$\frac{\pi a^3 b^3}{a^2 + b^2}$
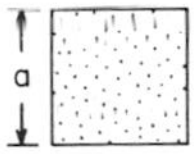	Square, with side a	$0.10406a^4$
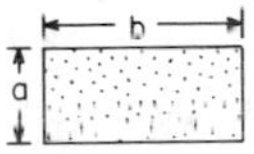	Rectangle, with small and large sides b and a $a{:}b$ = 1 2 4 8 c = 0.281 0.286 0.299 0.312	$\frac{ca^3b^3}{a^2 + b^2}$
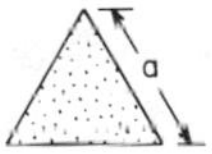	Equilateral triangle, with side a	$0.02165a^4$
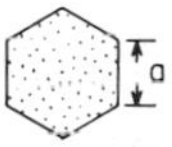	Hexagon, with side a	$1.03a^4$
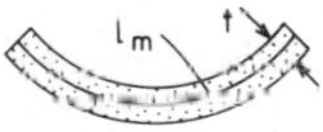	Thin-walled open section, with uniform wall thickness t and midwall perimeter length l_m	$\frac{l_m t^3}{3}$
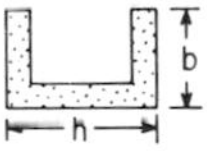	Channel section, with sides b, bottom h, and uniform thickness t	$\frac{t^3}{3}(2b + h)$
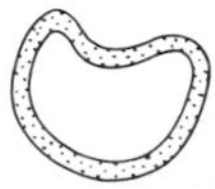	Thin-walled closed section, with uniform wall thickness t and midwall perimeter length l_m; A = area bounded by l_m	$\frac{4A^2t}{l_m}$
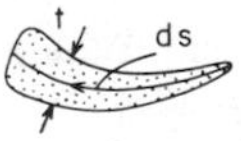	Thin-walled open section, with nonuniform wall thickness t	$(1/3)\int t^3\,ds$
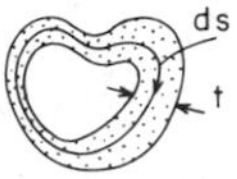	Thin-walled closed section, with nonuniform wall thickness t	$\frac{4A^2}{\int (ds/t)}$

length of shaft divided by the shear modulus; that is, if z is the axis of length, then $J = T(G\partial\theta/\partial z)$. The torsional constant depends on the cross-sectional geometry of a shaft. This is most easily seen for shafts with circular cross sections, where the torsional constant is equal to the polar second moment of area, symbolized by I_p, which equals the sum of the second moments of area measured in the two orthogonal planes through a cross section; the general solution for I_p for any solid cross section is given by the formula $I_p = I_{xx} + I_{yy}$, where I_{xx} is the second moment of area measured in the horizontal plane and I_{yy} is the second moment of area measured in the vertical plane. For example, the polar second moment of area of a solid circular cross section equals the sum of $\pi R^4/4 + \pi R^4/4$ or $\pi R^4/2$. Thus $J = I_p = \pi R^4/2$ and $C = G\pi R^4/2$. By the same token, the polar second moment of area of a hollow circular cross section equals the sum of $\pi(R_o^4 - R_i^4)/4 + \pi(R_o^4 - R_i^4)/4$ or $\pi(R_o^4 - R_i^4)/2$. Thus $J = I_p = \pi(R_o^4 - R_i^4)/2$ and $C = G\pi(R_o^4 - R_i^4)/2$.

For shafts with noncircular cross sections, however, $J < I_p$, and J must be determined analytically. Fortunately, the geometry of many plant organs can be approximated as circular cylindrical shafts and, for those that are not terete in cross section, closed-form expressions for J can be found (table 3.2). For this reason we begin the analytical treatment of torsion by considering shafts with solid circular cross sections. After this case, solid nonterete cross-sectional geometries (elliptical and triangular) will be considered. Regardless of the geometry being considered, however, our basic goal will be to find a quantitative expression for the maximum torsional shear stress and for the distribution of shear stress across each cross section.

When a circular cylindrical shaft with an external radius of R experiences a moment of torque T, the resulting torsional shear stress τ measured at any point is in the plane normal to the radius of the shaft (see fig. 3.8). The magnitude of the shear stress equals the product of the distance r measured along R, the angle of twist per unit length θ, and the shear modulus G:

$$\tau = r\,\theta\,G. \tag{3.18}$$

Since the shearing stress varies directly with the distance from the axis of the shaft, for a ductile material, plastic behavior in torsion is expected to begin just at the surface of a shaft. For materials that are weaker in shear longitudinally than transversely, such as wood, the first strains to appear are produced by shearing stresses operating along the axial section and appear on the surface in the longitudinal direction. Finally, for materials that are weaker in tension than in shear, a crack will appear along a helix inclined at 45° to the axis of the shaft. This helix is the result of pure shear (a state equivalent to one of tension in one direction and equal compression in the orthogonal direction; see chap. 2).

It is evident that $r\theta$ is an expression of the shear strain γ, since the shear modulus G equals the ratio of the shear stress to the shear strain ($G=\tau/r\theta=\tau/\gamma$). The shear stress measured at any point within each cross section has two components. If x and y denote the axes of any transection and if z denotes the longitudinal axis of the shaft, then these two shear stress components are given by the formulas

(3.19a) $$\tau_{xz} = -y\,\theta\,G$$

and

(3.19b) $$\tau_{yz} = x\,\theta\,G\,.$$

The elementary theory of torsion assumes that only pure shear occurs. That is, the bending stresses in the x, y, and z directions equal one another and also equal τ_{xy}, which in turn equals zero. This assumption provides for relatively easy solutions for the displacements u, v, and w, for any infinitesimally small element in each transection through a circular shaft measured in the x, y, and z directions:

(3.20a) $$u = -\theta_{yz}$$

(3.20b) $$v = \theta_{xz}$$

(3.20c) $$w = 0\,.$$

Since the torsional shear stress will increase as r increases, and since r has a limit of R, the maximum torsional shear stress τ_{max} will occur at the surface of the shaft (as shown in figs. 3.7 and 3.8). Also, τ and τ_{max} will increase as the angle θ of twist increases. For a shaft that is fixed at one end and torqued at the other, the shear strains will increase in magnitude from the base of the shaft to the tip (bottom illustration in fig. 3.8).

However, the elementary theory of torsion assumes that no contraction in the length occurs ($w=0$), which is counterintuitive; for example, if a washcloth is wrung out by twisting it, the cloth will shorten, indicating that rotational deformations occur and that $w\neq0$. True, depending on the magnitude of the torque, the contraction in length may be ignored, which is precisely what the elementary theory does. In the general case, however, it is evident that w must be a function of $\theta(x,y)$. To evaluate the torsional rigidity of a circular cylindrical shaft, we note that the net force on any cross section must be balanced by the applied moment of torque T:

(3.21)
$$T=\int \tau(2\pi r)(dr)r$$

or

(3.22)
$$T=\int rC\ (2\pi r)(dr)r\ ,$$

which is simplified by noting that $2\pi C$ is a constant: $T=2\pi C \int r^3\ dr=(\pi R^4/2)C=I_pC$. Solving for C yields

(3.23)
$$C=\frac{T}{I_p}\ .$$

Therefore the shear stress at any radius r in a circular shaft is given by the formula

(3.24)
$$\tau=\frac{r}{I_p}\,T=\frac{2\,r}{\pi R^4}\,T\ ,$$

while the maximum shear stress will occur when $r=R$:

(3.25)
$$\tau_{\max}=\frac{2\,T}{\pi\,R^3}\ .$$

These equations confirm the fundamental conclusions reached by the elementary theory of torsion, that the torsional shear stress increases from the center to the surface of the circular shaft (fig. 3.7). They also reveal that for a very long shaft, at a sufficient distance z from the fixed end, the shear stresses depend solely on the magnitude of the torsional moment and are essentially independent of whether the other end is fixed.

It is worth noting that eq. (3.24) is mathematically analogous to the equation for the tensile and compressive stresses that develop when a beam is bent. The bending stress σ at any distance d from the neutral axis is related to the bending moment M and the second moment of area I of a beam: $\sigma=Md/I$. Similarly, the shear stress τ measured at any distance r from the neutral axis of a shaft subjected to a moment of torque T is related to the torsional constant J, which for a circular shaft equals the polar second moment of area I_p; that is, $\tau=Tr/I_p$. Thus we can appreciate that T, r, and I_p are mathematically analogous to M, d, and I.

If the torsional constant is known, then the distribution and maximum intensity of shear stresses for any shaft can be calculated. For example, for an elliptical shaft whose semimajor and semiminor axes are a and b, $J = \pi a^3 b^3 / (a^2 + b^2)$. It is relatively easy to prove that the maximum shear stress occurs at the ends of the minor axis and that the absolute values of τ and τ_{max} are given by the formulas

(3.26a) $$\tau_{xz} = -\frac{2Ty}{\pi ab^3} \qquad \tau_{yz} = \frac{2Tx}{\pi a^3 b}$$

and

(3.26b) $$\tau_{max} = \frac{2T}{\pi ab^2}.$$

(Notice that when $a = b$, eq. 3.26b gives the τ_{max} for a circular shaft.) The displacement of any element in the shaft along its length is given by the formula

(3.27) $$w = T\frac{(b^2 - a^2)\,xy}{\pi\, a^2\, b^3\, G}.$$

Also, the shortening of the elliptical shaft as a function of the applied torque can be easily calculated, as when the petioles of the cottonwood (*Populus deltoides*), which have an elliptical cross section, torque in the wind. Further, the angle of twist θ for an elliptical shaft with a shear modulus G subjected to a moment of torque T is given by the formula $\theta = T(a^2 + b^2)/\pi a^3 b^3 G$. Thus, if we can measure the angle of twist and if we know the magnitude of the moment of torque, then we can compute the shear modulus of a biological structure (an elliptical petiole).

The torsional shear stresses developing within the triangular stems of sedges can be assessed based on the formula

(3.28) $$\tau_{max} = \frac{T}{8\,k\,a^3\,b},$$

where a and b are the depth and width of the triangular cross section and $k \approx 2.34$.

The relative ability of a member to resist bending versus torsion, based on its cross-sectional geometry, can be evaluated in terms of the dimensionless ratio of the second moment of area to the torsional constant, I:J. A ratio of one indicates that comparable geometric contributions are made to resisting bending and torsion, whereas ratios greater or less than unity reflect a geometric bias favoring torsion or bending, respectively. For solid or hollow circular cross sections with the same amount of material, $I{:}J = 0.5$, indicating that circular cross sections geometrically resist torsion twice as well as they resist bending. This bias can be altered by apportioning some material within each cross section farther away from the centroid axis, as in a square cross section ($I{:}J = 0.593$) or a solid equilateral triangular cross section ($I{:}J = 0.833$) or an elliptical cross section with a major axis twice that of the minor axis aligned in the plane of bending ($I{:}J = 1.25$). From a geometric perspective, a cylindrical tube provides the most efficient solution for using a given quantity of material to resist bending, elastic buckling, or torsion.

The biases that result from cross-sectional geometry are countered by the fact that, for most materials, $E = kG$, where $k>1$. Thus the ratio of flexural rigidity to torsional rigidity almost always tends to be greater than unity; that is, $EI{:}C = kGI{:}GJ = kI{:}J$, where $k>1$. For example, if we assume that members differing in their cross-sectional geometry are composed of the same isotropic material with a Poisson's ratio equal to 0.5, then $E = 2(1 + \nu)G = 3G$, and $EI{:}C = 3I{:}J$. Accordingly, for a circular member with a solid or hollow cross section, $3I{:}J = 1.5$, whereas $3I{:}J$ equals 1.78, 2.49, and 3.75 for members with square, equilateral triangular, and elliptical cross sections, respectively. Indeed, when the elastic and shear moduli of virtually any material are considered in conjunction with cross-sectional geometry, we conclude that virtually all members consisting of a homogeneous isotropic material resist bending as well as or better than they resist torsion.

Yet biological members like stems and petioles are not homogeneous in their material composition, nor are most plant tissues isotropic in their behavior. Further, many biological structures are mechanically designed to twist when subjected to wind pressure. The fluttering of leaves helps dissipate heat, reduces the projected areas (hence drag) of leaves, and disrupts boundary layers, thereby increasing the diffusion of gases into and out of the leaf lamina (Nobel 1983). Indeed, Shive and Brown (1978) report that the oscillations of cottonwood leaves increase the rate of O_2 flux through leaves by means of bulk airflow through the lamina. To some extent the design constraints imposed on petioles by the relation between E and G and by the requirement to resist bending but permit torsion are dealt with by simply placing more material within cross sections of petioles where a higher resistance to bending is required. Thus, flexural stiffness is increased by increasing the absolute magnitude of the second moment of area, regardless of the type of material used. Parenchyma and collenchyma are excellent materials for this strategy. Provided they are turgid, parenchyma and collenchyma have low compressibility and decent moduli of elasticity. Also, parenchyma can be apportioned developmentally wherever needed, since all tissues essentially begin their development as undifferentiated parenchyma. By contrast, the petioles of large pinnate leaves, as well as stems, typically resist torsion by placing stiff materials with high elastic moduli (like sclerenchyma) toward the perimeters of their cross sections. In this sense the petioles of pinnate leaves are mechanically analogous to stems: petioles mechanically support functionally photosynthetic units (leaflets and leaves) that are free to bend and deflect in the wind but are themselves very resistant to bending and torsion.

The truly elegant aspect of the mechanical design of petioles is seen in the way tissues differing in mechanical properties are spatially deployed within cross sections to accommodate the three stress distributions shown in figure 3.7, resulting from bending and torsion. Typically, tissues with relatively low shear moduli are placed just beneath the epidermis and at the center of cross sections taken along the midspan of petioles. This anatomy permits large peripheral torsional and centroidal bending shear strains, thereby allowing some circumrotation of petioles about their longitudinal axes. Since petioles must to some extent resist bending as well as excessive torsion, however, cables of elastic and relatively stiff vascular tissues are placed parallel to and at some distance from the centroid axis, which confers a reasonable bending stiffness. Indeed, when we come to look at a representative cross section of a petiole through the anatomically critical eye of a biomechanicist, we see that the composite tissue construction and spatial allocation of materials found in petioles reflect one of the most elegant expressions of evolutionary adaptation encountered in all of biology.

TORSION AND TORSIONAL BUCKLING OF CLOSED AND OPEN THIN-WALLED SECTIONS

Thus far we have assumed that when a support member buckles, deformations will occur in a plane of symmetry within a cross section. But support members with closed or open thin-walled cross sections, like the internodes and leaf blades of grasses, can buckle either by twisting or by a combination of bending and twisting, both of which can occur if the torsional rigidity of cross sections is small. Since the shear modulus of most plant materials is less than the elastic modulus, the possibility of torsional buckling in plant organs, as a consequence of either a bending or a torsional moment, cannot be entirely neglected. Fortunately, the geometric limiting case for torsional buckling for closed tubular cross sections is defined by the ratio of the thickness t to the outer radius R of a hollow cross section: when $t/R > 0.05$, the likelihood of torsional buckling is relatively low. This is somewhat reassuring, since $t/R \geq 0.20$ for most hollow plant organs (cf. Brazier buckling). By contrast, plant organs with open thin-walled sections, such as blades of grass, exhibit a considerable range in t/R, necessitating an appreciation of the influence of cross-sectional geometry on torsional buckling.

The torsional constant of a circular cylindrical member with a hollow cross section of uniform wall thickness t is given by the formula $J = (\pi/2)(R^4 - R_i^4)$, where R_i is the inner radius and R is the outer radius of the cross section (see table 3.2). Thus $C = G(\pi/2)(R^4 - R_i^4)$, from which we can readily appreciate that the ability of the circular shaft to resist torsion increases as the wall thickness $(R - R_i)$ increases. A similar conclusion is reached regarding thin-walled open cross sections, since in general $J = l_m\, t^3/3$ and $C = G(l_m\, t^3/3)$, where l_m is the length of the midwall perimeter. The formulas for J and C are prominent in the approximate mathematical solutions for simple torsion and torsional buckling of support members with thin-walled cross sections, either as a result of a compressive load or as a consequence of bending. The detailed derivations of these solutions, which are well beyond the scope of this book, are provided by Timoshenko and Goodier (1970) and Timoshenko and Gere (1961) and should be consulted, particularly with regard to their underlying assumptions. For example, torsion of thin-walled tubes is treated by assuming that the shear stress components are uniformly distributed throughout t within each cross section; that is, t is assumed to be very small compared with R. Accordingly, when a hollow tube is subjected to a moment of torque T, the shear stress within the wall will be predicted by the formula $T/2At$, where A is the area bounded by the inner and outer perimeters of the cross section. This formula indicates that the shear stress should be inversely proportional to the wall thickness and that within any cross section the maximum shear stress will occur where the wall is the thinnest. Significantly, when hollow plant organs undergo excessive twisting, they typically undergo torsional buckling at the point where their cross sections are thinnest.

In the simple case of a hollow circular cylinder with a uniform thickness defined by the inner radius R_i and outer radius R of the tube, the magnitude of the shear stress τ and the angle of twist θ are given by the formulas $\tau = T/[2\pi(R + R_i)t^2]$ and $\theta = (T/G)(l_m/4A^2t)$. The limits to the applicability of the first of these formulas can be seen by noting that, as the wall thickness increases, t approaches the value of R and R_i becomes zero. Thus τ is predicted to equal $T/2\pi R^3$. This predicted value is half the maximum torsional shear stress actually experienced within a solid circular cross section: $\tau_{max} = 2T/$

πR^3. The inaccuracy results from the stipulation that the shaft has a very thin cross section: $t/R<0.05$. As the wall thickness increases, this assumption is increasingly violated, and the predictions based on the previous formula show increasing error.

Buckling of thin-walled support members usually involves bending, and the critical load P_{cr} for global buckling calculated by the Euler column formula is always larger than the critical load the column can actually sustain. If the center of torsional shearing in each cross section coincides with the centroid axis of a column, then torsional buckling and buckling resulting from pure bending are assumed to be independent of one another. Thus the smaller of either the critical torsional load P_t or the critical load P_{cr} dictates the largest actual load that a column can sustain. A reasonable approximation of the critical torsional load is given by the formula $P_t = (A/I_p)(C + C_1\pi^2/l^2)$, where C_i is the warping rigidity of the column, which is the product of the elastic modulus E of the material and a parameter called the warping constant, which in turn is a complex mechanical parameter with units of length raised to the sixth power. Closed-form solutions for warping rigidities and warping constants of some cross-sectional geometries are provided by many engineering texts. Whenever $P_t > P_{cr}$, torsional buckling can be largely neglected, and the Euler column formula can be used with satisfactory results.

THE DEFLECTIONS OF CANTILEVERS

The bending of a cantilever under its own weight differs in kind from that of a vertical column. The bending of a cantilever is stable, whereas the global buckling of a column is not. Additionally, shear forces typically play a very significant role. This can be easily appreciated by holding a paperback book by the spine. When the front cover of the book is held parallel to the ground, the pages (which represent the material lines) of the solid (book) bend and slide. Thus, the material of the book has a low shear modulus G, and material lines are free to undergo laminar and rotational shearing as the book deforms by bending downward. The shear modulus of the book can be dramatically increased by restraining its material lines, increasing its modulus of shear. This can be seen by simply reversing the way we hold the soft-cover book. When we hold the side opposite the spine, the material lines (pages) are restrained by our fingers, the shear modulus is dramatically increased, and the book bends downward much less.

The stress distributions within each transection through a cantilever can be very complex (fig. 3.7). What is generally not appreciated, however, is that the magnitudes of these stresses differ as a function of where stresses are measured along the *length* of a cantilever. The maximum tensile and compressive stresses are achieved at the upper and lower surfaces of each transection, respectively, but they also increase toward the rigidly held end of the cantilever; the highest tensile and compressive stresses occur at the upper and lower surfaces of the transection nearest the anchored base. This is illustrated in plate 1, where tensile and compressive stresses are color coded along the length and depth of a cantilever viewed sideways. Bending shear stresses tend to have a longitudinal gradient counter to those of the tensile and compressive stresses. That is, bending shear stresses tend to increase toward the midspan of a cantilever. Although the cross-sectional geometry of a cantilever influences the distribution of tensile, compressive, and bending shear stresses, the longitudinal gradients of compressive and tensile stresses are similar for all cantilevers regardless of their cross-sectional geometry.

From a review of fracture mechanics (chap. 2) and a general appreciation that most plant tissues operate best in tension, the significance of the longitudinal gradients of tensile and compressive stresses within a cantilever takes on added meaning. Fractures typically propagate where tensile stresses are highest. Since this occurs at the base of a cantilever, any crack or imperfection whose length equals or exceeds the Griffith critical length will propagate with disastrous consequences. Plate 1 illustrates where and in what manner such a fracture is likely to occur. The fracture will almost invariably occur toward the anchored base and follow a curvilinear trajectory moving toward the bottom of the cantilever. The leading edge of the fracture is surrounded by a tensile stress field even as it passes through the lower portions of the cantilever that are dominated by compressive bending stresses. From this we can see that cantilevers are mechanically most vulnerable at the point where they are attached to some other structure.

The previous summary sheds some light on the functional significance of why branches are broader at the base and why branches and leaves break (and shear) at the base in windstorms. Taper and regional swellings on branches concentrate materials and regionally increase the second moment of area and the second polar moment of area. Hence they regionally increase the resistance of branches and petioles to bending and torsion at the base. Additionally, the taper of branches and the swollen bases of leaf petioles dissipate tensile force trajectories within larger volumes of material, reducing tensile stress concentrations and the likelihood that a fracture will develop (see Mattheck 1990). Indeed, branch collars—regional swellings of wood in which the grain of the wood differs from that elsewhere along the length of a branch—develop

CONCENTRATED LOAD AT ANY POINT

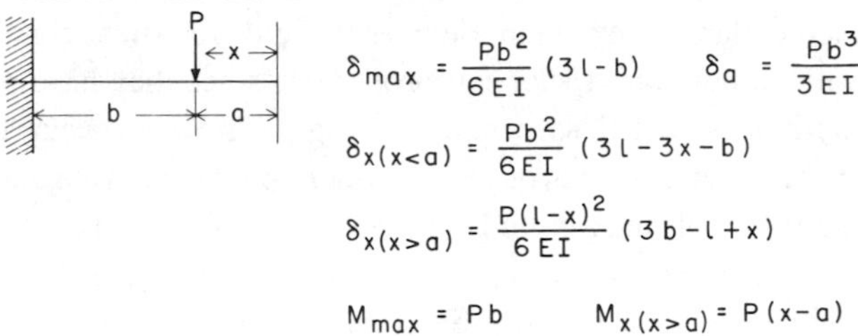

P
x
b
a
$\delta_{max} = \frac{Pb^2}{6EI}(3l-b)$
$\delta_a = \frac{Pb^3}{3EI}$
$\delta_{x(x<a)} = \frac{Pb^2}{6EI}(3l-3x-b)$
$\delta_{x(x>a)} = \frac{P(l-x)^2}{6EI}(3b-l+x)$
$M_{max} = Pb$
$M_{x(x>a)} = P(x-a)$

CONCENTRATED LOAD AT FREE END

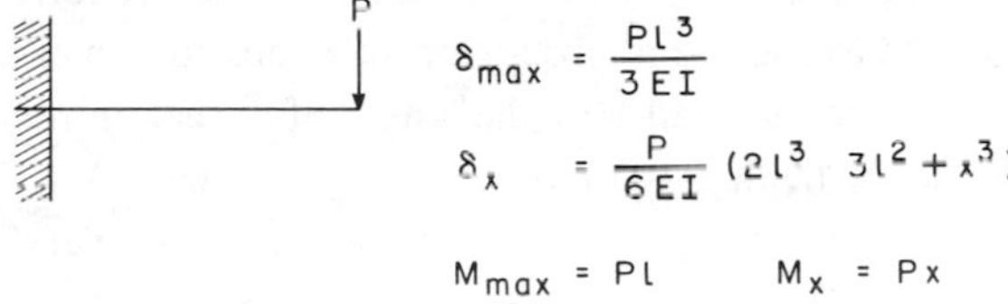

P
$\delta_{max} = \frac{Pl^3}{3EI}$
$\delta_x = \frac{P}{6EI}(2l^3 \;\; 3l^2 + x^3)$
$M_{max} = Pl$
$M_x = Px$

INCREASING (UNIFORM) LOAD TO FIXED END

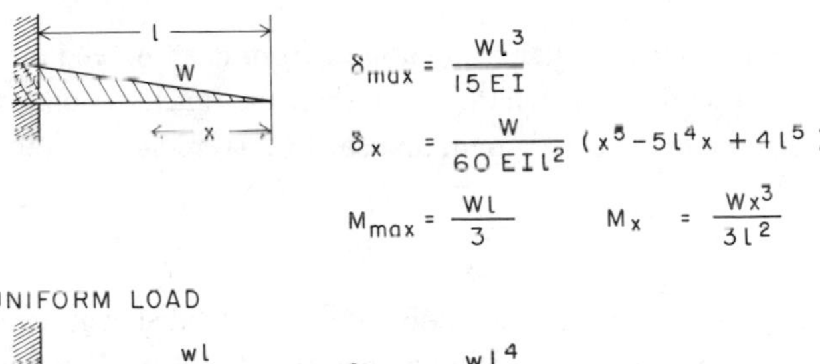

l
W
x
$\delta_{max} = \frac{Wl^3}{15EI}$
$\delta_x = \frac{W}{60EIl^2}(x^5 - 5l^4x + 4l^5)$
$M_{max} = \frac{Wl}{3}$
$M_x = \frac{Wx^3}{3l^2}$

UNIFORM LOAD

wl

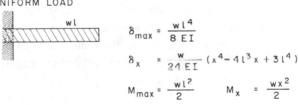

$$\delta_{max} = \frac{wl^4}{8EI}$$

$$\delta_x = \frac{w}{24EI}(x^4 - 4l^3x + 3l^4)$$

$$M_{max} = \frac{wl^2}{2} \qquad M_x = \frac{wx^2}{2}$$

FIGURE 3.9 Equations for predicting small deflections and bending moments for variously loaded cantilevered beams. Four loading configurations and three beam geometries are shown. In each case the cantilevered beam is rigidly fixed at one end (to the left of each diagram) and free at the other (to the right). The uppermost cantilevered beam has a concentrated load that can be placed anywhere along the length of the beam. The second cantilevered beam is loaded only at its free end. The hatching in the third and fourth cantilevered beams reflects the relative intensity of the loading along the lengths of the beams. P = point load, W = total weight, w = weight per unit length, EI = flexural rigidity. Deflections are measured in units of length; bending moments are measured in units of weight times length.

at the base of large branches where they are attached to tree trunks (Shigo 1990). Branch collars are excellent devices for reducing stress concentrations. They also reduce how far cracks can propagate, since they have a polylaminate construction. Nonetheless, when shearing failure does occur, it is almost always at the bases of branches or leaves, where bending tensile and compressive stresses reach their maximum levels.

We are now in a position to approach the subject of the deflections that occur in cantilevered beams resulting from bending and torsion under self-loading. Although exact solutions for large deflections are available, they are extremely difficult to compute, and it is much easier to treat the limiting conditions for small deflections that predict the onset of large deflections. Therefore the easier route is to use the approximate solutions for small deflections (those that are equal to or less than 10% the length of the cantilever). Some of the most useful equations dealing with this topic are summarized in figure 3.9.

The deflection δ measured anywhere along the length of a very long, thin cantilever sustaining a load P acting on its free end is given by the formula

$$\delta_{(x)} = \left(\frac{Pl^3}{3EI}\right)\left(1 - \frac{3x}{2l} + \frac{x^3}{2l^2}\right), \tag{3.29a}$$

where x is the horizontal distance measured from the free end. If we focus merely on the tip deflection, where $x = 0$, then the function contained in the second set of parentheses equals one, and eq. (3.29a) reduces to the formula

$$\delta_{(x=0)} = \frac{Pl^3}{3EI}. \tag{3.29b}$$

This equation describes the tip deflection of an end-loaded cantilever, a cantilever with a concentrated load at the free end (see fig. 3.9). Equation (3.29b) indicates that the deflection will decrease as the cantilever's flexural stiffness increases and will increase as the cube of the cantilever's length. All other things being equal, the length of the cantilever appears to be the single most important factor dictating mechanical behavior. A few simple experiments with strips of paper varying in length but not in width will confirm this conclusion.

However, eq. (3.29b) makes an implicit assumption in that it completely neglects the rigidity of the cantilever in terms of its shear modulus. Typically this assumption is justified, provided the bulk of the cantilever is made of a stiff material and that the ratio of the length to the depth of the cantilever is very large (≥ 20). But if the bulk of the cantilever is made of a material with a low shear modulus or if the aspect ratio of the cantilever is very small, then more complex equations must be used. To understand this, we must recognize that the total deflection δ at the tip of any cantilever is really the sum of the deflection resulting from bending δ_b and shearing δ_s. Thus, eq. (3.29b) should have the form

$$\delta = \delta_b + \delta_s = \frac{Pl^3}{3EI} + \frac{Pl}{AG}, \tag{3.30}$$

where A is the cross-sectional area of the cantilever. Since the shear modulus of an isotropic material is considerably less than the modulus of elasticity, and since most natural cantilevered organs are much longer than they are deep, the deflection resulting from bending is typically much more significant than that resulting from shearing. Thus, in most circumstances eq. (3.29b) is a very reasonable approximation.

If the cantilever is composed of two or more materials differing in their shear and elastic moduli, however, then shearing deflections cannot be neglected, particularly if there exists a large (by volume) inner core with a low shear modulus. For example, consider a cylindrical cantilever with an outer rind of stiff material and an inner core of a material with a low shear modulus (and radius R_i). When the appropriate second moments of area are substituted into eq. (3.30), we derive the formula

$$\delta = \frac{Pl}{\pi}\left[\frac{4l^2}{3E(R^4 - R_i^4)} + \frac{1}{GR_i^2}\right], \tag{3.31}$$

where R is the outer radius of the cantilever. Thus, as G decreases the second term in the parentheses increases. Also, as the radius of the inner core increases, the first term in the parentheses decreases. Incidentally, if a cantilever has a uniform cross section and consists entirely of a viscous material, then we can show from first principles that its deflection equals $Ptl^3/9\eta I$, where η is viscosity and t is time. Since $\eta = Ptl^3/9\delta I$, we can determine the viscosity of a viscous material molded into a cantilevered beam subjected to an end load. Indeed, this experimental approach is often used in the material testing of non-Newtonian fluids such as asphalt.

Caution should always be exercised in determining the elastic modulus of a cantilevered beam from deflection measurements, regardless of the formula used, since shearing effects become increasingly pronounced as the ratio of length to depth of a cantilever decreases, and because the supports used to artificially anchor one end of an organ can produce end-wall effects. When we are dealing with naturally long cantilevered organs, shearing artifacts in experimentally tested specimens can be reduced by examining specimens with ratios of length to depth ≥ 20 (the more anisotropic the material, the larger the aspect ratio required). To correct for end-wall effects, it is necessary to measure the elastic modulus of the same specimen clamped at different lengths. The true stiffness can be determined from the intercept of a plot of the reciprocal of the elastic modulus (the compliance) versus the reciprocal of length.

A recurrent theme in all our previous treatments is that geometry is critical to the mechanical performance of structures. This is true for cantilevers as well as for columns. The tapering of a cantilever can have profound consequences for its ability to sustain static loadings, and many plant organs are much more capable of modifying their geometry than the materials they are made of. Branches increase in girth from year to year and are typically thicker toward the base than the tip as a consequence of their secondary growth. The metabolic investment made in the cell walls of their woody tissues is amortized over many years, and since the bulk of wood is composed of dead cells, wood does not withdraw from the metabolic interest supplied by photosynthetic tissues elsewhere on the plant. Similarly, the petioles of many plant species are tapered along their length, but this tapering is typically achieved by an investment in thin-walled living tissues rather than in thick-walled dead tissues. When the volume fractions of different types of tissues within a petiole are computed and plotted as a function of their distance from the tip of a petiole, the volume fractions of thin-walled tissue types, such as parenchyma, disproportionately increase toward the petiole's base compared with the volume fractions of tissues with thicker cell walls, usually thought of as mechanical support tissues (xylem and phloem fibers). These thin-walled tissue types operate as hydrostats when their protoplasm is fully turgid. They provide bulk to a transection and can confer geometric stiffness with a minimum investment in cell wall material. This is a reasonable alternative to producing large amounts of woody tissue, particularly for an organ that typically has a limited functional lifetime. Yet the petioles of very large leaves, like those of many palm species, apportion a larger amount of the thick-walled tissues in their cross sections toward the surface than do the petioles of smaller leaves. This rind of thick-walled tissue surrounding an inner core of thinner-walled tissue confers great stiffness, particularly when it is placed in tension by a fully inflated core of hydrostatic tissue.

THE MECHANICAL ALLOMETRY OF PETIOLES

Petioles show remarkable mechanical design. For example, their flexural stiffness is often scaled to their size and the weight they must sustain. The petioles

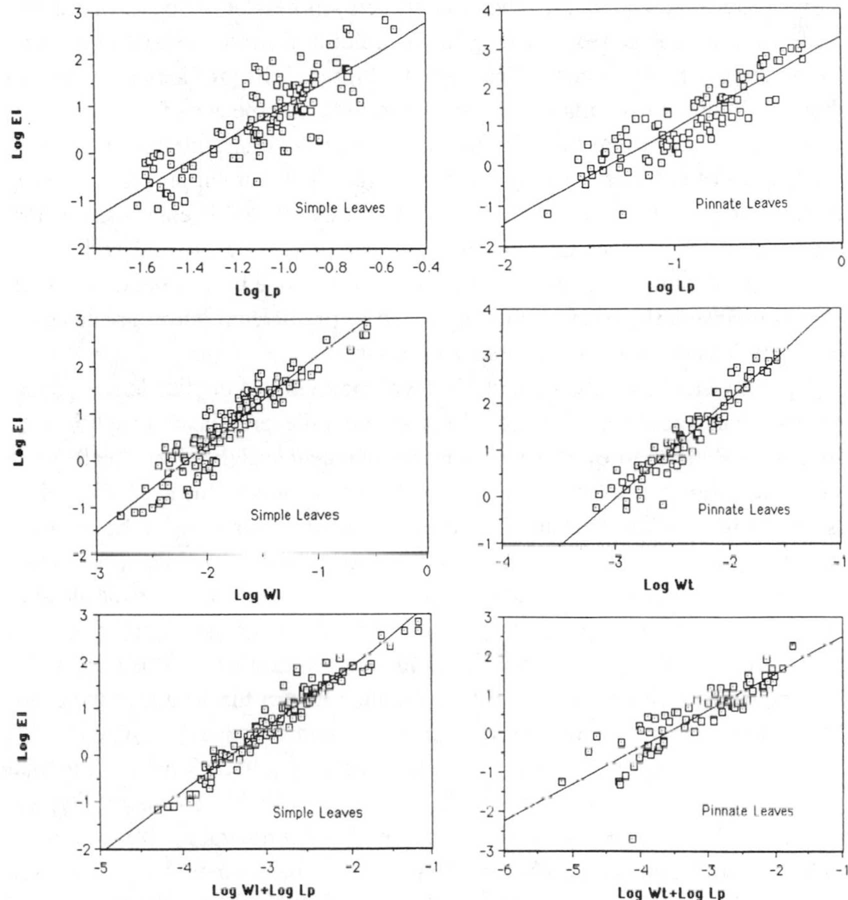

FIGURE 3.10 Allometry of flexural stiffness and leaf size for angiosperm and fern leaves. Log-log plots of the flexural rigidity (*EI*) versus the petiolar length *Lp*, lamina weight *Wl*, and total weight *Wt* of simple and pinnate leaves. *EI* disproportionately increases for each of the two leaf types as petiolar length increases or as *Wl* and *Wt* increase. For further details, see text.

of simple and palmate leaves are typically not tapered along their length, nor do they exhibit a longitudinal gradient in their elastic modulus. Thus they appear to be the mechanical equivalent of an end-loaded, untapered cantilevered beam, whose tip deflection ought to conform to the predictions of eq. (3.29). The point load *P*, in this case, is the weight of the leaf lamina sustained at the tip of the petiole. Equation (3.29) predicts that if the tip deflections among a large number of leaves differing in size (petiolar length or laminar weight) are to be equivalent, then the flexural stiffness of the petioles must be scaled to the cube of petiolar length. Also, flexural stiffness should be scaled to the weight (point load) of leaf lamina. Figure 3.10 provides data derived from mechanical tests made on twenty-two different species of plants (monocots, dicots, and some ferns). The data for the flexural stiffness of petioles, the lengths of petioles (*Lp*), and lamina weight (*Wl*) for simple leaves are log transformed. As we can see, there appears to be remarkable consensus in how these variables behave. Of particular interest is that the flexural stiffness of petioles from simple leaves is proportional to the cube root of petiolar length. This is precisely the relationship that would be predicted if leaves are designed to function as point-loaded cantilevered beams.

By contrast, the mechanical attributes of the rachis of pinnate leaves appear to very nearly conform to those of a tapered cantilevered beam (see fig. 3.9). In this case the load on the rachis is the total leaf weight (*Wt*). Leaflets are distributed along most of the rachis length, but the bulk of the total leaf weight is the weight of the petiole. Log-transformed data for pinnate leaves from monocots, dicots, and ferns are shown and compared with those from simple leaves in figure 3.10. The data for pinnate leaves indicate that the flexural stiffness of petioles is proportional to the square of rachis length. This does not conform to the deflection formula for tapered cantilevered beams, which predicts that *EI* should be proportional to the cube of the length of the cantilever. However, the flexural stiffnesses of the pinnate leaves examined in this study were disproportionately greater than those required to support the total weight of leaves—typically by a factor of three. Thus the rachis of a typical pinnate leaf is three times as stiff as the petiole of comparably sized simple or palmate leaves and need not scale length in the same manner as a petiole from a simple leaf. One explanation for this is that the rachis is the mechanical (functional) equivalent of a branch, supporting the loadings of individual leaflets much as a branch supports the weight of individual leaves.

Another hypothesis, which is not in opposition to the first, is that the rachis of pinnate leaves may be overbuilt to sustain its own weight and that of leaflets during periods of water stress. A reduction in the tissue water content of the rachis would in theory reduce the flexural stiffness, since thin-walled hydrostatic tissues would reduce in volume and the second moment of area of the rachis as a whole would decrease. This has been observed for the petioles of simple leaves, where experiments have shown that dehydration can dramatically reduce flexural stiffness (up to 90%), principally owing to a reduction in their cross-sectional area. When the reduction in flexural stiffness of the rachis of palm leaves is compared with the reduction in the flexural stiffness of petioles from simple leaves at comparable levels of water loss, the difference between the two mechanical designs is remarkable. The rachis of palm leaves reduces little in flexural stiffness, while the petioles of many simple leaves undergo mechanical failure when they reach their permanent wilting point. The reason appears to be that the palm rachis has an outer rind of thick-walled tissue that resists geometric distortion even when its inner core of thin-walled tissues shrinks as water is withdrawn from the leaf. Once again, geometry appears to be the vital component in the mechanical design of plant organs.

Before leaving this topic, an incidental observation I made while experimenting with leaves is worth noting. As petioles dehydrated, their elastic moduli actually increased, but not in a way that could compensate for the reduction in their second moments of area so that flexural stiffness was maintained. The increase in the elastic modulus could be accounted for in one of two mutually compatible ways. First, dry cell walls tend to have higher elastic moduli than their wetted counterparts. Thus, as water is withdrawn from a petiole or some other organ and as cell walls lose water, the elastic modulus of the organ would be expected to increase. But when the appropriate calculations are made concerning how much water would have to be withdrawn from petioles to affect the cell wall water content, it became obvious that this was not the likely mechanism. The second, and perhaps more common, way E could increase is that the volume fraction of thick-walled tissues is proportionally increased as the volume fraction of thin-walled tissues declines owing to water stress. Dehydration results in the collapse of thin-walled tissues and the densification of cell walls in general and those of thick-walled tissues in particular. Much work has shown that the mechanical strength of an organ is frequently a function of the relative volume fraction of cell walls within an organ. For example, Kokubo, Kuraishi, and Sakurai (1989) show that the strength of the stems of barley (*Hordeum vulgare*) cultivars placed in bending is highly correlated ($r = .90$) with the area of cell walls of sclerenchyma, as well as with the cellulose content of cell walls in general ($r = .93$). Similarly, Vincent (1982) showed that the strength of grass leaves is highly correlated with the volume fraction of sclerenchyma in leaf laminae. These two studies and others suggest that dehydration of plant tissues can effect changes in the material properties, such as E, of tissues and organs by influencing tissue geometry (the densification of cell walls per unit volume of tissue). Much more work in this area of plant biomechanics is needed, however, before any generalizations can be made.

In addition to supporting their weight against the force of gravity, many herbaceous plants with a rosette growth habit, like the bull thistle (*Cirsium pumilum*), have cantilevered leaves that mechanically compress the leaves and stems of plants growing under them. This compression is not simply a result of gravity operating on their weight. Rather, it results from growth stresses that induce a downward bending of the leaf, as pointed out by John Randall (unpublished data). Other examples of this type of loading can be found among some of the most common garden weeds (the broad-leaved plantain, *Plantago major,* and the dandelion, *Taraxacum,* to name just two).

Dynamic Beam Theory

Although alluded to in many of the previous sections, the consequences of dynamic loadings on the mechanical behavior of plant organs have not been treated with any quantitative rigor. Nonetheless, dynamic loadings and their mechanical consequences are very important, if for no other reason than that all terrestrial plants are subjected to some level of dynamic loadings from wind pressure, while for aquatic plants the movement of water has the same effect. At one extreme in their range of magnitude, these dynamic loadings can result in the mechanical failure of plant organs, as when trees uproot in storms. At the other end of the spectrum of dynamic loadings, plants are provided with a useful mechanism whereby pollen can be captured from or released into the air, as when the stalks of grass flowers sweep through the air and intercept airborne pollen or when stamen filaments oscillate and by inertia dislodge pollen from anther sacs. These movements occur because momentum is always exchanged between a moving fluid and an object that obstructs its movement (see chaps. 1 and 9). The exchange of momentum involves the change of kinetic energy from the fluid into potential energy stored within the solid in the form of strain energy resulting from elastic deformations. When the dynamic load is removed, the strain energy can be used to bring the elastically deformed object back to its original static equilibrium position. This typically involves a harmonic mode of mechanical behavior resulting from a cyclic exchange of potential energy and kinetic energy. The potential energy is released as elastic deformations are restored and the object attempts to recover its equilibrium, while the kinetic energy is associated with the movement (velocity) of the object. The rate of energy exchange between potential and kinetic energy is the natural frequency of vibration of the object, symbolized by f_i. In sum, resilient solids, like the strings of a cello or the petiole of a leaf, vibrate when they are struck, plucked, or otherwise moved out of their resting (equilibrium) position.

Plates

PLATE 1 Computer-simulated stress distributions in a bending cantilever (A) and in a cantilever through which a crack is propagated (B–E): (A) Tensile and compressive stresses developed within a cantilever fixed at one end (left) and subjected to a point load at its free end (right). The relative magnitudes of tensile and compressive stresses about the neutral axis (NA) are shown by color (deep blue = maximum tensile stress; amber-orange = maximum compressive stress). The highest tensile or compressive stresses occur at the surfaces of each transection through the cantilever and increase toward the fixed end of the cantilever. (B–C) Propagation of a crack initiated on the upper surface of a cantilever near its fixed end: (B) The advancing tip of the crack surrounded by high tensile stresses (deep blue; see D for details) relieves the remaining portions of the cantilever from most of the tensile stresses developed in bending. (C) As the crack migrates toward the lower surface of the cantilever, the magnitudes of the compressive stresses throughout most of the cantilever decrease but are still relatively high just beneath the crack's leading edge (amber-orange; see E for details). (D–E) Details of a crack propagating through the cantilever. The crack nucleus is surrounded by a region of high stresses that rapidly dissipate into higher compressive stresses. The arrow indicates the direction and the location of the crack migrating through the cantilever.

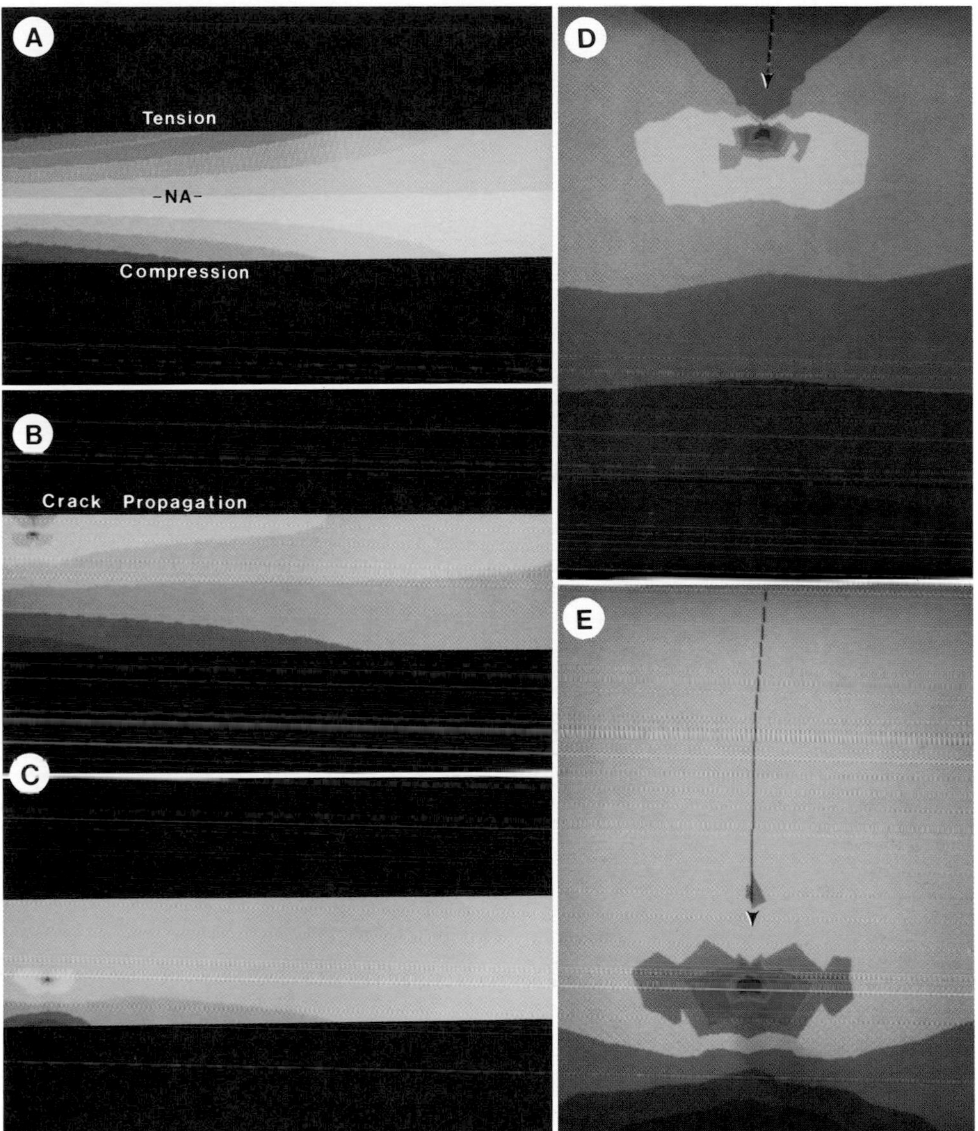
A
Tension
-NA-
Compression
B
Crack Propagation
C
D
E

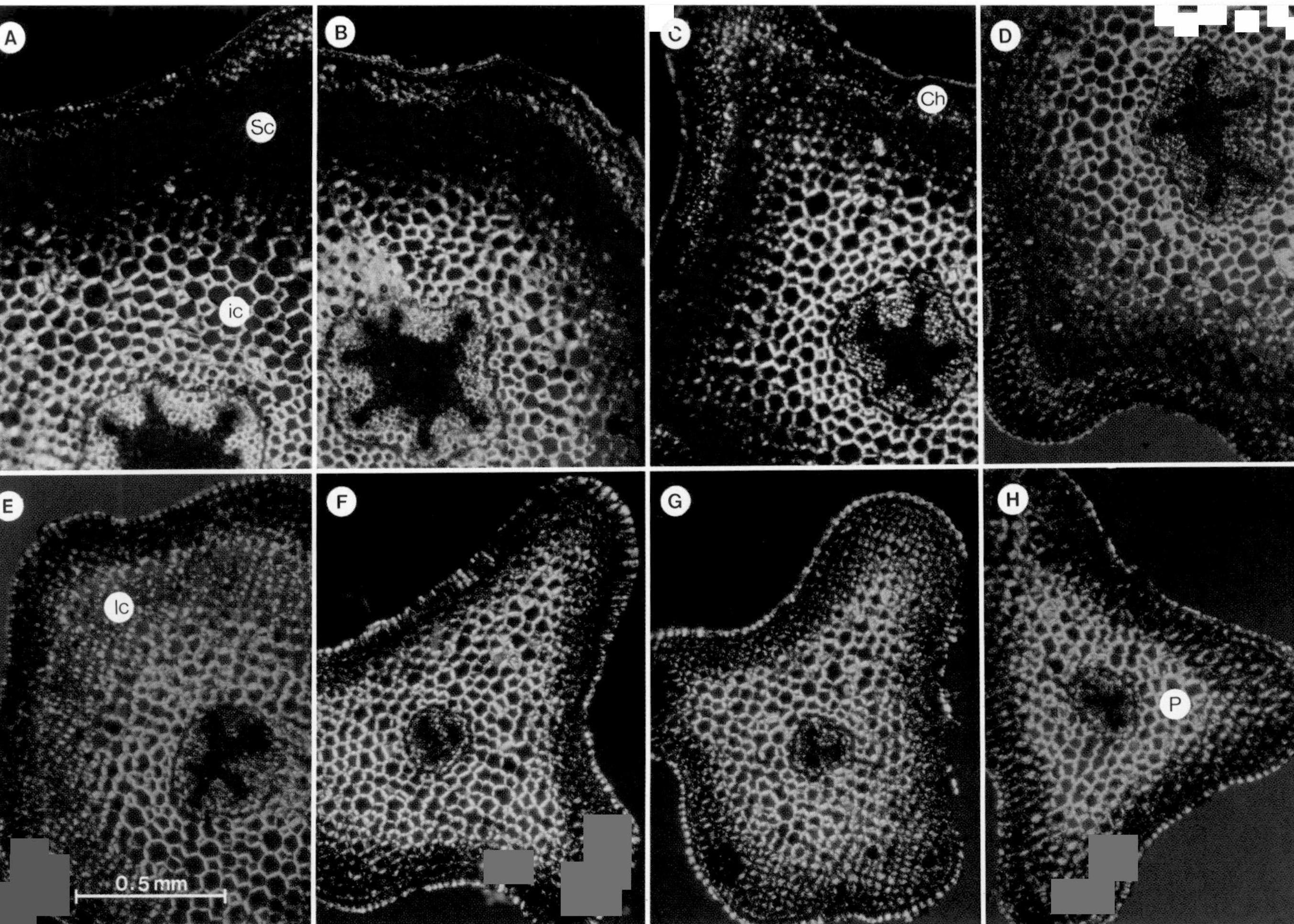

A
Sc
ic
B
C
Ch
D
E
lc
0.5 mm
F
G
H
P

PLATE 2 Representative transverse sections through branch elements of an aerial shoot of *Psilotum nudum* with eight levels of branching. Each section was stained with phloroglucinol to identify lignin (red) and phenolic constituents (yellow). The transections are arranged in ascending order of the level of branching (A is a transection through the basal branch element of the aerial shoot; H is a transection through one of the distalmost branch elements of the shoot). The relative proportions of sclerified cortex (Sc), lignified cortical parenchyma cells (lc), and cortical cells that have phenolic-rich cell walls (ic) decrease as the level of branching ascends, while the relative proportions of chlorenchyma (Ch) and unlignified parenchyma within the cortex increase. Other anatomical and morphological differences among the transections can be noted, such as changes in the relative proportion of vascular tissue and how lobed each transection appears.

PLATE 3 Computer analyses of the speed and direction of pollen grain trajectories around ovules of *Ephedra,* a gymnosperm. The direction of ambient airflow carrying pollen grains is from left to right in each picture. The outlines of ovules and the stems they are attached to are in white. These computer-generated pictures were produced by digitizing the shapes of ovules and stems into computer memory. The motion of pollen grains was determined by stroboscopic photographs (the sequential images of each pollen grain in its flight path were digitized). The computer analyses provide the trajectories of pollen grains (A–B), the average direction of pollen moving within each air space around ovules (C–D), and the average speed of pollen grains moving within each air space (E–F). (Relative speeds shown in all the figures are scaled with respect to the maximum speed of pollen grains; see scale in E.) From analyses such as those illustrated here, it is possible to determine some of the statistical attributes of the motion of airborne pollen grains around the reproductive structures of wind-pollinated plants.

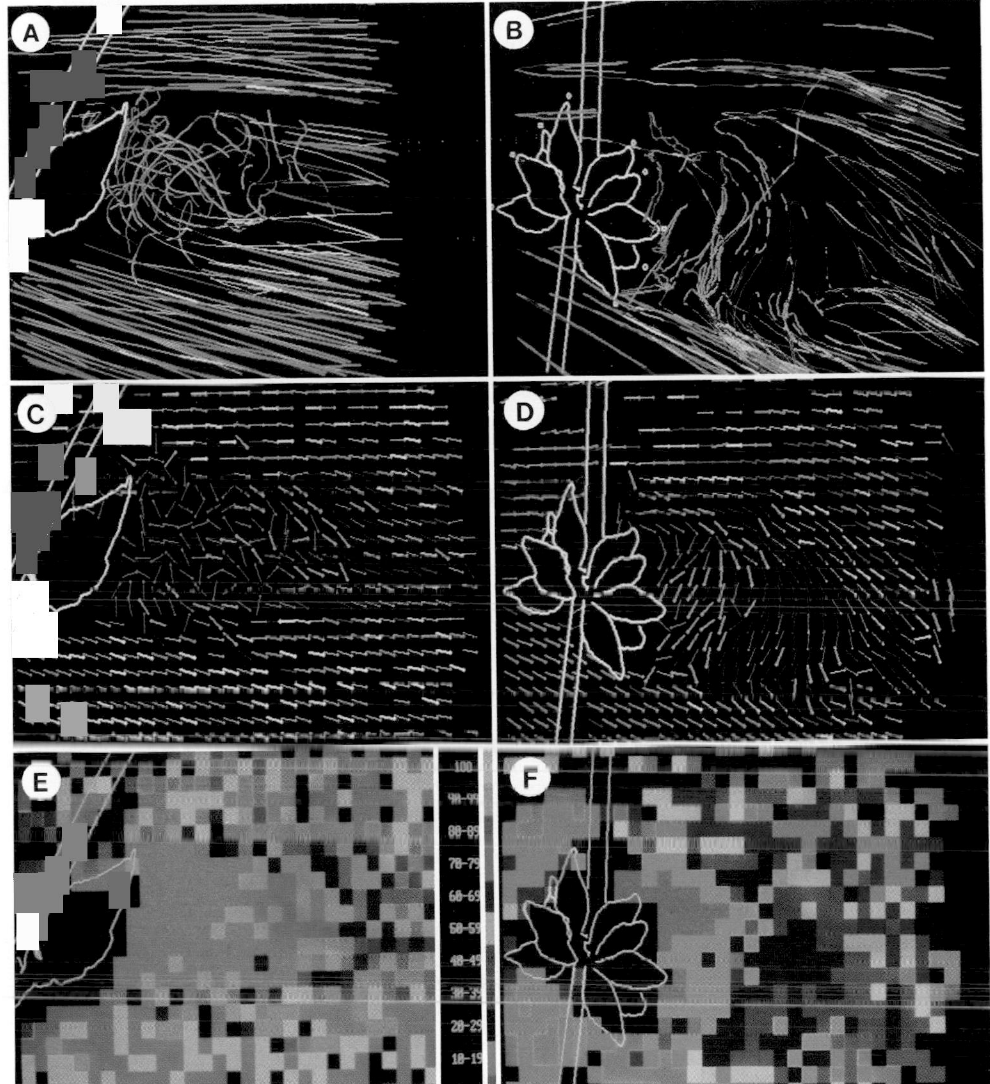
A
B
C
D
E
F
100
90-99
80-89
70-79
60-69
50-59
40-49
30-39
20-29
10-19
0- 9

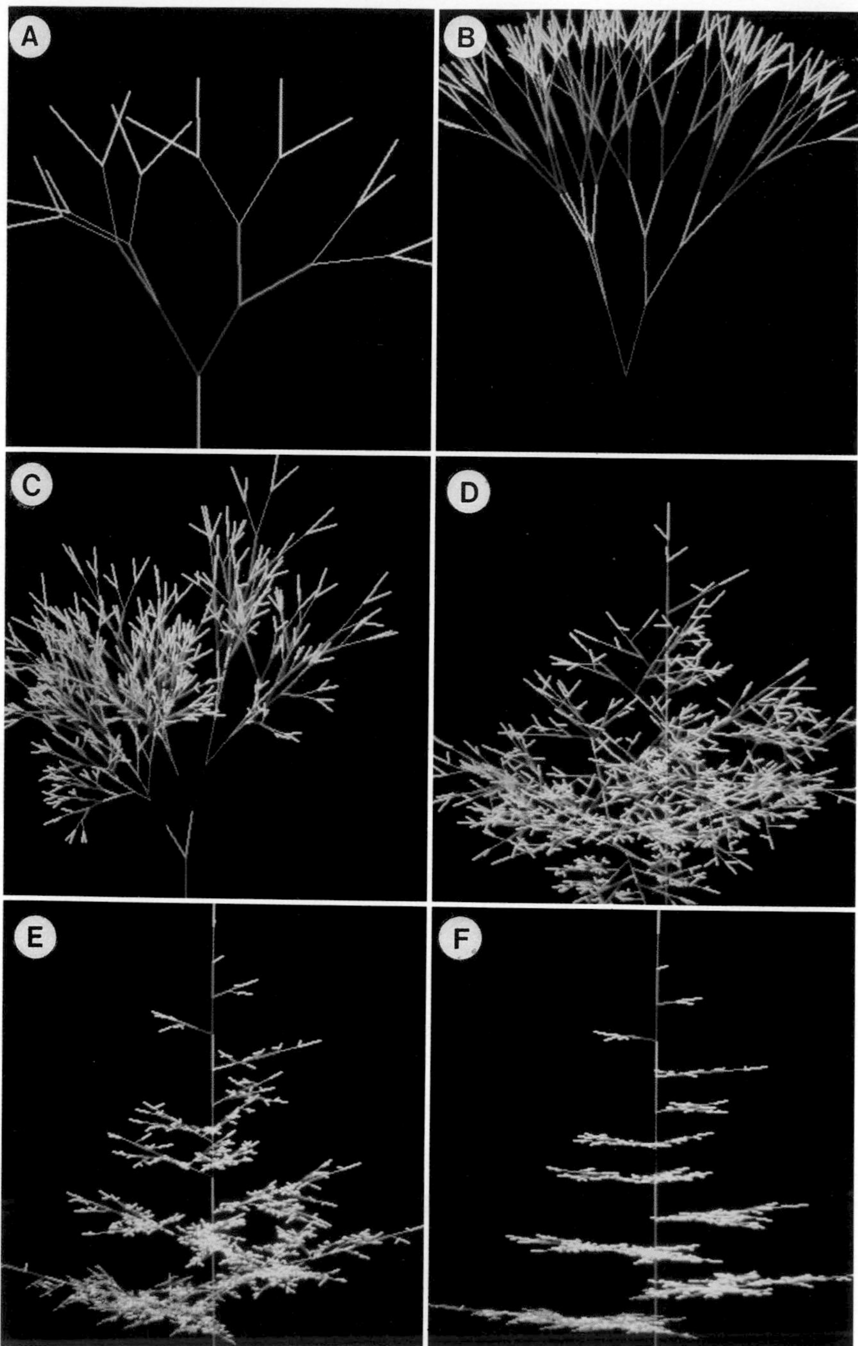

PLATE 4 Computer-generated branching patterns arranged in a sequence of increasing capacity to intercept sunlight and ability to sustain static mechanical loadings due to self-weight. These branching patterns were selected from within a three-dimensional "universe" of many thousands. The branching pattern shown in A was selected as the starting point for a computer program that assessed increasing efficiency in light interception and static loading. The computer program then searched for the nearest branching pattern to A within the universe that was more efficient (shown in B). This process was reiterated (C–E) until one of the most efficient branching patterns in the universe of possibilities was reached (F).

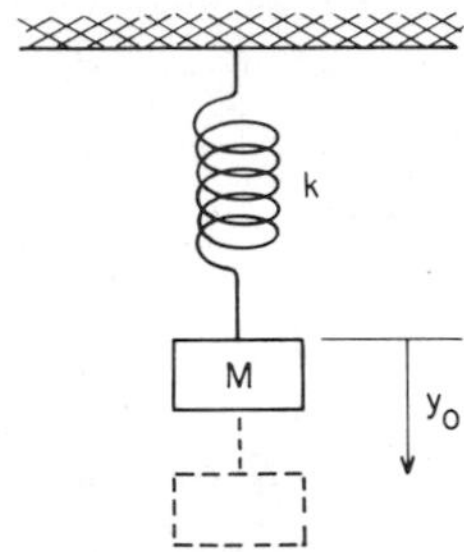

FIGURE 3.11 Displacement (y_o) of a mass M attached to a spring with a spring constant k. Equations relating the displacement and the spring constant to the potential and kinetic energies within the system can be used to derive the natural frequency of oscillation of the spring-loaded mass.

What is sometimes not so obvious is that if we know the geometry of an object and can measure its natural frequency of vibration, then we can determine its elastic modulus. This was first appreciated by Virgin (1955), who showed that the turgor pressure of plant tissues could be measured by means of the tissue's resonance frequency. More recently, Cosgrove and Green (1981) used the resonance frequency of cucumber seedlings to investigate changes in turgor and growth rate as affected by the rapid suppression of growth when seedlings are exposed to blue light. By the same token, if we know the material properties and geometry of an object, then we can predict its mechanical behavior in dynamic loading. Thus, natural frequencies of vibration give us a very sophisticated method to mechanically test a material, and at the same time they provide a way to predict the behavior of stems and leaves when subjected to wind or water pressure.

The relation between the natural frequency of vibration (f_i) of an elastic solid and the modulus of elasticity can be derived by considering a relatively simple physical system, such as a weight attached to the end of a spring (as shown in fig. 3.11). The task in considering this system is to find a relation between the natural frequency of vibration and what is known as the spring constant, because the spring constant is a measure of the material properties of the spring. We begin by recognizing that if the weight is displaced from its static equilibrium position by a distance y_o, the change in the potential energy of the system $\mathcal{PE}$ is given by the formula

$$\mathcal{PE} = \int_o^{y_0} F\, dy = \int_o^{y_0} ky\, dy = \frac{k\, y_0^2}{2}, \tag{3.32}$$

where k is the proportionality constant whose value we seek, the spring constant. If the vibration of the spring is harmonic in time t with a frequency f_i, then

$$y = y_o \sin (2\pi f_i)t \tag{3.33}$$

and

$$\frac{dy}{dt} = y_o (2\pi f_i) \cos (2\pi f_i)t . \tag{3.34}$$

Thus, the maximum kinetic energy $\mathcal{KE}$ of the weight with mass M is given by the formula

$$\mathcal{KE} = M (y_o^2) \left(\frac{2\pi f_i}{2}\right)^2 . \tag{3.35}$$

The foregoing equations indicate that the kinetic energy and potential energy are proportional to the amplitude squared; that the kinetic energy is a function of frequency; and that the potential energy is independent of frequency. If we assume that the mass at the end of the spring is massless and recognize that the kinetic energy equals the potential energy, then the frequency of vibration can be solved in terms of the spring constant and the mass of the spring:

$$f_i = \frac{1}{2\pi} \left(\frac{k}{M}\right)^{1/2} , \tag{3.36}$$

where the frequency of vibration has units of cycles per unit time (Hz). Equation (3.36) indicates that the natural frequency increases in proportion to the spring constant and decreases as the mass of the body (spring) increases. It is important to note that the natural frequency of vibration is independent of the amplitude—an assumption that is now verified mathematically.

Even though the vascular bundles running the length of the leaves of the polyanthus narcissus (*Narcissus tazetta*) and the stems of the inflorescences of the Neapolitan cyclamen (*Cyclamen hederifolium*) can mechanically operate as springs, causing these plant organs to condense in length as tissues within them age and lose water, harmonically oscillating springs are rarely relevant to botany, regardless of our possible fascination with the spiral chloroplasts of the alga *Spirogyra*. However, the previous equations can be extended to considering the mechanical behavior of columns and beams and provide a technique for measuring the dynamic modulus of elasticity of stems and leaves and roots from their natural frequencies of vibration.

All elastic beams have both mass and stiffness. When deflected from their

TABLE 3.3 Dimensionless Proportionality Factor λ_i for Untapered and Tapered Beams

		Mode Number *i*				
		$\lambda_{i=1}$	$\lambda_{i=2}$	$\lambda_{i=3}$	$\lambda_{i=4}$	$\lambda_{i=5}$
Untapered beam (clamped free)		$\lambda_1 = 1.8751$	$\lambda_2 = 4.6941$	$\lambda_3 = 7.8548$	$\lambda_4 = 10.996$	$\lambda_5 = 14.137$
Double linearly tapered beam (clamped free)						
$\alpha = 1.0$	$\beta = 1.2$	$\lambda_1 = 1.9279$	$\lambda_2 = 4.7344$	$\lambda_3 = 7.8778$	$\lambda_4 = 11.001$	$\lambda_5 = 14.149$
	1.4	1.9730	4.7689	7.8987	11.027	14.159
	1.6	2.0121	4.7989	7.9171	11.041	14.170
	2.0	2.0773	4.8496	7.9498	11.063	14.192
$\alpha = 1.2$	1.2	2.1331	5.0464	8.3060	11.572	14.853
	1.4	2.1819	5.0830	8.3283	11.589	14.866
	1.6	2.2245	5.1149	8.3493	11.606	14.879
	2.0	2.2953	5.1690	8.3839	11.632	14.899
$\alpha = 1.4$	1.2	2.3242	5.3351	8.7017	12.091	15.502
	1.4	2.3766	5.3738	8.7258	12.112	15.518
	1.6	2.4221	5.4075	8.7481	12.128	15.531
	2.0	2.4979	5.4646	8.7858	12.157	15.553
$\alpha = 1.6$	1.2	2.5041	5.6055	9.0730	12.617	16.109
	1.4	2.5597	5.6462	9.0980	12.598	16.125
	1.6	2.6081	5.6816	9.1219	12.617	16.140
	2.0	2.6887	5.7416	9.1629	12.645	16.165
$\alpha = 2.0$	1.2	2.8372	6.1041	9.7560	13.472	17.222
	1.4	2.8986	6.1484	9.7852	13.494	17.239
	1.6	2.9521	6.1869	9.8112	13.513	17.257
	2.0	3.0413	6.2524	9.8559	13.549	17.283

Note: α = ratio of representative cross-sectional dimension (measured in the plane of bending) at the bottom of the beam to the top of the beam. β = ratio of representative cross-sectional dimension (measured normal to the plane of bending) at the bottom of the beam to the top of the beam.

equilibrium position beams flex, and they alternately store potential energy in elastic bending and release potential energy in the form of transverse energy (see chap. 2 for a discussion of strain energy). Provided a few assumptions are made, a relatively simple mathematical relation can be derived relating the flexural stiffness *EI* of any beam to its natural frequencies of vibration f_i. (There can be more than one frequency of vibration: the fundamental frequency of vibration, denoted f_1, and higher harmonics, denoted $f_2, f_3, \ldots,$ f_n. Each of these frequencies can be mathematically related to *EI* by the appropriate proportionality constants.) These assumptions are (*a*) the beam is composed of a homogeneous, isotropic elastic material; (*b*) the beam has a uniform second moment of area; (*c*) the beam length *l* is much longer than beam radius $r(l{:}r \geq 10)$; (*d*) the beam is not end loaded. Some of these assumptions can be relaxed when dealing with plant organs that are heterogeneous both in their material properties and in their girth, as well as when dealing with organs that are loaded at their tips or along their sides; that is, there exist closed-form mathematical derivations for these conditions, and plant organs can be manipulated in the laboratory so that they conform to most of the assumptions listed above. A primary assumption that cannot be violated is *c,* since short and squat beams do not have easily measured natural frequencies of vibration. Another important consideration in applying dynamic beam theory is the extent to which the materials used to fabricate the beam evince viscoelasticity, since dynamic oscillations within viscoelastic materials are quickly dissipated owing to the viscous component of behavior (see Coleman, Gurtin, and Herrera 1967).

When assumptions *a–d* are valid, the natural frequencies of vibration of a dynamically loaded beam can be calculated from the general formula

$$f_i = \frac{\lambda_i^2}{2\pi\, l^2}\left(\frac{EI}{m}\right)^{1/2} \qquad (i = 1, 2, \ldots, n)\,, \tag{3.37}$$

where *l* is the length of the beam, *m* is the mass density of the beam, and λ_i is a dimensionless proportionality constant that depends on beam taper and the way the beam is allowed to vibrate (values for λ_i for a variety of beam geometries in a variety of loading conditions are available from the literature, e.g., Blevins 1984; some are given in table 3.3 for convenience). The mass density of the beam is the product of the density of the material the beam is made from and its cross-sectional area. Hence, for a cylindrical beam, eq. (3.37) takes the form

$$f_i = \frac{\lambda_i^2}{2\pi\, l^2}\left(\frac{E}{\rho}\right)^{1/2}\left(\frac{\pi\, R^4}{4\pi\, R^2}\right)^{1/2} = \frac{\lambda_i^2}{2\pi\, l^2}\left(\frac{E}{\rho}\right)^{1/2}\left(\frac{R}{2}\right). \tag{3.38}$$

This equation can be rearranged to solve for the dynamic elastic modulus,

$$E = \rho \left(\frac{4\pi\, l^2 f_i}{\lambda_i^2 R}\right)^2 . \tag{3.39}$$

(For an untapered, solid cylindrical beam, $\lambda_i^2 \approx 3.5$; therefore $4\pi/\lambda_i^2 \approx 3.6$.) The dynamic elastic modulus in eq. (3.39) reflects both the tensile and compressional elastic moduli of the material, because a rapidly vibrating beam simultaneously experiences both tensile and compressive loadings. Thus the dynamic elastic modulus is a realistic summation of the elastic moduli measured in tension and compression.

Equation (3.39) is very simple and useful. Provided the natural frequency of vibration and the length, radius, and unit density can be measured, the elastic modulus of any solid cylindrical plug of plant tissue or any solid, untapered cylindrical plant organ can be determined without recourse to tensile or compressional testing devices. All but the frequency of vibration can be measured with a ruler and a scale, and there are a number of relatively inexpensive (as well as very expensive) methods for measuring natural frequencies (see Virgin 1955; Cosgrove and Green 1981; Niklas and Moon 1988).

Equations (3.37) to (3.39) provide insight into the mechanical behavior of beamlike plant organs. For example, the frequency with which a cylindrical stem or leaf or stamen filament vibrates is directly proportional to the square root of its flexural stiffness and inversely proportional to the square of its length. Stiffer materials will have higher natural frequencies of vibrations than less stiff materials; the greater the ratio of length to radius of an organ, the lower its natural frequency of vibration will be. High frequencies of vibration are advantageous for shedding pollen from anther sacs, as well as for shedding air trapped within the boundary layers around leaf laminae. Accordingly, stamen filaments and leaf petioles should be stiff and not too long compared with their girth if their biological functions are dictated by dynamic loadings. By contrast, long, slender stems made of a very stiff material will have high frequencies of vibration that in some circumstances can produce shearing at the attachment sites of leaves. Yet experiments indicate that branching and the production of many leaves of different sizes dampen natural frequencies of vibration, thus reducing the magnitude of shearing and the strains within plants. Damping is a measure of a structure's capacity to absorb vibrational energy. It can be generated within the material of a structure (material damping), by the fluid surrounding a structure (fluid damping), or by the movement of joints within a structure (structural damping). Structural damping is evident in plants like papyrus (*Cyperus papyrus*), which has a cluster of strap-shaped leaves attached at the free end of a very long stem. Leaf fluttering induced by high wind pressures reduces the tensile and compressive stresses produced in the stem.

Equation (3.37) can be modified to accommodate many different beam geometries and loading conditions, such as untapered geometries or beams with a point load at the free end. For example, the natural frequencies of vibration of a hollow, tubular beam are given by the formula

$$f_i = \frac{\lambda_i^2}{8\pi l^2}\left[\frac{E\,(D_o^2 + D_i^2)}{m}\right]^{1/2}, \tag{3.40}$$

where D_o and D_i are the outer and inner diameters of the tubular wall. For slender cantilevers with a point load at the free end, like petioles, the following formula can be used:

$$f_i = \frac{\lambda_i^2}{2\pi l^2}\left[\frac{3EI}{(M + 0.24\,Mp)}\right]^{1/2}, \tag{3.41}$$

where M is the mass of the leaf lamina and Mp is the mass of the petiole.

The natural frequencies of vibration of a resonated beam can also be used to evaluate the shear modulus. As noted earlier, deformations can result from either flexure or shearing or both. Typically, flexural deformations dominate when very long, slender beams are resonated. Shearing may be important when short beams are dynamically loaded, however, as when short, rectangular pieces of gelatin are shaken. In general, the ratio of the deformations resulting from shear to flexure equals iD/l, where i is the mode number of resonance ($i = 1, 2, 3, \ldots, n$), D is beam diameter (or a typical dimension of the cross section), and l is beam length. When $iD/l \approx 1.0$, shear deformations cannot be neglected.

The shear modulus of an isotropic material molded into the shape of a beam can be determined by means of its resonance frequencies (see Blevins 1984) from the formula

$$f_i = \frac{\lambda_i}{2\pi l}\left[\frac{\kappa G}{m}\right]^{1/2}, \tag{3.42a}$$

where κ is the shear coefficient and m is the mass of the beam material. For a circular cylindrical beam composed of an isotropic material with a solid cross section, $\kappa = 6(1 + \nu)/(7 + 6\nu)$. When the mode number equals one, $f_i = 1$ and $\lambda_i = \pi$. Substituting these values into eq. (3.42a) and solving for G yields the formula

$$G = \frac{4ml^2f^2}{\kappa}. \tag{3.42b}$$

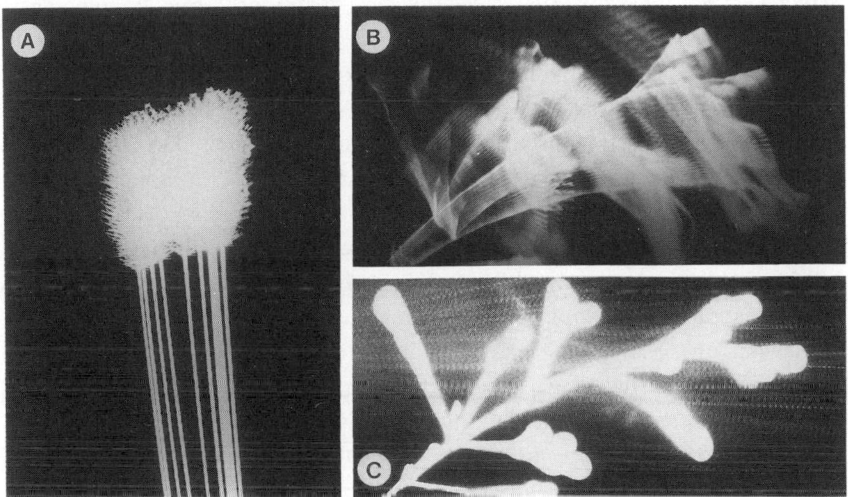

FIGURE 3.12 Harmonic springlike movements of grass infloresences as seen with stroboscopic photography: (A) Simple harmonic motion of an inflorescence of *Setaria*. The equilibrium position of the inflorescence lies at the midpoint of the extreme (left and right) displacements. Notice that the stem of the inflorescence slows down (the sequential images are closer together) as it approaches either of the two extreme displacement positions—the stem is resilient. (B) Complex harmonic motion of an inflorescence of *Agrostis*. Each branching element of the structure has its own frequency of vibration, dictated by its length and stiffness. (C) Pollen grains moving (from left to right) around an oscillating portion of an *Agrostis* inflorescence. Pollen grains are captured by flowers when grains collide with floral surfaces. The sweeping motion of the *Agrostis* inflorescence (B) increases the radius of search for airborne pollen grains.

For example, the Poisson's ratio of many types of parenchyma approaches 0.5. Thus $\kappa \approx 0.9$, and eq. (3.42b) takes the form $G \approx 4.4ml^2f^2$.

The relation between the shear modulus and the torsional frequencies of vibration of isotropic materials is approximated by the formula

$$f_i = \frac{\lambda_i}{2\pi l}\left[\frac{JG}{mI_p}\right]^{1/2}, \tag{3.43a}$$

where J is the torsion constant and I_p is the polar second moment of area. For a circular cylindrical beam, $J = I_p$. When the beam is fixed at one end and free to undergo torsional vibrations along its length, $\lambda = \pi/2$, and eq. (3.43a) reduces to the formula

$$G = 16ml^2f^2 . \tag{3.43b}$$

Equations (3.42) and (3.43) are applicable only to isotropic materials for which $G = E/[2(1 + \nu)]$. These equations are useful for testing the hypothesis that a material is truly isotropic, since they provide an extremely precise relation between G and the natural frequencies of vibration. Additionally, if the elastic modulus of a material suspected of being isotropic is known, then eqs. (3.42) and (3.43) can be used to calculate the Poisson's ratio of the material. Finally, these equations provide some guidance in treating biological members that consist of more than one material, when one predominates in relative volume fraction and is suspected of being isotropic; for example, petioles or stems with a large parenchymatous tissue component.

The response of plant organs to dynamic loadings can be very complex and somewhat unexpected. For example, when the inflorescences of grasses or other wind-pollinated species are dynamically loaded by wind pressure, they can oscillate much like a pendulum even when the airflow around them dramatically diminishes (or is artificially eliminated under laboratory conditions) (fig. 3.12). This oscillation can continue for as much as five minutes and, in fields of grass, appears as wavelike patterns spreading over much of the population. The wave is in large measure an artifact of the different natural frequencies of vibrations that occur in patches of grass stems that share nearly the same geometry and material properties but differ in these respects from neighboring patches. Although these patterns appear to be the immediate result of the movements of air currents, in many cases they are the lingering elastic response of stems striving toward their static equilibrium positions.

Under conditions of moving air, the oscillation of plant organs like the stems of grasses can result from the shedding of vortices of turbulent airflow generated along lateral and leeward surfaces of stems and flowers. The frequency with which these vortices are shed depends on a number of factors, especially the flexural stiffness of stems, the shape of surfaces obstructing airflow, and the wind speed and direction. Such oscillatory behavior can promote pollen capture. For example, the flowers of some grasses, such as *Setaria geniculata,* are clustered into dense inflorescences at the tips of stems. Airborne pollen is trapped aerodynamically in turbulent airflow eddies generated along the leeward surfaces of flowers. When moved from their static equilibrium position by a gust of wind, these inflorescences oscillate and essentially collide with the pollen moving in the airflow behind them. Other grasses, such as *Agrostis hyemalis,* have diffuse inflorescences where individual flowers or only a small number of flowers are produced at the tips of a highly branched inflorescence. When these species were studied in wind tunnels, it was found that the entire inflorescence oscillates as a complex pendulum where each branch had its own natural frequency of vibration. Each branch within the inflorescence was seen to sweep through the air column and trap airborne pollen by direct collision. In a sense, the dynamic behavior of the inflorescences of *Setaria* and *Agrostis* reflects two different "strategies" for utilizing strain energy to capture their airborne pollen.

Another way plants can use dynamic loadings involves the release of seeds and fruits attached to stems. When stems oscillate, they generate significant inertial forces. That is, they change direction and speed at the end of each of their oscillatory cycles. These forces can be used to shear the tissues that hold fruits to stems, so that the fruits are shed from the plant, or they can be used to eject seeds from their fruits much as a catapult ejects a stone. In wind tunnel experiments designed to measure the minimum wind speed required to dislodge the fruits of the goldenrod (*Solidago*), even very high speeds (20 $m \cdot s^{-1}$) failed to shear fruits from their attachment sites. When the same stems with fruits were subjected to gusts of wind (maximum speeds of 5 $m \cdot s^{-1}$), however, fruits were sheared free and dispersed.

PLASTIC DEFORMATIONS OF BEAMS

Since plastic deformations are essentially unrecoverable, and since these deformations use strain energy, it becomes obvious that plastic deformations reduce a structure's ability to return to its original static equilibrium condition after a dynamic load is applied. Hence, when a plastic beam is dynamically loaded it tends not to exhibit natural frequencies of vibration. This is not meant to imply that vibrations produced by the exchange of kinetic and potential energy will not occur. In fact, they do. But these vibrations will undergo material dampening and will show little or no periodicity. Plant materials can deform plastically when excessively loaded, and even tissues exhibiting elastic behavior when fully turgid will undergo plastic deformations as they wilt and their water content diminishes. Thus, plastic deformations in both static and dynamic loadings are biologically significant.

In chapter 2, the elementary theory of plastic behavior was reviewed briefly. At this juncture I shall rekindle this topic within the context of beam theory, where geometry is pivotal in determining mechanical behavior.

From the stress-strain diagram of a material, the moment and the deformation limits for any stage in bending can be predicted for any beam. The initial yielding of the extreme material fibers in each transection through the beam occurs when the maximum stress σ_{max} reaches the yield point f_y of the material. The corresponding bending moment M, therefore, is proportional to the product of the yield point and the second moment of area I. As the moment increases, the strain in the extreme material fibers is increased and a greater portion of the beam is subjected to the yield stress. That is, yielding spreads, and the beam can fail as a result of either compressive crushing or tensile rupture. For plants in which cellulose is the dominant elastic solid in cell walls, compressive failure is by far the more common mode of failure, particularly when the protoplast loses water and the cell wall is crushed under compression. Indeed, the osmotic pressure within tissues can be measured from the bending strength of columnar plant organs or cylindrical plugs of tissue. In an elegant and sadly neglected series of experiments, Lockhart (1959) measured the osmotic pressure of etiolated "Alaska" pea seedlings (*Pisum sativum*) by placing weights at the tips of stem segments and recording the deflections that resulted. His method was based on the fact that the degree of deformation (bending) of tissue equilibrated in a hypertonic solution is a linear function of how far the external osmotic pressure of the solution exceeds the osmotic pressure of the cell contents. By extrapolating the graph of deformation versus external osmotic pressure to zero deformation, Lockhart determined the osmotic pressure of the stem tissue at limiting plasmolysis. He also showed that the osmotic pressure determinations are independent of the magnitude of the applied loadings.

The deformed shape of a beam can be geometrically defined by the change in the angle of flexure ϕ per unit length z of the beam. If the ratio of this angle change ($d\phi/dz$) along the span of the beam is known, then the deflection of the beam at any point can be calculated. This is the method of the Elastica, where the deflection angle α equals the bending moment divided by the flexural rigidity ($\alpha = M/EI$). When plastic deformations occur, however, $\alpha \neq M/EI$. Rather, α is equal to the yield strain ε_y divided by the distance from the neutral axis. The corresponding bending moment is then given by the formula

$$M = F_y\,(2y) = F_e\left(\frac{4y}{3}\right), \tag{3.44}$$

where F_y is the normal force in the zone of yielding along the span of the beam and F_e is the normal force in the elastic zone. This equation provides the relation between the angle of bending and the bending moment from which the plastic deflections of a beam can be calculated.

The relation between the applied load and the deflection will be linear throughout the elastic range of a material's behavior. This explains why Lockhart got the results he did for pea seedlings osmotically stressed up to the point of incipient plasmolysis. Once the bending moment exceeds the elastic limit of the material, however, beams will continue to deflect even with no further increase in the applied load. Thus, plant tissues with thin walls that operate as hydrostats may continue to deform under their own weight once their protoplast reaches a critical low turgor pressure. This is particularly true for actively growing tissues, whose cell wall infrastructures are metabolically plasticized to permit the expansion of the protoplast when at full turgor pressure. Even woody tissues may exhibit plastic deformation when bending stresses exceed the elastic limits of secondary walls.

Clearly, many of the mechanical properties of plant tissues are dictated by cell and tissue water content. Indeed, the effects of water on plant biomechanics will occupy our attention throughout the next chapter.

Four

Plant-Water Relations

One of the nice things about water plants is that they never need watering.
Christopher Lloyd, *The Well-Tempered Garden*

This chapter examines how the flow of water through the plant body is hydraulically achieved and maintained. This topic is treated before the mechanical attributes of cells, tissues, organs, or the plant body are discussed, because the availability and transport of water influence every aspect of the survival, growth, and biomechanics of plants. The survival of terrestrial plants and even their individual organs in large part relates to a dependable water supply. The functional lifetime of a photosynthetic leaf depends on a positive average net photosynthesis and the maintenance of nonlethal temperatures. If the net photosynthesis is negative even for a comparatively short period, then leaves typically senesce, because there is no known mechanism to import nutrients like sugars in mature leaves. Provided light is not a limiting factor in the environment, net photosynthesis depends on the water balance of a photosynthetic organ, since water deprivation limits the capacity to exchange gases between plant tissues and the external atmosphere. When gases are exchanged with the atmosphere, water vapor is lost, cooling plant tissues that might otherwise achieve physiologically deleterious temperatures. Likewise, plant growth and the expansion of cells depend on a supply of water. The influx of water into cells that have metabolically reduced the yield stress of their cell walls causes the walls to expand and accommodate an increase in the volume of the protoplasts they envelop. An adequate supply of water is also necessary to maintain the stiffness of cells with mature (elastic) thin cell walls. Fully differentiated tissues composed of thin-walled cells mechanically operate as hydrostatic devices. The pressures generated by the protoplasts place cell walls in tension and reduce their deformation by externally applied stresses.

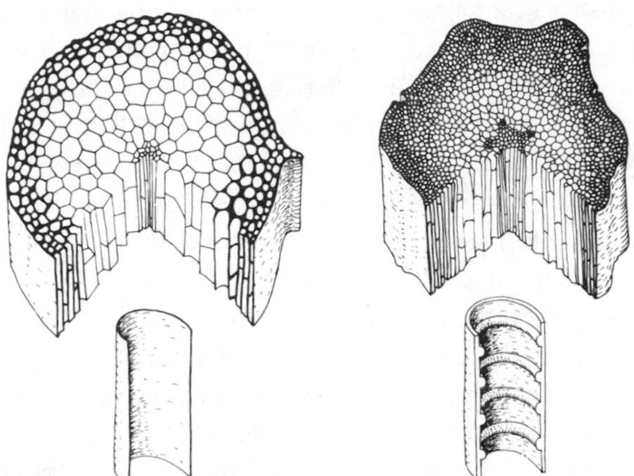

FIGURE 4.1 Morphological and anatomical parallels between nonvascular land plants (bryophytes) and vascular land plants. Among the many bryophytes, some have evolved a specialized tissue system for conducting water and cell sap. On the left a section from the axis of a moss gametophyte is shown, illustrating the centrally positioned water-conducting tissue composed of cells called hydroids (a section of a single hydroid is shown beneath the axial segment). Hydroids lack any internal wall thickenings. On the right a segment from the vertical axes of *Psilotum* is diagramed, showing the centrally positioned vascular strand and a segment of a single tracheid. Note that tracheids have internal secondary wall thickenings. Both the moss and *Psilotum* possess relatively thick cell walls in the peripheral tissues of their axes (shown by the relative density of each line drawing).

Terrestrial plants principally rely on water from the substrates on which or in which they grow. Dew or fog that has precipitated on aerial plant organs can be absorbed, providing a supplementary source of water, but the bulk of the water used by most vascular land plants is absorbed from the substrate, which in most cases is soil. Exceptions to this can be found, such as epiphytes (plants that grow upon other plants) and parasitic plants that are embedded within the hydrated tissues of their hosts. For example, the twining epiphyte *Dischida rafflesiana* produces pitcher-shaped leaves with very thick cuticles. The cavities of these cistern leaves are filled with highly ramified adventitious roots that absorb trapped water, mostly the condensed water vapor lost through transpiration. Indeed, the inner surface of a typical leaf has over twice as many stomata per square unit area as the outer surface. Thus little of the water absorbed by adventitious roots attached to the trees *Dischida rafflesiana* plants grow on is lost.

The vast majority of land-dwelling vascular plants absorb water from their substrates by means of a root system, and hydraulic continuity must be maintained between the root tissues and aerial photosynthetic tissues that constantly lose water when carbon dioxide is absorbed from the atmosphere. This continuity is achieved in large part by means of the xylem, which provides a continuous system of dead cellular conduits that offer comparatively little resistance to the flow of water because they lack living protoplasm. Xylem tissue is not essential to the survival of submerged aquatic plants, of course, since these organisms are continuously bathed in water. (The inessentiality of xylem is shown by the apparent reduction in its volume in many aquatic plants whose nearest relatives are terrestrial organisms possessing substantially greater volumes of this tissue.) Nor is xylem tissue essential to the survival of all plants on land, since the mosses and liverworts lack it. But evolutionary convergence among plant groups suggests the importance of producing a low-resistance pathway for water flow in plants occupying terrestrial habitats. For example, functional analogues to tracheids, called hydroids, are found in the mosses, which are remarkably successful in many terrestrial habitats (fig. 4.1). Hydroids are dead at maturity and transport water and dissolved nutrients. Additionally, relatively small nonvascular land plants like mosses rely on the thin layer of water that envelops their external surfaces. This outer blanket of water is drawn upward, much the way water flows over the surface of a wick, as moisture is lost by evaporation. The adhesion of water molecules to the external cell walls of mosses, as well as the cohesion among water molecules, provides for a hydraulic continuity with the soil water. In some mosses this external transport mechanism has become highly elaborated. For example, the outer cortical cells of the moss *Sphagnum* develop into large, thin-walled empty cells when mature. The empty cells frequently possess inner wall thickenings and are perforated, so water can pass through and over their cell walls. Similar cells develop on the leaflike phyllids. In 1873 Julius von Sachs commented: "These colourless cells both in the leaves and in the cortex . . . serve as a capillary apparatus to the plant to draw up the water of the bogs in which it lives and convey it to its upper parts; hence it is that plants of *Sphagnum,* which are continually growing taller, are filled with water like sponges up to their summits even though their beds are raised high above ground" (translated from the German in Goebel 1887, 183).

The conservation of water once it has been absorbed by land-dwelling vascular plants is largely achieved by means of the cuticle and stomata. When wetted, the cuticle is moderately permeable to water, as well as other substances, and permits the absorption of dew and fog by leaves, stems, and aerial roots. Nonetheless, the hydrophobic properties of the cuticle confer a

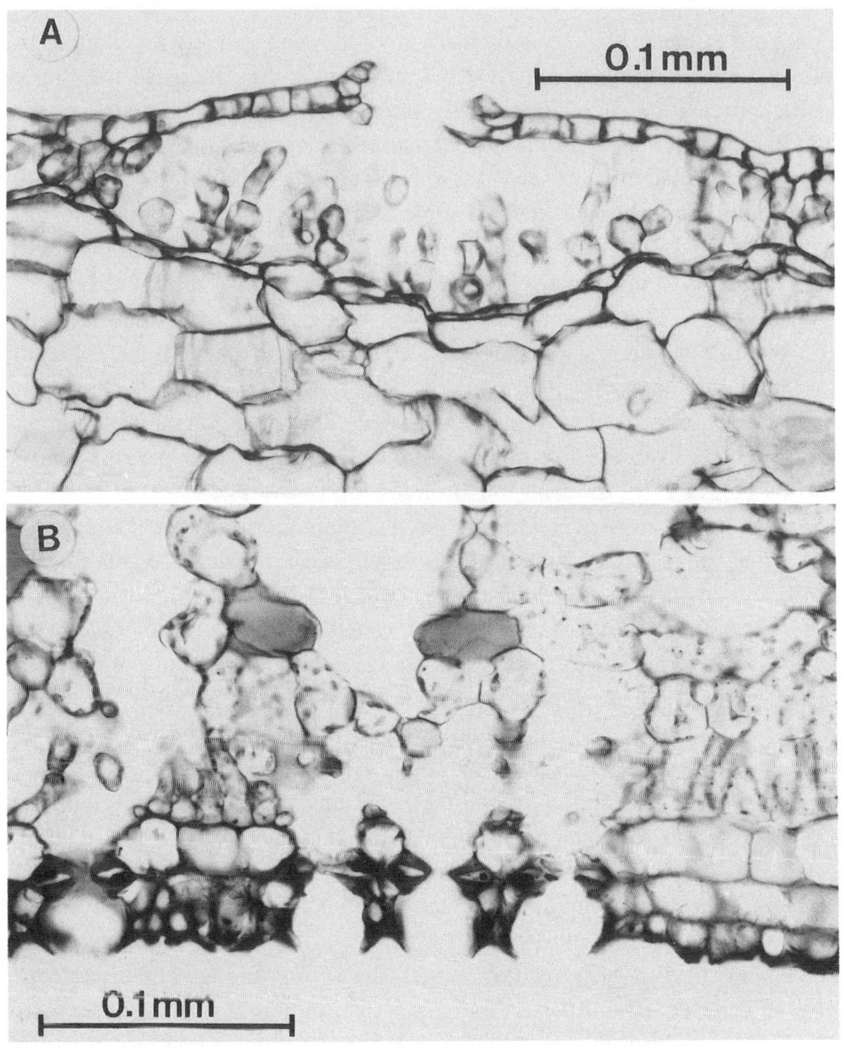

FIGURE 4.2 Comparison between the pore on the gametophyte of *Marchantia,* a bryophyte (A), and the stomata on the sporophyte of a vascular plant (B). Both pores and stomata provide openings to internal chambers within the plant body, through which gases can be exchanged with the external atmosphere. The pore of the bryophyte is flanked by cells that do not significantly alter the pore diameter as water is either gained or lost by their protoplasts. The guard cells that flank the stomatal opening alter the diameter of the opening as cell water content changes.

substantial resistance to the rapid loss of water. These properties depend on the chemical composition of the cuticle and to a more limited extent on its thickness, both of which can vary among plants and even on the same plant body. Thus subterranean roots typically have thin and sparsely distributed cuticles, whereas organs like leaves that are high in the canopy of trees have thicker cuticles that differ chemically from their more shaded counterparts. Since the presence of a cuticle or externally secreted mucilage limits the surface area over which water and atmospheric gases can be absorbed, virtually every land plant group has evolved some form of perforation that permits gas exchange (fig. 4.2). Many thalloid liverworts, for example, *Marchantia,* have pores that lead to invaginated chambers lined with highly specialized photosynthetic cells. These chambers represent topographically internalized surfaces that retain water vapor and so reduce the rate and extent of water loss to the external atmosphere. The pore diameters of these chambers, however, cannot be regulated to any significant degree. Thus, regardless of the rate or extent of water loss, the diffusion of water vapor through these pores is persistent. By contrast, the stomata found on the sporophytes of some nonvascular land plants (e.g., mosses) and most vascular plant species are highly specialized in that the cells that flank these perforations can regulate the pore diameter leading to the plant's surface. Each stomatal pore (opening) within the epidermis is flanked by specialized cells, called guard cells, whose geometry and relative size are sensitive to the water status of the plant and can mechanically deform to close the pore they flank, regulating further water loss when tissues become seriously dehydrated. Stomata are not evenly distributed over the surfaces of stems and leaves, and their frequency distribution and location can vary with the local environmental conditions attending the development of plant organs as well as from one plant species to another.

Although water is the most common material on the earth's surface, it is not available to the same degree to all plants, in all habitats, at all times. Accordingly, its availability is most probably the single factor most limiting to the growth and survival of terrestrial plants. The spatial and temporal heterogeneity of water access influences the global distribution of plant species, individual plant survival, the mechanical rigidity of tissues, and the rate at which individual cells grow. Thus the importance of water to plant life cannot be overemphasized, nor can we ignore the physical properties and physiological roles of water in our treatment of plant biomechanics.

In this chapter we will return to a number of concepts such as water potential (first mentioned in chap. 1). We will also introduce new concepts like osmotic potential and turgor pressure, which are essential to understanding how and when water molecules will move from one compartment (cell, tissue, or organ) within the plant body to another. A gradient in water concentration exists within the plant body, and how much this gradient is attenuated within the volume of the plant body depends on the availability of water to roots (or other absorptive organs or cells) and the rate of water loss from aerial photosynthetic tissues. To some extent plants can regulate or at least partially define their water gradients by metabolically regulating the amounts of solutes dissolved in their protoplasm. High concentrations of solutes decrease the local concentration of water molecules and provide the driving force for the movement of water molecules into compartments with less water. In turn, the relative quantity of water within each compartment of the plant body influences hydrostatic pressures in cells, tissues, and organs. As mentioned previously, these pressures provide a mechanism that regulates the stiffness of thin-walled cells and tissue systems within organs.

Let us begin our review of plant-water relations by examining some of the physical properties that make water unique. This will provide a context for understanding how solutes influence some of water's properties, such as vapor pressure, which in turn leads to the concepts of water potential and turgor pressure. Then follows a treatment of the movement of water from the soil into the root systems of plants. Toward the end of this chapter we will take a whole plant perspective in terms of the hydraulic continuity of water flow through the xylem tissue of the plant body by examining cohesion theory, which attempts to describe the ascent of xylem water through terrestrial plants. As its name implies, the cohesion theory draws on the cohesive properties of water molecules to explain how water is transported. Additionally, it relies heavily on the physical analogies between tracheary elements (tracheids and vessel members) and capillary tubes—long hollow tubes with very narrow bores that can draw water upward against the force of gravity by virtue of the adhesion of water molecules to the inner surfaces of the tubes and the cohesive forces among neighboring water molecules. The concepts of xylem tissue efficiency and safety will also be examined. Efficiency is treated in terms of the capacity of the xylem tissue to supply adequate amounts of water to leaves, while safety relates to the impairment of water flow when columns of water within the xylem tissue break under the high tensile stresses resulting from rapid translocation.

Throughout this chapter, we will be discussing pressure, so it is appropriate to note the units in which pressure is currently expressed (see table 2.1). In the past, vapor pressure was typically expressed in millimeters (mm) of mercury or millibars (mbar), while atmospheric pressure was expressed in bars (1 bar = 0.987 atmospheres). With the advent of the Système International, the primary pressure unit is now the pascal, symbolized by Pa. One bar equals 10^5 Pa, or 100 kPa, or 0.1 MPa. Thus, for example, 1.0 MPa equals 10 bars.

Accordingly, vapor pressure is now expressed in terms of kPa, while atmospheric pressure is now expressed by MPa, as are osmotic potential and water potential. For those who want to express pressure in terms of newtons per square meter, the conversion factor is 10^6 $N \cdot MPa^{-1} \cdot m^{-2}$. Thus 0.1 MPa is equal to 10^5 $N \cdot m^{-2}$. Note, therefore, that mechanical stresses and the elastic modulus of tissues can be given in units of MPa and that dimensionless ratios can be constructed between the hydrostatic pressures developed within cell walls and the tensile stresses they generate or the elastic moduli of entire tissues.

Physical Properties of Water

There is little doubt that water has the largest number of anomalous properties of any normally abundant material on earth, including one of the highest specific heats and the highest known heat of vaporization. The specific heat of any substance is the ratio of its thermal capacity to that of water at 15°C. The thermal capacity of any substance is the quantity of heat (expressed in calories) necessary to raise the temperature (in °C) of a unit mass (in grams). This high specific heat explains why large bodies of water stabilize the temperatures of islands and peninsulas. The equitable climate of the Hawaiian archipelago is largely due to the thermal capacity of ocean water. The only known substance with a higher specific heat is liquid ammonia—about 13% above that of water. Because of the high heat of vaporization (540 cal/g at 100°C), the evaporation of water has a cooling effect, whereas condensation can heat surfaces where liquid water accumulates. As mentioned earlier, the cooling caused by evaporation is essential for regulating tissue temperatures, particularly in leaves. During photosynthesis, as gases are exchanged between the plant body and the atmosphere, water vapor is lost and tissues that are exposed to the sun are cooled, also tempering the local atmosphere. (The shade cast by trees is only one advantage to a heat-stressed botanist.) When water is limited and stomata are closed to limit water loss, however, the temperature of leaves exposed to sunlight can increase many degrees and may eventually become lethal.

Water is only slightly ionized (roughly one out of every 56×10^7 molecules is dissociated) and has a very high dielectric constant, which accounts in part for its being called the universal solvent—a vital capacity that lets roots absorb minerals from the soil. Water has a high surface tension because the large cohesive forces among neighboring molecules are unbalanced at the water-air interface, where water molecules are attracted into the body of the liquid phase. Thus considerable energy has to be expended to draw water molecules out onto the surface of the liquid phase. Indeed, the energy per unit of new area *is* surface tension, which can be defined in terms of the force per unit length acting normal to the surface of a material. The surface tension developed at the air-water interface at 20°C is $7.28 \times 10^{-2}\ N \cdot m^{-1}$, equivalent to $73\ mJ \cdot m^{-2}$. The cohesive force of water, which causes its high surface tension, also provides for a relatively high tensile strength, essential for the ascent of water over the surfaces of mosses and within the conducting tissues of vascular land plants. Briggs (1949) reports that the tensile strength of water equals 22.6 MPa, roughly 10% the tensile strength of copper or aluminum. The tensile properties of water are temperature dependent, however, and many plants undergo "cold wilting" at 4°C, this being the temperature at which water achieves its maximum density instead of at the freezing point like most liquids. Since water expands on freezing, ice has a volume about 9% greater than the liquid water it is made from. Thus ice floats. If it did not, bodies of water as large as the Arctic Ocean would be completely filled with ice, with disastrous ecological and climatological consequences.

Other physical properties of water influence the survival of plants, particularly aquatic species, and may have played an important role in plants' colonization of the terrestrial habitat. Water is not transparent to the spectrum of photosynthetically available radiation (400–700 nm), symbolized by PAR. Although it appears colorless, water is actually a blue liquid that absorbs frequencies in the green and red regions of the spectrum; that is, the light absorption of water begins to rise as wavelengths increase above 550 nm and increases significantly in the red end of the spectrum. One beneficial consequence of the absorption spectrum of water is that it provides an excellent heat-absorption filter—a layer of pure water 1 m thick will absorb roughly 35% of the light at wavelengths greater than 680 nm (Kirk 1983, 50). Among its negative consequences are the overall reduction in the intensity of light as it passes through the water column and the selective attenuation of light in the red end of the visible spectrum so that most of the energy available to photosynthesis falls within a narrow window of frequencies (roughly between 450 and 600 nm). The in vivo absorption spectra of chlorophylls and carotenoids in a variety of algae extend the light-harvesting capacities of these plants into this window (Owens 1988, 125 and fig. 1). From an evolutionary perspective, the capacity of even moderately shallow columns of water to significantly attenuate light intensity, particularly the red end of the spectrum, combined with the capacity of particulates suspended in water to scatter and absorb light, may have provided the selective pressure to drive any green alga toward a terrestrial existence once it acquired the capacity to sustain periodic water deprivation. Ironically, once a roothold on land was achieved, the green canopies of taller plants filtered the quality and intensity of light for smaller plants growing in the understory in much the same way the vertical water column did for their algal ancestors.

The Molecular Structure of Water

The physical properties of water can be largely explained in terms of the molecular structure of water and the nature of the different types of electrostatic forces that hold its molecules together in the liquid and solid (ice) phases. The hydrogen and oxygen atoms within a water molecule share electrons, producing covalent bonds. As a result, the O-H bond in water is remarkably strong ($\approx$ 110 kcal/mol). (One mole, symbolized by mol, is the quantity of a substance that contains the Avogadro number of molecules of the substance. The Avogadro number, named in honor of Amedeo Avogadro, who delineated this concept in 1811, is roughly equal to 6.0225×10^{23} particles—atoms or molecules.) By contrast, the intermolecular forces, known as van der Waals forces or London forces, which operate among adjacent water molecules, are relatively weak (about 1 kcal/mol) and are effective only if molecules are very close together. Even in molecules that are on the average electrically neutral, van der Waals forces result owing to the temporary, instantaneous formation of dipoles (molecules that have two opposite electrical charges somewhere along their atomic structure). The presence of a dipole near a neighboring neutral molecule can spontaneously induce the formation of a temporary dipole, resulting in an instant but temporary attraction between the two molecules. Although van der Waals forces are weak, they plan an important role in accounting for physical properties of water (boiling point, surface tension, and heat of vaporization).

Hydrogen Bonding

The molecular structure of water, and hence many of its physical properties, depends on hydrogen bonding. Hydrogen-bonds have a binding force between 1.3 and 4.5 kcal/mol. They are stronger than van der Waals forces but considerably weaker than covalent bonds. Like van der Waals forces, hydrogen bonding develops as a consequence of the dipoles that form within substances like water. Hydrogen bonding results from the relatively weak electrostatic attraction that develops between partially charged (negative) oxygen and partially charged (positive) hydrogen atoms of neighboring molecules. The length of a hydrogen bond is about 0.177 nm. Although the charges that develop over a water molecule are asymmetric in their distribution, the hydrogen bonding within water can bind water molecules into a highly symmetrical crystalline lattice, particularly evident in the case of ice. The arrangement of water molecules in the lattice structure of ice has unusually wide intermolecular spacings, accounting for the lower density of ice than of liquid water. This is why ice floats. Surprisingly, when ice melts, only about 13% to 15% of its hydrogen bonds break, and only about 8% of the water molecules are free to move appreciably within the lattice. Nonetheless, as a consequence of the movement of water molecules, the low-density lattice structure of ice collapses, which accounts for the increase in density observed when the temperature of water rises from 0°C to 4°C. As the temperature rises more, the number of broken hydrogen bonds increases, further reducing the symmetry in the arrangement of water molecules and increasing the volume of liquid water as the kinetic energy within the system increases. Although there is no generally accepted picture of the molecular configuration of liquid water, most hypotheses view it as a three-dimensional infrastructure of hydrogen-bonded molecules with a statistical tendency for a tetrahedral arrangement. As much as 70% of the hydrogen bonding found in ice is retained in liquid water at 100°C. (Two hydrogen bonds must be broken for every water molecule that evaporates.) Thus, very large amounts of energy have to be applied to liquid water to break hydrogen bonding, which accounts for the high boiling point of water. If hydrogen bonding were not so extensive in water, then water would not exist as a liquid at room temperature. For its molecular weight, water should be a gas at room temperature—methane, which has the same molecular weight as water, boils at −161°C (in large part because it lacks any hydrogen bonding).

Hydrogen bonding is also responsible for water's unusually high viscosity and surface tension, because hydrogen bonding substantially increases the capacity of molecules to resist deformation (rearrangement). The viscosity of water can be increased by adding polar solutes, to which water molecules bind through hydrogen bonding, thereby increasing the structure of the system (the viscosity of a mixture of water and ethyl alcohol at 0°C is four times that of either water or ethyl alcohol alone). The addition of larger molecular species, such as proteins, also affects the structure of water solutions by binding water to the surfaces of dissolved molecules and lowering the free energy within the system. Indeed, the vapor pressure of pure water (0.61 kPa at 0°C and 101.3 kPa at 100°C) can be decreased by diluting it with solutes.

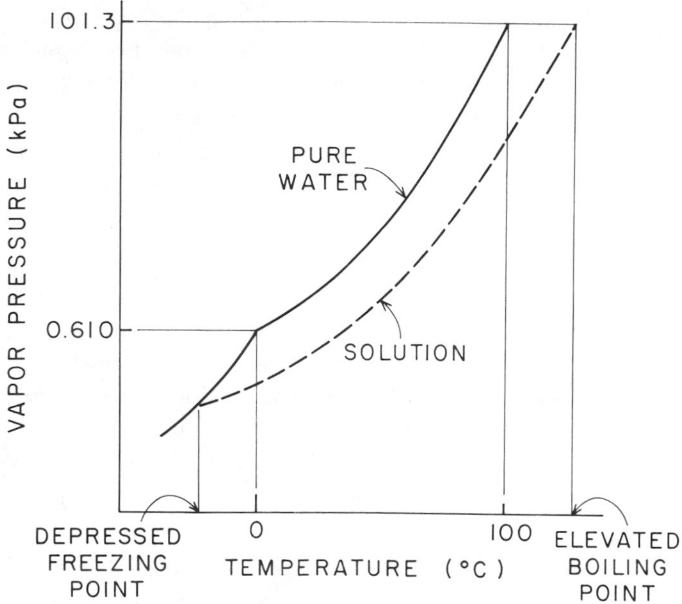

FIGURE 4.3 Vapor pressure of pure water and of a solution of water plotted as a function of temperature. Adding solutes to pure water depresses the freezing point of the solution compared with that of pure water. Adding solutes to pure water elevates its boiling point.

The capacity of small and large molecules to bind with water molecules is very important, since such binding effectively decreases the concentration of water molecules. In terms of plant-water relations, this decrease in the effective concentration of water establishes water gradients within tissues, organs, and the entire plant body. As we will see, water molecules move along gradients of decreasing water concentration. To understand this effect, we need to explore the consequences of solutes for the vapor pressure of water.

Vapor Pressure and Raoult's Law

The relation between the vapor pressure (e^o) of pure water and that of a solution of water (e) is given by Raoult's law, which makes the general statement that the vapor pressure of solvent vapor in equilibrium with a dilute solution is proportional to the mole fraction of solvent in the solution. Raoult's law is mathematically expressed by the formula

$$e = e^o \left(\frac{n_w}{n_w + n_s}\right), \tag{4.1}$$

where n_w is the number of moles of solvent (water) and n_s is the number of moles of solute. Although eq. (4.1) can be used only for very dilute molal solutions, it clearly indicates that the vapor pressure of water vapor in solution is proportional to the mole fraction of water in solution.

The presence of solutes in water raises its boiling point and lowers its freezing point (fig. 4.3). The effect of solutes on the boiling point of water solutions can be readily seen from eq. (4.1). Water boils when its vapor pressure is raised to that of the atmosphere. The presence of solutes lowers the vapor pressure of water. Thus, water containing solutes must be heated to a higher temperature than pure water to produce the necessary increase in vapor pressure to cause boiling. Adding solutes depresses the freezing point because it also decreases the solution's vapor pressure, thereby lowering the equilibrium temperature at which the liquid and vapor phases of water can coexist.

The influence of solutes on the freezing point of water has very meaningful biological consequences. As early as 1912, N. A. Maximov suggested that the primary cause of freezing injury in plants was the disruption of the plasma membrane by the formation of ice crystals within the cytoplasm. From Raoult's law, we can see that the metabolic accumulation of solutes within plant protoplasts provides a physiological mechanism to depress the temperature at which ice crystals form. It is not surprising, therefore, that the solute concentrations within plant cells typically increase as ambient temperatures gradually drop with the advent of winter. Solute concentrations within a cell can be increased by metabolically controlled processes involving the catalysis of large organic polymers or by processes that are not metabolically controlled. The latter involve the passive diffusion of water molecules. If the rate of cooling is relatively slow, water molecules can move across the plasma membrane in response to the external formation of ice crystals (which have a lower effective concentration of water because of the crystal lattice). The movement of water across the plasma membrane increases the solute concentration of the protoplasm (Steponkus 1984). The intracellular formation of ice requires seeding—the spontaneous aggregation of water molecules to form ice nuclei. A reduction in the concentration of water molecules within the cell reduces the likelihood that seeding will occur. By contrast, when the rate of cooling is very rapid, ice crystals may develop within the protoplasm and can mechanically perforate the plasma membrane or induce lesions within it, leading to the eventual death of the cell.

The efflux of water from a protoplast as a consequence of ice crystal formation outside the cell may seem counterintuitive, but the movement of water molecules across a permeable membrane is dictated by what is called a chemical potential. For thermodynamic reasons that will become readily apparent in the next section, water molecules will always move along gradients of decreasing concentration. When ice forms within water that contains solutes, the effective concentration of liquid water is decreased. Thus the water concentration within a protoplast surrounded by a mixture of liquid water and ice can be higher than the external concentration of liquid water, and water molecules move out of the protoplast. One consequence is that the buds of plants examined in winter are frequently coated with a layer of ice formed by water withdrawn from the bud tissues. This phenomenon is called extraorgan freezing. Water efflux from cells and the formation of ice inside gas-filled chambers within a tissue can also occur and constitute intratissue freezing. Both can be beneficial, since the high intracellular solute concentrations resulting from the efflux of water can reduce the temperature at which intracellular freezing injury will occur. There are also negative consequences to extracellular freezing, however. During the evacuation of water from cells, the surface area of the plasma membrane is reduced. When cells thaw and water is reabsorbed, the failure of the plasma membrane synthesis to keep pace with cell expansion can result in the death of protoplasts.

The Chemical Potential of Water

The efflux of water from a cell in response to the formation of ice outside the plasma membrane has been described in terms of chemical potential. The concept of chemical potential is very important to our consideration of plant-water relations, since it helps us understand why and when water molecules will move across membranes or from one part of the plant body to another. The chemical potential of any substance is a measure of its capacity to do work, which depends on a variety of factors, the most important from our perspective being the concentration of the substance. As the concentration decreases, the capacity of a substance to do work, and therefore its chemical potential, also decreases. Thus gradients in the chemical potential of water can be established by concentration gradients of solutes within the plant body, as well as by the simple fact that water enters and leaves a plant at different points along the plant axis. The extent to which the chemical potential of water in a solution is decreased by the presence of a solute is given by the formula

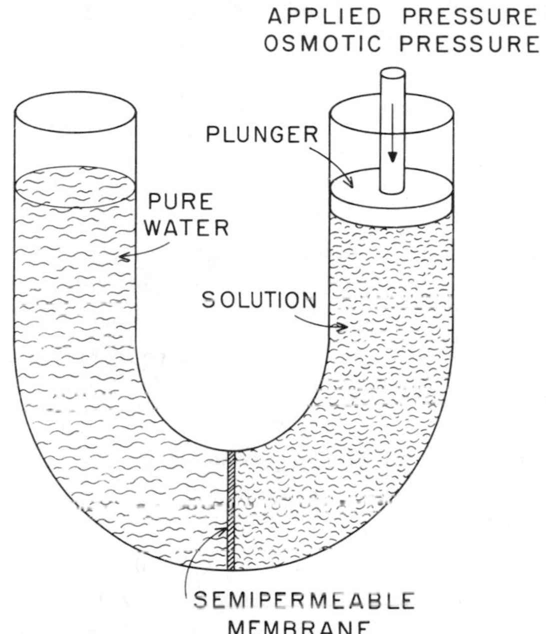

FIGURE 4.4 Osmotic pressure illustrated for a system composed of pure water and a solution of water separated by a membrane permeable to water and impermeable to the solute. The osmotic pressure is the pressure that must be applied to the solution (water and solute) to prevent a net movement of water molecules from the compartment containing pure water into the compartment containing the solution.

$$\mu_w - \mu_w^o = RT \ln \left(\frac{e}{e^o}\right), \tag{4.2}$$

where μ_w is the chemical potential of water in the solution, μ_w^o is the chemical potential of pure water at the same temperature and pressure, R is the gas constant (8.32×10^{-3} liter MPa per degree mol at 273°K), T is temperature (given in °K), and e/e^o is the relative vapor pressure (given in kPa). Equation (4.2) shows that when the vapor pressure of the solution (e) is the same as that of pure free water (e^o), $\ln(e/e^o) = 0$. Thus the difference in the chemical potentials of the solution and pure water ($\mu_w - \mu_w^o$) must be zero. This makes intuitive sense, since when e equals e_o, there is no gradient in water concentration, hence there can be no net movement in water molecules, and the chemical potential of pure water is taken as a standard equal to zero. Also, since the relative vapor pressure of any solution must always be less than that of pure water, the natural logarithm of the relative vapor pressure, $\ln(e/e^o)$, must always be negative. Therefore the chemical potential of any solution of water is less than that of pure water and is always expressed as a negative number.

Osmotic Pressure

From our discussion of the movement of water along gradients of chemical potential, we saw that if pure water is separated from a solution of water by a membrane permeable to water molecules but impermeable to the movement of solutes, then water molecules will move into the solution until the vapor pressures on both sides of the membrane are equilibrated (fig. 4.4). Indeed, this prediction comes directly from Raoult's law (eq. 4.1). The movement of water molecules across membranes is readily apparent when cells lacking cell walls are submerged in pure water. They swell and in some circumstances burst when the tensile strength of their plasma membranes is exceeded by the hydrostatic pressures that develop within them. Since the tensile strength of the primary cell wall (on the order of 10^2 to 10^3 MPa) is significantly greater than that of the plasma membrane, much greater internal pressures are required to burst cells with even very thin primary cell walls (Iraki et al. 1989). Since stresses can be expressed in terms of pressure, the pressure that must be applied to prevent the movement of water molecules across either a biological or an artificial membrane is called the osmotic pressure or osmotic potential, unfortunately often symbolized by π. The relation between π and the solute concentration in solution is given by Van't Hoff's law,

$$\pi = \left(\frac{n_s}{V}\right) RT \ , \tag{4.3}$$

where V is the volume of the water in solution (given in liters) and the fraction n_s/V is the solute concentration, symbolized by c. Thus $\pi = RTc$. As we can see, for equivalent temperatures, the osmotic pressure increases in direct proportion to the solute concentration.

The Van't Hoff equation provides reasonable predictions only when solutions are very dilute and provided nondissociating solutes are involved. When the solute concentration is high or when electrolytic ionization occurs, large deviations in osmotic pressure occur between predicted and observed values. For example, eq. (4.3) predicts that the osmotic pressure of a molal solution of sodium chloride should be 2.27 MPa, when in fact it is almost twice this value: 4.32 MPa. This discrepancy arises because solutes can dissociate in solution, releasing more particles into solution than the solute concentration would have us believe. The Van't Hoff equation also fails to yield satisfactory predictions even when some nondissociating solutes are used, because many types of nondissociating molecules become hydrated—they bind water to them, reducing the effective concentration of water within the solution. For example, a single sucrose molecule can bind six water molecules. Thus the osmotic pressure that must be exerted to prevent water molecules from moving across a membrane is higher than that predicted by Van't Hoff's law.

Clearly, the number of water molecules in a solution does not change as the solute concentration increases. Thus the glib phrase effective concentration is somewhat misleading and must be clarified. In reality, adding solutes to water decreases the free energy of the water molecules in the solution. Free energy is the energy available to do work at any constant temperature and pressure. The change in free energy, symbolized by ΔG, equals the change in the total energy of the system, symbolized by ΔE, plus the change in pressure and the volume of the system, symbolized by $\Delta \mathcal{P} V$, minus the product of the absolute temperature and the change in the entropy of the system, symbolized by ΔS; that is, $\Delta G = \Delta E + \Delta \mathcal{P} V - T\Delta S$. The addition of any molecular species, or the presence of any surface that binds water molecules, reduces the free energy of water. This means that all dissociating solutes, most nondissociating solutes, and all water-binding surfaces (called matrices) reduce the chemical potential of water in solution. Similarly, any change in the pressure or temperature of a water solution will change the chemical potential of water. The influence of solutes, matrices, and pressure on the chemical potential of water can be quantified provided the chemical potential is expressed in terms of units of energy per unit volume, as discussed in the next section.

WATER POTENTIAL

If both sides of eq. (4.2) are divided by the partial molal volume of water, symbolized by V_w (expressed in units of cm^3 per mole), then chemical potential is expressed in terms of the units of energy per unit volume. This is called the water potential, symbolized by ψ_w:

$$\psi_w = \frac{\mu_w - \mu_w^o}{V_w} = \frac{RT \ln (e/e^o)}{V_w} . \tag{4.4}$$

From eq. (4.4), we once again see that the water potential of pure liquid water is zero, while the water potential is lowered by the addition of solutes. Equation (4.4) can be altered to include terms that reflect the influence on water potential of water-binding surfaces or matrices and pressure, as well as solutes. The fundamental relation among these three features is given by the formula

$$\psi_w = \psi_s + \psi_m + \psi_p , \quad (4.5)$$

where ψ_s is the solute potential, ψ_m is the matric potential, and ψ_p is the pressure potential. The solute potential expresses the depression in the free energy of water owing to the consequences of solutes within the protoplast, while the matric potential expresses the effects of cell surface areas, colloids, and capillarity effects on the free energy of water. Both the solute potential and matric potential are expressed as negative numbers. The pressure potential describes the consequences of fluid pressures developed within or externally applied to the system. It usually has a positive value, reflecting the influence of ambient atmospheric pressure, although ψ_p can be negative, as for xylem tissue experiencing rapid transpirational loss of water through the surfaces of leaves. From eq. (4.5) we can see that any solution of water will always have a water potential expressed as a negative number. Since water will always move along gradients of decreasing concentration, it will always flow from a solution with a higher (less negative) water potential to one with a lower (more negative) potential.

Some doubt has been cast on whether the matric potential should be included in the water potential equation (Passioura 1980; Tyree and Karamanos 1980), because there is some ambiguity in its definition; the matric potential includes the effects of colloids and of the microcapillarity of cell walls. How much the matric potential can be ignored in terms of the water potential of a cell or tissue, however, appears to depend on the relative volume fractions of the vacuole and the volume fraction of the cell wall of a single cell or the walls within a tissue. When the volume fraction of vacuoles is very high compared with the volume fraction of cell walls, as in parenchyma, the matric potential can be largely neglected. But in thick-walled tissues the matric potential appears to have a very real effect on measurements of water potential.

From the foregoing, we can see that osmotic pressure should really be viewed as an osmotic potential, and that the osmotic potential and the solute potential are expressions of the same phenomenon. Indeed, for very dilute solutions, osmotic potential π and solute potential are the same thing. For example, when treating the behavior of xylem water (which is a very dilute aqueous solution of organic and mineral solutes), the water potential equation may be written as

$$\psi_w = -RTc + \psi_m + \psi_p , \quad (4.6)$$

where the osmotic potential π is expressed as $-RTc$. The negative sign is a

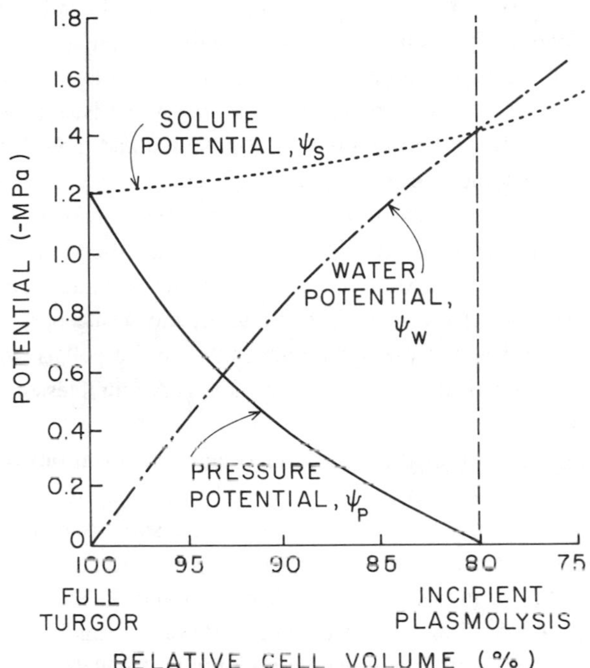

FIGURE 4.5 Changes in the solute potential (ψ_s), pressure potential (ψ_p), and water potential (ψ_w) of a cell with a very thin, elastic cell wall plotted as functions of the relative cell volume. Turgor pressure is equal to ψ_p. The value of ψ_s is always negative; the value of ψ_w is equal to or less than zero. A fully turgid cell (full turgor) has a ψ_w equal to zero. As water exits a cell and as the solute potential becomes more positive, the water potential and the turgor pressure become more negative until incipient plasmolysis occurs ($\psi_w = \psi_s$). (From Kramer 1983.)

convention, since osmotic potential will reduce the water potential. We will have other reasons for modifying the equation for water potential, particularly when we examine the influence of gravity on xylem water ascending to the tops of trees, but for the time being we can use eq. (4.5) to derive a very important physiological concept—turgor pressure.

TURGOR PRESSURE

Turgidity refers to how fully protoplasts within cells are hydrated. In the case of turgid protoplasts, the value of the positive pressure potential equals the sum of the negative solute and matric potentials, and the cell water potential equals zero. As the cell water potential becomes more negative, the turgidity of the cell decreases: the cell becomes flaccid and ultimately plasmolyzes. Since matric potentials rarely change significantly for a mature cell or tissue, the influence of the matric potential on turgidity can be largely neglected. In these circumstances, turgor pressure is equal to ψ_p; that is, turgor pressure is defined as the difference between the water potential and the solute potential, or $\psi_p = \psi_w - \psi_s$. As the cell water potential decreases (becomes increasingly more negative), the turgidity of the cell decreases (fig. 4.5). For example, if the pressure potential of the cell is +3.0 MPa, then the value of the solute potential must equal −3.0 MPa, since the cell water potential of a fully turgid cell is zero ($\psi_w = 0$). Likewise, if the value of the solute potential is *more negative* than the positive pressure potential, then the cell is only partially turgid. In a flaccid cell the pressure potential is very much less than the value of the solute potential.

Turgor pressure is biomechanically important because it profoundly influences the tensile stresses generated within cell walls and the mechanical stiffness of thin-walled cells and thin-walled tissues, such as parenchyma. The influence of turgor pressure on the mechanical parameters of cells and tissues will be dealt with in greater detail in chapters 5 and 6. Note, however, that when the protoplasts within thin-walled cells are fully turgid and appressed to the cell walls, they exert a hydrostatic pressure that places the cell walls in tension. The primary cell walls of plants are uniquely capable of dealing with extremely high tensile stresses (chap. 5). Significantly, the apparent stiffness of cells and tissues whose cell walls are placed in tension is greater than the stiffness of these structures when their cell walls are not in tension. The inflated protoplasts of fully turgid, thin-walled cells reduce the freedom of cell walls to move (and buckle) under a compressive stress. By contrast, changes in turgor pressure tend to have little effect on the stiffness of thick-walled cells and tissues. The volume of thick-walled cells changes little (2% to 5%) with a decrease in the turgor pressure of their living protoplasts. Indeed, as water is lost from a thick-walled tissue, the elastic modulus of the tissue can increase as a result of dehydration of the cellulosic components in the cell wall as well as densification of the volume fraction of cell wall material in the tissue. (The elastic modulus of dry cellulose is higher than that of wet cellulose.) This property can be seen particularly well in leaves of wheat and *Rhododendron,* which get much stiffer as they dehydrate. These leaves can store as much as 30% of their total tissue water content in the cell walls or in intercellular spaces within the cell wall infrastructure of their leaf tissues. This extracellular water may also provide a reservoir for cells to draw on when the soil water content becomes low.

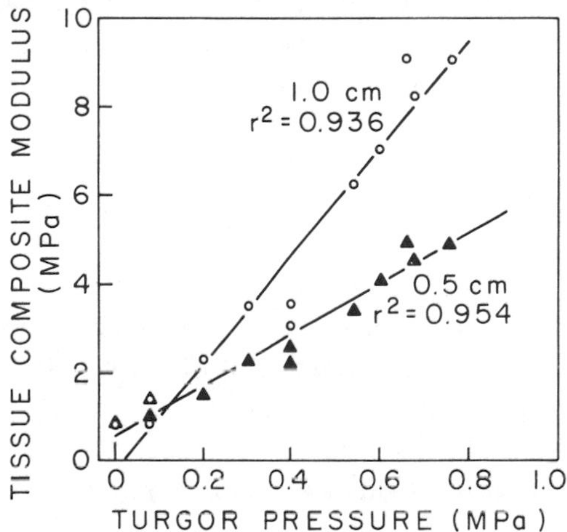

FIGURE 4.6 Relation between the tissue composite modulus (the elastic modulus of a tissue) and the turgor pressure of two plugs of parenchyma (differing in cross-sectional diameter, 0.5 and 1.0 cm) isolated from potato tubers. As the turgor pressure of the tissue increases, the elastic modulus of the tissue samples increases linearly (note the high correlations, $r^2 \geq 0.9$).

The influence of turgor pressure on the mechanical properties of thin walled plant tissues is illustrated in figure 4.6, where the apparent tissue elastic modulus or tissue composite modulus (expressed in units of MPa) of parenchyma from potato tubers is plotted as a function of the turgor pressure. Cylindrical plugs of parenchyma, differing in diameter (either 0.5 or 1.0 cm), were removed from potato tubers and submerged in varying concentrations of mannitol to alter tissue turgor pressure. Two important points are illustrated by the relation between the apparent elastic modulus of the tissue samples and tissue turgor pressure. First, the apparent elastic modulus of the tissue increases linearly as a function of increasing turgor pressure, which is explicable in terms of the hydrostatic (turgor) effects of the tissue's protoplasts on the stiffness of the tissue. Second, for comparable turgor pressures above roughly 0.2 MPa, the apparent elastic modulus is higher for plugs of tissue with a greater diameter, since the number of cells influences how much cell walls within the tissue are free to bend and deform when subjected to a mechanical stress. Up to a point, the more cells, the less relatively thin-walled cell walls are free to mechanically deform. These data reveal an often unappreciated feature of plant tissues—unlike most engineered materials, in plant tissues elastic parameters are not necessarily constant but can be altered by how fully tissues are hydrated and by the absolute dimensions of the tissue.

Turgor pressure is also extremely important with respect to cell enlargement. Actively growing cells can metabolically decrease the yield stress of their cell walls. As water moves into the protoplast and the protoplast expands, the internal hydrostatic pressure results in biaxial tensile stresses within the cell wall that can deform the wall, permitting an increase in cell volume. If the cell wall is not metabolically altered and remains elastic, then reducing the tensile stresses developed within the cell wall will restore the original dimensions of the cell. A permanent enlargement of growing cells may be effected either by the synthesis of new cell wall material or by permanent plastic deformations of the original cell wall material with concomitant thinning of the wall (see chap. 5). Accordingly, the extent to which cell enlargement progresses and is maintained is determined by the maintenance of positive turgor pressures and by how much the physical properties of the cell wall are metabolically altered.

The elastic properties of plant cell walls are critical to cell expansion attending growth, as are a number of features of water movement. Accordingly, we must understand the relation between the change in the internal pressure of a cell $\Delta\mathcal{P}$ and the change in the volume of a cell ΔV as water flows into it. This relation can be quantified by the volumetric elastic modulus, usually symbolized by ε; that is, $\varepsilon = \Delta\mathcal{P}/(\Delta V/V_o)$, where V_o is the original volume of the cell. (Recall from chap. 2 that $\Delta V/V_o$ is cubical dilatation and that the bulk modulus K of a material equals $\Delta\mathcal{P}/[\Delta V/V_o]$. Accordingly, the volumetric elastic modulus and the bulk modulus are identical.) The volumetric elastic modulus increases as a function of cell wall thickness, and empirically determined values of ε have a significant range, 1 to 50 MPa, indicating that cells can change in volume by as little as 0.2% to as much as 10% for each 0.1 MPa change in internal pressure. A useful association can be derived relating the volumetric elastic modulus to the initial internal osmotic pressure and the initial volume and surface area of cells in terms of a time constant t_e for the resulting volume change: $t_e = V_o/[AL_w(\varepsilon + \pi_i)]$. The time constant is the time required to complete all but $1/e$ of the decrease or increase in volume (about 37%), while the half-time required to change from the initial to final volume equals the product of ln 2 and t_e. For any given value of initial internal osmotic pressure, the relations among V_o, A, L_w, and ε mean that the change in volume is slow for a cell with a high volumetric elastic modulus resulting from a very rigid cell wall. Also, the change in cell volume will be slow when the cell has a large ratio of surface area to volume (S/V), a high permeability to water flow (L_w), or both. The formula for the time constant can be significantly simplified because in most cases the value of π_i is very small (on the order of 0.5% to 2%) compared with the volumetric elastic modulus. Thus $t_e = V/[AL_w(\varepsilon + \pi_i)] \approx V/AL_w\varepsilon$. Consider a spherical cell with a surface area-to-volume ratio equal to 1×10^{-2} μm and a volumetric elastic modulus of 20 MPa. If $L_w = 10^{-12}$ m·s^{-1}·Pa^{-1}, then $t_e = (100 \times 10^{-6}$ m)/[(10^{-12} m·s^{-1}·Pa^{-1})(20×10^6 Pa)] = 5 s, illustrating that the change in volume of such a cell can be very rapid. Clearly, the magnitude of L_w or ε is significant in calculating the time constant. For example, if a value of 10^{-13}m·s^{-1}·Pa^{-1} had been taken for L_w, then the time constant would increase by an order of magnitude (50 s). Nonetheless, it is clear that time constants on the order of a few seconds or minutes are very important to a variety of physiological processes, such as stomatal closure and opening.

Small gradients of water potential may be sufficient to enlarge cell volume. These differences can easily be calculated from Fick's first law (introduced in chap. 1) provided we know the permeability of water flow at the cellular level: the volume flux density of water J_w equals the product of the permeability of water flow L_w and the difference in the water potential between the outside of the cell and the inside $(\psi_{wo} - \psi_{wi})$: $J_w = L_w \Delta\psi_w$ (see eq. 1.7). Solving for the difference in water potential yields $\Delta\psi_w = J_w/L_w$. Consider a spherical cell (with unit radius 5×10^{-4} m), submerged in a dilute aqueous solution of pond water, which grows 1% in total volume every twenty-four hours (8.64×10^{-4} s). For a cellular system such as this one, we will take L_w to be on the order of 10^{-12} m·s^{-1}·Pa^{-1}. Some additional calculations reveal that the surface area A and volume V of the cell is 3.41×10^{-6} m^2 and 5.24×10^{-10} m^3, respectively. Since

$$(\psi_{wo} - \psi_{wi}) = \frac{J_w}{L_w} = \frac{\frac{1}{A}\left(\frac{dV}{dt}\right)}{L_w} = \frac{\frac{1}{(3.41 \times 10^{-6}\ m^2)}(1\%)\left[\frac{(5.24 \times 10^{-10}\ m^3}{(8.64 \times 10^{-4}\ s)}\right]}{(10^{-12}\ \mathrm{m \cdot s^{-1} \cdot Pa^{-1}})},$$

the internal water potential (ψ_{wi}) required to sustain the influx of water necessary to accommodate a 1% increase in cell volume per day is roughly 19 Pa less than the external water potential (ψ_{wo}). Assuming that the external water potential of a dilute aqueous solution of pond water is roughly −7,000 Pa, the internal water potential of the cell need only be 0.27% less than the external water potential. One must clearly appreciate, however, that the difference between the external and internal water potentials is *not* the driving force per se for cell volume enlargement. The driving force is the solute potential of the protoplasm, since it is the solute potential within the cell that causes water to move into the cell protoplast, which in turn exerts hydrostatic pressure on expandable cell walls.

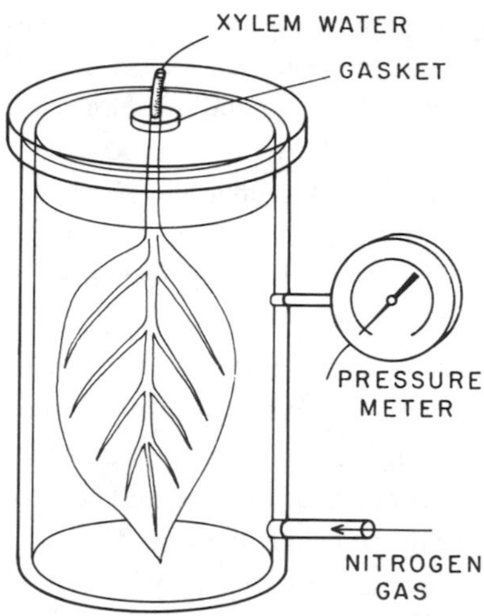

FIGURE 4.7 Diagram of a leaf inside a pressure bomb used to measure its water potential. A gas (nitrogen) is fed into the chamber to increase the internal pressure. At some applied pressure, xylem water is exuded from the exposed cut end of the leaf's petiole. The applied pressure that results in the exudation of xylem water is equal to the water potential of the leaf.

WATER POTENTIAL MEASUREMENTS AND THE VOLUMETRIC ELASTIC MODULUS OF TISSUES

The level of hydration of plant tissues and organs, such as leaves, influences many physiological processes. Therefore plant physiologists have long sought convenient and dependable methods for measuring the water potential of plant organs. For leaves, one of the most convenient and simple ways is the pressure chamber technique, which capitalizes on the fact that when a leaf is cut off a plant, the xylem water is suddenly at atmospheric pressure while the leaf as a whole is at some negative water potential. Thus the xylem water rapidly moves into the cells of the leaf. The water potential of the leaf can be determined by measuring the magnitude of external pressure (uniformly applied over the surface of the leaf) required to drive water from living cells into the xylem tissue. Pressure is applied by placing the entire leaf in a pressure chamber so that the cut tip of the petiole is exposed to view (about 5 mm) (fig. 4.7). (It is essential that water loss through transpiration be reduced, because this will lower the water potential of the leaf. Therefore pressure chamber measurements typically are done quickly. Also, it is desirable to place the leaf lamina in a plastic bag before cutting it off the plant to reduce transpiration.)

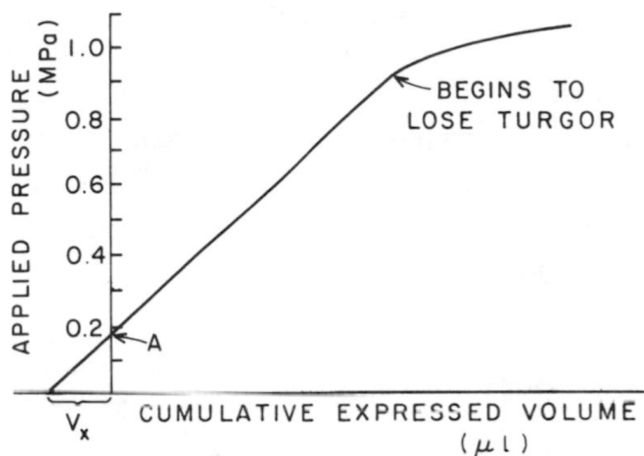

FIGURE 4.8 Pressure-volume curve used to calculate the elastic modulus of a plant organ. The organ is placed in a pressure bomb (see fig. 4.7), and the internal pressure of the bomb is slowly increased. The amount of water (expressed volume of water, abscissa) is recorded as a function of the applied pressure (ordinate). The pressure-volume curve is initially linear but becomes curvilinear when the tissue loses turgor. The slope of the linear portion of the pressure-volume curve is used to compute the elastic modulus of the organ's tissues. See text for further details.

The pressure within the chamber is gradually increased (typically at a rate $\leq$ 0.03 MPa$\cdot s^{-1}$) until the xylem water meniscus surfaces at the cut end of the leaf (this can be seen with the aid of a hand lens or microscope). The water potential of the leaf equals the negative value of the applied pressure within the chamber.

The volumetric elastic modulus of the leaf (or any other organ that can be examined in this way) can be determined from a pressure-volume curve (Melkonian, Wolfe, and Steponkus 1982). As the pressure within the chamber increases, water accumulates at the cut end. The volume of water expressed at the cut end is collected with capillary tubes (or any other method that allows the amount to be determined, by either weight or volume; 1.0 ml of water = 1.0 mg) and recorded as a function of the applied pressure. The applied pressure is then plotted as a function of the expressed water (fig. 4.8). The pressure-volume curve is initially linear and typically occurs for pressures $\leq$ 1.0 MPa, but how long the relationship remains linear depends on the nature of the plant material being examined. Beyond its initial linear portion, the plot becomes hyperbolic when the turgor of tissues is lost. The slope of the linear portion of the pressure-volume curve equals the ratio of change in the applied pressure to the change in the volume of expressed water, $\Delta\mathcal{P}_a/\Delta V_e$. The volumetric elastic modulus ε (which is the bulk modulus, K) can be calculated from the formula

$$\varepsilon = \Delta \mathcal{P}_a (\Delta V_e / V_o)^{-1}, \tag{4.7}$$

where $\mathcal{P}_a$ is the applied pressure, V_e is the corresponding amount of expressed water, and V_o is the total amount of water in the organ at full turgor. Four data points are typically required to ensure a reasonable determination of *E,* but only data collected for turgid tissues can be used; for most leaves and stems, the applied pressures should not be less than 0.8 MPa.

Osmotic Adjustment

Before we explore the way water potential influences the flow of water into and through the plant body, we will briefly treat an effect called osmotic adjustment. Plant physiologists and agronomists have known for many years that the osmotic potential of tissues in some species can be physiologically lowered when plants are water stressed, thereby maintaining cell and tissue turgidity. Removing water from a cell or a tissue can decrease the osmotic potential simply as a result of the reciprocal increase in the concentration of solutes within the protoplast, but in osmotic adjustment there is a *net* increase in the concentration of solutes, as opposed to a passive increase caused by water loss. The capacity for osmotic adjustment provides for cell enlargement and growth under water-potential conditions that would otherwise inhibit growth. It also permits stomata to remain open, allowing photosynthesis to continue. Osmotic adjustment can be viewed as a "strategy" that permits plants to tolerate drought. Nonetheless, the capacity for osmotic adjustment does not persist indefinitely, nor does it allow for the full maintenance of all physiological functions (Paleg and Aspinall 1981). Typically, the solutes involved in osmotic adjustment are inorganic ions (e.g., Na^+, K^+, and Cl^-), carbohydrates, and a variety of organic acids (Ford and Wilson 1981).

When some cells are osmotically stressed, they divert metabolites from the construction of their primary cell walls to solutes needed for osmotic adjustment, reducing the tensile strength of their cell walls. Iraki et al. (1989) grew tobacco cells in a solution of 428 millimolar NaCl to induce severe osmotic stress (−2.3 MPa) and observed a reduction in growth evinced by slower cell expansion and cell volumes that were only one-fifth to one-eighth those of untreated cells. The tensile strength of the primary cell walls of stressed cells was reduced two- to fivefold. This reduction in tensile strength was highly correlated with a reduction in the proportion of crystalline cellulose in the primary cell wall, as was the amount of the amino acid hydroxyproline in the insoluble protein fraction. Hydroxyproline is an important component

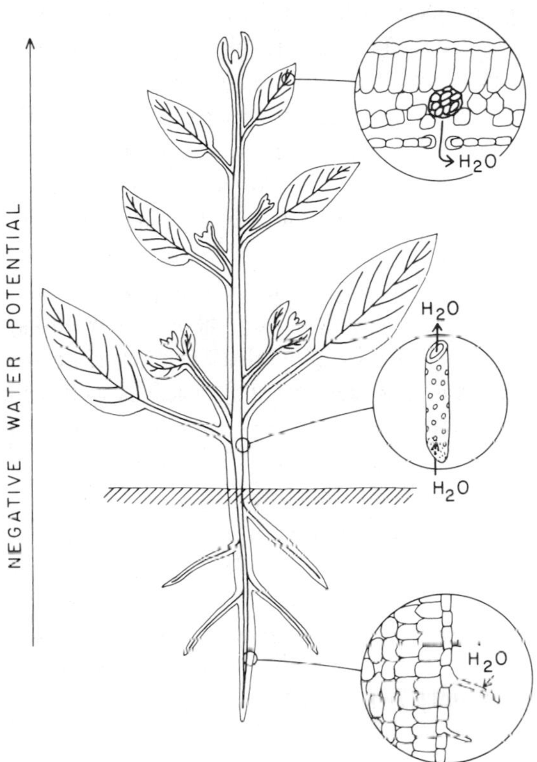

FIGURE 4.9 A negative water potential gradient (decrease in water potential) established in the plant body. When the water potential of the roots of the plant is less than that of the soil, water flows into the epidermal cells of the root and passes into its xylem tissue. Water moves through the tracheary elements of the stem and leaves and exits from the leaves as water vapor. The gradient of water potential is maintained by the supply of water to the roots from the soil and by the decrease in leaf tissue water content. Stomatal closure reduces water loss from the plant body.

in extensin—a protein within the primary cell wall that influences how far walls can stretch. The findings of Iraki et al. (1989) indicate that the cellulose-extensin framework within the cell wall influences the tensile properties of the primary cell wall and that osmotic stress affects the synthesis of these cell wall components.

WHOLE PLANT-WATER RELATIONS: SOME GENERAL POINTS

The general aspects of whole plant-water relations are treated here in the contexts of the absorption of water by roots and also the movement of water throughout the plant body. In the latter context, it is important to recognize that eq. (4.4) can be used to calculate the direction water will flow within a single organ or an entire plant provided gravitational effects are neglected. Water ascending through vertical pipes is subject to a gravitational potential.

Recall that water will always flow from regions with a high water potential to regions with a low water potential (fig. 4.9). This tells us that if water is to flow from the soil into a root, then the root must have a lower water potential than the soil. The water potential of roots can be kept at a lower (more negative) value than that of the soil by the concentration of solutes within root tissues. Similarly, if water is to flow from the root into leaves, then the water potential of leaves must be lower than that of the root. Losing water through evaporation can reduce the water potential of leaves well below that of other tissues within the plant, so that a steep gradient of water potential is established throughout the plant. The site of evaporation of water is the liquid water-air interface on the cell walls of the spongy mesophyll within leaves. Menisci of water within the fibrillar cell walls release water vapor into the air-filled chambers percolating through each leaf. Neglecting the diameter of conducting cells, we could estimate the maximum height a tree could reach in terms of the hydraulic limitations on drawing water vertically against the force of gravity, provided we can estimate the pressure difference across these menisci. The pressure difference $\Delta\mathcal{P}$ across a spherical meniscus is given by the formula

$$\Delta\mathcal{P} = \frac{2T_s}{K} , \tag{4.8}$$

where T_s is the surface tension coefficient and K is the radius of curvature of the meniscus (Rand 1978). The surface tension coefficient for a water-air interface (at 20°C) is 73×10^{-9} MPa·m. The spacings among adjoining cellulose fibrils within the primary cell walls of spongy mesophyll average 1×10^{-8} m. Therefore $K = 0.5 \times 10^{-8}$ m, and from eq. (4.8), we can calculate that $\Delta\mathcal{P} = 29.2$ MPa. If we neglect the matric potential, we can recast the equation for water potential in terms of parameters that have a more immediate and apparent meaningfulness to the ascent of xylem water in tall plants:

$$\psi_w = \mathcal{P} - RTc + \rho gh , \tag{4.9}$$

where $\mathcal{P}$ is hydrostatic (turgor) pressure, R is the gas constant, T is temperature, c is solute concentration (assumed to be dilute), ρ is the density of water, g is the acceleration of gravity (9.80 $\text{m} \cdot \text{s}^{-2}$), and h is height. (Note that $-RTc = \psi_s$ and that $\mathcal{P} = \psi_p$.) The third term (ρgh) is called the gravitational potential, symbolized by ψ_g. Some simple arithmetic will show us that if water moves 10 m vertically in a tree, then the gravitational contribution to the water potential is roughly 0.1 MPa. Since 0.1 MPa is equal to a gravitational head of about 10 m, we can readily appreciate that the pressure difference generated by the evaporation of water within leaves (given by eq. 4.8) is more than sufficient to account for the ascent of water even in the tallest trees, such as the General Sherman tree, a specimen of *Sequoia sempervirens* measuring 275 feet (83.9 m) in height, or the coast redwood (*Sequoiadendron*), which can reach heights of 102 m.

The limiting hydraulic factor in the vertical ascent of xylem water appears to be more closely defined by the tensile strength of water, which dictates the physical limits in height on the continuity of vertical columns of water ascending through the xylem tissue of the plant. The maximum theoretical tensile strength of a perfect column of water has been calculated to exceed 30 MPa (Hammel and Scholander 1976), which on first inspection suggests that plants could grow to a theoretical height of 3,000 m before the columns of water within them snapped from their own weight. However, defects on the inner surfaces of the conduits water flows through or the presence of dissolved gases within the water column greatly reduce the practical limits on the tensile strength of water. It is not entirely clear whether plants can grow a vascular system containing water relatively free of dissolved air. Scholander, Love, and Kanwisher (1955) found that the xylem water of grape (*Vitis vinifera*) is fully saturated with atmospheric nitrogen and partially saturated with oxygen. Accordingly, there is some evidence that the tensile strength measurements made for water from which gases have been removed incorrectly estimate the tensile strength of water in functional xylem. Clearly, imperfections on the inner surfaces of xylem-conducting cell walls, together with rapid transpirational rates and limits on the availability of water to roots (as well as resistance to water flow within roots), reduce how tall plants can grow before water columns in xylem tissues break under tensile stress. The breakage of a column of water involves cavitation—the formation of bubbles of water vapor. Indeed, with the appropriate acoustic equipment, the formation of water vapor bubbles in xylem tissue can be heard as a snap (see Ritman and Milburn 1988 for a comparison between two techniques for detecting cavitation acoustically). As we shall see, the formation of water vapor bubbles within tracheids and vessels is an important feature in considering the design factors of plants subjected to very cold temperatures.

When the root system withdraws sufficient water from the soil around it that the water potential of the soil is less than that of the root tissues, water ceases to flow into the plant. Subsequent dehydration of aerial tissues results in stomatal closure so that further water loss from photosynthetic tissues is dramatically reduced, but there are consequent effects on photosynthetic rates and tissue temperatures (net photosynthetic rates approach zero and tissue temperatures can rise, since the evaporative cooling of tissues owing to water loss is essentially eliminated). As water percolates through the soil and the water potential of the soil approaches zero, water flow into the root system is restored, plant tissues regain their turgidity, stomata reopen, and photosynthesis is reactivated in the presence of sunlight.

With this general picture in mind, it is useful to consider some of the previously mentioned aspects of plant-water relations in greater detail. Two are of particular interest. One is the interaction between the soil and the root systems of plants; the second is the ascent of water within the conducting tissues of large plants. As mentioned previously, eq. (4.5) can be used for plants of relatively low stature. When considering large vertical plants such as trees, however, or when dealing with the movement of water within a soil's profile, we cannot neglect the effect of gravity. For example, the upward movement of water within the conducting tissue (xylem) must overcome the gravitational force of 0.01 $MPa \cdot m^{-1}$. Likewise, gravity causes water to drain downward in soils. I will treat the soil-root interface and the ascent of water in the next two sections.

THE SOIL-ROOT INTERFACE AND THE ABSORPTION OF WATER

As we have seen, the driving force behind the movement of water is the difference in the chemical potential, which can be expressed in terms of the basic equation for water potential (eq. 4.5). In the contexts of soil-water relations and the ascent of water in large plants, an additional term must be added to eq. (4.4). As mentioned previously, this term is the gravitational potential, symbolized by ψ_g. Thus, eq. (4.5) takes the form

$$\psi_w = \psi_s + \psi_p + \psi_g \,. \tag{4.10}$$

Each of the three terms to the right of eq. (4.10) can be viewed as a driving force operating in the context of water absorption from the soil by the roots and subsequent translocation through the stems. In soils that are wetter than field capacity, gravity causes water to drain downward through noncapillary pore spaces until the matric forces within the soil balance the tension developed in the water column within the soil. Height below ground level is designated $-h$, and the gravitational potential ψ_g, which equals $-\rho g h$, results in a reduction in the water potential of the soil. In soils that are drier than field capacity, water movements occur principally along gradients of decreasing matric potential produced by the evaporation of water from the soil's surface. As the water content of a soil decreases, the difference between the water potential of roots and soil decreases; there is an increase in the resistance of water flow through any of the soil-root compartments, and there is a decrease in the conductance to water flow.

Attempts have been made to predict the radial flow of water into a single root of infinite length to gain insight into the way the water content of the soil varies with radial distance from the root, and into the rate of soil water depletion within the absorptive zone of a root. The equation for radial flow into a root is $(\partial\theta/\partial t) = (1/r)[\partial(rD\partial\theta/\partial r)/\partial r]$, where θ is the volumetric soil water content at a distance r from the root surface and D is the soil water diffusivity, which equals the ratio of the hydraulic conductivity to the water capacity of the soil. The solution of the formula is difficult because soil water diffusivity is not an analytical term. If D is taken as a constant, however, then well-known solutions to this formula exist. For example, Gardner (1960) studied time and distance scales for water absorption by a single root in an effort to determine the influence of the initial matric potential (wetness) of the soil on the subsequent radial distribution of the matric potential under the condition of a constant uptake of water by the root. As expected, he found that larger radial gradients near the root occur as the matric potential of the soil decreases. In a subsequent paper, Gardner (1968) reported that the drop in the potential from the soil to the root is negligible provided the matric potential of the water in the bulk soil is near zero (≥ -1.0 MPa). The drop in potential was calculated to be about 100 kPa when the matric potential of the bulk soil is -100 kPa and 10 MPa when the bulk soil is at -1.5 MPa. Other authors have suggested much lower values for the drop in the matric potential from the soil to the root (see Marshall and Holmes 1988, 311–12). It appears that if root systems are well developed, then the drop in the matric potential is relatively unimportant until the potential decreases below -1.5 MPa. Thus, -1.5 MPa appears to be a reasonable lower limit for the availability of water to most plant species. Drought-tolerant plants, however, can absorb water from soils with matric potentials as low as or even lower than -1.5 MPa over long periods because the osmotic potential of their leaf tissues may be as low as -20 MPa, whereas that of most common agricultural plants is -1 to -2 MPa.

Kramer (1983, 198) provides the following formula for the principal forces involved in water absorption A_b by roots:

$$A_b = \frac{(\psi_m + \psi_s)_{soil} - (\psi_p + \psi_s)_{root}}{r_{soil} - r_{root}}, \quad (4.11)$$

where resistance r is given in units of $sec \cdot cm^{-1}$. In moist, well-aerated, warm soils, the roots of moderately transpiring plants have lower water potentials than the water potential of the soil. This is achieved by the accumulation of solutes within the living cells of root tissues. As a consequence, water flows passively into the root, and roots develop what has been called root pressure—a positive pressure is exerted on the columns of water that are sustained within the xylem tissue. High positive root pressures are frequently reflected in guttation—xylem water (which contains solutes) appears as droplets on the surfaces of leaves. The absorption of water from the soil by roots as a consequence of root osmotic potential alone has been called osmotic absorption. This is in contrast to what has been called pressure mass flow, which can occur in plants experiencing high transpirational water loss. When transpirational rates are high, mass flow of water through the roots develops as a consequence of the large volumes of water lost by leaf tissues when stomata are open. The mass flow of water through the xylem tissue of roots dilutes the osmotic potential, whereupon the absorption of water by roots becomes predominantly controlled by the pressure potential in the xylem water resulting from the mass flow of water. Hence the description of this phenomenon as pressure mass flow. Kramer (1983) gives values for the pressure potential within roots experiencing mass flow of water between -1.5 and -2.0 MPa. These values for ψ_p show that a much steeper gradient in water potential across the soil-root interface can develop as a result of rapid transpirational water loss effects than that resulting from the strictly osmotic absorption of water by roots. Before leaving eq. (4.11), we should explore the relative magnitudes of the root and soil resistances. How much root resistance exceeds or is less than soil resistance to the flow of water depends on the water potential gradient. Within the range of -0.02 to -1.1 MPa for the water potential of the soil, root resistance appears to be greater than soil resistance. In part, the high root resistance may result from a decrease in the soil-root contact area rather than from an intrinsic physiological alteration of root tissues, although it has been suggested that water could move across gaps created within the soil and be absorbed by roots in the form of water vapor. In general, root resistance appears to be higher than the resistance of any other compartment of the plant body (root, stem, or leaf). Jensen, Taylor, and Wiebe (1961) reported that root resistance in the sunflower, *Helianthus,* is more than 1.5 times that of the leaves and more than 3.8 times that of the stem. Similar relative resistances among the three types of plant organs have been reported for a variety of herbaceous plant species. The principal resistance to water flow within roots appears to result from the endodermis, a specialized more or less cylindrical layer of cells whose anticlinal cell walls are suberized and therefore resistant to hydration. Thus, water must move through the living protoplast of the endodermis, which offers a high resistance to water flow. The absorptive surface areas of roots are greatly increased by the presence of mycorrhizae, symbiotic fungi whose hyphae penetrate root tissues and extend into the fabric of the soil.

As the water potential of the soil decreases owing to the absorption of water by roots, the roots' capacity to take up water, even by the effects of mass flow of water through the xylem tissue, decreases and eventually equals zero when there is no gradient in water potential across the soil-root interface. The drying of the soil may also decrease the contact area between the soil and root surfaces, aggravating the roots' inability to absorb water from their subterranean environment.

The circumstances involving the transition from the route of osmotic to pressure mass flow absorption are predicted by the formula

$$J_v = Lp\,(\Delta\psi_p + \omega\Delta\psi_s)\,, \tag{4.12}$$

where J_v is the total volume flow, Lp is the hydraulic conductance, $\Delta\psi_p$ is the difference between the pressure potential of the root xylem tissue and the soil, ω is the reflection coefficient, and $\Delta\psi_s$ is the difference between the osmotic potential of the xylem water and the soil. When the difference between the pressure potential of the xylem within the root and that of the soil approaches zero ($\Delta\psi_p \rightarrow 0$), absorption is primarily by means of the osmotic route,and $J_v \approx Lp\omega\Delta\psi_s$. Conversely, as the difference in the pressure potential increases and the osmotic potential of the xylem water is reduced ($\Delta\psi_s \rightarrow 0$), the difference in the osmotic potentials of the root and the soil increases, water absorption becomes dominated by pressure mass flow, and $J_v \approx Lp\Delta\psi_p$.

Clearly, the depth and horizontal spread of the root system, as well as the structure of the soil, play an important role in the capacity of plants to absorb water. The morphology of the root system is largely dictated by the genetic characteristics of the species, although considerable variation in root systems within a species results from differences in soil texture and water availability within the growth habitat. Additionally, the spacing among neighboring plants may have a profound effect on the morphology of the root system through root-root competition. For example, the root systems of sparsely planted trees tend to have a more extensive horizontal spread than those of densely planted trees. As the number of plants rises, the depth of their root systems increases.

The production of mucigel by root caps reduces friction between the growing apex and the surrounding soil. Mucigel is a highly viscous secretion that acts as a lubricant and can also nourish soil bacteria that may produce secondary growth products beneficial to root growth. Mucigel has a complex chemical composition consisting of carbohydrates, amino acids, organic acids, enzymes, and vitamins. It is not clear how many of these components (or in what volume fractions) are produced exclusively by root tissues versus the microbial fauna associated with the root-soil interface.

Provided no other parameters become limiting, the strength of the soil is the principal factor controlling the distribution of roots within the soil profile. In compact soils, root growth is often confined to preexisting soil fissures and cracks, particularly in soils dominated by clays. When the growth in length of newly formed roots is impeded by a compacted layer of soil, their tips expand laterally. This lateral expansion propagates fissures (by means of simple fracture mechanics), opening up avenues of less soil resistance through which roots can continue to grow longitudinally. Additionally, fissures within clayey soils can be generated by the lateral expansion of older roots through secondary growth. Subsequent lateral root formation along the lengths of older roots can capitalize on these fissures, and young roots that absorb water more efficiently than older roots can reinvade previously occupied subterranean territories.

In the absence of fissures or cracks within the soil profile, hydrostatic pressures developed within the growing tips of roots provide a mechanism for penetrating the soil. Hydrostatic pressures of 0.5 to 1.0 MPa are not uncommon within roots whose progress has been impeded by soil compaction. This impedance-growth response may involve ethylene production. The role of ethylene is much more clearly defined and agreed upon in terms of roots' tolerance to flooding, which reduces the diffusion of oxygen into growing roots. Acclimatization to flooded soil conditions appears to be the result of ethylene-induced modifications of root anatomy, primarily in terms of the lysis of cortical cells, leading to the formation of extensive, longitudinally interconnected gas-filled spaces within roots.

The Ascent of Water

The evolution of tissue systems specialized to conduct water and sap and to simultaneously support the weight of tissues and organs above the ground permitted plants to achieve great vertical stature. As plant stature increases, however, the demand for water tends to increase as well. For example, over 4 liters of water per day is required for the growth of a typical corn plant, and over 200 liters per day is not atypical for a large tree. Accordingly, the elevation of photosynthetic tissues above the ground leads to high transpirational rates. For example, with a xylem cross-sectional area of 0.2 cm^2, a transpirational rate of 200 g·hr, and a total leaf area of 2 m^2, the velocity of water flow at the base of a corn plant is estimated to be about 10 m·hr (Russell and Woolley 1961). Greenridge (1958) measured the rate of water flow through the trunk of the American elm, *Ulmus americana,* at between 4.3 and 15.5 m $\cdot hr^{-1}$, and Kuntz and Riker (1955) reported a flow rate of between 27 and 60 $m \cdot hr^{-1}$ for the oak species *Quercus macrocarpa*. Although flow rate measurements are difficult and sometimes misleading, those reported for a variety of herbaceous and tree species indicate that high transpirational demands are often placed on plants and that the rate of transpiration increases with the height of the plant. Indeed, the diameters of tree trunks can decrease significantly (as much as 6%) when transpiration rates are high, owing to a decrease in the water content (hence volume) of cells near the xylem tissue.

The mechanism by which xylem water reaches the tops of tall trees has been the object of considerable research and theorizing (Zimmermann 1983 provides an excellent review of this topic). By the end of the nineteenth century, it was well understood that the ascent of water in the xylem tissues did not necessarily require the presence of living cells. For example, Strasburger (1891) had shown that water could rise through stems that had been killed by heat. From these experiments, Strasburger concluded that transpiration was sufficient to pull water through the xylem. This research foreshadowed what has become known as the cohesion theory for the ascent of water in xylem. There are three essential aspects to the cohesion theory:

1. The high cohesive force of water. When confined to tubes with small diameters, considerable tensile force (possibly as high as 10 MPa) must be applied before the water column ruptures.
2. The continuity of water within the plant. Water columns within the xylem are part of a continuous, uninterrupted system ranging from the water-saturated walls of cells within the roots to the water-saturated walls of cells in the leaves.
3. A gradient in water potential within the plant body. Evaporation of water from photosynthetic tissues in leaves reduces the water potential, causing water to be pulled through xylem cell lumens, thereby reducing the pressure potential within the xylem tissue, which in turn is transmitted to the root system. Hence water absorption is subject to a lag resulting from tissue capacitance but is essentially controlled by the rate of transpiration.

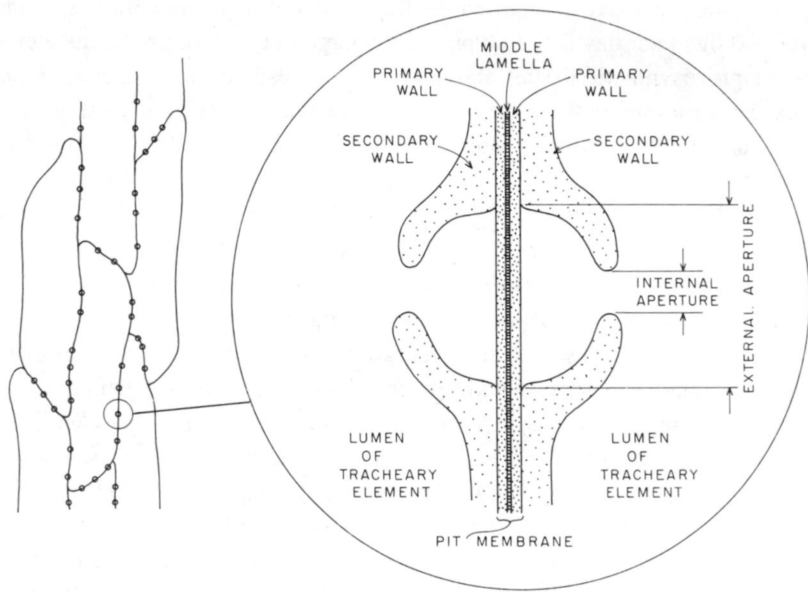

FIGURE 4.10 Diagram of tracheary elements with bordered pits (left) and two bordered pits with the intervening pit membrane between two adjoining tracheary elements (right). The pit membrane consists of the middle lamella shared by the two cells and the primary cell walls of the two cells. The pit membrane offers resistance to the flow of water molecules laterally across the pit membrane (from one cell to the other).

I have spent some time discussing the continuity of the water phase from the soil to the photosynthetic tissues within the plant body, as well as the nature of the water potential gradient. Thus it seems appropriate to dwell for a moment upon the first feature of the cohesion theory—the tension and capillarity of water in xylem tissue.

Although the cohesive force of water is difficult to measure, since the geometry of the container and the adhesive stresses developed between water molecules and the material of the container surface play a critical role in how well columns of water resist tensile stresses, Briggs (1949) demonstrated that water columns could withstand centrifugal forces of over 22 MPa before they snapped by means of cavitation. Similarly, Zimmermann and Brown (1971) estimated that a maximum tension of from 0.015 to 0.020 MPa was required to lift water up the trunks of rapidly transpiring trees. It is reasonable to surmise that a pressure of roughly 2.0 MPa is sufficient to overcome the force of gravity and the resistance to flow in a column of wood 100 m tall. Kramer

(1983, 283) discusses the xylem water potentials measured in trees and shows that the gradients in water potential are clearly greater than those required to lift water to considerable heights.

This lift is aided by the relatively small diameters of the conducting cell types (tracheary elements) found in xylem, which can be dealt with in terms of the laws of capillary flow. The tracheary elements within xylem tissue that have become evolutionarily specialized to conduct water are called tracheids and vessel members. Tracheids are found in species distributed among all vascular land plant groups. Vessels are found in some pteridophyte and gymnosperm species and most angiosperm species. Tracheids are characterized by having a complete primary wall, lacking any perforations, with tapered end walls. The secondary walls of tracheids possess variously thickened regions, whose geometry and wall thickness vary with the developmental age of the xylem, the local environment attending growth, and the genetic capacity of the species. When mature, tracheids lack a living protoplast and have a cell lumen that can be occupied by water. Water flows from tracheid to tracheid, both longitudinally and laterally, through pits where the secondary wall layers are lacking (fig. 4.10). The portion of the two adjoining primary cell walls (with their intervening middle lamella) that spans each pit in the secondary wall is called the pit membrane. Since the pit membrane resists water flow, there is considerable resistance to the flow of water from one tracheid to another. Vessel members are also dead at maturity and possess an inner secondary cell wall. Unlike tracheids, however, vessel members have end walls that are perforated—the end walls lack primary cell walls and the intervening middle lamella—and are stacked end to end to form vessels. Because water is free to move longitudinally from one vessel member to another without an intervening pit membrane, individual vessels have much lower resistance to water flow than do tracheids. Nonetheless, the lateral movement of water from one vessel to another involves flow through pits and pit membranes. Accordingly, the lateral flow of water in xylem tissue with vessels may have the same resistance as the lateral movement of water in xylem tissue dominated by tracheids.

Vessels (and to a much more limited degree tracheids) are structurally and functionally similar to capillary tubes—which are very long compared with their diameter and have very small inner diameters. For example, the average vessel diameter and length in the latewood of the red maple, *Acer rubrum,* are 45 μm and 1.2 cm, while the vessels of the holly, *Ilex verticillata,* can be as long as 1.3 m. (One of the simplest ways to crudely determine the lengths of vessel in a stem is to dry the stem and force smoke through one of the cut

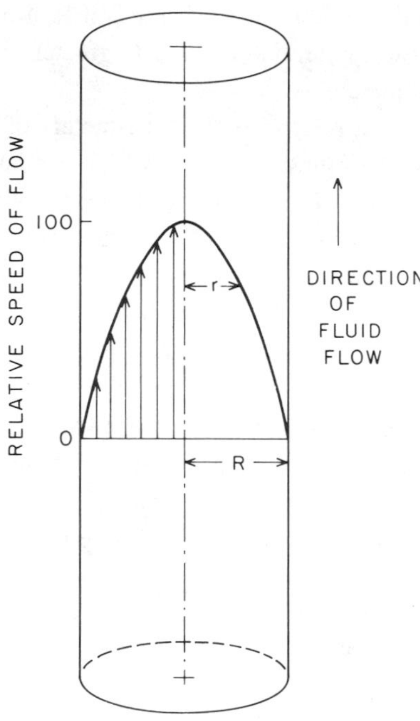

FIGURE 4.11 Diagram of fluid (liquid) flow through a vertical capillary tube. The direction of flow is from the bottom to the top of the figure. The velocity profile of the flow is parabolic; the maximum speed of flow is at the center of the capillary tube ($r = 0$), while the speed of flow of the liquid at the liquid–tube wall ($r = R$) interface is zero.

ends. By snipping off portions of the stem and watching for smoke to exit, one can make a very rough estimate of vessel length. It must be emphasized that the frequency distribution in vessel length can be very skewed [leptokurtic]—more than 90% of any of the vessels found in any particular plant tend to be significantly shorter than the maximum vessel length.) Unlike capillary tubes, however, which have a smooth inner wall surface, the inner walls of vessels and tracheids are ornamented with secondary wall thickenings. Clearly, the geometry of the inner secondary walls plays a role in dictating the flow patterns of water—they can generate turbulence when flow rates are increased. Nonetheless, we can readily appreciate why many plant physiologists have modeled water flow through xylem in terms of the flow of water through capillary tubes.

The physics of capillary flow was treated initially by Gotthilf Hagen and

Jean Poiseuille, who independently investigated nonturbulent flow in very long but narrow-bored glass tubes. By carefully injecting dyes at various locations within the cross-sectional opening of a tube, both Hagen and Poiseuille discovered that water flowed more rapidly at the center of the cross section than toward the periphery of the tube's bore and that the flow rate diminished parabolically toward the inner surface of the tube's opening (fig. 4.11). Based on their analyses, Hagen and Poiseuille proposed the following formula, now known as the Hagen-Poiseuille equation, to describe the relation among the hydraulic conductance *Lp*, the radius of the tube's bore *R*, and the viscosity μ of the liquid flowing through the capillary tube:

$$Lp = \frac{\pi R^4}{8 \mu} . \qquad (4.13)$$

This equation will be discussed in the context of plant evolution (chap. 10), after we have discussed fluid dynamics in some detail (chap. 9), but for the time being it is important to see that the parabolic flow of water through a capillary tube is the reason the hydraulic conductance through the tube is proportional to the fourth power of the tube's inner radius. (For now, the variations in the viscosity of xylem water owing to temperature and the concentration of dissolved materials can be ignored.) This geometry of flow results in large part from the adhesion of water molecules to the surfaces of the capillary tube. The adhesive forces generated within narrow tubes are difficult to measure, but they play a vital role, together with the cohesive forces of water, in drawing water up a narrow tube against the force of gravity. Clearly there is a trade-off involved between the hydraulic conductance of a capillary tube, which can be maximized by increasing the diameter of the tube's bore, and the extent to which adhesive forces can assist in the upward migration of water, which depends on the ratio of the inner surface area to the inner volume of a tube. Tubes with small bore diameters have low *Lp* but can draw water upward; tubes with large bore diameters have comparatively high *Lp* but draw water upward very little. Needless to say, however, if positive pressure is applied from below the column of water, flow will occur in any tube with any bore diameter. This pressure head in plants is supplied in the form of root pressure, discussed earlier in the context of the osmotic absorption of water by roots from the soil.

The flow rate of water through a tube, expressed as the change in volume V per unit time t (or dV/dt), is related to the hydraulic conductance *Lp* of the tube (given by eq. 4.13) and the applied pressure gradient (the change in the pressure $\mathcal{P}$ per unit length l) according to the formula

(4.14a) $$\frac{dV}{dt} = Lp \frac{d\mathcal{P}}{dl}$$

or

(4.14b) $$V = \frac{\pi R^4}{8\mu} \frac{\mathcal{P}}{l}.$$

For the mathematically minded, the derivation of eq. (4.14b) can be easily comprehended if the parabolic profile of the liquid, under a pressure difference of $\mathcal{P}$, flowing through the tube with length l and radius R, is viewed as consisting of a series of concentric sheaths, much like an extendable telescope. At a distance r from the center of the tube, the rate of change of velocity v with r will be dv/dr. From fluid mechanics, the velocity v equals $\mathcal{P}(R^2 - r^2)/4\mu l$. The volume of each sheath is given by the product of its length l, its circumference $2\pi r$, and its width dr. The total volume of flow V through the capillary tube per second is the integral of the velocities and volumes of all these concentric sheaths, from 0 to R; that is, $V = \int_0^R v\,(2\pi r)\cdot dr = (\pi\mathcal{P}/2\mu l) \int_0^R (R^2 - r^2)r\cdot dr = (\pi\mathcal{P}/8\mu l)(2R^4 - R^4) = (\pi R^4/8\mu)(\mathcal{P}/l)$. The maximum velocity of flow v_{max} (at the very center of a capillary tube) is given by the formula

(4.14c) $$v_{max} = \frac{R^2}{\mu V} \frac{\mathcal{P}}{l}.$$

Thus the flow rate of water through a single capillary-like vessel ought to be dramatically influenced by the average vessel member diameter as well as by the pressure gradient. Indeed, Zimmermann (1983, 14) provides a splendid discussion of this topic and shows (provided all other things are equal) that if the relative diameter of a vessel is increased by a factor of 4, then the relative flow rate is increased by a factor of 246 and the percentage flow rate is increased by more than 93%. Based on eqs. (4.13) and (4.14), Zimmermann made the following prediction: "Whenever evolution brought about a slight increase in tracheary diameter, it caused a considerable increase in conductivity" (Zimmermann 1983, 14). (Recall that tracheary element refers to both tracheids and vessel members.) This prediction has been substantiated in part by the increase in the average tracheid diameters reported for successively geologically younger plant fossils (chap. 10). Zimmermann appreciated that water has to flow laterally from tracheid to tracheid or vessel member to vessel member and that the Hagen-Poiseuille equation is only a very crude approximation, particularly for imperforate tracheary elements (tracheids). He also specifically emphasized that the inner secondary wall thickenings of vessel members can generate turbulent flow during episodes of high transpirational water loss, diminishing the relevance of the Hagen-Poiseuille equation. Nonetheless, that equation provides a very reasonable first-order approximation for drawing broad comparisons among species differing in xylem morphometry. It also provides a very useful pedagogic basis for pointing out the importance of the frequency distribution of vessel and tracheid diameters in xylem. That is, the flow rates of water through xylem tissue have to be approximated, at the very least, in terms of the collective diameters of tracheids and vessels. The bulk of the water that flows longitudinally through xylem tissue will be carried by the vessels with the larger diameters, but all unobstructed vessels will transport water.

XYLEM EFFICIENCY VERSUS SAFETY

The flow of water within plants is often discussed in terms of the trade-off between efficiency and safety. In this context, efficiency is treated as the relation between the size (typically, the cross-sectional area) of the xylem tissue and the leaf area to which water must be supplied, while safety refers to the factors that influence the continuity of water columns within the xylem, which can be broken as a result of mechanical or freezing injury. B. Huber (1928; see also 1956) was one of the very first workers to discuss the efficiency of the xylem. He considered two parameters. The first was the specific conductivity of the xylem, which is measured as the volume of water moved per unit time under a given pressure through a stem segment of specified length and cross-sectional area. (Provided length and cross-sectional area are standardized, specific conductivity is essentially expressed by the relationships given in eq. 4.12.) The other measure of xylem efficiency Huber proposed is the relative conducting surface, which is the ratio of xylem cross-sectional area to leaf surface area. (When the ratio of xylem cross-sectional are to leaf fresh weight is taken, the analogue to the relative conducting surface is called leaf-specific conductivity, or LSC. The relative conducting surface is sometimes referred to as the Huber value; see Zimmermann 1983, 66–67.) Specific conductivity is typically given in units of $ml \cdot hr^{-1} \cdot cm^{-2} \cdot MPa^{-1}$, whereas the relative conducting surface or Huber value is dimensionless. Based on his surveys of plants, Huber reported values for specific conductivities of 20 $ml \cdot hr^{-1} \cdot cm^{-2} \cdot MPa^{-1}$ for conifer species, from 65 to 128 $ml \cdot hr^{-1} \cdot cm^{-2} \cdot MPa^{-1}$ for deciduous broadleaf trees, and values as high as 1,273 $ml \cdot hr^{-1} \cdot cm^{-2} \cdot MPa^{-1}$ for some vine species. Even higher values were reported for roots. Further, Huber found that the specific conductivity of twigs and branches was

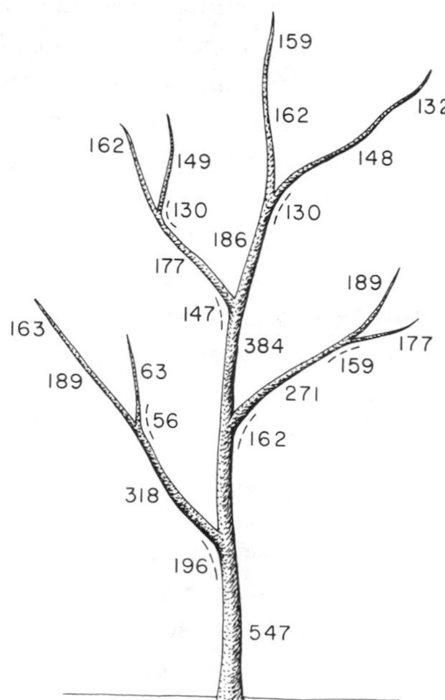

FIGURE 4.12 Hypothetical distributions of leaf-specific conductivities (LSC) as measured along the trunk and branches of a stylized tree. Branch-trunk and branch-branch junctions (shown by dashed lines) have lower LSC than areas proximal or distal to the junctions. There is a general trend for a distal decrease in LSC along the length of the trunk and the lengths of branches.

typically lower than that of their trunks and that the specific conductivity could vary along the length of a trunk. In terms of the relative conducting surface or Huber value, plants manifest a significant range. For example, Huber reported that the relative conducting surface value of trees averaged 0.5, while the values of herbaceous and desert succulents were 0.2 and 0.1, respectively. Certain aquatic plants had even lower values (Nymphaeaceae = 0.02), while the relative conducting surface value of bog plants was found to be not unlike that of some desert plants.

The values of relative conducting surface are somewhat suspect because the entire cross-sectional area of the xylem is used in their computation. Huber (1928) was painfully aware that not all of the xylem in a tree trunk or branch is hydraulically functional. Indeed, only the sapwood typically conducts water, while the heartwood of the trunk and branches functions primarily to mechanically sustain loadings. Accordingly, the observation that the specific conductivity of a trunk decreases acropetally might change significantly if specific conductivity was calculated only in terms of the cross-sectional area of the sapwood. For example, Kaufmann and Troendle (1981) found comparable ratios of sapwood area to leaf area in various portions of tree crowns for three of the four species they examined.

Zimmermann (1983 and references cited therein) has provided the most comprehensive survey of changes in the leaf-specific conductivities (LSC) of trees. Most of his findings come from diffuse-porous tree species (those in which vessels are found in both spring and summer wood and are of comparable diameter), such as poplar (*Populus*), birch (*Betula*), and maple (*Acer*), and are summarized in figures that plot LSC on the branching patterns of trees (Zimmermann 1983, figs. 4.3 to 4.6). Zimmermann found that LSC increased from the base to the top of the trunks of two young specimens of maple, while it decreased acropetally in the trunk of a specimen of birch. No particular trend was discernible in the trunk of a poplar or of one specimen of maple. Despite this interspecific variability in LSC, the leaf-specific conductivities of lateral branches were found to be consistently lower than those measured for the trunks (fig. 4.12). Finally, Zimmermann found that branch junctions always showed the lowest LSC values measured along any specific path of water flow. It is this last finding that is of particular interest to us here.

Branch-trunk junctions appear to be hydraulic bottlenecks, called hydraulic constrictions, as evinced by their locally low values of LSC. The decrease in the LSC value across a hydraulic constriction is not the result of a local drop in the Huber value—that is, the localized decrease in LSC is not caused by a regional decrease in the cross-sectional area of functional xylem. Indeed, branch junctions actually bulge, perhaps because of regionally high cambial activity stimulated by dynamically induced mechanical stresses. Rather, the hydraulic constriction appears to result from a localized decrease in average vessel diameter just above a branch junction as well as an increase in the frequency of vessel endings. Also, many branch vessels below the branch junction are found to be occluded with gums (Zimmermann 1983, 74). Hydraulic constrictions appear to operate as design factors in trees, limiting the extent to which water vapor bubbles that have formed in cavitated vessels distal to the hydraulic constrictions can enter the trunk.

The safety features of tracheids versus vessel members become readily apparent when xylem water is subjected to very low temperatures. The passage of an embolism within an individual tracheid is greatly restricted by the small inner diameters of pits and the presence of the pit membrane. By the same token, vessel members of small diameter can restrict the passage of vapor bubbles through the length of an individual vessel, reducing the chance that bubbles will coalesce—the smaller the bubble, the greater the probability that it can redissolve when xylem water thaws. The increased frequency of vessel endings at branch junctions is yet another design factor, since vessel endings confine embolisms and limit the passage of an embolism from one vessel to another through pits.

STOMATA

As mentioned in the introduction to this chapter, land-dwelling plants have evolved a number of ways to conserve water once it is absorbed by their subterranean parts. One of the most important is regulating the rate of water vapor loss from the wetted surfaces of photosynthetic surfaces within the leaf, which involves stomata and their guard cells (see fig. 4.2B). A water deficit within the plant body leads to stomatal closure, followed by a recovery of tissue turgor and a reopening of stomata. This section offers a brief review of stomatal structure and function.

Typically two guard cells flanking the future opening, the stomatal pore, develop from the transverse division of a single protodermal cell. The middle lamella (a pectinaceous material between adjoining primary cell walls) shared by the guard cells disintegrates and the two walls separate, forming the stomatal pore. The walls bordering the pore are typically cutinized and become thicker than the rest of the cell wall.

The degree of opening of the stomatal pore depends on the turgor of the guard cells and adjacent cells, which in turn affects how much the cell walls of guard cells elastically deform. Cooke et al. (1977) considered the elastomechanics of guard cells by modeling the behavior of cell walls as linearly elastic, thin-walled shells. They found that an increase in hydrostatic pressure in the guard cell and a decrease in the hydrostatic pressure of neighboring cells cause opening of the stomatal pore, while an increase in the hydrostatic pressure of neighboring subsidiary cells and a loss of pressure in the guard cells tend to close the pore. Notice that the model used by Cooke et al. did not require a differential thickening in the guard cell walls. Previously, it had always been thought that the thickenings in the cell walls flanking the stomatal pore were mechanically essential for stomatal opening and closure. By contrast, Cooke et al. (1977) found that the elliptic shape of the guard cells was the single most important geometric parameter dictating the hydrostatic opening and closing of stomata. Other features, such as wall thickenings and radial stiffening of the guard cell walls, aided opening and closing but were not essential. If the two guard cells were modeled as a circular torus, however, an increase in hydrostatic pressure caused the stomatal pore to close rather than open.

Experimental observations have shown that the width of the stomatal pore often oscillates, typically with a period of ten to fifty minutes. The gaseous flux through the stomatal pore appears to operate in a feedback mechanism responsible for stomatal oscillation. Delwiche and Cooke (1977) examined this effect. Their analysis was subsequently extended by Rand et al. (1982). From these studies, the following picture concerning stomatal oscillations has emerged: Water leaves the wet mesophyll and subsidiary cell walls and diffuses through the stomatal pore. This water loss is compensated for by the efflux of water from the guard cells to the subsidiary cells. The resulting decrease in hydrostatic pressure within the guard cells causes the stomatal pore to close. The diminution in the stomatal pore diameter increases the resistance to water vapor loss, and the water potential of mesophyll and subsidiary cells increases as water is supplied by the xylem tissue. The stomatal pore opens as the hydrostatic pressure of guard cells is reestablished, and the entire cycle of water loss is reiterated. Stomatal oscillation in an autonomous system that can be described by two first-order ordinary differential equations, involving the pressures within the guard cells and subsidiary cells (Delwiche and Cooke 1977). Rand et al. (1982) found that stomatal oscillations within the surface of a leaf are propagated as waves that pass over the leaf surface. Thus, stomatal oscillations exhibit stable, spatially uniform, and synchronized behavior, but the period of oscillation might be expected to vary among species differing in their xylem conductance and in the geometry of their subsidiary cells. Sadly, the relations among these physiological parameters have not been explored.

Is there a Darwinian explanation for stomatal oscillations? Apparently so. Upadhyaya, Rand, and Cooke (1988) found that when typical physiological values were modeled in terms of stomatal oscillations, the metastable state of oscillatory behavior confers an advantage to water conservation under dry atmospheric conditions. Some stomatal pores are open, permitting gas exchange and photosynthesis; others are closed, permitting the reacclimation of hydrostatic (turgor) pressure within the leaf. Conserving water in this manner reduces the rate of carbon dioxide assimilation, but not as much as if all the stomata were closed; that is, stomatal oscillations increase the ratio of carbon assimilation to water loss.

Five
Plant Cell Walls

> I take it therefore, that the cylindrical cell of *Spirogyra,* or any other cylindrical cell which grows in freedom from any manifest external restraint, has assumed that particular form simply by reason of the molecular constitution of its developing wall or membrane; and that the molecular constitution was anisotropous, in such a way as to render extension easier in one direction than another.
>
> Sir D'Arcy W. Thompson, *On Growth and Form*

This chapter treats a highly specialized subject, but one essential to the more general treatment of plant biomechanics—the biomechanics of plant cell walls. This focus is justified because some features are so impressive in their importance to a group's biology. The cell wall infrastructure of their tissues sets plants apart from all other organisms. An additional reason for treating the cell wall in a separate chapter is that its behavior during and after cell growth illustrates how we must simultaneously understand a number of levels of biological organization to identify the relations between form and function, as well as comprehend these relations in the context of dynamic growth processes.

With rare exceptions, such as gametes, plant cells are enveloped by a rigid or semirigid wall that is secreted by the protoplast external to its plasma membrane (Esau 1977). In mature cells, the cell wall is a mechanical barrier against sudden distensions of the protoplast by the rapid influx of water caused by a steep osmotic gradient. Thus it represents the limits of the size and shape of a mature cell and in large part defines the texture and mechanical properties of mature tissues and organs (see chap. 6). Additionally, in many tissues the protoplast of some cell types dies in a developmentally controlled manner, and the cell walls that are left behind remain functional, contributing to mechanical strength and providing conduits for water and other nutrients. Tracheids, vessel members, fibers, and sclerids are functional when their protoplasts die and are removed from cell wall lumens. By contrast, in living cells, cell walls can remain relatively thin (as in parenchyma) or manifest differential thickening (as in collenchyma). Turgor pressure within the cells of these tissues varies (between 1 and 8 bars, in normal circumstances) and provides a hydrostatic mechanism for altering the stiffness of living tissues. In growing tissues or isolated cells, the osmotic gradient is the driving force for cell expansion and elongation, and depending on the physiological influence of the protoplast, the cell wall can be fluidlike and extensible in living and growing cells and tissues or stiff and rigid in nongrowing cells. This versatility in the material properties of cell walls is a function of both cell wall architecture and chemistry.

One of the most important concepts concerning the mechanical behavior of plant cell walls is the behavior of polymers, because cellulose, which chemically dominates the essentially fibrous infrastructure of cell walls, is a polymer. In chapter 2 we briefly reviewed the distinction between an extended polymeric solid and a molecular solid, but in this chapter we will gain a direct appreciation of the importance of this distinction in terms of cell growth and morphogenesis. In general, polymers such as cellulose have higher tensile moduli and greater tensile strength when stressed along their length than when stressed normal to their length. An expression of this is seen when the modulus of elasticity of cell walls isolated from wood is measured along and transverse to the longitudinal axis of cells. The elastic modulus measured along cell length can be as high as 3.5×10^{10} N·m^{-2}, while the elastic modulus measured transverse to cell length can be as low as 1.0×10^{10} N·m^{-2}. Accordingly, we see that there can be over a threefold difference in the elastic behavior of cell walls depending on the direction in which material properties are measured. Thus the cellulosic fibrous infrastructure of plant cell walls confers an anisotropy in mechanical behavior that is the cornerstone of our understanding both of why tissues exhibit different mechanical behaviors depending on the direction in which they are stressed and of how plant cells can achieve a variety of shapes and sizes as they grow. In terms of the latter, the living protoplast deposits strands of cellulose in geometric configurations that prefigure the orientations of the future axial elongation and lateral expansion of the cell. As we will see, however, there are competing theories to explain how the depositional patterns of cellulose fibers are achieved and how these patterns, once established, constrain cell morphogenesis. Although the validity of these theories is of biological concern, the extent to which they use mechanical principles and are internally consistent with these principles will be our primary focus.

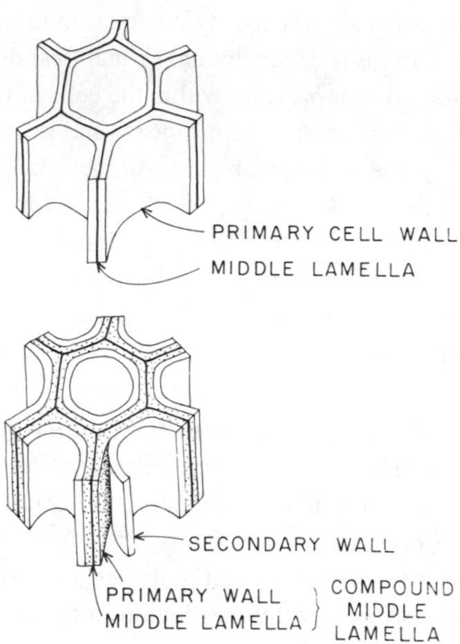

FIGURE 5.1 Plant cell wall formation through apposition. Primary and secondary cell wall layers are deposited in sequence so that the first-formed layers are farthest from the living protoplast. The primary cell wall is formed first and is in contact with the middle lamella (shown by dark lines in both diagrams), a pectinaceous layer that binds neighboring cells together. The secondary wall is deposited after the primary cell wall is secreted by the protoplast. During secondary wall formation, the middle lamella and the cell wall layers of both the primary and secondary cell walls are chemically impregnated with lignin. (The chemical alteration of the primary cell wall is shown by stippling.) As a result of chemical impregnation, the middle lamella and the primary cell wall are difficult to distinguish from one another and are collectively referred to as the compound middle lamella.

THE CELL WALL IN THE CONTEXT OF TISSUES

All cell walls are formed through apposition. That is, the wall is synthesized and deposited by the protoplast in discrete layers, each one originally in contact with the plasma membrane but subsequently pushed toward the outside of the cell wall as new layers are secreted by the protoplast. Thus the oldest cell wall layer is farthest from the plasma membrane. When a cell wall is formed, primary cell wall layers are deposited in apposition during the expansionary phase of cell growth. Additional layers composing the secondary cell wall are deposited in some cells after elongation and expansion have essentially ceased (see fig. 5.1). The material properties of both the primary and the secondary cell walls can be altered by the protoplast after volumetric growth has ceased by means of chemical impregnation with compounds like lignin. The young primary cell walls of actively growing cells typically exhibit viscoelastic behavior, however, since expansion and elongation of the protoplast encased within the primary cell wall require that the cell wall distend ("flow") while manifesting some elasticity (recoverable deformations). By contrast, secondary cell walls tend to operate mechanically as rigid, elastic solids when loaded within their proportional limits.

In tissues, the primary walls of adjoining cells are separated by a pectinaceous layer, called the middle lamella (fig. 5.1), that acts essentially as a cementing agent. Experiments with parenchymatous tissues at high turgor show that the tensile fracture of these tissues can result from cell wall rupture, while at low turgor cell-cell debonding can occur by the propagation of fractures within and parallel to the middle lamella (Lin and Pitt 1986). Thus the mechanical behavior of tissues cannot be determined solely by considering the material properties of cell walls but also depends on other factors, among which the physical properties of the middle lamella and the physiological status (tissue water content) of the protoplast are important. Additionally, the age of the tissue must be accounted for, since the material properties of both the cell wall and the middle lamella can be altered by chemical impregnation with compounds such as lignin. When this occurs the distinctions among the primary cell walls of adjoining cells and the intervening middle lamella are lost, and the three layers are then referred to as the compound middle lamella (fig. 5.1). The impregnation of the middle lamella and adjoining cell walls reduces the probability of cell-cell debonding, and when failure in tension occurs it typically takes the form of cell wall rupture as fractures propagate across cell walls. The mechanical behavior of tissues will be treated in chapter 6; for now it is important to remember that the mechanical behavior of isolated cell walls and those embedded within tissues is distinct and that cell walls rarely operate in isolation from neighboring cell walls.

THE CHEMISTRY OF PRIMARY CELL WALLS

Cell walls have a fibrous infrastructure embedded within what has traditionally been viewed as a more or less amorphous matrix. The fibers constitute a polymeric extended solid chemically dominated by polysaccharides, cellulose and the hemicelluloses being the most important (see fig. 5.2). As discussed in chapter 2, the physical properties of extended solids are dictated by the length of the polymers composing the solid and by how and how extensively

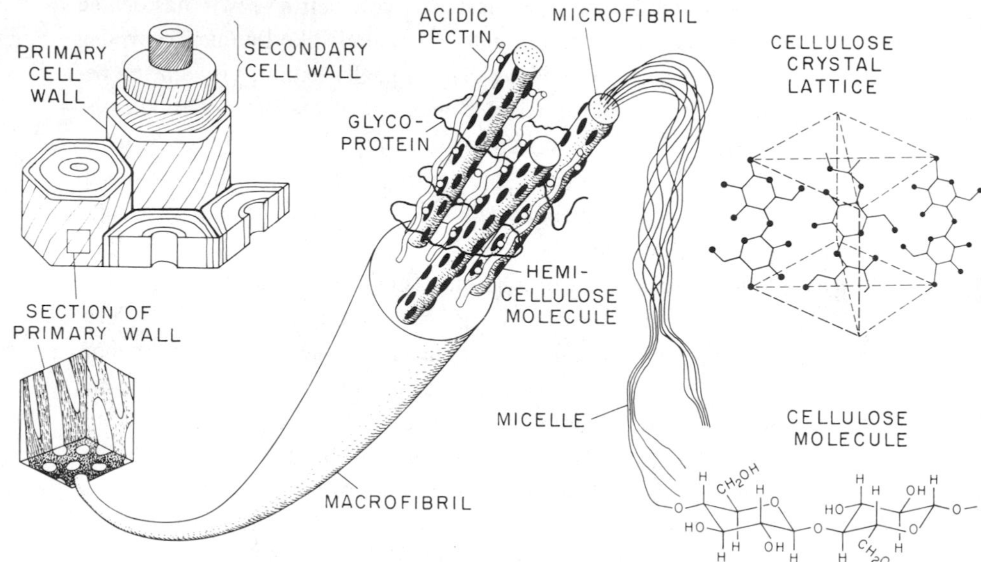

FIGURE 5.2 Various levels of organization within plant cell walls. The cell wall consists of sequentially deposited cell wall layers that have a network of cellulosic fibrils whose orientation angles within the cell wall varies from layer to layer (upper left of diagram). Two different size categories of fibrils can be distinguished. Macrofibrils are visible at the level of the light microscope (lower left of diagram). These in turn consist of microfibrils that are visible at the level of transmission electron microscopy. The microfibrils are bound within a matrix of glycoproteins, hemicelluloses, and acidic pectins (middle of diagram). In turn, each microfibril consists of a number of cellulose molecules organized into paracrystalline and crystalline (micellar) regions. Cellulose molecules, in which cellobiose is the repeating disaccharide unit (lower right of diagram), have a crystal lattice composed of antiparallel chains stabilized by interchain hydrogen bonding (upper right of diagram). (After Esau 1977.)

the polymers are cross-linked. Accordingly, the physical properties of primary cell walls are profoundly influenced by the chemistry of cellulose and, to a lesser extent, the chemistry of other primary cell wall constituents. To understand many aspects of the physical biology of cell walls, research has focused on the chemistry of cellulose, particularly the polymeric configuration of cellulose as it exists in the cell wall, referred to as native cellulose.

Cellulose is highly resistant to many solubilizing agents, whereas hemicelluloses are more easily extracted from cell walls. Consequently, early attempts to isolate cellulose from cell walls often resulted in chemical artifacts from which the physical properties of cell walls were erroneously deduced (reviewed by Preston 1974). The stability of cellulose is due in large part to the β-1-4-linkage between its monomeric units, glucose. In addition, molecules of cellulose are long chained and tend to form crystalline regions separated by paracrystalline regions. Although glucose is the monomeric unit within the cellulose molecule, cellobiose is the repeating disaccharide, since molecules of glucose within the chain are alternately rotated 180°. Thermodynamic theory indicates that cellulose can take on two different polymeric configurations, though only one is assumed by native cellulose within cell walls. One of these, the extended-chain configuration, is that of native cellulose, while the folded-chain configuration occurs when cellulose is chemically extracted from cell walls and when polymerization and crystallization do not occur simultaneously. The distinction between these two configurations is important, since the elastic tensile modulus and tensile strength of cellulose depend on the polymeric configuration of the cellulose in the polymeric extended solid constituting the fibrous component of the cell wall.

In its native crystalline state, cellulose exists as an extended-chain polymer; the unit cell within the crystal lattice consists of antiparallel chains that are stabilized by interchain hydrogen bonding (see fig. 5.2). The unit cell has been measured and has the dimensions of $0.835 \times 1.03 \times 0.79$ nm. This polymeric configuration is lost when cellulose is extracted and recrystallized. Under these conditions the cellulose molecule takes on the folded-chain configuration, which is substantially less strong when placed in tension. The differences in the material properties of the extended-chain and folded-chain configurations of cellulose and the failure to appreciate which of these two configurations is represented by native cellulose led early workers to erroneous conclusions concerning the physical biology of plant cell walls (see Preston 1974 for a comprehensive review of this topic).

In contrast to cellulose, the hemicelluloses are a group of polysaccharides characterized by possessing acidic groups (usually D-glucuronic or D-galacturonic residues, or the 4-0-methyl esters of the former). Hemicelluloses can dominate the dry weight composition of some primary cell walls (e.g., they compose 38% to 53% of the dry weight of *Avena* coleoptiles) and can vary among plant groups or even among tissues from a single organ.

The polymeric nature of cellulose confers a chemical anisotropy on the molecule, while the polymeric extended solid constituting the fibrous infrastructure of the cell wall confers a mechanical anisotropy on the cell wall as a whole. Polymers characteristically resist tensile forces along their length but tend to deform appreciably when forces are exerted perpendicular to their length. Cellulose is no exception to this general rule. Hence the fibrillar network of cellulose polymers within the cell wall provides a physical system in which tensile and compressive forces can be preferentially accommodated. That is, the stressed cell wall contains multidirectional but nonequivalent strain lines. As I mentioned previously, the source of the stress in actively growing cells is the exertion of hydrostatic pressure internal to the cell wall as the protoplast contained within expands. The protoplast can be viewed in this context as an essentially incompressible amorphous and highly viscous fluid that in itself has no preferred direction(s) for expansion. By virtue of its direct control on the orientation of cellulose fibers within the cell wall, however, the protoplasm manifests cell morphogenesis (change in shape) as it grows. The cell wall therefore, in both its macroscopic and microscopic infrastructure, is a direct physical manifestation of the much less observable but nonetheless very real developmental anisotropy within the protoplasm.

Although these views have remained relatively unaltered since they were originally proposed, more recent evidence suggests that the cell wall matrix may play an equally important role in dictating the mechanical behavior of cell walls. (In particular, the matrix must be sheared as cellulosic components within the cell wall are extended or reoriented as the cell wall expands or extends. Therefore the shear modulus of the cell wall matrix could conceivably be very important.) In retrospect, the perception of the significance of the matrix should come as no surprise, since the physical properties of cell walls are dramatically altered when the matrix is impregnated with compounds like lignin. But regardless of its lateness in coming to the attention of researchers working on the mechanics of cell walls, even the primary cell wall is now increasingly viewed in terms of being a composite material in which the role of the matrix cannot be ignored.

The Fibrillar Network of Cell Walls

The fibrillar nature of cellulose within walls is evident at the level of observation with both the light and the electron microscope. Accordingly, the fibrils within the cell wall are traditionally classified according to size. The largest fibrils, which are visible with the light microscope, are called macrofibrils and can measure on the order of 100 to 250 nm in width. The smallest fibrils, visible only with the electron microscope, are called microfibrils and measure on the order of 3.5 to 8.5 nm in width (Frey-Wyssling 1954; Preston 1974; Esau 1977). (Microfibrils can coil around one another, giving the cablelike appearance of macrofibrils seen with the light microscope.) Each microfibril is composed of a number of cellulose molecules, organized into crystalline regions called micelles and separated by less orderly paracrystalline regions

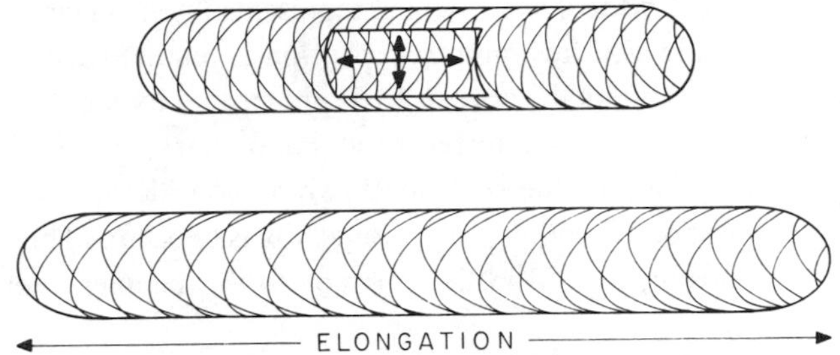

FIGURE 5.3 Restraining influence of microfibrillar orientation on the preferred axis of cellular elongation. An internal (turgor) pressure (indicated by orthogonal arrows within the cell lumen shown at the top) is resisted by transversely oriented microfibrils. The preferred axis of elongation is perpendicular to the axes of microfibrils (as shown at the bottom).

(these relationships are diagrammed in fig. 5.2). Hemicelluloses are attached to the surface of microfibrils by means of hydrogen bonds, and acidic and neutral pectin molecules cross-link hemicelluloses among neighboring microfibrils. This cross-linking is thought to involve glycoproteins. In primary cell walls, the hemicelluloses, pectins, and glycoproteins constitute the principal molecular species of the cell wall matrix. The matrix can be hydrated to varying degrees, thereby changing the distances between neighboring microfibrils and presumably altering the nature and extent of chemical bonding. Consequently the matrix is not without a molecular structure and organization, nor is it unresponsive to physiological changes in the cell wall as cells grow. Unfortunately, its relative plasticity, its ductility, and its behavior in shear under different conditions of water content or enzyme activity are largely unknown.

As I mentioned earlier, the variation in the orientation of microfibrils in different layers within the cell wall confers the capacity to resist extension along fibril length. Therefore, in theory, preferential microfibrillar orientations provide a visible manifestation of potential preferred directions in cell elongation (Preston 1974; Taiz 1984). Simple shell theory can be used to predict the differences in the lines of strain required to manifest spherical versus cylindrical cells when an internal pressure with equivalent directional lines of force is exerted within a shell. Multidirectional but nonequivalent strain lines within the cell wall are not necessary for cells that grow in an isodiametric manner, since turgor pressure is equidirectional in magnitude. Theory indicates that for a spherical shell, representing the isodiametric cell wall, with an internal fluid pressure $\mathcal{P}$, the stress exerted in all directions should equal $r\mathcal{P}/2t$, where r is the radius of the shell and t is the shell-wall thickness. However, for an essentially cylindrical shell, which represents the geometry of an elongating cell, some physical restraint to expansion is required, since a ratio of the transverse to longitudinal stress of 2:1 is anticipated by theory; that is, the transverse stress, sometimes referred to as hoop stress, in such a cylinder would equal $r\mathcal{P}/t$. Clearly, the shell must exhibit considerable anisotropy in its capacity to resist deformation if cell elongation is to occur. (Thus the words of D'Arcy Thompson, quoted in the epigraph to this chapter, seem particularly apt.)

Cellulose microfibrils, which tend to be helically arranged within cell walls, provide an obvious mechanical solution to the requisite for anisotropy in the material properties of differentially expanding cell walls. Fibrils, originally oriented preferentially in transverse direction, would operate mechanically to restrain expansion while permitting elongation (as shown in fig. 5.3). An alternative mechanism, operating along the same lines, would be to preferentially loosen specific microfibrils through metabolically controlled processes. The protoplast would decrease the yield stress at which hydrostatic pressure would cause the cell wall to flow. Note, however, that any model purporting to explain the morphogenesis of elongated cells in this manner is accepting a number of underlying assumptions, and one—the 2:1 ratio of transverse (hoop) to longitudinal stresses—is critical. Two other assumptions are relevant. First, it is assumed that the cell wall is uniform in thickness, relatively thin, and (except for the cellulosic microfibrils) isotropic in its material properties. Second, it is assumed that elongation creates a nonlinear strain gradient across the cell wall. That is, the most recently deposited layers would experience little or no straining, while the first formed (outermost) layers would be stretched by a factor proportional to the overall elongation of the cell. Since cellulose microfibrils resist tension along their lengths, we would anticipate that fibrils within wall layers would be reorientated parallel to the direction of maximum strain (cell length) and that this reorientation would be proportional to the strain experienced within each layer across the cell wall. What is often neglected in this scenario is that there exists an instantaneous stress gradient in the reverse direction created by the hydrostatic pressure exerted by the turgid protoplast. Thus cell elongation must result in strains of sufficient magnitude to countermand the stress gradient produced by turgor pressure.

The Turgor Stress Gradient

The stress gradient across a cell wall at any instant in the growth of a cell

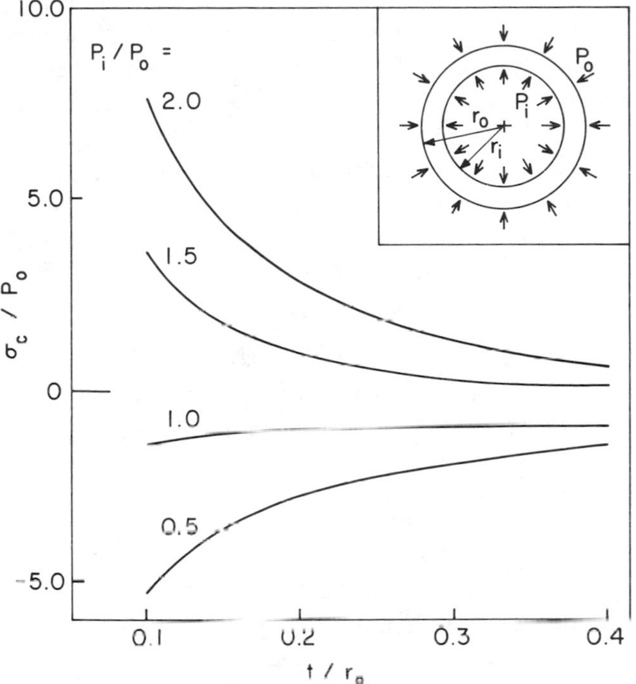

FIGURE 5.4 Theoretical relation among circumferential (hoop) stresses (σ_c) generated within a tubular cell with wall thickness t and outer radius r_o subjected to an inner (turgor) $\mathcal{P}_i$ and an outer pressure $\mathcal{P}_o$. (The inserted diagram to the upper right of the graph shows a cross section through the cylindrical cell wall with an inner radius of r_i.) The relations among pressure, circumferential stress, and cell wall dimensions are expressed in terms of three dimensionless ratios (t/r_o, $\mathcal{P}_i/\mathcal{P}_o$, and $\sigma_c/\mathcal{P}_o$). The ratio of circumferential stress to the external pressure ($\sigma_c/\mathcal{P}_o$) is plotted as a function of the ratio of cell wall thickness to cell wall outer radius (t/r_o) for four different ratios of turgor to external pressure ($\mathcal{P}_i/\mathcal{P}_o$). As the thickness of the cell wall increases (as t/r_o increases), the predicted ratios of circumferential stress to the external pressure ($\sigma_c/\mathcal{P}_o$) converge for any ratio of turgor to external pressure ($\mathcal{P}_i/\mathcal{P}_o$); for any wall thickness (t/r_o is a constant), the ratio of circumferential stress to external pressure ($\sigma_c/\mathcal{P}_o$) increases as $\mathcal{P}_i/\mathcal{P}_o$ increases. These relationships indicate that when the wall thickness is equal to or exceeds roughly 20% of the cell radius, the turgor pressure plays little or no role in determining the circumferential stresses within the cell wall. (From Niklas 1989c.)

depends on the difference between the inner turgor pressure, $\mathcal{P}_i$, and the outer external pressure, $\mathcal{P}_o$ (fig. 5.4). For an isolated cell, not flanked by neighboring cells, the internal (turgor) pressure can equal or far exceed the ambient pressure. Indeed, turgor pressure can equal 10 bars. Thus, as a crude approximation (for illustration only), $\mathcal{P}_o$ can be set equal to one, while $_i$ would be larger by many factors. By contrast, for a cell embedded within a tissue and surrounded by comparable cells, the condition $\mathcal{P}_o = \mathcal{P}_i$ could be accepted as a reasonable approximation. Close-form solutions are available from engineering theory to treat the general condition; these two are only examples. This solution provides for the quantification of the circumferential and radial stresses, σ_c and σ_r, that would develop within the wall of a pressurized cylindrical shell with closed caps. (A treatment of the longitudinal stresses developed in cylinders may be found in Roark and Young 1975, 504.) The close-form solution is given by the formulas

(5.1) $$\sigma_r = \frac{r_i^2\, r_o^2}{r_o^2 - r_i^2}\left[\frac{\mathcal{P}_o - \mathcal{P}_i}{(r_i + d)^2}\right] + \frac{\mathcal{P}_i\, r_i^2 - \mathcal{P}_o\, r_o^2}{r_o^2 - r_i^2}$$

(5.2) $$\sigma_c = \frac{r_i^2\, r_o^2}{r_o^2 - r_i^2}\left[\frac{\mathcal{P}_o - \mathcal{P}_i}{(r_i + d)^2}\right] - \frac{\mathcal{P}_i\, r_i^2 - \mathcal{P}_o\, r_o^2}{r_o^2 - r_i^2},$$

where r_i and r_o are the inner and outer radii of the cylindrical wall and d is the distance of a wall layer within the wall measured from r_i. The relation between the circumferential or radial component of stress within the cell wall and the distance d within the cell wall at which they are measured is easily seen with the aid of these two equations. As d increases, all other things being equal, these two stresses decrease. Consequently, in the case of an isolated cell with an internal (turgor) pressure greater than the ambient external pressure, the highest stresses within the cell wall will occur within the innermost (youngest) cell wall layers. (For a detailed treatment of the response of a mechanically anisotropic cylindrical cell to multiaxial stress, see Sellen 1983 and Carroll 1987.)

Equations (5.1) and (5.2) provide additional insight. For an isolated cell, the maximum circumferential stress when the internal pressure is much greater than the external pressure is given by the formula

(5.3) $$\sigma_c^{\max} = \mathcal{P}_i\left[\frac{(r_o - t)^2 + r_o^2}{t(2r_o - t)}\right],$$

where t is the wall thickness. Equation (5.3) indicates that the circumferential stress is always greater than the turgor pressure and approaches the value of the turgor pressure as the outer radius of the cell increases. Significantly, the maximum circumferential stress can never be reduced below the internal pressure, regardless of the wall thickness. This can be seen by calculating the circumferential stress for the inner wall layer σ_c^i (where $d = 0$) and the outer wall layer σ_c^o (where $d = t$) and taking the difference between the two:

(5.4) $$\sigma_c^i - \sigma_c^o = \frac{\mathcal{P}_i}{t\,(2r_o - t)}\,[(r_o - t)2 + r_o^2] - \frac{2\mathcal{P}_i\,(r_o - t)^2}{t\,(2r_o - t)} = \mathcal{P}_i\,.$$

However, there exists a critical wall thickness at which the pressure differential between the inside and outside of an isolated cell will not influence the circumferential stresses developing within the cell wall. This critical wall thickness is roughly 20% of the radius of the cell. This is shown in fig. 5.4, which plots the ratio of the circumferential wall stress to external (ambient) pressure as a function of the ratio of the wall thickness to the outer radius of the cell for four different ratios of internal to external pressure. As can be seen, for thin walls, as the internal pressure increases (as when the inner pressure is twice the external pressure, $\mathcal{P}_i/\mathcal{P}_o = 2$), the circumferential stress increases, but as the wall thickness increases, the circumferential stresses produced by different ratios of internal to external pressures become asymptotic to the value of the internal (turgor) pressure.

In contrast to an isolated cell, when a cell is embedded within a tissue the internal turgor pressure and the external pressures exerted by neighboring cells are equivalent or nearly so. Thus, from eq. (5.2) we can derive the following formula:

(5.5) $$\sigma_c = \frac{\mathcal{P}\,(r_o - t)^2 - r_o^2}{r_o^2 - (r_o - t)^2} = \frac{\mathcal{P}\,(t - 2r_o)}{2r_o - t}\,,$$

where $\mathcal{P}$ is the turgor pressure of the tissue. Thus no stress gradient across the cell wall will be produced.

THE IMPORTANCE OF THE FIBRILLAR NETWORK

The importance of the fibrillar network of microfibrils within the cell wall matrix is illustrated by the theoretical relations among the circumferential (hoop) and radial stresses within a cylindrical isotropic wall, the modulus of elasticity E, the Poisson's ratio of the wall material ν, and the circumferential (hoop) strain ε_c:

(5.6) $$E = \frac{\sigma_c - \nu\,\sigma_r}{\varepsilon_c}\,.$$

This equation indicates that the sum of the circumferential and radial stresses is constant throughout the thickness of the cell wall and that the two stresses produce uniform extension or contraction in the longitudinal direction of the cylindrical cell. Thus, without the microfibrils, the cross sections perpendicular to the longitudinal axis of the cell will remain in plane stress; each transection does not interfere with the deformation of its neighboring transections. Microfibrils running through cross sections of the cell wall will translate circumferential stresses diagonally from one transection to another. The extent to which these stresses are translated diagonally through the cell wall will depend on how the fibrils are anchored within the cell wall matrix. If they are arranged as "hoops" or are anchored at both ends to other fibrils, much like the wires embedded in the rubber latex of industrial vacuum tubing, then tensile strains will occur. If they are free at their ends, however, much like strands of pasta in a bowl of sauce, then shearing strains between the fibrils and the matrix will develop, and the shear modulus of the cell wall matrix will be critical in dictating the physical properties of the wall as cells grow.

The Potential Importance of the Matrix

Before describing the two models currently used to treat the biomechanics of cell wall extension, it is instructive to consider the likelihood that microfibrils can "slip" past one another within the wall matrix. The behavior of collenchyma cells provides some circumstantial evidence that microfibrils can slip within the cell wall matrix, since their cell walls swell and shrink depending on how fully the matrix is hydrated. When submerged in pure water, the primary cell wall thickness of collenchyma cells can swell to as much as 150% of that observed in fresh tissue. When placed under tensile stress, cell walls can extend to as much as 170% of their original length. Clearly, the pectin within the cell walls of this tissue can behave as a gel, and the high extensibility of the entire wall suggests that microfibrils can significantly move past one another when the matrix is hydrated, placed in tension, or both. The slippage of microfibrils within the gelatinous matrix appears significant in determining the physical attributes of collenchyma, which in organs provides mechanical support while not inhibiting longitudinal growth. Additional, albeit highly circumstantial, evidence for the slippage of microfibrils and the importance of the material properties of the matrix under a shearing load comes from rheological investigations showing that collenchyma is a nonlinear viscoelastic material whose shear modulus varies with the age of the tissue (see chap. 6 for further details).

In any circumstances, biomechanical models of cell wall stretching must take into account the way microfibrils are anchored within the cell wall, as well as the strain gradients across cell walls that differ for isolated cells and cells within a tissue. If a model views changes in microfibril orientation within each wall layer caused by cell growth as being the result of a nonlinear stress gradient, then it must acknowledge that there exists another nonlinear, hydrostatic (turgor) stress gradient in the opposing direction. It must also deal with the absence of a hydrostatic stress gradient across cell walls embedded within tissues. Unfortunately, no current model deals adequately with all of these features.

TWO MODELS FOR CELL WALL EXTENSION

There are two competing models that attempt to deal with cell wall extension: the multinet model asserts that new microfibrils are deposited in a predominantly transverse orientation to the future axis of cell elongation, and the helicoidal model asserts that microfibrils are arranged in parallel arrays but that the orientation angles of successively deposited wall layers shift in an alternating or rotational pattern. These two models share many of the same underlying assumptions, principally that tensile stresses reorientate microfibrils; both argue that each cell wall layer experiences a different strain depending on its distance from the plasma membrane and the extent of cell elongation; and both models essentially ignore the implications of microfibril anchorage within the matrix as well as the importance of the cell wall's modulus in shear. The two models differ in only one essential way—the initial orientation of microfibrils within newly deposited cell wall layers. Unfortunately, both models are inadequate to the task of defining a rigorous conceptual framework in which the biophysical features of cell elongation and morphogenesis can be understood, but each provides an internally consistent logic that can assist in experimental design and can be used to review what is currently known about cell growth and morphogenesis.

THE MULTINET MODEL

The multinet model of cell wall growth asserts that new microfibrils are deposited in a predominantly transverse orientation to the future axis of cell elongation and that, as elongation proceeds, each sequential layer of microfibrils is stretched longitudinally and circumferentially so that the microfibrils in successively older wall layers are progressively reoriented in the longitudinal direction (Roelofsen 1951; see Preston 1974, 385–96). Also, since the volume of each layer is assumed to remain constant after the layer has been synthesized, each layer becomes progressively "thinner" as the cell elongates and expands. Most of these precepts are illustrated in figures 5.5 and 5.6,

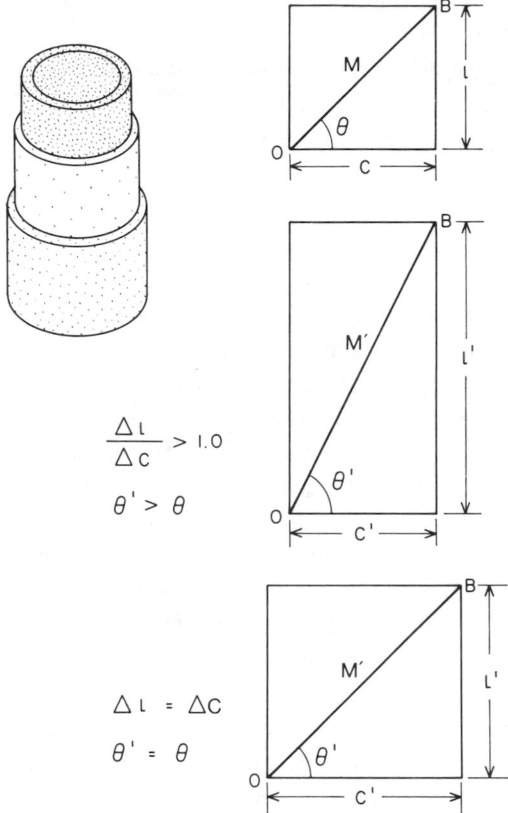

FIGURE 5.5 Hypothetical reorientations of a microfibril *M* of a cell wall layer within an elongating or uniformly expanding cell (upper left diagram). The cell wall layer can be envisioned as a splayed-out square with a length *l* and circumferential width *C*. The orientation of a microfibril *M* (with a length equal to the diagonal line O–B) is denoted by the angle θ, such that $\tan \theta = l/C$ (upper right diagram). If the cell wall preferentially elongates but does not expand ($C' = C$), then the microfibril will be stretched to a length l' and will be reoriented with an angle θ′ greater than θ (middle diagram). If the cell elongates and expands at the same rate (the change in *l* equals the change in *C*), then the microfibril will retain its original orientation ($\theta' = \theta$) but will be stretched (*M′* is greater than *M*, lower diagram).

which provide a graphic basis for reviewing the multinet model in further detail.

The reorientation of microfibrils within a cell wall layer as the cell elongates is diagramed in figure 5.5. The original cell wall layer is drawn as a cylinder, whose circumference *C* and length *l* are equal, cut along its length

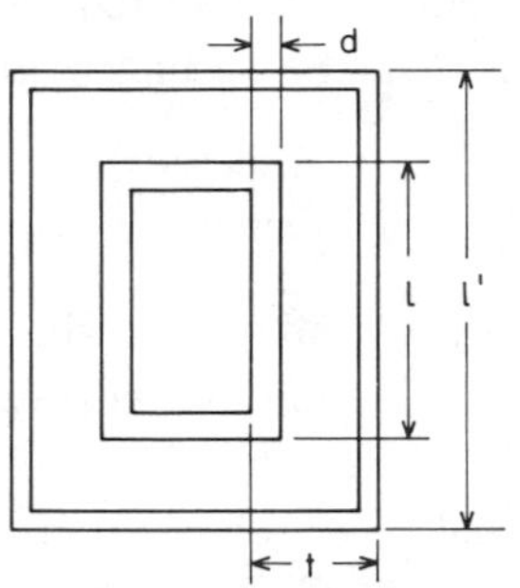

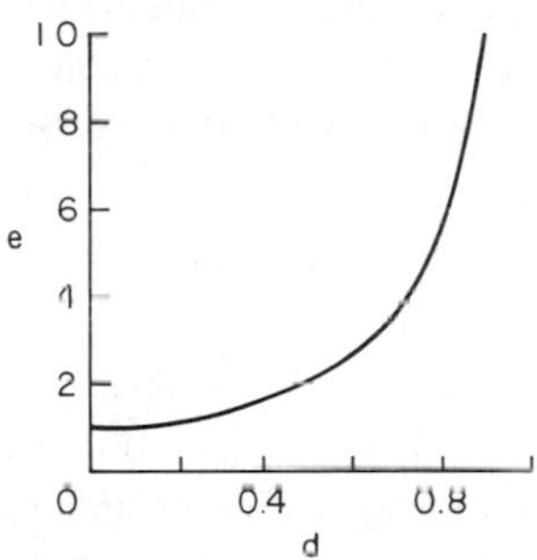

FIGURE 5.6 Nonlinear strain gradient across the cell wall predicted by the multinet model. The innermost cell wall layer with length l and distance d from the plasmalemma will be pushed outward, with a concomitant alteration in length l' (top diagram). The proportional strain e is predicted to increase rapidly as d increases (lower graph).

and unrolled. A single microfibril M, representative of many others, is also shown with its original orientation angle θ. (For simplicity, the microfibril is drawn as a diagonal line with a relatively steep orientation angle, whereas the multinet model would argue that the microfibril should lie more or less in the horizontal direction. A horizontal orientation, however, is not essential to illustrate the predicted future deformations of the microfibril as predicted by the model.) From figure 5.5, the longitudinal and circumferential strains, ε_l and ε_c, can easily be calculated from the simple trigonometric relationships

(5.7a) $$\varepsilon_l = \frac{l' - l}{l} = \frac{l'}{l} - 1 = \frac{C' \tan \theta'}{C \tan \theta} - 1$$

(5.7b) $$\varepsilon_c = \frac{C' - C}{C} = \frac{C'}{C} - 1 = \frac{l' \tan \theta}{l \tan \theta'} - 1 ,$$

where the prime notation indicates the altered (deformed) dimension. Also, the strain on the microfibril M can be calculated according to the formula

(5.7c) $$\varepsilon_m = \frac{l' \sin \theta}{l \sin \theta'} - 1 .$$

Notice that if the microfibril was allowed to slip within the matrix, then its strain would be significantly different from that calculated by eq. (5.7c), yet it would still be reoriented as the matrix of the wall layer was strained. Hence the model cannot discriminate from observations of microfibrillar reorientations the extent to which microfibrils are anchored within the cell wall layer. Also note that the orientation angle of a microfibril will not change in a cell whose expansion and elongation are equivalent, as in the case of an isodiametric cell that increases in volume while maintaining its original spherical shape (shown at the bottom of fig. 5.5). Nonetheless, the microfibrils within such a wall layer will experience tensile strains, since they must increase in length as the wall layer is "pushed" outward.

The multinet model predicts that the strain gradient across the cell wall will be nonlinear, as shown in figure 5.6. Microfibrils embedded within the innermost (youngest) wall layers will experience little or no tensile strain; those embedded in the outermost (oldest) wall layer will experience the highest tensile strains; and those within the intervening wall layers will experience exponentially higher tensile strains as a function of their distance d from the innermost wall layer. The proportional strains e across the entire cell wall can be calculated in relation to the maximum strain experienced at the outermost wall layer as a function of the wall thickness t and the distance d:

(5.8) $$e = \frac{t}{t - d} .$$

This equation indicates that at the plasma membrane ($d = 0$) the proportional strain e equals one and that as d approaches the external surface of the cell wall e increases dramatically to a value defined by ε_l and ε_c (see eq. 5.7a,b). Calculations indicate that if the cell maintains a constant overall thickness, then the inner half of the wall receives proportionally much less strain than the outer half of the cell wall. Assuming that the original orientation of microfibrils in each wall layer is transverse, the distribution of strains across the wall is not sufficient to change the net orientation of microfibrils to the longitudinal direction. Consequently, even with reorientations of microfibrils in the outermost layers of the cell wall, the elongating cell will continue to be "girdled" by predominantly transverse microfibrils.

Some of the predictions of the multinet model have been shown to be consistent with experimental observation, but others have not. The passive realignment of fibrils has been demonstrated experimentally by physically clamping the cells of the green alga *Nitella* and placing them in tension (Green 1960; Gertel and Green 1977; Metraux, Richmond, and Taiz 1980). Stretching alters the alignment of microfibrils in the outermost wall layers of these algal cells but does not alter the orientation angles of microfibrils in the inner cell wall layers. Experiments of this sort involve the application of external tensile stresses that operate over a short time, however, and the behavior of cell walls considered by the multinet model must be evaluated in the context of long durations of stress applications, owing to the viscoelastic nature of growing cell walls. That is, the reorientation of microfibrils under abrupt tensile stress need not reflect the phenomenology of steady-state, relatively slow, cellular growth. Similarly, experiments using isolated cell walls cannot detect the effects of protoplasmic metabolic activity on the mechanical behavior of cell walls. Finally, patterns of microfibrillar orientation developing during growth often fail to conform to those predicted by the multinet model. For example, the pericyclic fibers of the asparagus plant (*Asparagus officinalis*) have transversely oriented microfibrils on inner wall layers deposited only in the early stages of elongation (Sterling and Spit 1957). During subsequent elongation (these cells can reach twenty times their original length), microfibrils within the inner wall layers have been observed to be organized into sets of helices with a slope of 60° to the cell axis, while microfibrils on the outermost wall layer lie mostly in the transverse direction. This is precisely the opposite of the orientation predicted by the multinet model. Since these fibers have apical intrusive growth, one might argue that their growth pattern, which differs from that of most cells in tissues, may not be especially relevant to the multinet model. But careful ultrastructural observations indicate that the primary cell walls of isolated cells and of cells aggregated into tissues consist of parallel arrays of microfibrils that differ in direction among successive wall layers, giving rise to a characteristic successively arched pattern when cell walls are sectioned obliquely. These and other observations suggest that one of the assumptions made by the multinet model (that all newly formed microfibrils are oriented transversely to the future plane of cell elongation) may be in error, or at least that there are enough exceptions that the assumption must be qualified. The characteristic arched patterns, which give a herringbone appearance to oblique sections through some plant cell walls, are the observational basis for an alternative that has been called the helicoidal model. This model shares many of the assumptions made by the multinet

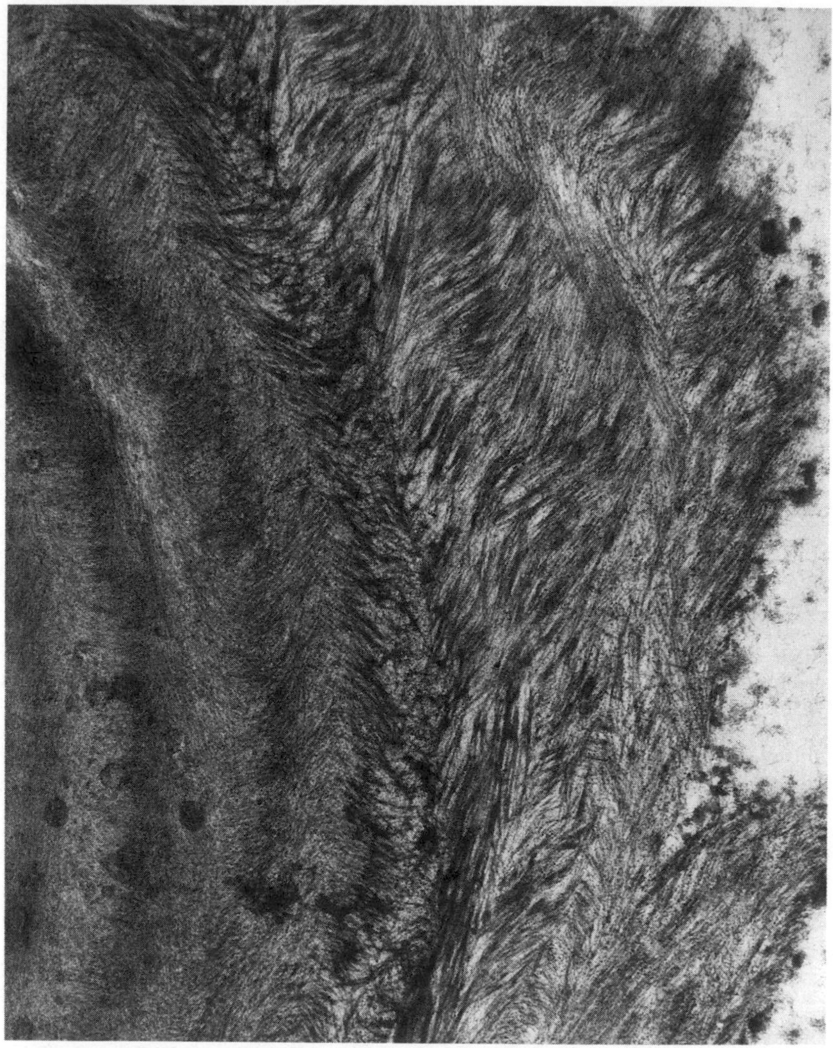

FIGURE 5.7 Patterns of microfibrillar orientations in sequential cell wall layers predicted by the helicoidal model. The helicoidal model states that a more or less constant angle is maintained between the orientations of microfibrils from one cell wall layer to the next. When the cell wall is obliquely sectioned, revealing progressively more internal wall layers, the constant mutual angle results in a herringbone pattern—obliquely sectioned walls will appear to have "arcs" of ascending and descending fibrils.

model, however, and should be considered an amended version of its intellectual predecessor.

THE HELICOIDAL MODEL

The helicoidal model asserts that the cell wall is synthesized in successive wall layers, each containing microfibrils arranged in parallel arrays, but with the microfibrillar orientation angle shifting among them in a clockwise rotating pattern (Neville 1985, 1986). Each shift at its initial (synthesis) stage involves a constant mutual angle. This results in an arched, herringbone pattern of microfibrillar orientations when cell walls are sectioned obliquely. Each arch corresponds to a rotation of 180° that brings microfibrillar parallel arrays into an antiparallel orientation. Consequently, two adjoining arches correspond to a 360° rotation and a parallel realignment of microfibrils (this is shown diagrammatically in fig. 5.7).

The helicoidal model can be visualized by considering the orientation of microfibrils as parallel to the minute hand of a clock. As the hand rotates, so does the orientation of successive arrays of microfibrils. If each layer within the cell wall requires equivalent time to be deposited, then a constant mutual angle will occur between successive wall layers. If the hand of the clock moves sporadically or unevenly (because different wall layers require different amounts of time to be deposited), then the angles between successive wall layers will be unequal. This can be visualized by drawing parallel lines on mobile, superimposable disks of paper. Rotating each disk with respect to another is equivalent to changing the orientation angle in neighboring sets of microfibrils. Varying the angle generates a variety of arched patterns, many observed in real cell walls.

As initially postulated, the helicoidal model differs from the multinet model only in the initial orientation of microfibrils in the youngest cell wall layer. Both models envisage a passive reorientation of microfibrils as the cell elongates, and the matrix extensibility and extension of microfibrils within the matrix are intrinsic features of both. Additionally, both models share the concept of a nonlinear strain gradient across the width of the cell wall, as well as the notion that the oldest (outermost) wall layers will be "thinned" as elongation proceeds. The helicoidal arrangement will be maintained in the innermost wall layers and distorted into a crossed polylamellate (herringbone) pattern in the older cell wall layers. As in the multinet model, spherical cells are predicted to retain undistorted arched patterns of microfibrils, since growth in width and length is isometric.

The problems with the helicoidal model are numerous, but two (one conceptual, the other procedural) deserve attention. First, if a constant mutual orientation angle is maintained and if this angle is small, then cell wall expansion and elongation are physically impossible given the assumption that microfibrils are anchored at their ends and cannot slide through the matrix or deform laterally. Second, the herringbone patterns seen when cell walls are sectioned obliquely may be artifacts resulting from the shrinkage of cell walls and the resulting concertina-like distortions that can occur in wall layers. Some very simple physical models (superimposed sheets of cellulose acetate, on which parallel arrays of lines are drawn, yield herringbone patterns when they are crumpled and sectioned obliquely) confirm the possibility that the helicoidal arrays seen in real walls may be artifacts of tissue preparation. The first of these two problems is not too damaging, since the helicoidal model does not have to assume that the orientation angles are small, nor must it absolutely insist that the angles are constant. Slight modifications of these assumptions yield a model that provides for the physical possibility of cell wall elongation and expansion. The second problem, however, is potentially serious enough to warrant concern. Procedures exist for eliminating cell wall shrinkage or at least ensuring that it has not occurred, but in my opinion, insufficient attention is currently paid to problems of artifact.

Refinements of the Models

As previously noted, refinements of both the multinet and helicoidal models are required for full understanding of the phenomenology of cell wall expansion and elongation. Some of these have already been accomplished; others are still absent from the literature. Since the helicoidal model and the multinet model are two variants on a single theme, the following discussion applies to both.

One refinement that is absolutely essential to understanding cell wall biophysics is to note that the cell wall is a composite material consisting, at the very least, of two components—a fibrous, polymeric (extended solid) microfibrillar component, and a more ductile, perhaps amorphous matrix. The high modulus of elasticity of cellulose (roughly 30 to 40 $GN \cdot m^{-2}$), which can account for as much as 20% of the dry weight of primary cell walls, and the much lower modulus of the cell wall matrix (from 10^7 to 10^8 $N \cdot m^{-2}$) are compatible with this suggestion, as are most ultrastructural observations. Further, it would be desirable to precisely define the category of composite material that best suits the empirical data on cell walls. That is, there are a variety of ways a composite material can be constructed. One that is known as a periodic microstructured composite, in which fiber-fiber interactions dictate and induce deformations when the composite is stressed (Walker, Jordan, and

Freed 1989), appears particularly relevant. Periodic microstructured composites are typically composed of a metal (ductile) "matrix" that is structurally reinforced with "fibers." When the fibers occupy a large volume fraction in such a composite material, deformations within one fiber induce deformations in neighboring fibers owing to the transmission of shear stresses through the matrix. When the fibrous component within such a composite material dominates in terms of volume fraction (the critical percentage is 20%), fiber-fiber interactions become the dominant feature in the constitutive formulations required to predict mechanical behavior.

The mathematical description of the mechanical behavior of periodic microstructured composites is well beyond the scope of this chapter, but there exists a substantial body of literature on cell wall biophysics that the botanist can refer to. For example, the stress-strain history in each unit periodic "cell" (fiber) is temperature dependent, and the inelastic and viscoelastic behavior of the constituents in the unit periodic cells are mathematically predictable (see Weng and Chiang 1984; Tandon and Weng 1986 for treatments of this subject). Further, Nemat-Nasser and his colleagues have developed elastic, plastic, and creep constitutive models for the behavior of these composite materials (Nemat-Nasser, Iwakuma, and Hejazi 1982; Nemat-Nasser and Taya 1981; Nemat-Nasser and Iwakuma 1983), while Walker, Jordan, and Freed (1989) have formulated equations to deal with nonperiodic microstructured composites exhibiting viscoelastic behavior. Research in this area has progressed to treat the strains developing in the "matrix" phase of composites resulting from the strains developing in the "fibers." Also, the behavior of laminated composites has been approached mathematically and empirically. It is likely that the same procedures could be applied to our understanding of the mechanical behavior of primary cell walls. Among the many benefits of such an approach would be a deeper appreciation of how cell walls are metabolically controlled in the context of the yielding of cross-linkages among neighboring microfibrils (fiber-fiber interactions) and the influence of the material properties of the matrix on mechanical deformations among neighboring microfibrils. For example, when periodic microstructured composites are placed in tension, the fibrous infrastructure progressively distorts and eventually disperses into the flowing matrix. When the tensile stresses are removed, the fibrous infrastructure can reassemble, much like some of the effects envisaged in real cell walls. Sadly, in light of what is known about the behavior of fabricated composite materials, the complexity in the biophysics of plant cell walls is most likely highly underestimated.

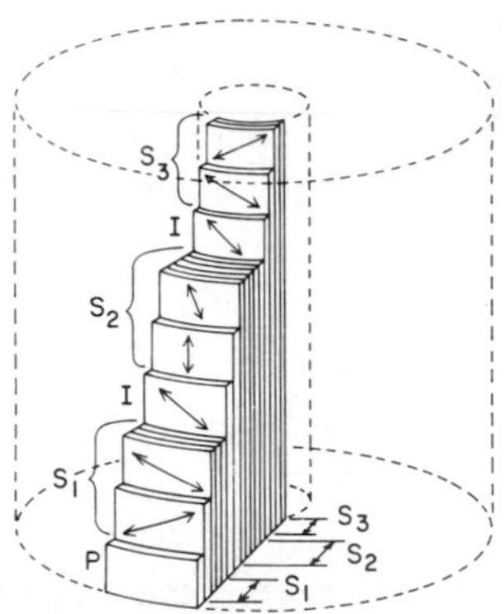

FIGURE 5.8 Sequential wall layers (S1–S3) and average microfibrillar orientation (arrows) within the secondary cell wall. The layers are numbered from the first (S1) to the last (S3) formed (S1 is in contact with the primary cell wall P). An intervening wall layer (I) exists between S1 and S2 and between S2 and S3. The microfibrillar orientation within each I layer introgrades between the set of its flanking S wall layers.

SECONDARY CELL WALLS

The protoplasts of some cells secrete a secondary cell wall internal to the primary one (see fig. 5.1). The initial deposition of the secondary cell wall may or may not coincide with the terminal phase of cellular expansion and elongation, but because of chemical alterations that occur throughout the cell wall during the deposition of subsequent secondary wall layers, we can view the bulk of the secondary wall as a secretory product of nonelongating cells. The secondary cell wall is a prominent feature of cell types found in both primary and secondary tissues (sclerids, fibers, collenchyma, xylem, and phloem).

Data indicate that the architecture and spatial arrangement of microfibrils, together with the potential for the chemical impregnation of the entire cell wall, dictate the physical properties of isolated cells and tissues subjected to mechanical forces. For example, the mechanical properties of tissues largely lacking secondary cell walls are highly dependent on the physiological status and water content of the protoplast, whereas the mechanical properties of tissues in which the secondary cell walls are highly developed appear less variable in tissue water content. Tissues consisting of thin-walled (primary cell wall) cells are stronger in tension than in compression, but they are susceptible to cell-cell debonding modes of mechanical failure. The reverse is typically true for tissues composed of cells with secondary cell walls. A clear distinction, however, must be maintained between the physical properties of cell walls and tissues, since in the latter case we are dealing with the compound middle lamella and its adhesive effects on neighboring cell walls.

The sequential walls deposited within the secondary wall are designated S1, S2, and S3 (see fig. 5.8). The letter S is the traditional shorthand for "secondary," while the numbers indicate the temporal sequence of deposition. Thus the S1 layer is the outermost (first formed) secondary cell wall layer, while the S3 layer is the innermost (youngest) layer. Some cells, particularly those formed in tension wood (to be considered in chap. 8), secrete a gelatinous layer, denoted by G (which in this case is not the shear modulus), between the S3 layer and the plasma membrane. The G layer has been shown to be rich in the enzyme lactase.

Evidence is accumulating that the secondary wall has a helicoidal structure, much like the one hypothesized for the primary cell wall. It has been recognized that the S1, S2, and S3 layers have different microfibrillar orientation angles—the microfibrils in the former two layers are more perpendicular to the longitudinal axis of the cell than the microfibrils in the S3 layer. But detailed studies of transmission electron micrographs reveal that each S layer is composed of many thinner layers with variations in microfibrillar orientation angles. Also, recent studies show that there exist intervening wall layers between the S1 and S2 layers and the S2 and S3 layers. The microfibrillar orientation angles of these intervening layers intergrade with those of the two S layers that flank them. Accordingly, the secondary wall appears to have a helicoidal architecture that may be continuous with that of the primary cell wall. Clearly, however, since the S layers are secreted after the bulk of cell elongation is completed, the differences in the microfibrillar orientation angles seen in the secondary wall layers cannot be the result of passive realignments induced by mechanical growth stresses. Also, unambiguous evidence for a complete continuity of the "helicoid" across the primary to the secondary wall interface is not available. The outermost lamination of the S1 layer may contain transversely oriented microfibrils. If so, then it is possible that in cells producing secondary walls the helicoidal "clocklike" pattern producing mutual angles between adjoining wall layers continues in an uninterrupted sequence. Alternatively, if evidence was found for an intervening layer in which the mutual angle was interrupted, then the clock of the helicoid would be "reset" with the advent of an ontogenetically distinct phase of wall synthesis.

Regardless of the way its layers are deposited, the secondary wall is neither spatially nor chemically homogeneous. In secondary xylem ("wood") produced late in the growth season, the S2 layer is typically thicker than the S1 layer, while the S3 can be either very thin or entirely lacking. In early spring wood, cells have an S2 layer that is very much thinner than either the S1 or the S3 layer. Hence the heterogeneity of wall layer thicknesses depends in part on the development of the individual cell and on the developmental status of the tissue. Additionally, secondary wall deposition retroactively alters the chemistry of the entire cell wall. Lignin, which is deposited in the cell wall when the S layers are synthesized, is typically found in higher concentrations in the compound middle lamella (Bailey 1936; Preston 1974, 288–91). Lignin can be found concentrated at the corners of cell walls, within radial walls as opposed to tangential walls, or in concentric lignin-rich layers alternating with cellulose-rich layers within the cell wall. Significantly, delignification produces little or no change in the X-ray diffraction patterns of cell walls, which are dictated by the crystalline cellulose network. Thus lignification occurs in sites removed from physical contact with the plasma membrane and involves a chemical impregnation process through the matrix of the cell wall that does not disrupt the architecture and pattern of deposition of cellulosic fibrils.

Other chemical constituents within the cell wall show nonlinear gradients in concentration. For example, arabinan is largely confined to the compound middle lamella, while galactan is absent in the S2 and S3 layers. Cellulose is higher in concentration within the S layers than in the primary cell wall. These "countergradients," together with the infiltration of lignin, provide a chemical basis for interpreting the temporal pattern of overall cell wall stiffening as cells mature. The material properties of the cell wall temporally shift from those dominated by the ductile matrix (young, expanding cells) to those dominated by a rigidifying microfibrillar network (old, mature cells). Lignification, which is superimposed on this temporal shift in the abundance of polysaccharides, retroactively stiffens and waterproofs the entire cell wall. Although not a mechanically useful material in its own right, lignin (which has little resistance to tensile stresses) lets less water infiltrate cell walls. Since the tensile modulus of cellulose decreases as a function of hydration, the importance of lignin to the mechanical behavior of cell walls is that it stabilizes the mechanical properties of the fibrillar cellulose network within the cell wall. Essentially, lignin provides a safety factor, and it can increase the compressive stresses that cell walls can sustain before they fracture.

Wood Fracture Mechanics

The foregoing explanation provides some insight into how tissues with secondary cell walls can mechanically fail. For example, microfibrillar orienta-

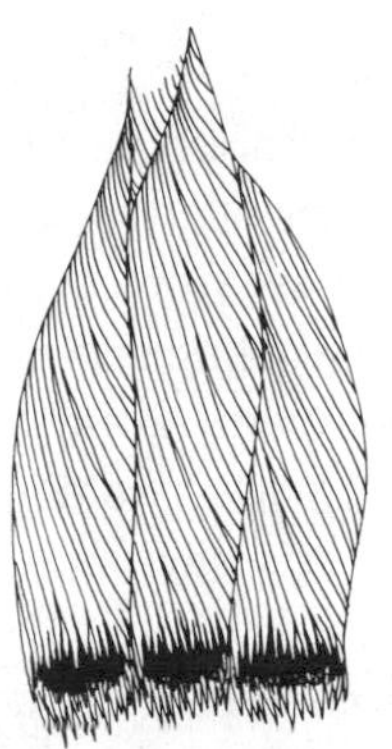
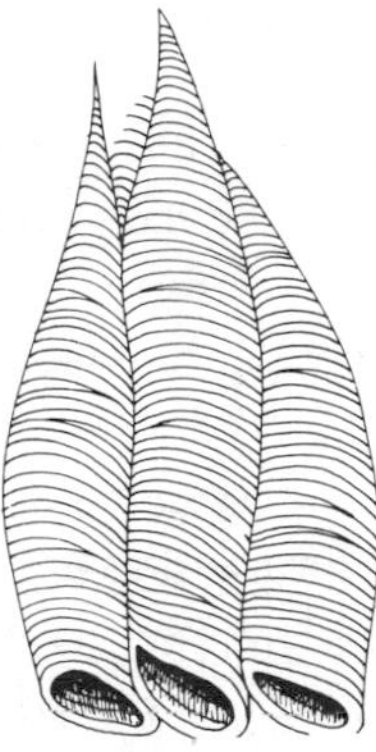

FIGURE 5.9 Two modes of tracheid wall fracture dependent on the orientation of microfibrils in S2 (see fig. 5.8). (The dominant microfibrillar orientation in the S2 layer is indicated by the hatching on the surface of the cells shown in this figure.) Tough fracture occurs when a cell is placed in tension along its length and when the S2 wall layer has a steep microfibrillar orientation angle (A). This mode of failure involves the cleavage of primary covalent bonds of cellulose chains, and cellulosic microfibrils break. Brash fracture occurs when microfibrils have a low orientation angle (B) and secondary bonds between neighboring microfibrils break, resulting in the separation of microfibrils in the S2 layer after the mechanical failure of the S1 layer.

tion within the primary and secondary cell walls profoundly influences the way tissues fracture. This was elegantly demonstrated by Mark (1967), whose studies indicate that wood fracture is rarely initiated within the compound middle lamella. Rather, the S1 layer is the first to undergo mechanical failure, usually because of shearing. Mark also showed that the shear stresses in the S1 and S2 layers are typically opposite in direction, as might be anticipated from the difference in their microfibrillar orientation angles (see fig. 5.8). The consequent separation of these two layers produces a second point of fracture. Depending on the orientation angle of microfibrils, two forms of fracture can occur, as shown in figure 5.9. One form, known as tough fracture, involves the breaking of primary covalent bonds within the cellulose polymers and occurs when microfibrils are more or less aligned with the direction of cell length. This mode of fracture occurs relatively slowly and obviously requires considerable energy. The second form of fracture, called brash fracture, involves the breaking of secondary bonds, allowing microfibrils to separate along their length, and occurs when microfibrils are predominantly aligned circumferentially to the longitudinal axis of cells. Brash fractures occur suddenly and require less energy than tough fractures.

MODELING CELL WALL FAILURE

The modeling of isolated cells provides a reference point concerning the adjustments that occur at the molecular level involving the absorption of strain energy when cells are deformed in either tensile or compressive loading. Modeling is useful here because a number of vital aspects of cell wall failure cannot be directly observed. Modeling also provides a way to compare anticipated results based on assumptions that reflect what we think we know with observed empirical results that are immune to arbitration. For instance, no straightforward empirical studies are available to show whether microfibrils slip when an isolated cell is placed in tension. If they do, however, then the wall can be modeled as an array of helical springs arranged in parallel. Such a helical spring model would provide a basis for relating the properties of a spring to those of the fibrillar cellulose network within the cell wall. Such a model would predict that the modulus of elasticity would increase as the orientation angle decreases. A number of workers have shown a relation between the modulus of elasticity measured for isolated cells and the fibrillar (spring) orientation angle θ. Working with cotton hairs, Spark, Darnborough, and Preston (1958) showed that when $\theta = 10°$ the modulus of elasticity was thirty times higher than when $\theta = 50°$, while Balashov et al. (1957) showed that stretching decreases the orientation angle in isolated cell walls, just as a spring model would predict. Hearle (1958, 1963) proposed a theoretical relation between the effective modulus of elasticity, E_θ, measured parallel to cell length, and the orientation angle *q:*

$$E_\theta = \frac{E\ F(\theta)\ [K\ (1 - 2\cot^2 q)^2]}{E\ F(\theta) + K\ (1 - 2\cot^2 q)^2}, \tag{5.9}$$

where K is the bulk modulus (which is assumed to be three times that of the cell wall matrix; the cellulosic component is assumed to be incompressible) and $F(\theta)$ is a function of the pitch angle of fibers based on the theory of twisted yarns. This equation is not accurate for low orientation angles (1°–18°), but when the modulus of elasticity is plotted against the orientation angle of microfibrils in plant fibers and compared with predictions from Hearle's analysis, a remarkably good fit is seen. The correlation between predictions from a spring model and the observed value of E is even more remarkable when we consider the assumptions underlying Hearle's theory and appreciate the variety of experimental formats used to accumulate data on fibers.

Six

The Mechanical Behavior of Tissues

> Let me try to illustrate this by a few examples picked somewhat at random out of thousands, and possibly not just the best ones to appeal to a reader.
>
> Edwin Schrödinger, *What Is Life?*

This chapter treats plant biomechanics at the interface between cell wall architecture and the mature plant organ—the tissue level of organization. From a biomechanical perspective, this level of organization is distinct because the mechanical behavior of a tissue involves two components: the protoplast (when it is retained in a mature tissue) and the internal scaffold of cell walls (secreted by the protoplast), whose geometry differs from tissue to tissue. Each type of tissue can be defined based on the metabolic condition of its protoplasm and the three-dimensional arrangement of its cell walls, and each has a structure and mechanical properties that cannot be deduced solely from either of its two components.

Nonetheless, the tissue level of organization may appear to be an intellectual abstraction resulting from a reductionist view of plants. This criticism is valid up to a point. From previous chapters we appreciate that the size, shape, and development of an organism influence the magnitude and distribution of mechanical stresses that will occur within each of its tissues. Hence the mechanical behavior of a tissue is subservient to and integrated within the individual plant body and must be viewed in that context. But familiarity with plants should persuade us that in many cases the whole organism can be remarkably uniform in its cellular construction, while many plant organs (e.g., potato tubers, thorns, spines) consist predominantly or entirely of a single type of tissue (storage parenchyma, sclerenchyma). In the case of the plant body, many algal species are parenchymatous or pseudoparenchymatous in their tissue construction. The former, consisting of a living protoplast incompletely partitioned by cell walls through which plasmodesmata interconnect all neighboring cells, not infrequently lacks evident cellular specialization. Likewise, the pseudoparenchymatous construction, which consists of intertwining branched or unbranched filaments of cells in which only some cell-to-cell contact surfaces are interconnected by plasmodesmata, may exhibit little or no cellular specialization. Nonetheless, both types of tissue construction can be used to construct remarkably complex morphologies, particularly in an aquatic milieu where there exists little or no intrinsic limitation on the ultimate size of an organism. Indeed, specimens of marine kelp are some of the largest plants in the world. For these plants, a biomechanical perspective based on the tissue level of organization is useful and biologically meaningful. Additionally, there are pedagogic reasons for focusing on the mechanical behavior of individual tissue types. Different tissues manifest different mechanical properties, and understanding the mechanical behavior of an organ, much less an entire plant, is often contingent on our understanding the differences among types of tissues. A rough estimate of the number of different types of plant cells (aggregates of which can constitute a single tissue) is twenty-five. If only two types of tissue are combined to form an organ, then there are a minimum of roughly 2^{25} or 10^7 possible combinations. This calculation is not intellectually vacuous. It forces us to appreciate that cells and tissues are not combined randomly or with equiprobability. Among the many possible combinations, only a relatively few have been used by plants, and some are used more frequently than others. By now the reason for this should be immediately apparent—only a few are structurally compatible with the functions essential to plant growth and survival.

Any useful perspective on plant tissues must give insight into both plant anatomy and tissue biomechanics—it should provide an appreciation of the diversity seen in cell size, shape, and function (anatomical distinctions among tissues) and permit the quantification of how biomechanical performance relates to cell size, shape, and function. We can gain such a perspective by combining the precepts of the organismic theory with those of the theory of cellular solids. The organismic theory was treated in chapter 1. It argues that the entire plant body, however large and internally complex it may be, consists of a single protoplasmic phase that controls growth and development. At one extreme this protoplasm may secrete a single cell wall; at the other extreme it may partition itself into a number of cells, albeit incompletely, by retaining cytoplasmic continuity among neighboring cells in the form of plasmodesmata. The multicellular organism secretes an internal system of cell walls whose geometry may vary from one part of the organism to another. This geometric heterogeneity is reflected in the criteria anatomists use to designate types of tissues that have different physical and functional attributes depending on whether the protoplasm functionally persists or dies, as well as on the fabric of the cell wall infrastructure. The cellularity of the protoplasm—how far the protoplast is variously partitioned—provides for mechanical as well as physiological and reproductive specialization. Although the latter kinds of specialization are important, this chapter will emphasize the mechanical roles of tissues in terms of the architecture of their cell walls and the metabolic condition of their protoplasm.

This emphasis would not be quantitatively meaningful without the theory of cellular solids. This branch of engineering theory treats the behavior of composite materials that contain a solid phase (consisting of walls, struts, beams, or columns) that variously partitions a fluid phase. Plant tissues have a system of cell walls consisting of walls or struts or some other partitioning geometry. The fluid phase in tissues may be a liquid, such as the living protoplasm found in parenchyma, or a gas, such as the air-filled lumens of the dead cells found in cork and wood, or a combination of both types of fluids, such as in aerenchyma, which has living cells with a solid and a fluid phase arranged to form gas-filled chambers and canals. Tissues like aerenchyma, cork, and wood are manifestly cellular solids, as are parenchyma and collenchyma. The distinction between a living and a dead tissue largely depends on whether the solid phase (cell walls) permeates a liquid phase (the protoplast) that can exert transient changes in hydrostatic pressure or whether the solid phase permeates a gaseous phase that varies little in pressure owing to its equilibrium with the external atmosphere.

Although the mechanical perspective on tissues provided by the theory of cellular solids may as yet be unfamiliar, it is extremely useful. It affords an understanding of why living plant tissues deform (wilt) when they lose water, and it helps us understand why wood can get stronger when it is compressed. It also gives ecological insights into the mechanical advantages of possessing dead thick-walled tissues in habitats that experience periodic water stress versus possessing living thin-walled tissues in an aquatic habitat. No single theory can explain all the anatomical and morphological diversity seen in plants, but the theory of cellular solids does more than a little to clarify the functional significance of different types of tissues.

MERISTEMS AND TISSUE SYSTEMS

Our treatment of plant tissues begins with a review of the meristematic origin of the various tissue types. Those who are already familiar with this topic can

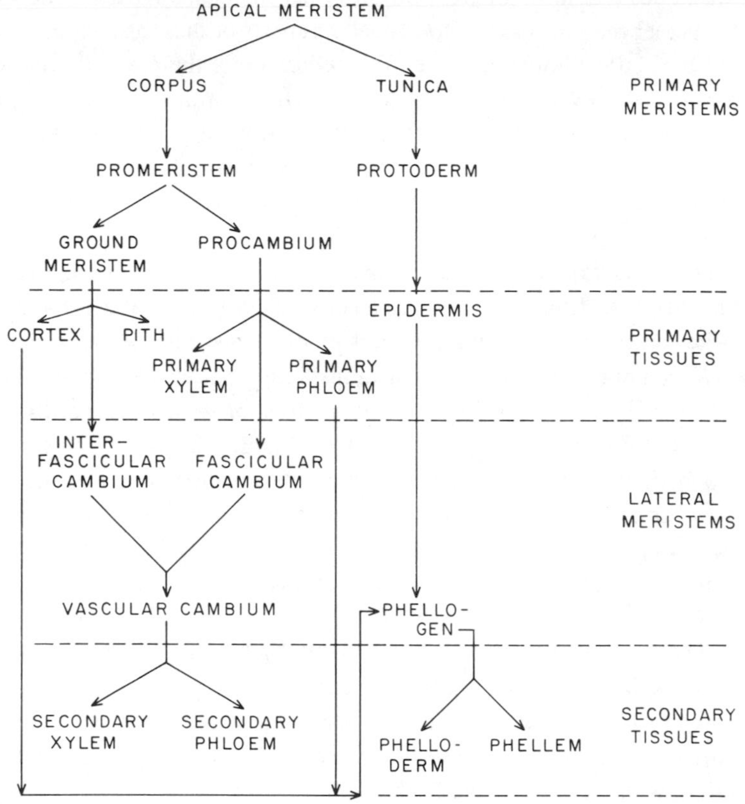

FIGURE 6.1 Flow diagram illustrating some of the developmental relations among the various tissue systems that are produced by the shoot apical meristem of a dicot. The diagram indicates the course of development from the tip toward the base of the shoot. The apical meristem is viewed here as comprising at least two components, a tunica and a corpus. The tunica, which consists of one or more layers of cells that cover the corpus, gives rise to the protoderm, a meristematic layer of cells that gives rise to the epidermis. The corpus, which consists of a more or less lenticular mass of cells, gives rise to the promeristem, which differentiates into the procambium and the ground meristem. The procambium gives rise to primary xylem and phloem; the ground meristem gives rise (in dicots) to the pith and the cortex. In some plants, secondary growth occurs after primary ontogeny. This involves the production of two lateral meristems; the innermost is called the vascular cambium. It develops from fascicular and interfascicular cambia (which develop in and between the primary vascular bundles, respectively). The second, more external lateral meristem is called the phellogen (sometimes referred to as the cork cambium), and its origin may be from subepidermal, cortical, or primary phloem parenchyma cells. The meristematic cells with the phellogen divide to produce the phellem externally and the phelloderm internally. Collectively, the phellem and the phelloderm are called the periderm. The "bark" of a tree trunk or branch consists of all the cell layers external to the vascular cambium.

move on to the following section (The Apoplast and the Symplast).

Plant anatomists have categorized plant tissues based on cellular morphology, function, and development. Although differences in morphology and function are clearly important, developmental criteria are more comprehensive in distinguishing among the various types of tissue, because different tissues may converge in their cellular appearance and functional roles even within the same plant. Perhaps more important, many of the anatomical and functional features used to diagnose tissues are prefigured in the growth regions giving rise to them, thereby emphasizing the developmental integration that exists at the whole plant level and the need for an organismic perspective on plant anatomy (fig. 6.1).

Cellular regions within the plant body that retain an embryonic capacity to provide new and undifferentiated cells are called meristems. Meristematic cells divide to give rise to derivative cells that subsequently differentiate and mature into the various types of cells and tissues found within the plant body. There are two types of plant meristems (see fig. 6.1): apical meristems, found at the tips of roots and stems, give rise to derivative cells that prefigure the primary tissues of the plant body; lateral meristems, found as more or less cylindrical layers of meristematic cells running the length of some stems and roots, give rise to derivative cells that will differentiate and mature into the secondary tissues of the plant body. Apical meristems provide the potential for stems and roots to increase in length indefinitely by means of primary growth. Lateral meristems permit stems and roots to increase in girth by means of secondary growth. If the primary growth of a plant is such that apical meristems continue to produce derivative cells unless they are traumatically killed, then the plant is said to exhibit indeterminate growth. When growth is developmentally truncated, the plant exhibits determinate growth.

Three primary tissue systems are traditionally distinguished by their primary meristematic origins (Esau 1977)—the dermal tissue system, the primary vascular system, and the ground tissue system. The dermal tissue system protects stems, leaves, and roots as well as regulating aeration and water loss by means of stomata. It consists of the epidermis and traces its meristematic origin to the protoderm, which in turn is the product of the tunica (see fig. 6.1). The primary vascular system and the ground tissue system trace their meristematic origins to the promeristem, specifically to the procambium and the ground meristem, respectively. Both the promeristem and the ground meristem found in stems are derived from the corpus of the shoot apical meristem. The primary vascular system comprises the primary xylem and the primary phloem. The former contains cells specialized to conduct water; the latter contains cells specialized to conduct cell sap. The ground tissue system is typically characterized by relatively undifferentiated parenchymatous cells. In dicot stems, the ground tissue system internal to the primary vascular tissue is called the pith, while the ground tissue system external to the primary vascular tissue is called the cortex. Since the vascular bundles are scattered in cross sections of monocot stems, no distinction can be drawn between pith and cortex.

Two types of lateral meristems are responsible for secondary growth: the vascular cambium and the phellogen (the cork cambium). The vascular cambium gives rise to secondary xylem ("wood") and secondary phloem. The phellogen gives rise to phellem (or cork) and the phelloderm, which collectively are referred to as periderm. The origin of the initials within the phellogen is variable among plant species, but the phellogen is typically derived in part from cells within the epidermis.

A noteworthy feature of the relations summarized in figure 6.1 is that the functions of the dermal and primary vascular tissue systems are supplanted in older portions of the same plant by functionally analogous tissues whose meristematic origins differ. The protective function of the epidermis is transferred to the periderm, while the hydraulic functions of the primary vascular tissues are transferred to the secondary vascular tissues in older portions of the plant.

FIGURE 6.2 (*facing page*) Idealized three-dimensional diagram of sections through a typical dicot stem (with no secondary growth) illustrating the locations and general morphologies of various cell and tissue types. (The anatomical relations shown here are one of many anatomical configurations found among dicot species and differ in many respects from those of a monocot.) Going clockwise (starting at the bottom of the figure), the outermost layer of the stem is the epidermis (shown in surface view), which typically has stomata (openings in the epidermis flanked by modified cells called guard cells). Beneath the epidermis, a region of collenchyma is shown. In transection, collenchyma cells appear isodiametric or nearly so; in longitudinal section they may be elongated. The vascular bundles, consisting of primary xylem and phloem, are arranged in a more or less concentric pattern. The xylem tissue (found toward the inside of vascular bundles) contains cells (tracheids or vessel members or both) that are dead when mature and that conduct water. The end walls of vessel members are perforated, lowering their resistance to water flow. The phloem tissue contains living cells, called sieve tube members, specialized to conduct cell sap. The end walls and lateral walls have sieve areas, each consisting of a highly aggregated collection of plasmodesmata. The ring of vascular bundles delimits the ground tissue of the stem into an outer cortex and an inner pith consisting of more or less isodiametrically shaped cells. The cortex and the pith are typically composed of parenchyma. Fibers are generally dead at maturity and are elongated parallel to the longitudinal axis of the stem. In cross section, the outlines of fibers are angular and polygonal. Fibers are typically found in association with both the xylem and the phloem (phloem fibers are illustrated in this diagram).

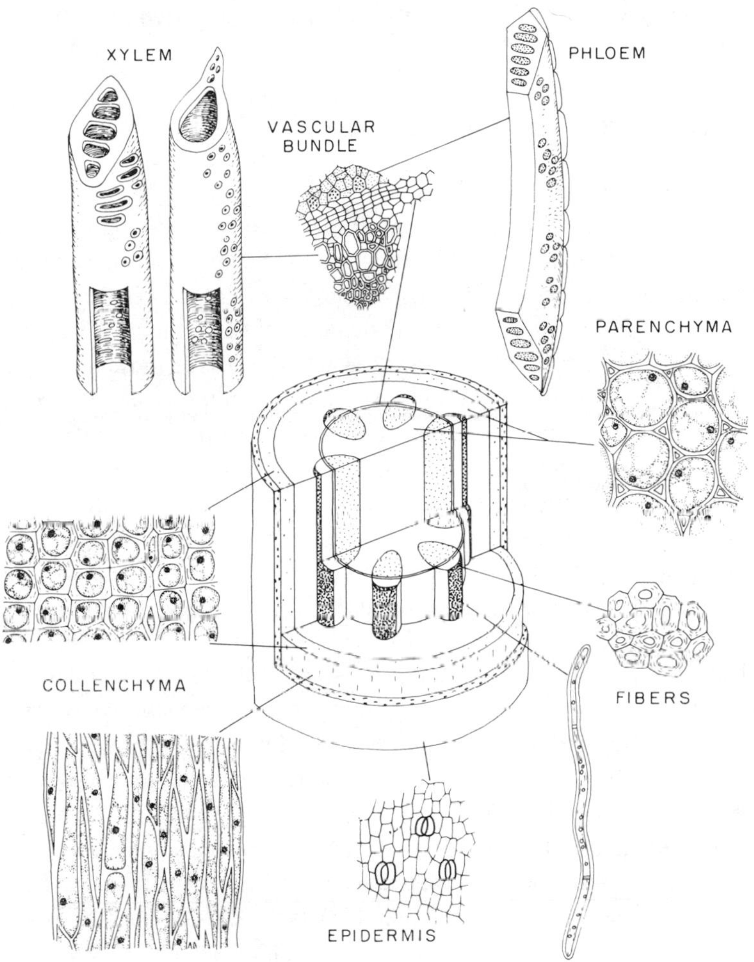

Clearly, the cells within the plant body are ultimately derived either directly or indirectly from cells tracing their ontogenetic origins to the primary meristems. Thus, although the distinctions between cells and tissues produced by the apical meristems versus the lateral meristems are useful, from one philosophical perspective the meristematic capacities of initials within the lateral meristems are the residual expression of the legacy of apical meristems. Nonetheless, the distinction between the primary and secondary ontogenetic phases in the development of a plant, as well as the transfer of functions from primary to secondary tissues, is biologically meaningful, if for no other reason than that the mechanical properties of primary and secondary tissues are often distinct.

THE APOPLAST AND THE SYMPLAST

Regardless of their meristematic origins, all primary and secondary tissues originally possess a living protoplasm responsible for secreting the cell wall infrastructure. Subsequent developmental modifications can lead to the regional death of the protoplasm, as in secondary xylem and phellem, in which case the cell wall infrastructure of the tissue is all that remains. Thus the way the cell walls are geometrically arranged within a tissue (how they differ in size, shape, and thickness) and the extent to which the protoplasm remains alive vary among cell and tissue types and provide for the anatomical diversity of tissues seen in the plant body (fig. 6.2).

Nonetheless, we must recognize that the plant body as a whole consists of two biological components whose physiological and mechanical properties differ substantially. In the physiological literature, these two components are referred to as the symplast and the apoplast. The symplast is the entire living protoplasm within the plant body. The apoplast consists of everything within the plant body other than the protoplasm—that is, the cell walls and inter- and intracellular spaces. Physiologically, the apoplast provides an avenue for transporting water and dissolved nutrients. In the primary and secondary xylem, the apoplast is specialized to conduct water. Some xylem cells (tracheids and vessel members) are preferentially elongated and have differentially thickened cell walls to resist implosion from rapidly moving liquids. Additionally, the cell walls of tracheids and vessel members are chemically treated with lignin to resist hydration. Vessel members lack walls at each end and are aligned end to end to produce long conduits (called vessels) running parallel to the longitudinal axes of stems, roots, and leaves, dramatically reducing their resistance to the flow of water.

The apoplast and symplast can operate mechanically in very sophisticated ways, as shown by the capacity of some tissues to regain positive turgor without additional uptake of water. Levitt (1986) reports that the midrib portions of wilted, detached leaves of cabbage (*Brassica oleracea* var. *captitata*) can regain positive turgor pressure and restiffen. Although the precise mechanism by which this happens remains unknown, it appears to be related to the capacity of different populations of cells to unload into the apoplast solutes that are later taken up by the symplast of a more distant population of cells, permitting them to differentially absorb water, regain hydrostatic pressure, and thereby

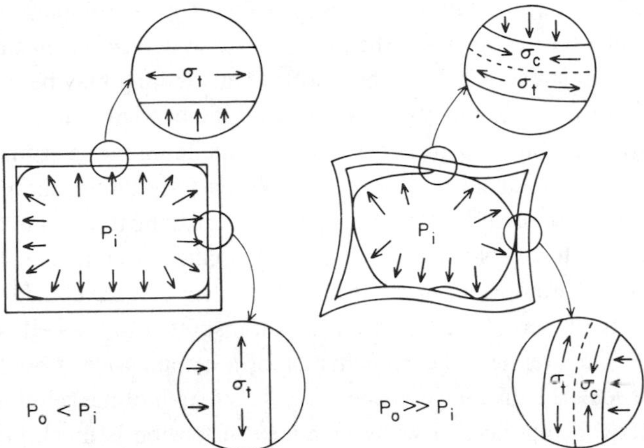

Figure 6.3 Theoretical effect of turgor pressure $\mathcal{P}_i$ on the distribution of tensile (σ_t) and compressive (σ_c) stresses within the cell wall of a hydrostatic (thin-walled) cell. When fully inflated and appressed to the cell wall, the protoplast will place the cell wall in overall tensile stress and will restrain it from bending when a load is applied externally (diagram to left of figure). When the internal turgor pressure of the cell drops below the external pressure, the protoplast deflates and may pull away from the cell wall (highly exaggerated in the diagram to the right). When an external force is applied to the cell, the cell wall is free to bend and buckle, and it experiences both compressive and tensile stresses as it deforms.

stiffen the tissue (Weisz, Randell, and Sinclair 1989). Perhaps not too surprisingly, the donor population of cells is in the lamina of the leaf, while the recipient population is associated with the vascular tissue of the midrib.

The symplast is the incompressible liquid phase of the plant body. Together with the apoplast, which in large part is the solid phase of the plant body, the symplast can mechanically operate as a hydrostatic device. As it expands in volume owing to growth or undergoes a transient increase in volume by the influx of water, the symplast can exert hydrostatic pressure on its apoplastic system of cell walls, which can be many times the atmospheric pressure (see chap. 4). When water is removed from cells and tissues, however, turgor pressure can drop well below atmospheric pressure. How much turgor pressure changes is important to the biomechanical properties of cells and tissues with thin walls. When water is supplied to these tissues, the resulting increase in the hydrostatic pressure the symplast exerts against its cell walls can increase the flexural stiffness of a tissue as a whole. Conversely, when the hydrostatic pressure within tissues is low, stems and leaves wilt because tissues and organs can no longer sustain their own weight owing to a transient decrease in their stiffness. During wilting the protoplasm shrinks in volume, and the cell walls within a tissue are free to bend or otherwise deform when subjected to the mechanical stresses induced by gravity. Thus wilting may be due to local buckling in response to the compressive force of a bending stem.

The hydrostatic role of the symplast in maintaining tissue stiffness is diagramed at the cellular level in figure 6.3. When fully turgid, the protoplasm within a cell is appressed to its enveloping wall, placing the wall in tension. If compressive or tensile stresses are externally applied, then the resulting compression of the protoplasm increases the tensile stresses developed within the cell wall—the externally applied stress is transmitted to the cell wall in the form of tensile stresses. The transmission of a compressive stress to the cell wall is particularly advantageous because the strength of the cellulosic microfibrillar network within cell walls measured in tension is much greater than the strength measured in compression. When the volume of the protoplasm within a cell wall is reduced by dehydration, cell walls submitted to compressive or tensile loadings are free to deform (fig. 6.3), depending on how much the volume of the protoplasm has been reduced. Obviously a limit exists on how much dehydration the protoplast can sustain and still remain physiologically viable. Permanent wilting occurs when the protoplasm is no longer capable of resuming its hydrostatic (mechanical) role.

The apoplast is either the dominant or the only material phase in some tissue types, such as secondary xylem (wood) and phellem (cork). When fully mature, the symplast within these tissues is either much reduced or totally absent. Tissues like wood and cork cannot operate mechanically as hydrostatic devices. Rather, they operate as gas-filled solids. Since gases are compressible, the solid phase (the apoplast) within these tissues exclusively provides mechanical support. It is not surprising, therefore, that the apoplast in tissues like wood is extensively reinforced with secondary cell walls that are chemically impregnated with lignin. Lignin functions as a bulking agent that can increase the compressive strength of cell walls and reduce the extent to which water infiltrates and consequently reduces the elastic and shear moduli of cell walls.

When placed under compression, the gas phase within tissues like cork and wood is expelled, and the solid phase densifies owing to the crushing and appression of cell walls. During densification, the elastic modulus of these tissues can increase to the limiting value of the elastic modulus of the cell walls. Accordingly, the material properties of tissues like wood and cork depend on the magnitude of the external stresses they sustain, just as the material

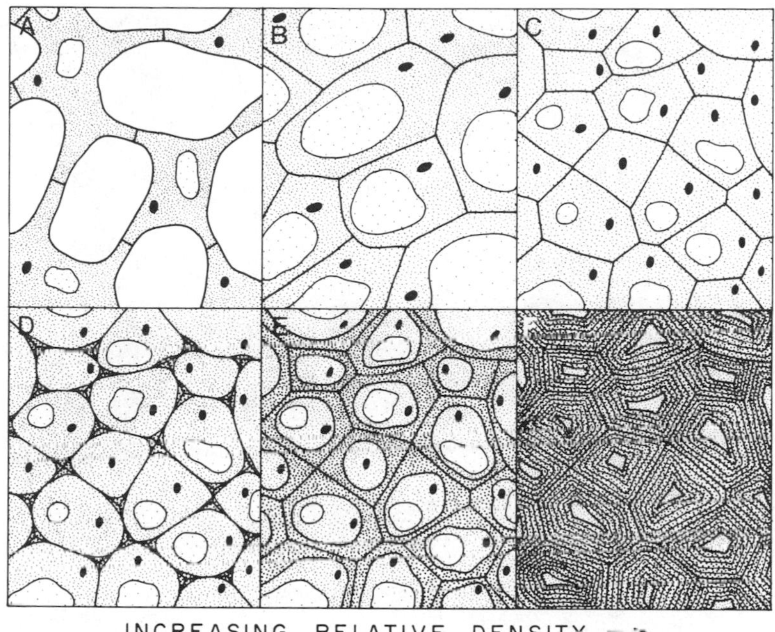

FIGURE 6.4 Variations in the relative volume fractions of apoplast, symplast, and cell wall thickness within some tissue types. The apoplast is the nonliving portion of a tissue (cell walls and areas not occupied by the protoplasm); the symplast consists of the living protoplasm. Primary cell walls are shown as black lines or, if thick, as densely stippled areas; secondary cell walls are darkly stippled, polylaminated areas, protoplasm is shown as densely stippled, nonlaminated areas; vacuoles are lightly stippled areas; nuclei are black ovales. Depending on the volume fraction of the symplast within a tissue, the tissue will mechanically operate either as a hydrostatic tissue, where the symplast dominates and cell walls are thin (A–E), or as a cellular solid, where the apoplast dominates and cell walls are thick (F). Cellular solid tissues may have living protoplasm or may be dead at maturity. (A) Tissue configuration typical for aerenchyma or spongy mesophyll with large intercellular, air-filled spaces. (B–C) Tissue configuration of parenchymatous tissues. (D–E) Tissue configuration of collenchyma (seen in transection). (F) Tissue configuration of sclerenchyma or secondary xylem (seen in transection).

properties and mechanical behavior of hydrostatic tissues depend on the water content of the symplast.

THE SYMPLAST VOLUME FRACTION

Hydrostatic tissues with a liquid phase, such as parenchyma, and cellular solid tissues with a gaseous phase, such as wood or cork, are extremes in a

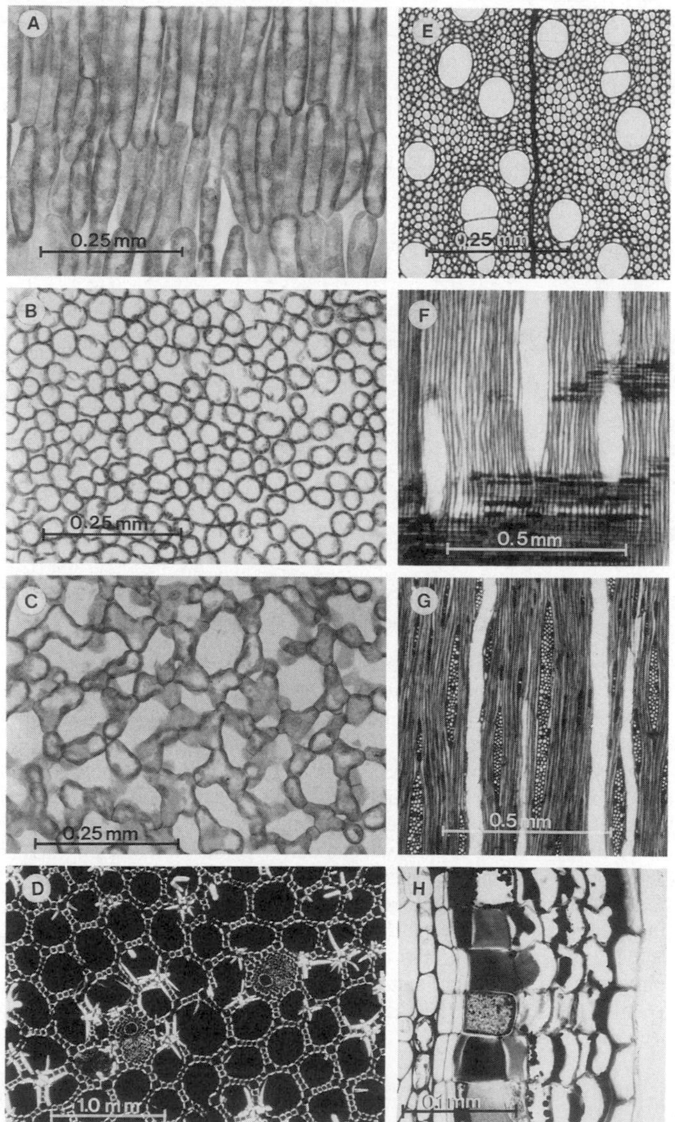

FIGURE 6.5 Examples of tissues that mechanically operate as hydrostats (A–B), pressurized cellular solids (C–D), or gas-filled cellular solids (E–H): (A) Transection of pallisade mesophyll. (B) Paradermal section of pallisade mesophyll. (C) Spongy mesophyll. (D) Aerenchyma (dark field, polarized light). (E) Transverse section through secondary xylem (wood). (F) Radial section through secondary xylem. (G) Tangential section through secondary xylem. (H) Transection through phellem ("cork"). For further details, see text.

continuum of tissues differing in the volume fractions of the symplast and the apoplast (fig. 6.4). In general, as the symplast volume fraction decreases, tissues are increasingly less likely to mechanically operate as hydrostatic devices and increasingly more likely to exhibit the mechanical behavior of nonhydrostatic gas-filled solids.

This trend holds even for aerenchymatous tissues that have a low symplast volume fraction by virtue of their chambered construction (fig. 6.5). The mechanical behavior of aerenchyma is relatively indifferent to tissue water content because how far the neighboring cells hydrostatically reinforce one another is a function of the cell-to-cell contact area, which in aerenchyma can be very small. Accordingly, although aerenchyma is mechanically weak and provides little stiffness to terrestrial plants, it most likely serves a physiological function analogous to that of the spongy mesophyll by providing large internal surface areas through which gases can diffuse. Nonetheless, aerenchyma can lower the overall weight of an organ and thereby reduce self-loading. In aquatic plants, aerenchyma can aid the aeration of submerged organs and provide buoyancy.

By the same token, as the volume fraction of gas-filled spaces within spongy mesophyll decreases, this tissue may provide mechanical support to leaves as well as provide aeration. The sun leaves produced in the upper canopy of trees have comparatively less spongy mesophyll than the shade leaves produced lower down within the canopy. The relative proportion of spongy mesophyll affects physiological processes but may also provide different mechanical strategies. Sun leaves experience greater water deprivation and larger mechanical stresses caused by wind than shade leaves do. Since they have a lower volume fraction of spongy mesophyll, the flexural stiffness of sun leaves may vary less as a function of tissue hydration than does that of shade leaves. To my knowledge, no one has examined sun and shade leaves for biomechanical differences, though such differences very likely exist.

Differences in the relative volume fraction of the symplast within plant tissues probably influenced the evolution of land plants. The aquatic antecedents of the first land plants were most likely parenchymatous in their tissue construction and therefore relied on turgor for mechanical support. This plant body construction is cheap and efficient because water is not a limiting factor in an aquatic habitat and a hydrostatic design can maximize the tissue volume fraction within an organism that is both mechanically and photosynthetically competent. As plants evolved into the terrestrial environment and progressively radiated into drier habitats where water deprivation is episodic, organisms relying exclusively on a hydrostatic design for mechanical support would have been at a disadvantage. Selective advantages would have been conferred on plants whose protoplasts could vary the symplast volume fraction throughout the plant body. The capacity to produce tissues with thick walls in addition to thin-walled tissues would permit specialization in physiological, as well as mechanical, function. The evolutionary transition from an exclusively hydrostatic plant body to one that could also rely on thick-walled tissues for mechanical support suggests that the evolutionary appearance of different tissue types would roughly conform to the sequence diagramed in figure 6.4.

Ultimately, the innovation of tissues whose symplast died at maturity would have provided for highly specialized tissues systems for mechanical support and hydraulic transport. This evolutionary scenario predicts a sequence of appearance of different tissue types not too unlike that documented in the fossil record for the early evolution of land plants. Additionally, it is not incompatible with the anatomical differences found among hydrophytes, mesophytes, and xerophytes seen in present-day flora. In general, plant species occupying wet habitats (hydrophytes) mechanically sustain their static and dynamic loadings by means of hydrostatic tissues (e.g., *Impatiens*), whereas species adapted to dry habitats (xerophytes) typically have substantial portions of their organs composed of thick-walled, dead tissues.

Cellular Solids and Density

The behavior of commercially fabricated cellular solids provides a conceptual framework within which to evaluate the mechanical behavior of plant tissues with a gaseous phase. I begin this review by considering tissue density, recognizing that the bulk density of tissues is much lower than the density of their constituent materials (cell walls and protoplast). The difference between the bulk density of a cellular solid and the density of its constituent materials plays an important role in the mechanical behavior of tissues and commercially fabricated cellular solids.

A variety of primary and secondary tissues are characterized by large volume fractions of gas-filled spaces, which can be intercellular, as in aerenchyma and spongy mesophyll, or intracellular, as in wood and cork (fig. 6.5). Inter- and intracellular spaces profoundly influence the mechanical behavior of a tissue. In living tissues they provide an unoccupied volume into which cellular fluids can evacuate when the protoplast is subjected to a large external stress; in living and essentially dead tissues, stresses, whether slow or rapid, can deform (buckle and bend) cell walls into empty spaces. Additionally, gas-

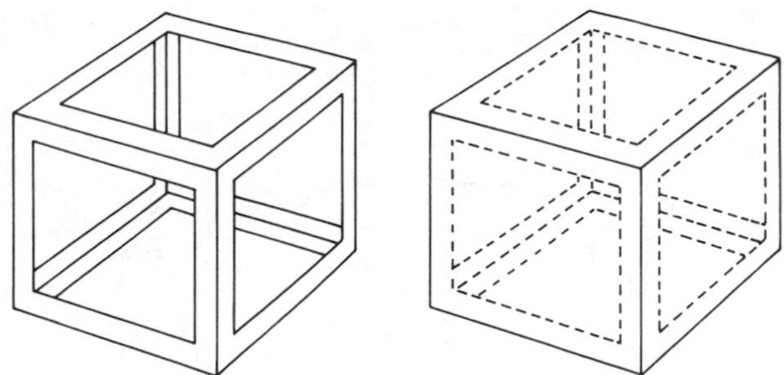

FIGURE 6.6 Two general types of cellular solids. The solid phase within a cellular solid can be arranged three dimensionally to produce a strutted, open-walled configuration, or it can be arranged three dimensionally in the form of solid walls enclosing chambers. The former (shown to the left) are referred to as open-walled cellular solids, while the latter (shown to the right) are closed-walled cellular solids. In these two diagrams, only one "unit" within either the open-walled or closed-walled cellular solid is shown. Each type of cellular solid would have an infrastructure composed of many of either type of unit. The interconnected air-filled spaces within aerenchyma and spongy mesophyll give these two types of tissue the appearance of an open-walled cellular solid (see fig. 6.5C–D). The completely enclosed cell lumens of secondary xylem and phellem give these tissues the appearance of a closed-walled cellular solid (see fig. 6.5E–H).

filled volumes reduce a tissue's relative density—the ratio of the density of the whole tissue to the density of the material used to construct its cell wall infrastructure. The relative density of plant tissues is less than one—thus they float in water even though the density of their cell wall constituents (lignin, cellulose, and hemicelluloses) is greater than that of water.

The anatomy of many gas-filled plant tissues closely parallels the structure of a variety of commercially fabricated and economically important materials that are technically called cellular solids—for example, polyvinyl chloride, polystyrene, and rubber latex foams. Cellular solids have two general types of internal construction. Their solid phase can consist of open strutlike beams interconnected at their ends or of complete walls (fig. 6.6), and they are referred to as open-walled or closed-walled cellular solids. Aerenchyma and spongy mesophyll are examples of biological, open-walled cellular solids (see Kraynik and Hansen 1986). Cork and wood are closed-walled cellular solids. Commercial cellular solids are used when insulation, padding, or lightweight but stiff structures are required. Many plant tissues provide the same functions—cork is an excellent thermal insulator and can sustain high impact loadings because of its great resilience, and wood is one of the strongest materials for its density.

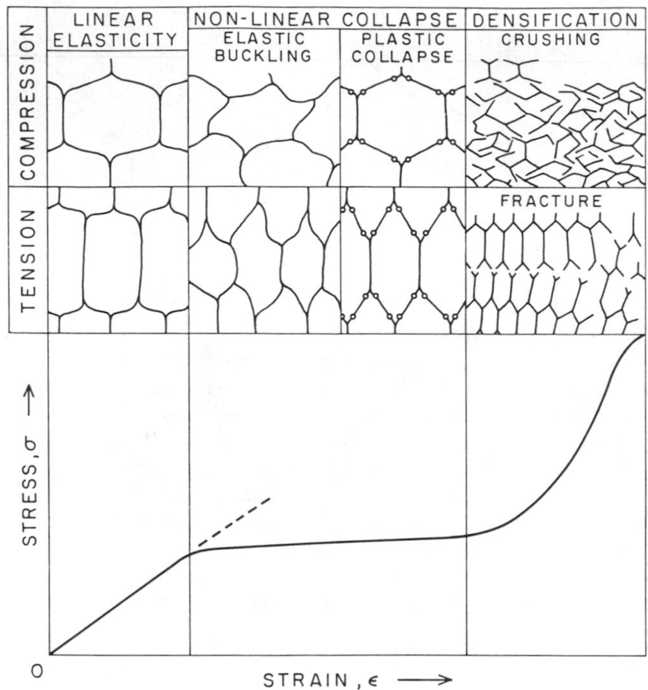

FIGURE 6.7 Geometric distortions with a typical honeycomb cellular solid subjected to compression (above) and a typical stress-strain diagram under compressive loading (below). Transections through a honeycomb cellular solid reveal the types of geometric distortions that are likely to occur when the solid is subjected to compression. These distortions can involve elastic buckling, plastic collapse (the likely locations of plastic hinges are shown as small circles), or crushing (in compression). The stress-strain diagram resulting when a cellular solid is placed in compression typically shows three regions (linear elasticity, nonlinear collapse, and densification). The features of the stress-strain diagram result from deformations occurring within the cellular voids of the solid (upper panel). (From Niklas 1989c.)

The relative density of cellular solids is the single most important parameter influencing their mechanical behavior. In commercial cellular solids, relative density is the ratio of the density of the cellular solid ρ to the density ρ_s of the material used to construct the solid phase: ρ/ρ_s. The relative density of commercially fabricated cellular solids typically ranges between 3×10^{-1} and 3×10^{-2}. This range compares favorably with that of many plant tissues; for example, the ρ/ρ_s of aerenchyma can be as low as 10^{-3} (leaf and stem aerenchyma from the sedge *Juncus*), while that of cork ($\rho/\rho_s = 0.09$) or balsa wood ($\rho/\rho_s = 0.13$) is well within the range of commercially fabricated cellular solids. However, the relative densities of heartwood can range well above those of commercial cellular solids (between 0.09 and 0.94). The cell wall lumens in heartwood can be obstructed with phenolic residues and cell wall protrusions. The mechanical behavior of plant tissues with relative densities approaching unity, like some heartwoods, should not be modeled according to the theory of cellular solids. Thus, based on their relative densities, we could infer the mechanical behavior of plant tissues like aerenchyma and cork and some wood species from the theory of cellular solids and would expect it to be largely dictated by cellular structure rather than by the material properties of cell walls. By contrast, the mechanical behavior of heartwood would be influenced less by cellular structure and more by the material properties of the cell wall solid phase. Nonetheless, as we will see, the mechanical behavior of even heartwoods is not impervious to the concept of cellular solids.

The Mechanical Behavior of Cellular Solids

When a fabricated cellular solid is subjected to compressive stresses, its stress-strain diagram typically shows three characteristic responses (diagramed in fig. 6.7): a linear elastic response under low stress; an elastic or plastic deformation response under constant stress (which appears as a plateau on the stress-strain diagram); and a densification response associated with the crushing of the solid phase within the cellular solid. Each of these three phases within the stress-strain diagram can be phenomenologically explained by means of elementary strength of materials theory (treated in chap. 3), because each strut in an open-walled cellular solid or each wall in a closed-walled cellular solid is more or less free to deform along most of its length, like a beam placed in bending or a column in buckling. Thus the stress-strain diagram differs from that of a true solid material because it reflects the mechanical response of a structure as well as the material properties of the solids from which the structure is constructed. For much the same reason, when plant tissues like cork or wood or aerenchyma are mechanically tested, the anatomy of the tissue profoundly affects the response to a given stress level. Indeed, because the cellular geometry of a tissue sample can differ as a function of the plane in which a tissue is sectioned (see fig. 6.5), the material properties measured for a tissue sample will differ depending on the direction in which stresses are applied. Accordingly, anatomically complex tissues are mechanically anisotropic. The mechanical anisotropy of plant tissues is capitalized on by plants in terms of how the geometries of tissues are oriented within the plant body with respect to naturally occurring types of loading. The mechanical response of a tissue to some externally applied mechanical stress, such as a high-velocity impact loading or a static compressional loading from the tissues above, will depend on the direction of the load's application to the struts, beams, or cantilevers within the tissue. The layers of cork in the bark of a tree can sustain very high impact loadings directed toward the center of the tree, but substantially less vertical compressive loadings along the length of the tree, whereas wood can sustain high compressive stresses applied along the length of its grain compared with the stresses it can bear laterally.

As I already noted, the relation between stress and strain for a cellular solid is not constant over the entire range of loadings, and when densification occurs, a cellular solid can actually get stiffer, deforming proportionally less as the level of stress increases. In plants this can confer many advantages, as when the wood in a tree trunk or branch is placed in compression or tension, because the material of the organ (which is really a structure) can resist deformations even as the stress levels increase beyond the proportional limits of the solids used to construct cell walls within the tissue. Since wood is essentially dead tissue, small microstructural (cell wall) deformations have little effect on the metabolism of the plant as a whole.

A more quantitative treatment of the mechanical behavior of cellular solids will provide greater insight into these anecdotes. This treatment requires the application of beam theory, where the walls or struts within a cellular solid operate as mechanical devices. For example, the geometry of an open-walled cellular solid (like aerenchyma) consists of many beamlike struts interconnected end to end. From simple beam theory, we know that, when compressed, the nonvertically aligned struts will deflect δ under an applied force F. We also know that the stress is proportional to F divided by the square of the average beam length l, while the strain is δ divided by l. The relative density of the cellular solid must be proportional to $(t/l)^2$, where t is the average beam thickness, while the second moment of area of the section through each beam is $t^4/12$. (In a closed-walled cellular solid, like wood or cork, the relative density is proportional to t/l and the second moment of area is $t^3l/12$.) The deflection δ of each beam within the cellular solid must be equal to the product of $Fl^3/12\,E_sI$ and some proportionality constant, where E_s is the elastic modulus of the solid from which the beam is constructed. Accordingly, the elastic modulus of the cellular solid E as a whole is proportional to E_st^4/l^4. Thus, within the linear elastic response range of a cellular solid with an open-walled construction, $E = (\rho/\rho_s)^2\,k_1\,E_s$. This association can be rearranged to yield a dimensionless expression relating the comparative density to the comparative elastic modulus of the cellular solid operating within its linear elastic range of behavior (see Ashby 1983, 1758):

$$\frac{E}{E_s} = k_1 \left(\frac{\rho}{\rho_s}\right)^2. \tag{6.1}$$

Experimental data for fabricated cellular solids conform remarkably well to the theoretical relationship given in eq. (6.1). The equation predicts that the elastic modulus of the cellular solid E will decrease sharply as the interstitial volume increases. For many commercial cellular solids, the relative density averages 10^{-1} and the relative elastic modulus averages 10^{-2}. For cellular solids with a relative density of 10^{-2}, E/E_s drops to 10^{-4}.

The linear elastic behavior of cellular solids occurs with small strains (about 5% in compression and between 5% and 8% in tension). Nonlinear deformations will occur under higher strains. The principal form of nonlinear elastic deformation is elastic buckling of the struts or walls within the cellular solid, for which Euler's column formula is reasonably applicable (see chap. 3). This formula gives the critical load P_{cr} for which buckling occurs in relation to beam length and the flexural stiffness of the beam:

$$P_{cr} = k_0 \left(\frac{E_s I}{l^2}\right), \tag{6.2}$$

where k_0 is a proportionality factor. For an open-walled cellular solid, the stress σ_e at which this occurs is proportional to P_{cr}/l^2 and can be computed from the formula

$$\sigma_e = k_2\, E_s \left(\frac{\rho}{\rho_s}\right)^2, \tag{6.3}$$

where k_0 is another proportionality factor. Equation (6.3) holds for cellular solids with relative densities less than 0.3. When the relative density is higher than this value, as in some heartwoods, the gas-filled spaces within the cellular solid are too small or sparsely distributed and the struts or cell walls within it are too thick or short to buckle elastically as predicted by the Euler equation.

If the solid phase within the cellular solid is a plastic material, as in the growing cells of aerenchyma, then compressive or tensile stresses can induce plastic yielding of the solid phase. The yield stress at which yielding will occur σ_{p1} is proportional to the plastic moment of the struts or walls within the cellular solid and inversely proportional to beam length:

$$\sigma_{p1} = k_3\, \sigma_y \left(\frac{\rho}{\rho_s}\right)^{3/2}, \tag{6.4}$$

where σ_y is the yield stress of the plastic material constituting the solid phase of the cellular solid.

Higher compressional stresses produce densification in which walls or struts are crushed. In tension, walls may fail through the propagation of fractures. The crushing or rupture stress σ_{rup} is related to the relative density and the modulus of rupture of the solid phase σ_r:

$$\sigma_{\text{rup}} = k_4 \, \sigma_r \left(\frac{\rho}{\rho_s}\right)^{3/2}, \tag{6.5}$$

while the plain stress toughness K_{IC} in tension is given by the formula

$$K_{IC} = k_5 \, \sigma_r (\pi l)^{1/2} \left(\frac{\rho}{\rho_s}\right)^{3/2}, \tag{6.6}$$

which predicts the relation between relative density and the rupture stress of the solid phase in terms of the propagation of a crack. Equations (6.5) and (6.6) are particularly useful when dealing with tissues whose relative densities are equal to or less than 0.3. For example, eq. (6.5) reveals why aerenchyma is not a good mechanical tissue—the relative density of this tissue is so low that it is easily crushed in compression or pulled apart in tension. Similarly, eq. (6.6) indicates that if the size of a crack equals or exceeds the average dimensions of the cells within the cellular solid, then the rupturing of a single cell can radiate outward from this nucleation site and propagate across the entire specimen. Accordingly, low-density woods (which have relatively thin cell walls) are predicted to be susceptible to fracture—a prediction testified to by rueful carpenters who use low-density woods to make bookshelves or by the failure of willow (*Salix*) trees in storms. What saves most woods from mechanical failure is their polylaminate construction. Sequential growth layers of secondary xylem, deposited by the vascular cambium within the trunk or branches, have slightly different grain orientations than their immediate neighbors. Thus a layer-layer interface can act as a barrier to the propagation of fractures. Indeed, the deposition of denser secondary xylem at the end of each growth season provides a tissue heterogeneity even within each growth layer of wood, so fractures may terminate within a single growth layer.

The deformations resulting from nonlinear elastic buckling, plastic collapse (due to yielding), and compressive crushing are visible as a plateau in the stress-strain diagram of the cellular solid once the stress level exceeds the linear elastic limit of the cellular solid (see fig. 6.7). Actively growing plant tissues can exhibit this response, particularly if they are deprived of water: they begin to wilt (which is partially due to plastic deformations within the

TABLE 6.1 Equations for Some Mechanical Properties of Open-Walled Cellular Solids

Mechanical Property	Open-Walled (foamlike)
Linear elasticity	$\frac{E}{E_s} = k_1 \left(\frac{\rho}{\rho_s}\right)^2$
Elastic buckling	$\sigma_e = k_2 E_s \left(\frac{\rho}{\rho_s}\right)^2$
Plastic collapse	$\sigma_{pl} = k_3 \sigma_y \left(\frac{\rho}{\rho_s}\right)^{3/2}$
Rupture	$\sigma_{rup} = k_4 \sigma_r \left(\frac{\rho}{\rho_s}\right)^{3/2}$
Fracture toughness	$K_{IC} = k_5 \sigma_r (\pi l)^{1/2} \left(\frac{\rho}{\rho_s}\right)^{3/2}$

Source: Data from Ashby (1983, 1764, table 3).
Note: Definitions and units of symbols are given in the text.

cell wall infrastructure) and deform under their own weight.

Some of the equations that predict the mechanical behavior of closed-walled and open-walled cellular solids are given in table 6.1. The following provides a brief summary of the attributes of cellular solids:

1. Regardless of the characteristics of the solid phase, cellular solids will have three regions in their stress-strain diagrams: a region of linear elastic behavior, followed by a plateau region of elastic, plastic, or brittle deformation, and finally, at higher stresses, a region of densification.

2. There is a unique elastic modulus for the linear elastic regime of loading, however.

3. Under high compressive stresses, the apparent elastic modulus will increase dramatically (to the limiting case of E_s) before the solid fails. This results from out-gassing of the cellular solid as walls or struts crush together. The solid begins to fail at the start of the plateau for elastic-plastic and elastic-brittle materials.

4. The relative density of the cellular solid profoundly influences mechanical behavior. It in turn depends on the anatomy of the cellular solid—the thickness, length, and girth of cell walls or struts or any other solid phase microstructure.

5. When the relative density of a cellular solid exceeds roughly 0.3, the cell walls within the solid are too bulky to undergo bending and Eulerian buck-

ling. Thus plant tissues with very high relative densities, like heartwoods, are likely to behave mechanically more like true solids than are cellular solids.

Empirical studies are always the final arbiter for any theory, and with the theory of cellular solids they provide substantial support. For example, Easterling et al. (1982) examined the elastic modulus and crushing strength of balsa (*Ochroma lagopus*), one of the strongest woods for its density, in the three principal orthogonal planes to the grain of the wood (radial, tangential, and longitudinal). They found that the mechanical properties of this tissue depend on the material properties of the cell walls *and* on the geometry (cell size and shape) of the tissue viewed in the direction of the applied force, which differ in the radial, tangential, and longitudinal planes of section. Comparing the theoretical predictions based on the theory of cellular solids with the empirical results leaves little doubt that balsa behaves as a closed-walled cellular solid exhibiting marked mechanical anisotropy. An elegant feature of this research was the use of the scanning electron microscope (SEM) to visualize deformations. Samples of balsa (to which displacement transducers were attached) were mechanically tested and simultaneously photographed within the SEM. Thus, deformations in the cellular structure of wood samples were documented as a stress-strain diagram was recorded.

Hydrostats and Pressurized Cellular Solids

The theory of cellular solids can be applied to a variety of living plant tissues. In many other living tissues, however, cells are so closely packed and the interstitial volume is so small that beam theory cannot be legitimately used to treat cellular structure in terms of beams, columns, or cantilevers. Perhaps the best example of a living tissue with relatively little interstitial space is storage tissue parenchyma, such as that found in the tubers of potato. Although spaces exist among neighboring cells in this tissue, their volume fraction is relatively small ($< 5\%$) compared with that found in aerenchyma (up to 50%) or spongy mesophyll or secondary xylem and cork. To treat tissues such as parenchyma, another conceptual framework is required—the hydrostat, a thin-walled, inflatable structure whose mechanical behavior is in part dictated by the differential between the internal and external pressures exerted on the thin wall or membrane composing the hydrostat's outer surface. Hydrostatic devices are commonly used by engineers—for example, air-filled tires and balloons. By the same token, plants use hydrostats in the form of turgid, thin-walled cells and tissues. The basic design advantage of a hydrostatic device is that the wall or membrane is placed in tension (and so stiffens) as the internal pressure $\mathcal{P}$

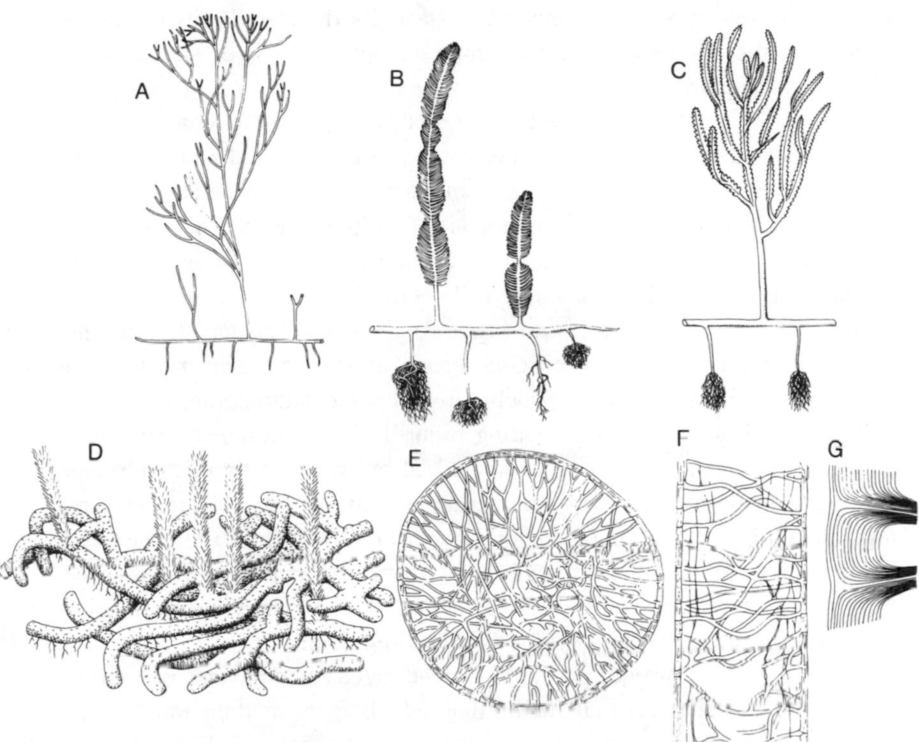

FIGURE 6.8 Morphological (A–D) and anatomical (E, G) complexity in the siphonaceous algal genus *Caulerpa:* (A) *C. fastigiata.* (B) *C. taxifolia.* (C) *C. cupressoides.* (D) Horizontal "rhizomatous" elements of the thallus covered with "rhizoid-like" cellular extensions. (E) Transection through vertical component of a cell showing internal "strutlike" extensions of the primary cell wall (called trabeculae). (F) Longitudinal section through a cell showing trabeculae. (G) Details of the primary cell wall layers interconnecting trabecular extensions within the cell (to right) and the external cell wall (to left).

increases. Indeed, hydrostats are remarkably cost efficient, since the tensile stresses σ_t developed in the hydrostat's wall increase as the outer radius of hydrostat r_o increases and as the wall or membrane thickness t decreases (see Hettiaratchi and O'Callaghan 1978):

$$\sigma_t = \left(\frac{\mathcal{P}\lambda}{2}\right)\left(\frac{r_o}{t}\right), \tag{6.7}$$

where λ is the wall stretch ratio (the tensile strain). Since stress divided by strain is the elastic modulus, eq. (6.7) shows that the hydrostat will increase in stiffness ($E = \sigma_t/\lambda$) as r_o/t increases. Naturally there are limits to this design, since the thinner the wall, the more susceptible the hydrostat is to perforation and rupture.

Perhaps the most remarkable examples of plant hydrostats come from the siphonaceous algae. These plants consist of a single cell and a multinucleate protoplast. Significantly, these organisms can achieve considerable size and morphological complexity without the aid of multicellularity. The algal genus *Caulerpa* can grow to twenty meters long and as much as one meter tall, yet individual plants consist of a single cell—a hydrostat with an inflatable protoplast and an extendable cell wall (fig. 6.8). Such a mechanical design is not without its limitations, however. *Caulerpa* is an aquatic organism whose capacity for mechanical support benefits from the buoyant medium it grows in and because water is never a limiting factor in the environment. Species of *Caulerpa* that grow in habitats characterized by waves or by relatively high flow regimes have trabeculae—an internal reticulum of cell walls transversely and longitudinally spanning the single external cell wall (fig. 6.8). The trabeculae operate as an internal scaffold of struts and beams and cantilevers through which solutes can be transported, as well as providing mechanical support against local shearing forces and bending moments.

Large hydrostats are not particularly good mechanical devices in a terrestrial habitat, where water may be limiting and where the medium (air) vertical plant organs grow through provides no buoyancy whatever. If we think of parenchyma as a biphasic material, however, with a living protoplast and an internal scaffold of cell walls, then a meaningful analogy can be drawn between this multicellular tissue and the *Caulerpa* cell. The cell wall infrastructure of parenchyma and the trabeculae of *Caulerpa* serve the same mechanical function, and the entire mass of parenchyma operates as a single macrohydrostat composed of smaller, inflatable cellular hydrostatic units.

There is another logical component to our understanding of plant hydrostats—they are pressurized cellular solids. The cell wall infrastructure within a living tissue such as parenchyma is a solid phase; the symplast is a liquid phase. How much the hydrostatic pressure of the liquid phase influences the mechanical behavior of the cellular solid depends on the wall thickness of the solid phase. Parenchyma is a thin-walled hydrostatic tissue—its mechanical behavior is very much affected by its hydrostatic pressure. At the opposite end of the spectrum, wood has no liquid phase in terms of a living symplast. It is a nonpressurized cellular solid whose mechanical behavior is influenced by both the geometry and, to a limited extent, the chemistry of its cell wall infrastructure. Between the two extremes (parenchyma and wood) there lies a broad range of tissue types, some with thin walls, like parenchyma, others with relatively thicker cell walls. Within this middle ground, tissues operate as pressurized cellular solids whose mechanical behaviors are influenced both by the cell wall infrastructure and by hydrostatic pressure.

In chapter 5, the cutoff point was mathematically predicted for the influence of cell wall thickness on the circumferential stresses developing within the wall of a cylindrical cell. The limiting cell wall thickness was 20% of the cell radius. For thicker cell walls, the difference between the internal (turgor) pressure and the external (ambient) pressure exerted on the cell plays little or no role in dictating the stresses developing within the wall. For thinner cell walls, the internal and external pressure differential profoundly influences cell wall stresses. If we extrapolate this to plant tissues, then we can predict that tissues whose average cell wall thickness significantly exceeds 20% of their average cell wall radius will not exhibit hydrostatic mechanical behavior. Conversely, when the average cell wall thickness of a tissue drops well below 20%, a tissue will operate as a hydrostat. When the average cell wall thickness approaches 20% of the average cell radius, a tissue should operate as a pressurized cellular solid—a mechanical hybrid between the two extremes of behavior.

Significantly, in most cases thick cell walls typically have secondary cell wall layers that are usually lignified. Thus, even though we can construct an intellectually satisfying bridge (pressurized cellular solids) between hydrostatic and nonhydrostatic cellular solids based on a geometric parameter (cell wall thickness), we cannot neglect the biology of our system, which involves the chemical alterations attending secondary wall deposition. Lignification not only influences the extensibility of cell walls but, as we have seen, can influence the extent to which water can infiltrate a cell wall and so alter the modulus of elasticity of the microfibrillar cellulose network. Accordingly, cell wall thickness and chemistry influence the mechanical behavior of cells and plant tissues. At first glance the possible permutations of cell wall architecture and chemistry that may be encountered among different plant tissues may appear hopelessly complex, but in reality they are not. Indeed, many potential combinations of wall thickness and chemistry are never biologically expressed in plants, possibly because some are mechanically disastrous. Regardless of the reasons, plant tissue types fall into relatively circumscribed categories of mechanical behavior, which I will discuss at the end of this chapter. Before generalizations are made, however, it is advisable to become familiar with the major plant tissue types, noting their similarities and dissimilarities as well as the advantages and disadvantages they confer to the survival, growth, and evolution of plants.

PARENCHYMA

If wood and cork represent outstanding examples of plant cellular solids, then parenchyma is the archetypal hydrostatic tissue. Although secondary cell wall layers can occur in this type of tissue, parenchyma cells typically have thin walls that consist exclusively of primary wall layers, which in themselves confer little resistance to bending stresses. When the symplast of this tissue is at or near full turgor, however, externally applied compressive or tensile forces can be transferred to the cell wall infrastructure as tensile stresses, stiffening the tissue as a whole. Accordingly, the mechanical behavior of parenchyma depends on its water content.

By the same token, how far parenchyma is utilized within the plant body for mechanical support also depends on how closely cells are packaged together. The mechanical role of parenchyma is maximized when there is little interstitial volume and when cell wall contact area is high, because the thin cell walls are relatively free to deform into intercellular spaces. This provides some insight into the anatomical apportionment of different types of parenchyma within plant organs. The cells of storage parenchyma are roughly isodiametric and multifaceted, with a geometry approximated by an orthotetrakaidecahedron (a fourteen-sided polyhedron with eight hexagonal and six quadrilateral facets). The interest in the geometric similarities between parenchymatous cells and liquid films traces its intellectual legacy to Joseph Plateau, who in 1873 explored the minimum surface areas that liquid films assume when suspended from wire frames. Lord Kelvin developed this concept further in establishing what his critics referred to as *his* tetrakaidecahedra. In 1894 Kelvin showed that a thirty-six-edged structure consisting of eight curved hexagonal faces and six plane quadrilateral faces could be stacked endlessly without interstices. (The geometry is further characterized by having a 120° angle between adjacent faces and an angle of 109° 28′16″ between its edges. This has been called the minimal tetrakaidecahedron.) In 1923 Lewis examined the three-dimensional geometry of one hundred cells isolated from the pith of the elder, *Sambucus canadensis,* and found that the average cell had 13.97 faces instead of the predicted 14. Calculations indicate that the fourteen-sided cell geometry maximizes the capacity to tessellate cells into any given volume, thereby minimizing interstitial volume and maximizing cell wall to cell wall contact area (see Matzke 1950). Not too surprisingly, densely packaged parenchyma is found in anatomical locations within stems and leaves where high compressive loadings or bending and torsional shear stresses are anticipated, such as toward the center and periphery of aerial stems lacking secondary growth and the petioles of dicot leaves. Additionally, meristematic tissues tend to be parenchymatous in their construction. Experimentally determined values for their elastic moduli can be surprisingly high, from 19 to 40 $MN \cdot m^{-2}$ (see Kutschera and Kende 1988). By contrast, some specialized forms of parenchyma contain large volume fractions occupied by air, such as the parenchyma composing the spongy mesophyll found in the leaves of many dicot species (see fig. 6.5). An extreme case of this condition is seen in aerenchyma, where over 50% of the tissue volume is filled with air. In these circumstances the hydrostatic capacity of the protoplast to stiffen the cell wall infrastructure is greatly reduced. Therefore spongy mesophyll and aerenchyma provide little mechanical support to organs in themselves; for example, the elastic modulus of aerenchyma isolated from the leaves of *Juncus effusus* is estimated to be 2.26 $MN \cdot m^{-2}$. Thus, although it is common to speak of parenchyma in a generic sense, the mechanical properties of this type of tissue depend on how densely cells are packaged.

One would anticipate that the stiffness (the modulus of elasticity) of parenchyma, a hydrostatic tissue, would increase as a function of tissue water content. This expectation was first investigated empirically by Virgin (1955), who found that the apparent elastic modulus of parenchyma isolated from the pith of potato tubers was directly proportional to the tissue turgor pressure. This finding was subsequently verified and elaborated on by Falk, Hertz, and Virgin (1958). The maximum elastic modulus these authors reported for potato tuber parenchyma with a turgor pressure of 0.67 MPa is 19 $MN \cdot m^{-2}$, which falls in the lower portion of the range of E reported for actively growing meristematic tissue: 19–40 $MN \cdot m^{-2}$. At a turgor pressure of 0.31 MPa, the maximum elastic modulus that Falk, Hertz, and Virgin (1958) report is roughly 8 $MN \cdot m^{-2}$. In a companion paper, Nilsson, Hertz, and Falk (1958) proposed a model to explain these results. Their model assumed that cell walls are linearly elastic under small deformations, such that the cell wall extension in length $\Delta l/l$ could be approximated by the formula

$$\frac{\Delta l}{l} = \frac{F}{E_c \, wt} , \tag{6.8}$$

where E_c is the elastic modulus of the cell wall, F is the applied (hydrostatic) force, w is cell wall width, and t is cell wall thickness. Since the transverse contraction in cell wall width accompanying cell wall elongation is related to Poisson's ratio of the wall material, such that

$$\frac{\Delta w}{w} = -\upsilon \left(\frac{\Delta l}{l}\right), \tag{6.9}$$

Nilsson, Hertz, and Falk (1958) concluded that the original cell wall radius r_o would increase to Δr as turgor pressure $\mathcal{P}$ increases according to the formula

$$\Delta r = \frac{1-\upsilon}{E_c\,t}\left(\frac{\mathcal{P}\,r_o}{2}\right). \tag{6.10}$$

In an extremely thoughtful and elegant manipulation of these assumptions, Nilsson, Hertz, and Falk (1958) derived a formula that predicts the apparent elastic modulus of parenchyma E for any turgor pressure:

$$E = 3\left[1 + \frac{(7-5\upsilon)}{20\,(1+\upsilon)}\right]\mathcal{P} + \left[\frac{3(7-5\upsilon)}{10(1-\upsilon^2)}\right]\left(\frac{E_c t}{r_o}\right). \tag{6.11}$$

The first term in eq. (6.11) expresses the contribution to the elastic modulus of the tissue resulting from turgor pressure; the second term expresses the influence of the material properties and the geometry of the cell walls within the tissue.

There are two aspects of eq. (6.11) that have a direct bearing on the mechanical behavior of parenchyma in particular and plant tissues in general. First, the elastic modulus of the tissue is not a constant (nor is it even of the same magnitude as the elastic modulus of the solid cell walls within the tissue). Unlike true solids that have clearly defined material properties, as shown by the nature of their stress-strain diagrams, the material properties of plant tissues are physiologically and developmentally variable. Thus the elastic modulus of parenchyma varies as a direct function of turgor pressure as well as the material properties of cell walls that can change with the aging of the tissue. Second, the structure of a plant tissue—the geometry of the cell wall infrastructure (cell wall thickness) as well as its material properties (cell wall elastic modulus)—plays a vital role in dictating mechanical behavior. This illustrates a point made repeatedly throughout this book—that plant tissues are not materials in the strict sense of the word. They must be viewed, at the very least, as composite materials or, with greater conceptual rigor, as structures. This was seen when wood and cork were treated as nonhydrostatic cellular solids, and it is seen here in terms of parenchyma as a hydrostatic tissue.

As I mentioned in chapter 4, one important way plant tissues differ from true solids is that the elastic modulus of tissue samples sometimes depends on the cross-sectional area of the sample used in mechanical testing, even when loading forces are normalized with respect to that variable. Recall (from chap. 2) that stress is the applied force divided by the cross-sectional area through which the force is applied. Thus, in theory the stress-strain diagram for a solid is independent of the dimensions of the sample, affording us the opportunity to measure the material properties of the material examined. In their study of parenchyma, however, Falk, Hertz, and Virgin (1958) showed that the elastic modulus of parenchyma increases as the diameter of the tissue sample increases. This effect has been confirmed and is not an artifact (see Niklas 1988a). The explanation for these observations is that the number of cells in the cross-sectional area through which either compressive or tensile forces operate influences the mechanical behavior of the material—the greater the number of cells, the less cell walls are free to deform. Thus the greater the number of cells, the stiffer the parenchyma sample. This has obvious implications for the dependence of the material properties of parenchyma on the absolute volume of parenchyma tissue within a plant organ. Essentially, bulky parenchymatous organs are stiffer than their smaller counterparts, even though they are made of the same material. This strategy may have been employed during the early phase of land plant evolution, when tissue differentiation was less evident than in geologically younger plant species and when parenchyma was one of the few mechanical materials developmentally available for constructing vertical plant organs.

Indeed, the mechanical behavior of parenchyma is much more complex than eq. (6.11) leads us to believe. In general, parenchyma behaves as a nonlinear elastic material—its stress-strain diagram does not typically exhibit a linear relation between applied stress levels and strain. Additionally, parenchyma exhibits short-term elastic recovery, long-term plasticity, stress relaxation, and creep. As discussed in chapter 2, stress relaxation and creep are characteristic of viscoelastic materials, while short-term elastic and long-term plastic behavior are seen in many solids. Clearly, parenchyma is neither a viscoelastic material nor a solid (elastic or plastic). Rather, it is a structure that manifests properties that parallel those of viscoelastic or elastic or plastic materials. Parenchyma cell walls flatten in the plane perpendicular to the direction of an externally applied load, and cell fluids (principally water) evacuate from protoplasts as cell walls undergo plastic deformations at strain rates that depend on the permeability of the plasma membrane. We must understand all these properties to evaluate the mechanical role of this tissue.

The expulsion of cell fluids from the protoplast explains much of the mechanical behavior of parenchyma. The slope of the stress-strain diagram increases with the strain rate, because higher strain rates give fluids less time to cross the plasma membrane. With higher strain rates, the tissue demonstrates less compressibility and the apparent elastic modulus of the tissue increases. In turn, stress relaxation is the result of pressure stabilization as turgor pressure within cells increases. Loading followed by unloading produces recoverable cell wall elastic deformation and unrecoverable plastic deformations owing to the loss of fluids from cells. Creep is also the result of fluid evacuation. Finally, tissue failure under compression occurs when the applied stress equals or exceeds the rupture stress of cell walls. Thus the mechanical behavior of parenchyma depends on the rate of stress application, cell turgor, plasma membrane permeability, and the stiffness of cell walls. Additionally, it depends on the ductility of the pectinaceous middle lamella and the size, shape, and number of cells in the plane through which the stress is applied.

During histogenesis many of these features can change (e.g., cell wall stiffness, cell size and shape), while cell turgor can undergo transient physiological modification even in mature tissues. For example, the amount of insoluble pectin in the middle lamella decreases in the storage parenchyma of maturing fruits. This results in fruit softening, since soluble pectins hydrate and reduce how much the middle lamella binds together neighboring cell walls. Pectin is a more or less amorphous material that transmits shear. It is highly viscoelastic in its mechanical behavior and supports little tensile stress. The relative plasticity of the pectinaceous middle lamella contributes to the mechanical behavior of parenchyma by permitting deformations in cell shape under compressive or tensile loadings. Cell wall rupture is the principal mode of mechanical failure when the middle lamella is stronger than the primary walls. If shear stress in the middle lamella exceeds the strength of the pectinaceous material and the strength of cell walls is high, however, then cell-cell debonding will occur. Thus, when parenchymatous fruits ripen, their mode of tissue failure changes from one dominated initially by cell wall rupture to one dominated by cell-cell debonding. This makes mature fruits with a high volume fraction of parenchyma easily chewed or ruptured by impact loadings (as when they fall to the ground), exposing less digestible seeds and permitting their dispersal by frugivores.

By the same token, the mode of tissue failure is also influenced by cell turgor, which can change over short intervals. When the dominant mode of tissue failure is cell wall rupture, as in unripened parenchymatous fruit tissue, high turgor pressure reduces the strength of parenchyma. Any externally applied stress is added to the tensile stresses produced in cell walls by the turgid protoplast, so less external stress is required to rupture cell walls. When the dominant mode of tissue failure is cell-cell debonding, as in ripened fruits, high turgor pressure increases the strength of tissues because it inflates protoplasts and increases the cell wall to cell wall contact area.

The mode of failure in parenchyma for any given turgor pressure and strain rate depends on the relative strengths of intercellular bonding (the cohesive

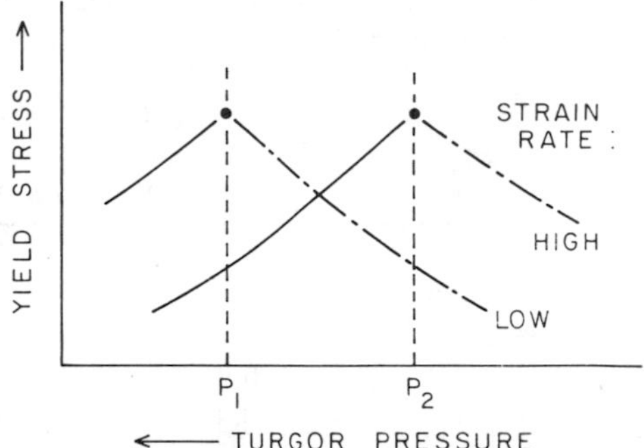

FIGURE 6.9 Theoretical effects of turgor pressure $\mathcal{P}$ and strain rate on the mode of mechanical failure and yield stress of parenchyma. Two strain rates are shown: a "high" strain rate (graph to right) and a "low" strain rate (graph to left). Two modes of tissue failure can occur: cell-cell debonding (dashed lines) or cell wall rupture (solid lines). The transition between the two modes of failure (black dots) in a tissue sample depends on the turgor pressure of the sample and on the strain rate. High turgor pressure decreases the yield stress at which cell wall rupture occurs when a high strain rate is applied to the tissue. High turgor pressure also increases the yield stress at which the transition between cell wall rupture and cell-cell debonding occurs when a low strain rate is applied to the tissue. See text for further details.

strength of the middle lamella) and cell walls (Lin and Pitt 1986). If the applied stresses that can cause cell-cell debonding and cell wall rupture are denoted by σ_d and σ_r, respectively, then the tissue yield stress σ_y is the minimum of either σ_d or σ_r. Thus, $\sigma_y = \min(\sigma_d, \sigma_r)$ (Lin and Pitt 1986). If $\mathcal{P}_1$ and $\mathcal{P}_2$ are the turgor pressures at which a transition between the two modes of tissue failure occurs for high and low turgor pressures, then when the turgor pressure is less than $\mathcal{P}_1$, higher strain rates increase σ_y. If the turgor pressure is greater than $\mathcal{P}_2$, however, high strain rates decrease σ_y. Finally, if the turgor pressure lies between $\mathcal{P}_1$ and $\mathcal{P}_2$, then a change in the strain rate causes a shift in the mode of failure and may increase or decrease σ_y. These relationships are summarized in figure 6.9. Lin and Pitt (1986) found that the probability distribution of cell wall strengths within parenchyma tissue samples statistically conforms to a Weibull frequency distribution with a coefficient of variation of 30%, so even within tissue samples isolated from the same population of plant organs there exists considerable variation.

An often neglected aspect of tissue structure is the mechanical role of intercellular spaces. If a significant portion of the tissue is occupied by gas-filled spaces, then the Poisson's ratio will be low, since stresses will cause cells to deform into these spaces and the tissue will behave as a compressible material

TABLE 6.2 Mechanical Behavior of Collenchyma (under Uniaxial Tension) Isolated from Lovage (*Levisticum officinale*)

Stress ($MN \cdot m^{-2}$)	Extended Length (m)	Unrecovered Deformation ($\times 10^{-3}$ m)	Engineering Strain ($\times 10^{-3}$ m)
4.079	0.1893	0	1.59
8.169	0.1896	0	3.17
12.26	0.1900	0.1	5.29
16.38	0.1902	0.2	6.35
20.41	0.1904	0.4	7.41
24.53	0.1906	0.6	8.47
28.65	0.1909	0.8	10.1
32.67	0.1912	1.0	11.6
36.78	0.1914	1.1	12.7
40.91	0.1918	1.2	14.8
44.93	0.1922	1.3	16.9
49.05	0.1924	1.5	18.0
53.17	0.1927	1.8	19.6
57.19	breakage	—	—

Source: Data from Ambronn (1881, 521–23).

(see chap. 2). As the volume of interstitial spaces decreases, the tissue will increasingly behave as a compressible material, with a Poisson's ratio convergent on a maximum value of 0.5. This explains why storage parenchyma and spongy mesophyll or aerenchyma operate mechanically as very different materials. Additionally, cell fluid evacuation through intercellular spaces can result in flow rates with various degrees of viscous resistance. Therefore, in theory, the volume of intercellular spaces in a tissue can have significant effects on mechanical behavior. Unfortunately, few experimental data are available to indicate how much these factors influence the mechanics of plant tissues in general.

Parenchyma provides an excellent example of a thin-walled hydrostatic plant tissue, but there are many other tissues important to plant biomechanics that have wall thicknesses intermediate between parenchyma and the thick cell walls found in wood and other nonhydrostatic cellular solids. Let us now turn to these tissues.

COLLENCHYMA

Unlike parenchyma, which has been the focus of considerable theoretical and

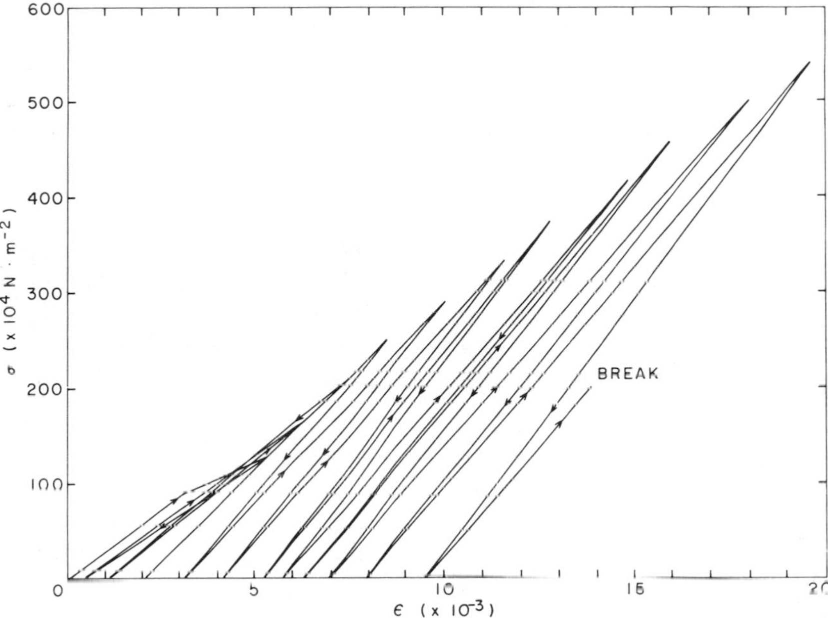

FIGURE 6.10 Plot of stress versus strain for collenchyma. Data (from Ambronn 1881) are given in table 6.1.

empirical research, relatively little is known about the elastic parameters of collenchyma. Nonetheless, collenchyma was one of the first plant tissues to intrigue botanists with its mechanical role in growth and development. It is often a prominent tissue in rapidly elongating stems and leaves, where it is frequently found as a component of vascular bundles, particularly within petioles—for example, celery (*Apium graveolens*) and lovage (*Levisticum officinale*), members of the Umbelliferae (Apiaceae). H. Ambronn (1881) was the first to suggest that collenchyma provides mechanical support to rapidly growing organs and that its low elastic range and high plasticity prevent it from hindering elongation. He carefully removed strands of collenchyma from the vascular bundles of leaves from a variety of species and measured their elastic range and breaking stress. The data from lovage shown in table 6.2 are from this seminal study. The original length and transverse area of the collenchyma strand are 18.93 cm and 1.2×10^{-3} cm^2, respectively. (Ambronn reported that only half of the cells exposed in transection were filled with protoplasm. The rest were empty of cell contents. Hence he reported a transverse area of 0.6×10^{-3} cm^2 for the tissue sample. When calculating stress, however, the

TABLE 6.3 Relative Tensile Breaking Stress of Collenchyma and Primary Xylem Strands Isolated from the Petioles of Celery (*Apium graveolens*)

Collenchyma			Primary Xylem		
Breaking Load (N)	Area (10^{-7} m^2)	Breaking Stress (MN·m^{-2})	Breaking Load (N)	Area (10^{-7} m^2)	Breaking Stress (MN·m^{-2})
11.77	5.32	22.12	2.55	4.91	5.19
12.75	9.02	14.13	0.93	3.22	2.89
14.72	5.83	25.21	1.37	4.13	3.33
14.72	9.51	10.30	1.47	4.14	3.55
12.75	7.85	16.29	1.52	4.00	3.81
11.48	4.75	24.13	1.67	3.28	5.08
12.56	4.97	25.31	2.21	3.48	6.35
14.72	4.91	20.01	1.47	3.16	4.66
7.85	2.66	29.53	0.93	2.66	3.50
10.79	3.96	27.27	1.28	3.24	3.93
11.38	3.26	34.92	1.28	3.36	3.79
11.77	4.27	27.57	1.28	3.14	4.06
7.85	2.91	26.98	0.79	2.51	3.13

Source: Data from Esau (1936, 457, table 2).

total transverse area must be used.) Ambronn loaded the strand of collenchyma with 50 g weights at intervals of 5 minutes and measured the extended length of the strand, as well as the plastic deformation (unrecovered extension), after the weights were removed. His data are graphed in figure 6.10. (This protocol was exactly like the one Robert Hooke used to measure the stress and strain of metal wires, but Ambronn appreciated the importance of recording the rate of strain.) Unfortunately he did not record the change in the transverse area after each loading-unloading cycle, so we cannot calculate the instantaneous stress from his data, but only the nominal stress.

Ambronn found that the collenchyma sample could elongate from 2% to 2.5% before breaking. The data indicate a maximum extension of 0.37 cm (roughly 1.96% of the strand's original length). By contrast, fibers removed from the primary xylem of lovage typically extend less than 1.5% before they break under tension. The elastic range of the collenchyma Ambronn examined is low compared with that of fibers, however. For example, his sample of collenchyma underwent plastic deformations at stresses higher than 8.33 MN·m^{-2}, whereas fibers regain their original lengths after tensile stresses of 150 to 200 MN·m^{-2}. Within its elastic range of behavior, collenchyma has an elastic modulus roughly equal to 22 MN·m^{-2} (table 6.2). Clearly, in collenchyma plastic deformations at relatively low stress levels are advantageous, since they permit growth in length, whereas plastic deformations in primary vascular tissue fibers would be disadvantageous, since they would confer a low resistance to bending moments in mature organs.

Esau (1936) examined the breaking stresses of collenchyma strands isolated from the petioles of celery and compared them with the breaking stresses of the primary xylem strands removed from the same vascular bundles as the collenchyma samples. This was an elegant experiment and involved great technical finesse. Representative data from this experiment are shown in table 6.3, which reveals that the average breaking stress of collenchyma is roughly 5.7 times that of the associated primary xylem (roughly 23.3 compared with 4.1 $MN \cdot m^{-2}$). The average breaking stress of these collenchyma samples (from celery) was roughly 44% that of Ambronn's collenchyma samples (from lovage), indicating that the mechanical properties of this tissue cannot be extrapolated from one species to another. Indeed, a 40% change in the breaking stress of collenchyma is possible owing to changes in tissue water content. Also, Esau was able to show that variations in breaking stresses were correlated with the age of the tissues (based on the age of the leaves they were isolated from). Higher breaking stresses were typical for older and more mature leaves. From this she concluded that cell wall thickening attending histogenesis increases the strength of collenchyma (Esau 1936, 459). Thus the material properties of collenchyma vary as a function of the developmental age of the tissue.

The effect of aging on the mechanical attributes of collenchyma is beautifully shown by the research of Jaccard and Pilet (1975), who examined young and senescent collenchyma from celery. When subjected to equivalent stresses, senescent collenchyma exhibited total strain roughly 17% less than that of young tissue. Also, the ratio of plastic to elastic strains decreased as the tissue aged. Accordingly, young collenchyma is more plastic in its behavior than older collenchyma. Seen from the perspective of the trade-offs between the requirements for plastic deformations in elongating portions of a plant and the requirement for elastic rigidity in mature portions of the same plant, these data are very satisfying. By varying the magnitude of the loading, Jaccard and Pilet (1975) also demonstrated that young collenchyma is a nonlinear viscoelastic material and that the theoretical limit of linear viscoelasticity is smaller than the force that would be required to stretch this tissue in growing leaves. Thus the driving force of organ elongation is sufficient to cause collenchyma to flow.

Unfortunately, the data for this study were derived from tissue samples that had been boiled in methanol to inhibit metabolic alterations during mechanical testing. Although this treatment is typical for studies examining cell wall extensibility, it very likely obscures many mechanical features of the primary cell walls of living tissues. One of these features is the extent to which externally applied loads may cause protoplasmic fluids to evacuate into the cell walls of collenchyma, which have a relatively high volume fraction of gelatinous matrix. As seen in parenchyma, fluid evacuation from the protoplast affects the mechanical properties of tissues. This may be even more pronounced for collenchyma, since the primary cell walls of this tissue typically swell and contract as a function of cell wall hydration.

The Epidermis

Tissues within an organ can grow at different rates, and these differences can cause tensile and compressive growth stresses among neighboring cells or adjoining tissues. Accordingly, we might anticipate that the epidermis, which is a continuous layer of cells over the entire plant body (see fig. 6.2), would experience mechanical stresses resulting from the differential rates of growth occurring within the plant body it envelops. Gregory Kraus (1867) was one of the first biologists to follow this line of reasoning. He showed that the epidermis of growing stems is in longitudinal tension by slicing an internode (a segment spanning the interval of a stem separated by the attachment sites of two successive leaves) into longitudinal strips each consisting of a single tissue and comparing the length of each strip with the original length of the internode. His measurements showed that the outer strips of tissue shrank from their original lengths while the inner tissues expanded. Thus Kraus concluded that the epidermis normally experienced tensile growth stresses, while tissues toward the center of the internode normally experienced compressive growth stresses. Subsequent work by Schüepp (1917) and Schneider (1926) confirmed the earlier work of Kraus and also showed that longitudinal tension on the epidermis decreases from the tips of growing stems toward their bases. In retrospect this should come as no surprise; since growth in length diminishes away from a shoot apex, longitudinal growth stresses should also decrease.

All normal materials experience a lateral contraction when they are extended in length. Conversely, when released from longitudinal tension, materials expand laterally. Thus, when cylindrical segments of young epidermal tissue are excised from stem segments, they would be expected to expand in girth and shrink in length—precisely what is typically seen (Veen 1971). Yet attempts to measure the mechanical properties of the epidermis from the behavior of excised specimens placed in uniaxial tension would be futile owing to Poisson's effect. Recall (from chap. 2) that materials tested in uniaxial tension tend to have lower elastic moduli and higher ultimate strains in uniaxial tension than in biaxial tension. The epidermis naturally operates under biaxial tension, and the strains within the attached epidermis are not the same as those produced by uniaxial loading experiments. You could measure the elastic moduli in uniaxial tension and use the following relations to get the biaxial behavior: $\varepsilon_1 = (1/E)(\sigma_1 - \nu\sigma_2)$ and $\varepsilon_2 = (1/E)(\sigma_2 - \nu\sigma_1)$ (as kindly pointed out to me by Lorna Gibson). Another way to measure the mechanical properties of the epidermis is to vary the hydrostatic pressure within intact fruits by injecting varying quantities of an isotonic solution. Bernstein and Lustig (1985) and Lustig and Bernstein (1985) investigated the tendency of the epidermis of grapes to split by assuming that the fruit was spherical and that the epidermis was isotropic. These assumptions allowed them to use a simple relation between the tensile stresses σ_{-} developed within the grape skin as a function of hydrostatic pressure $\mathcal{P}$, that is, $\sigma_{-} = (\mathcal{P}R)/2t$, where t is the thickness of the skin and R is the radius of the fruit. These authors found a fairly consistent relation between their experimental results and the tendency of grape varieties to split under field conditions. Thus two-dimensional strains clearly dictate the mechanical properties of the epidermis.

The apparent elastic modulus of epidermal tissues could be easily measured by trapping epidermal specimens between a pair of cylindrical tubes fitted with rubber O-rings. By varying the pressure in one of the two tubes and measuring the deflection δ at the center of the bulging hemispherical specimen, the elastic modulus can be calculated from the formula $E = (12/64)[(5 + \nu)(1 - \nu)(\mathcal{P}D^4)/(\delta t^3)]$, where D is the diameter of the disk of epidermis. A reasonable estimate of the Poisson's ratio would be 0.3, in which case $E \approx (0.698\mathcal{P}D^4)/(\delta t^3)$. The strength of such a specimen could be calculated from the stress of rupture, which for an isotropic material equals $[3(5 + \nu)\mathcal{P} D^2]/8t^2$. (These formulas are valid provided the epidermis is truly isotropic. The assumption of anisotropy could be tested by comparing the magnitudes of orthogonal strains measured as the displacements of small resin markers placed on the surface of the hemispherical specimen.)

How reasonable is the assumption that the epidermis is isotropic? The answer appears to depend on the age of the tissue. From chapter 5, we might expect that the mechanical behavior of the epidermis was in some way associated with the orientation angles of cellulose microfibrils within cell walls. For example, according to the multinet model, extension in length would be permitted by the epidermal cells if the dominant orientation angle within the cell walls of this tissue was oriented transverse to the longitudinal axis of internode extension. Conversely, in older epidermis, which is found on stems that have essentially stopped growing in length, the dominant orientation angle might be aligned more with the longitudinal axis of the stem. Accordingly, an ontogenetic change in the dominant orientation angle of microfibrils within the epidermis would be anticipated. Takeda and Shibaoka (1981a,b) have shown that in cowpeas (*Vigna*) the dominant orientation of microfibrils changes as the epidermis matures. In young epidermis, a transverse orientation of microfibrils dominates the ultrastructure of the innermost wall layers of the external epidermal cell walls (Takeda and Shibaoka 1981a). In medium-aged epidermal cells, transverse, oblique, and longitudinal microfibrillar alignments are present with similar frequency. But in older epidermal cell walls, the dominant microfibrillar orientation angle is longitudinal to the axis of stems. These findings appear particularly significant with regard to the preferred direction of cell wall extensibility and its consequences on the directionality of the principal stresses experienced in epidermal cell walls. Recall that most polymers resist extension along their longitudinal axis. Thus, cell walls dominated chemically by cellulose microfibrils ought to expand preferentially perpendicular to the longitudinal axis of cellulose microfibrils. As such, young epidermal cells will resist widening but permit elongation; adolescent cell walls ought to have little or no preferred directionality in extension, and mature cell walls would permit expansion but resist elongation. This pattern of permissible deformations in epidermal cell walls is consistent with the notion that the epidermis imposes a limiting anisotropic boundary condition to growth in the size and shape of stems as well as other organs. It is also entirely compatible with the organogenetic changes of stems, in which elongation is initially favored over expansion in girth, followed by a cessation in growth in length and an increase in girth.

In addition to the ultrastructural features of cell walls, the epidermis has other macrostructural features that may be equally important in dictating the mechanical behavior of this tissue. The epidermis is a layer of cells (characteristically with little interstitial volume) rather than a three-dimensional solid. Its cells are closely spaced and frequently look like pieces of a jigsaw puzzle when seen from above. Accordingly, epidermal cells have a very high cell wall to cell wall contact area. The middle lamella, which binds adjoining cell walls, is known to have a high shear modulus in parenchyma. Thus there is reason to suspect that the middle lamella of the epidermis may be important in determining the mechanical behavior of this tissue when placed in bending and torsional shear.

Research has also shown that epidermal cell walls are not mechanically homogeneous. The peripheral cell walls of the epidermis are more rigid—they have a lower plastic extensibility than the cell walls that make contact with the tissues they envelop (Cosgrove 1989). The peripheral cell walls can be up to five times as thick as the inner cell walls (Haberlandt 1909; Ray 1967) and thus can withstand significantly higher stress.

Significantly, applying exogenous growth hormones to the epidermis alters the mechanical behavior of this tissue. By inference, this alteration has been interpreted within the developmental context of stem elongation. Under the influence of auxin, the cut halves of an actively growing pea internode exhibit less curvature than those of an untreated internode (Masuda and Yamamoto 1972). Tanimoto and Masuda (1971) have concluded that auxin induces stem elongation by removing the restraining influence of the epidermis. Measurements of epidermal extensibility and stress relaxation parameters by Masuda and Yamamoto (1972) are consistent with this interpretation. At water saturation and at reduced turgor pressures, stem elongation appears to be correlated primarily with the mechanical properties of epidermal cell walls. Isolated epidermis does not respond to 2, 4-D, however, though subepidermal tissue layers do. This suggests that several layers of cells adjacent to the epidermis are involved in the response of the epidermis to auxin. The pioneering work of Kraus (1867), who observed that the outermost layers of the cortex also contract when they are excised from stems, lends credence to the notion that the epidermis mechanically operates as the outermost layer of cells adjacent to more internal layers of tissue that collectively manifest a gradient of mechanical and physiological responses to growth.

Regarding the contribution the epidermis makes to the bending stiffness of cantilevered plant organs, such as the petiole, Julian F. V. Vincent (pers. comm.) makes the following suggestion: Remove a turgid petiole with a ratio of length to depth $\geq$ 20 and orient it as a simple cantilever (one fixed end). Load the free end of the cantilever with weights P, measure the deflection from the vertical, and determine the elastic modulus of the petiole with the aid of the formula $\delta = Pl^3/3EI$ (see eq. 3.29 and fig. 3.9). Carefully remove the adaxial layer of epidermis, remeasure I, determine E with the stripped surface oriented uppermost and then lowermost on the cantilever, and compare the difference between these measurements and the original E of the intact petiole. When large, fleshy petioles, such as those of rhubarb, are examined in this fashion and the reduction of the second moment of area resulting from the removal of the epidermis is noted in the calculations, one can see that the epidermis contributes ≤ 70% of the total bending stiffness yet contributes from ≥ 5% to 10% of the total cross-sectional area. Clearly, the epidermis provides a comparatively stiff outer rind to the relatively less stiff inner parenchymatous core of some petioles. Vincent goes on to note that the contribution to bending stiffness made by other structural components in the petiole could be determined by the successive surgical removal of other tissue systems such as subepidermal collenchyma and phloem fibers.

THE VASCULAR TISSUE SYSTEM: SOME GENERAL COMMENTS

Unlike parenchyma and collenchyma, where each tissue essentially consists of a single cell type, the vascular tissue system is composed of a variety of developmentally homologous but structurally and functionally divergent cell types (see fig. 6.2). For example, depending on the species, the mature primary vascular tissue system may possess a significant volume fraction of parenchyma, or this tissue may be entirely lacking. Likewise, the secondary vascular tissues of some species may be dominated by tracheids or fibers and may or may not contain a significant proportion of xylem and phloem ray parenchyma. As a result of the heterogeneity in cell types within the vascular tissues of a single species, the mechanical properties of mature vascular tissues typically show marked anisotropy. Similarly, the mechanical properties of the vascular tissue system may differ substantially across species boundaries.

An additional consideration we should bear in mind when treating the vascular tissue system is the need to distinguish between primary and secondary vascular tissues. Although both tissues may contain the same cell types, their proportion, spatial distribution, and mechanical properties may differ significantly. Thus, a primary phloem fiber may not have the same mechanical attributes as a secondary phloem fiber. The potential for these differences should engender some skepticism in accepting generic statements about the mechanical performance of a given cell type within the vascular tissue system. This caveat is most probably not needed for those who are familiar with the way plants develop and grow, since ontogenetic differences in cell size and shape and in cell wall texture, chemistry, and thickness are evident within any tissue type, highlighting the need to precisely define the developmental age of any cell or tissue type. Nonetheless, biomechanical treatments of the vascular tissue system are particularly prone to errors of generalization and oversimplification.

Perhaps the most important distinguishing feature among the various developmentally homologous cell types found in vascular tissues is their marked

TABLE 6.4 Elastic Moduli of Wood Samples from Six Species of Trees

Genus and Species	Moisture Content (%)	Elastic Modulus E_L ($GN \cdot m^{-2}$)	E_T/E_L	E_R/E_L
Ochroma lagopus	9	3.92	0.015	0.046
Betula alleghaniensis	13	14.91	0.050	0.078
Pseudotsuga menziesii	12	16.29	0.050	0.068
Picea sitchensis	12	12.07	0.043	0.078
Liquidambar styraciflua	11	10.10	0.050	0.115
Liriodendron tulipifera	11	10.38	0.043	0.092
		11.3 ± 4.4	0.04 ± 0.014	0.08 ± 0.023

Source: Data from American Society of Civil Engineers (1975, 35, table 1.10).
Note: See table 2.3 for additional values of E.

axial asymmetry. Cells such as fibers, tracheids, and vessel members are typically elongated in the direction of the axis of organ elongation. This geometry contributes significantly to the anisotropic behavior of primary and secondary vascular tissues. Several early studies demonstrated that the elastic modulus of isolated fibers correlates with the dominant fibrillar orientation angle within secondary cell wall layers, which in turn correlates with cell length. In general, the elastic modulus E increases as the orientation angle θ decreases (as microfibrils are oriented more and more along the length of the cell). For example, the elastic moduli of phloem fibers whose microfibrillar orientation angles are 42° and 5° are 3 and 3.5 $GN \cdot m^{-2}$, respectively. Since longer fibers typically have lower microfibrillar orientation angles, tissues with longer fibers will have higher elastic moduli than tissues with shorter fibers. In many species, fiber length increases with the age of organs. Hence older organs should be stiffer and therefore more capable of resisting bending moments. Experience confirms this expectation.

Secondary xylem shows a pronounced mechanical anisotropy that correlates with the geometric asymmetry seen in this tissue. The elastic modulus measured along the grain of wood samples E_L is significantly higher than when measured tangentially or radially to the grain E_T and E_R, as shown in table 6.4. For the species listed, the average value for E_L is 11.3 $GN \cdot m^{-2}$, while E_T/E_L and E_R/E_L average 0.04 and 0.08, respectively. Accordingly, for the species listed, the average value for E_T is 0.487, and that for E_R is 0.926 $GN \cdot m^{-2}$. Thus there can be as much as two orders of magnitude difference in the stiffness of wood depending on the direction this tissue is mechanically stressed.

The relations among the relative values of the elastic moduli measured in the three principal orthogonal planes of wood samples shown in table 6.4 hold true regardless of the water content of the wood samples, but the absolute values of E typically decrease as a function of tissue water content up to fiber saturation (roughly 27%). The relation between the elastic modulus E_O measured at M_O% moisture content and the elastic modulus E_M adjusted to a higher moisture content, M%, is given by the formula

$$E_M = E_O\,[1 + H\,(M - M_O)], \tag{6.12}$$

where H is the coefficient of moisture effect—the change in the elastic modulus with 1% change in moisture content. Typically, for most wood species, $H = -0.02$. It is worth noting, however, that the increase in the elastic modulus of wood resulting from a decrease in tissue moisture content reduces the strength and the extensibility of tree branches. Indeed, observations tend to support the notion that dry branches are more brittle and fracture more easily than their moistened counterparts.

Equation (6.12) demonstrates that the strength of secondary xylem within trees will change as a function of how much the apoplast of a plant is hydrated. Typically the most recent secondary xylem layers within a tree trunk are more hydrated than the older (heartwood) secondary growth layers. Thus there usually is a radial gradient in the stiffness of wood so that the outer portions of a tree trunk are less stiff than the inner portions. The profile of this gradient may also show seasonal fluctuations depending on the availability of water to roots. The changes in the stiffness of wood as a consequence of changes in its moisture content are not trivial, as calculations indicate. For example, at 12% moisture content, the elastic modulus of Sitka spruce (*Picea sitchensis*), measured along the grain, equals 12.1 $GN \cdot m^{-2}$, whereas at 24% moisture content the elastic modulus measures 9.2 $GN \cdot m^{-2}$. This is a 23.6% reduction in stiffness. As the moisture content of the wood increases, there is likely to be an increase in the water content of tissues and plant organs (leaves) above the trunk. Hence as the elastic modulus of the tree trunk decreases, the compressive loading imposed on it by its canopy of branches and leaves is likely to increase. From the Euler column formula (described in chap. 3) we can see that this is not particularly advantageous in terms of bending strains, especially when we consider that in most deciduous trees new leaves, produced when wood moisture content is seasonally high, impose considerable drag in the wind. In fact, however, most trees have a considerable design factor in their construction (see chap. 8).

Data from isolated primary phloem fibers indicate a similar dependency of

TABLE 6.5 Formulas for Approximating Material Properties of Wood Based on the Specific Gravity (g) of the Wood Sample

Oven-Dried Specimens	
In static bending	
Fiber stress at proportional limit	$(5.75 \times 10^7)g^{1.15}$
Modulus of rupture	$(9.98 \times 10^7)g^{1.15}$
Modulus of elasticity	$(1.36 \times 10^{10})g$
Compression parallel to grain	
Fiber stress at proportional limit	$(2.98 \times 10^7)g$
Maximum crushing strength	$(3.88 \times 10^7)g$
Modulus of elasticity	$(1.68 \times 10^{10})g$
Compression perpendicular to grain	
Fiber stress at proportional limit	$(1.73 \times 10^7)g^{2.25}$
Air-Dried Specimens (12% Moisture Content)	
In static bending	
Fiber stress at proportional limit	$(9.62 \times 10^7)g^{1.15}$
Modulus of rupture	$(1.48 \times 10^8)g^{1.15}$
Modulus of elasticity	$(1.61 \times 10^{10})g$
Compression parallel to grain	
Fiber stress at proportional limit	$(5.04 \times 10^7)g$
Maximum crushing strength	$(7.03 \times 10^7)g$
Modulus of elasticity	$(1.95 \times 10^{10})g$
Compression perpendicular to grain	
Fiber stress at proportional limit	$(2.67 \times 10^7)g^{2.25}$

Note: All formulas yield units of $N \cdot m^{-2}$.

the elastic modulus on cell wall hydration and draw attention to the role of lignification in both primary and secondary xylem and phloem fibers. The elastic modulus of wet phloem fibers measured in tension is roughly 19 $GN \cdot m^{-2}$; when they are air dried, E measures 51 $GN \cdot m^{-2}$, and when kiln dried, E averages 60 $GN \cdot m^{-2}$. Thus, just as in secondary xylem, the elastic modulus of phloem fibers increases as water is withdrawn from cell walls. The strength of dry delignified fibers is greater than that of wet delignified fibers, indicating that the strength of predominantly cellulosic cell walls increases as water is withdrawn. Regardless of cell wall moisture content, however, lignified cell walls are less strong than delignified cell walls. Thus, at any moisture content lignin appears to reduce the strength of cell walls. Although lignification appears to be a disadvantage, it is not. The presence of lignin in cell walls permits less water to infiltrate and weaken the microfibrillar infrastructure of cell walls. Thus it stabilizes the mechanical properties of the cell wall from transient fluctuations in water content. Additionally, data indicate that the high concentration of lignin in the compound middle lamella reduces the potential for cell-cell debonding when wood is subjected to tensile stress. Mark (1967) estimated that the shear modulus of the compound middle lamella is roughly 77 $GN \cdot m^{-2}$, higher than that of secondary cell walls by a factor of three. As discussed in chapter 5, under tensile stress, when xylem cell walls do fracture, the incipient fracture occurs in the S1 wall layer. This propagates to the remaining secondary wall layers, which are estimated to have lower shear moduli but higher elastic moduli. Thus the compound middle lamella, largely by virtue of its lignified condition, provides wood with a gluelike material stronger than the cell wall infrastructure it binds together.

The mechanical properties of clear- and straight-grained samples of wood can be approximated from formulas provided the specific gravity and moisture content of a sample are known (U.S. Forest Products Laboratory, Madison, Wisconsin), although when the wood sample is not clear- or straight-grained, the influence of specific gravity is much diminished. In the absence of detailed and extensive tests, the approximate relations between specific gravity and mechanical properties for a green wood sample can be obtained from table 6.5, where g is the specific gravity of an oven-dried wood sample, based on the volume measured when the sample was green (formulas yield units of $N \cdot m^{-2}$). The values of all these parameters will be substantially larger for a sample of wood that is air dried to 12% of its moisture content (table 6.5).

MECHANICAL FAILURE OF SECONDARY XYLEM

Since secondary xylem (wood) is the principal mechanical support tissue for many plant species, and since the mechanical failure of this tissue involves macroscopic as well as microscopic deformations, the behavior of wood when it mechanically fails deserves special mention.

Under compression, the walls of tracheids and vessel members buckle or concertina (Preston 1974, 348–52), while whole samples of wood densify. Robinson (1920) first noted microscopic slip planes as tissue failure begins. A slip plane is a shear zone running through any material. In wood these result from the buckling of microfibrils and can be visualized by polarized light microscopy, where slip planes appear as bright lines running obliquely across cell walls at angles between 61° and 69.5° to the axis of cell length (Dinwoodie 1968; Kisser and Steininger 1952). If the walls were isotropic, then a 45° angle would be anticipated. Hence the angle of slip planes confirms the anisotropic nature of xylem cell walls. Slip planes caused by compression begin to form under applied loads that are roughly one-half the breaking load

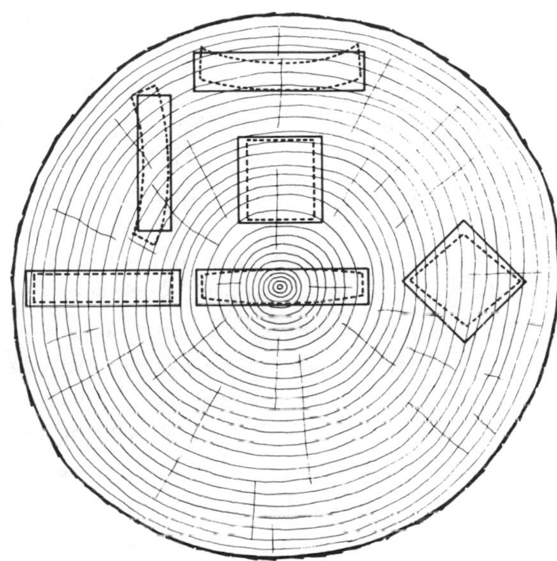

FIGURE 6.11 Diagram of a transverse section through a tree trunk illustrating the deformations that result when blocks of secondary xylem (wood) are taken out and allowed to dry. The in situ geometry of each block of wood is shown by solid lines; the bent outline of each block, once it is removed from its original location, is shown by dotted lines.

of wood.

At higher stresses, deformations propagate from regions where slip planes develop, resulting in a compression crease. Slip planes are microscopic points of stress concentration that radiate locally, producing macroscopic compression creases. Dinwoodie (1974) found that 71.6% of the variation in the slip plane angle could be accounted for by differences in fibril orientation angle in the S1 layer and by the ratio of longitudinal to transverse stiffness in wood samples. By contrast, only 48.4% of the variation could be explained by the fibril orientation angle alone.

Slip planes and compression creases can develop naturally within wood compressionally loaded well below its breaking load. These slip planes result from differential dehydration when the surfaces of wood are exposed to air. As a result of the anisotropy in the mechanical properties of secondary xylem, shrinkage is not uniform in the three principal orthogonal planes, and compressive and tensile stresses can develop, as shown in figure 6.11. These localized stresses can reach magnitudes approaching the breaking load, and slip planes and compression stresses can rapidly propagate through the entire trunk.

Additionally, secondary xylem experiences growth stresses much like those discussed in the context of the epidermis and the growth stresses developing within primary stems. For example, Jacobs (1938, 1945) showed that a plank cut from the median longitudinal section through a tree trunk may split (explosively) along the line of the pith, with the resulting halves bending away from the pith. This rapid longitudinal shearing is accompanied by an increase in the length of the wood toward the center of the trunk and a decrease in its length toward the periphery. Thus the wood toward the center of the trunk is evidently placed in compression by subsequent growth layers of wood, while wood toward the periphery experiences tensile growth stresses. At the cellular level, newly differentiated tracheids (toward the periphery of the trunk) undergo longitudinal contraction that places older tracheids toward the center under increasing compressive loading as the tree grows laterally. When the plank is cut, the tensile stresses toward the perimeter of the plank are eliminated and the cells toward the center are relieved of their compressive load, resulting in a potentially violent mechanical response. This effect is mechanically similar to the curvature of split stem segments lacking secondary growth, where the epidermis is mechanically analogous to the newly differentiated tracheids in wood. Another mechanically analogous system is seen in the common onion. When sliced transversely, the concentric segments of the leaf bases forming the onion bulb are tightly appressed to one another. When cut to form two hemispheres, however, the older leaf bases (toward the outside of the onion bulb) reflex outward and straighten, while the younger leaf bases (toward the center of the bulb) more or less retain their original curved geometry.

Although the growth stresses that can develop within secondary xylem can reach levels theoretically high enough to cause tissue failure, Boyd (1950) has argued that these stresses are never fully achieved owing to the formation of minute compression failures (in the form of slip planes and compression planes) that locally relieve the strain energy stored within the trunk as a whole. Detailed theoretical treatments of growth stresses have been developed and are in reasonable agreement with the empirical data (Gillis 1973). Thus it appears that localized mechanical failure in secondary xylem is part of the normal process of the growth and development of trees. The only significant departure between early formulations of the theory of growth stresses and experimental results is the magnitude of curvatures occurring on strips of wood cut from median longitudinal sections. The radial gradient of the longitudinal stress is generally less than would be expected from theory. More recently, however, Gillis and Hsu (1979) have modified early theoretical treatments to deal with the mechanical behavior of wood as a consequence of elastic-plastic behavior (rather than as purely elastic behavior). This modification has drawn empirical data and predictions based on theory into even greater agreement. Gillis and Hsu have concluded that the core of the tree trunk typically behaves as if it had a significant plastic component. Significantly, all treatments of this subject have indicated that the longitudinal growth stress distribution is uncoupled from either of the distributions of radial or circumferential stress (Kübler 1959; Archer and Byrnes 1974; Gillis and Hsu 1979).

Structures versus Materials

In this chapter we have seen that plant tissues are structures, not materials in the true sense. This distinction has been repeatedly emphasized because the mechanical behavior of a structure can be understood only when its geometry and the material properties of its constituent solids or fluids are considered together. The geometric features of a plant tissue provide the basis for evaluating the mechanism by which it operates. True, the material properties of the solids or other substances a tissue is made from are equally important, as we have seen in considering how plant cell walls help determine the mechanical behavior of tissues. But as discussed in the context of nonhydrostatic and hydrostatic cellular solids, the geometry of the cell wall infrastructure within a plant tissue may in some cases supersede in importance the material properties of the infrastructure. Thus we can never dissociate the mechanical behavior of a tissue from its anatomical configuration.

Despite the innumerable differences among the various plant tissues reviewed in this chapter, some basic patterns in their mechanical behavior provide insights into fundamental ecological and evolutionary questions. Perhaps the most basic of these patterns is expressed by the range of variation in the ratio of the symplast to the apoplast among tissue types. When the various types of tissues are compared based on this single criterion, two extremes immediately become evident—the gas-filled cellular solid and the liquid-filled hydrostat (see fig. 6.4). Between these two extremes lies a spectrum of plant tissue types that, for lack of a better term, I have collectively defined as pressurized cellular solids. Wood, cork, spongy mesophyll, and aerenchyma were discussed in terms of the mechanical theory underlying the behavior of commercially fabricated cellular solids. These artifacts and their biological counterparts share the feature of having a solid phase dispersed (in various geometric ways) within a fluid phase. In the case of wood and cork, the solid phase is the cell wall infrastructure, while the fluid phase is the air within these tissues. Aerenchyma and spongy mesophyll differ only in that their solid phase consists of the living symplast in addition to the cell walls within this tissue. Despite the presence of a living symplast, aerenchyma and spongy mesophyll typically operate as cellular solids because of the limited cell-cell contact area within these tissues.

The opposite of the cellular solid is the hydrostatic tissue type, such as parenchyma, whose thin walls are easily deformed but are hydrostatically bolstered by an incompressible symplast. The mechanical properties of hydrostatic tissues are largely dictated by the turgor pressure of the living symplast and by the ratio of the cell wall thickness to cell radius. As this ratio increases, the mechanical effects of hydrostatic pressure are diminished, and the tissue becomes more and more like a cellular solid. Between the gas-filled cellular solid and liquid-filled hydrostat lie the vast majority of plant tissues—collenchyma, epidermis, and others not treated here. The mechanical attributes of these tissue types depend on both the cell wall infrastructure and the hydrostatic pressure, often in very complex ways.

Clearly this spectrum in mechanical behavior is somewhat contrived and must be suspect, given its all too obvious simplicity. As already seen, each of the tissue types reviewed here has unique characteristics, perhaps as an evolutionary specialization for a particular mechanical function. Thus collenchyma permits the elongation of organs by its plastic and viscoelastic properties; the epidermis appears to influence (or at least reflects) the preferred axes for organ elongation and expansion by virtue of the microfibrillar architecture within the entire fabric of its cell walls, much like a macroscopic version of the individual microscopic cell wall. Finally, even wood, which has been presented as the archetypal cellular solid, demonstrates mechanical properties that one would never infer strictly from mechanical theory.

Nonetheless, the categorization of plant tissues in terms of cellular solids, hydrostats, or hybrids between the two is not without merit. Indeed, it provides some basis for understanding why plant species that have adapted to aquatic or very wet habitats typically have the bulk of the plant body composed of tissues with very high ratios of symplast to apoplast (with due acknowledgment that in the case of aerenchyma the gas-filled spaces must be added to the apoplast side of the equation). These high ratios reflect the mechanical architecture of a hydrostat—a structure that is remarkably strong provided its internal pressure can be maintained. In aquatic habitats water is not a limiting growth factor, and hydrostatic tissues are ideally suited to rapid and mechanically robust growth. By contrast, the bulk of many terrestrial, perennial species reflects the mechanical architecture of a cellular solid. Wood and cork are relatively lightweight yet remarkably strong materials whose mechanical behavior is relatively insulated from the vagaries of water availability. Thus the two extremes in tissue construction, the liquid-filled hydrostat and the gas-filled cellular solid, provide two potential ecological "strategies" for constructing the plant body.

By the same token, the ratio of the symplast to the apoplast indirectly reflects the chemical nature of a plant tissue, since tissues with a very low ratio typically have thickened cell walls that are lignified. Lignin is often regarded as a material of little or no mechanical significance, owing to its low tensile elastic and shear moduli. As we have seen, however, lignin indirectly influences the mechanical behavior of the cellulosic microfibrillar network within cell walls. It prevents water from infiltrating cell walls and therefore reduces the loosening of cellulose microfibrils by water. Additionally, lignin provides a bulking agent to cell walls, increasing the capacity of the cell wall infrastructure of lignified tissues to sustain compressive loadings.

THE CRITICAL ASPECT RATIOS OF TISSUES

The extent to which tissues operate mechanically as either hydrostats or cellular solids and the capacity of a given type of tissue to construct a cylindrical, vertical plant organ are illustrated in figure 6.12. In this figure, the average or maximum elastic modulus of representative samples of tissue is plotted on the abscissa, while the maximum ratio of length to radius of a cylindrical plug of each tissue type based on a buckling criterion is plotted on the ordinate. For comparison, the elastic modulus of cellulose is also plotted. This value was determined from dry cotton fibers. The other data reflect the elastic modulus of fully turgid hydrostatic tissues (parenchyma and collenchyma) and, with the exception of balsa wood (*Ochroma*), fresh or green cellular solid tissues (removed from living plants). The elastic modulus was measured by multiple resonance frequency spectra (Niklas and Moon 1988). This technique utilizes dynamic beam theory by vibrating cylindrical tissue samples, measuring their natural frequencies of vibration, and calculating the dynamic elastic modulus (see chap. 3). Recall that the dynamic elastic modulus reflects the mechanical behavior of a material placed alternately in tension and compression. Therefore it is an elastic parameter that treats the mechanical properties of a material in both static and dynamic bending.

Figure 6.12 is useful because it shows the theoretical relation between the material properties of a tissue and the maximum height to which a cylinder fabricated from each tissue can grow before it deflects from the vertical under its own weight. Cellulose, the strongest material for its density, provides a baseline for comparison. The universe of elastic and geometric possibilities is rather limited: the elastic modulus spans five orders of magnitude, while the ratio of length to radius (the critical aspect ratio) spans only three. Nonetheless, the range in the critical aspect ratio is significant. Parenchyma, as a material, can be used to construct a cylindrical plant organ that is roughly fifteen times as long as it is wide, while sclerenchyma and the wood from Douglas fir (*Pseudotsuga*) can be used to construct a cylinder roughly six hundred times as long as it is wide. However, the commitment to the cell wall infrastructure of the tissue that has to be made to achieve this fortyfold increase in the critical aspect ratio is significant. The bulk of parenchyma is the living protoplast; the bulk of sclerenchyma and wood is dead. The hydrostatic tissue types (parenchyma and collenchyma) are metabolically competent tissues; most gas-filled cellular solids (sclerenchyma, cork, wood) are the sarcophagi of the once living symplast.

Some features shown in figure 6.12 are worth additional comment. First, the trajectory of increasing elastic modulus does not partition into primary and secondary tissues. The value of E used for sclerenchyma was taken from a primary tissue, yet this sclerenchyma has a higher elastic modulus than either balsa or Douglas fir wood. Similarly, bast fibers have an elastic modulus comparable to that of the secondary xylem of the two tree species. Thus, in theory sclerenchyma can be used as a structural, mechanically supportive tissue with equal or perhaps greater efficiency than wood. Indeed, some plants, like the palms and tree ferns, have achieved very large critical aspect ratios (heights) based on this primary tissue type.

Second, there is a disparity of two orders of magnitude between elastic moduli of the hydrostatic and the gas-filled cellular solid tissue types. Parenchyma and collenchyma are clearly separated from the other materials plotted in figure 6.12 in terms of their material properties. Likewise, the critical aspect ratios of the hydrostats and the cellular solids are clearly delineated and significantly different. This is interesting from an evolutionary perspective. The earliest land plants were essentially hydrostatic mechanical devices—owing to their limited capacity for tissue specialization and ancestral parenchymatous tissue construction, the maximum heights they could have grown to before reaching elastic disequilibrium must have been quite small and, in addition, susceptible to the vagaries of water availability. By contrast, with the evolutionary advent of thick-walled tissue types (but not necessarily second-

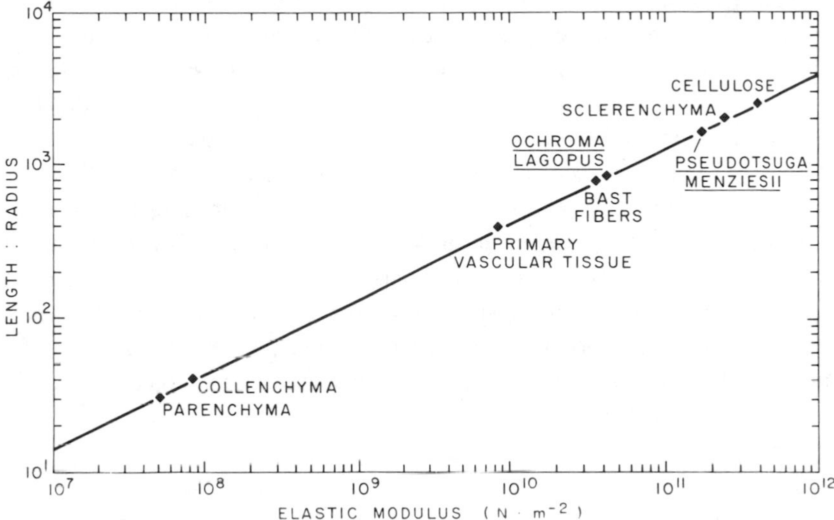

FIGURE 6.12 Maximum aspect ratios and elastic moduli for cylinders made from various plant tissues and materials (hydrated cellulose). The maximum aspect ratio (length divided by radius) is calculated from the Elastica and reflects the maximum length (height) that can be achieved before the cylinder begins to deflect from the vertical under its own weight.

ary tissues), plant height could have been substantially increased and made somewhat immune to the mechanical effects of tissue dehydration. Thus plant height, tissue specialization, and habitat are inextricably bound together.

Third, there appears to be an intrinsic compromise between lignified and nonlignified cell walls in terms of the elastic modulus. Notice that, in terms of the critical aspect ratio, cellulose is a much better material for fabricating a cylinder than are lignified cell walls. The safety factor that lignification confers on cell walls (a stabilization of the elastic modulus) is bought at a price (a reduction in the elastic modulus). Nonetheless, the slight lowering of the elastic modulus resulting from lignification is probably minimal compared with the benefits. Thus it is not surprising that lignification was an early evolutionary achievement of many plant groups. Lignin has been reported from the algae and the bryophytes as well as the vascular plants. In the former two plant groups, lignin appears to have little or no mechanical role but may deter herbivores. Accordingly, lignification may be viewed as an exaptation to mechanical support tissues. Regardless of the selection pressures associated with the evolutionary appearance of lignification, the potential for mechanical support is obvious and was quickly achieved during early land plant evolution.

What is not illustrated in figure 6.12 is that plant organs and their critical aspect ratios are mechanically governed by the selective apportionment of different tissues to produce histologically heterogeneous structures. Stems, leaves, and roots are not typically composed of a single tissue type, although some (like spines and thorns) may be dominated by a single tissue. As such, each plant organ is a mechanical composite, and its mechanical behavior is governed by the spatial deployment of its many tissue constituents. The anatomy of plant organs that serve similar mechanical functions is convergent not by accident, but by the design constraints imposed by the reconciliation of these functions in terms of physical laws and processes. Thus the plant anatomist, ecologist, physiologist, and paleontologist share a common interest, and through the medium of biomechanics, they share a common language.

Seven

The Mechanical Attributes of Organs

> And thus, in contrast to the mere gaze, which by scanning organisms in their wholeness sees unfolding before it the teeming profusion of their differences, anatomy, by really cutting up bodies into patterns, by dividing them into distinct portions, by fragmenting them in space, discloses the great resemblances that would otherwise have remained invisible; it reconstitutes the unities that underlie the great dispersion of visible differences.
>
> Michel Foucault, *The Order of Things*

This chapter treats the mechanical behavior of individual plant organs. Its primary focus is on various modeling procedures that can be used to approximate the morphology of organs, as well as on those that treat the composite material properties resulting from a heterogeneous tissue construction. The former modeling procedures largely stem from solid mechanics theory as it relates to the performance of columns and cantilevered beams (see chap. 3). With the aid of solid mechanics, the consequences that the morphology of an organ has for mechanical behavior can be approximated. How far the predictions of theory coincide with observation, however, largely depends on how well the morphology of an organ coincides with the geometry of a column or a cantilevered beam. Morphologically complex organs must be approached by combining columns and cantilevers into more sophisticated mechanical devices. This approach was explored in chapter 6, where the behavior of numerous interconnected beams, configured into a strutted network, was used to infer the behavior of wood and cork in terms of the theory of cellular solids. A similar approach can be taken to approximate the vascular network of fibers and thick-walled cells in dicot leaves as a strutted network. Likewise, the behavior of a parallel array of elastic beams interwebbed by a more ductile membrane could be used to model the mechanical behavior of a grass leaf, while the phloem and xylem fibers within the vascular bundles running the length of stems could be likened to guy wires or elastic restraint beams running through a homogeneous foundation material comparable to the ground tissue system in a stem.

Clearly, however, physical models must be used with discretion. The theory underlying the mechanical behavior of an artifact is always based on assumptions concerning geometry and the boundary conditions of loading that may or may not be appropriate when applied to a particular plant structure. For instance, the Euler column formula, discussed in chapter 3, treats the small deflections of axially compressed columns assumed to have uniform flexural rigidity. Stems and columnar leaves can be examined by means of the Euler formula only under very limited and often highly contrived laboratory conditions. Under natural field conditions, the types of loading conditions stems and leaves experience do not comply with the assumptions underlying the formula. Nonetheless, the Euler column formula provides many qualitative insights, and regardless of the level of intellectual or computational sophistication, biomechanical analyses typically yield nothing more than first-order approximations of reality. Within any given population of plants, there exists an intrinsic morphological and anatomical variability, which for various reasons is never fully explored, despite the best intentions. Consequently, even the most sophisticated modeling procedure provides only a passing estimate of biological complexity. Provided this estimate is conservative, many difficult practical situations can be analyzed with comparative ease. This is as true for engineering as for biology.

In contrast to the relative ease with which the geometric attributes of plant organs can be approximated by the solid mechanics of ideal columns and cantilevered beams, the ability to predict the composite material properties of organs resulting from the juxtaposition of tissues differing in their individual material properties is much more circumspect. In this chapter we will review some of the fundamental models treating composite materials. Among them are the Voigt and the Reuss models. Although limited in their applicability to plant organs, as well as many engineered composite materials, these models provide crude estimates for the material properties of plant structures having a heterogeneous tissue construction. Unlike simple solids, whose material properties are immune to geometry, the material properties of composite materials are profoundly influenced by geometry. The locations of elastic, strong materials within a composite material influence the material properties of the composite as a whole. Therefore, in treating the material properties of composites, the geometry of the constituent materials is surprisingly important. In this sense we cannot speak of the material properties of a composite material without referring to the shape, size, and location of its constituent materials.

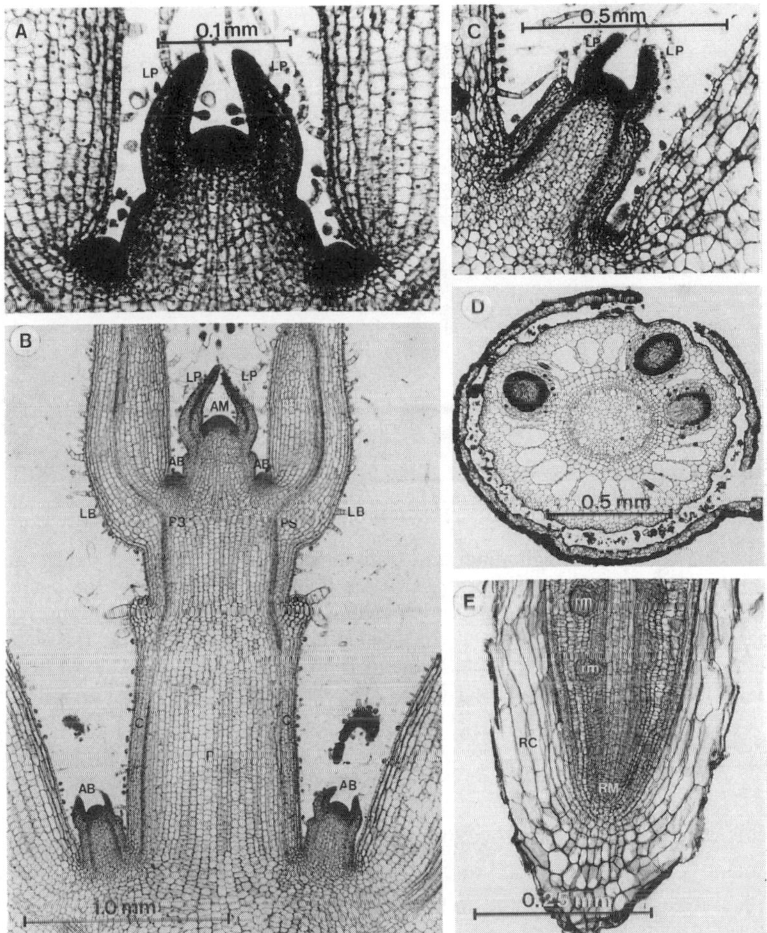

FIGURE 7.1 Shoot and root anatomy: (A) High magnification of shoot apical meristem shown in B, showing two leaf primordia (LP) flanking the apical dome. (B) Longitudinal section through the distal portion of a dicot (*Coleus*) shoot, showing shoot apical meristem (AM), leaf primordia (LP), leaf bases of more mature leaves (LB), axillary buds (AB), procambial strands (PS), pith (P), and cortex (C). (C) Higher magnification of one of the axillary buds shown in A, revealing leaf primordia (LP) flanking the apical dome. The axillary bud develops exogenously from superficial meristematic cells derived from the shoot apical meristem shown in B. (D) Transverse section through the root of the aquatic plant *Pistia,* revealing three endogenously developing root apical meristems extending into the root cortex. (E) Longitudinal section through the root of *Pistia,* showing root cap (RC), root apical meristem (RM), and endogenously developing root meristems (rm).

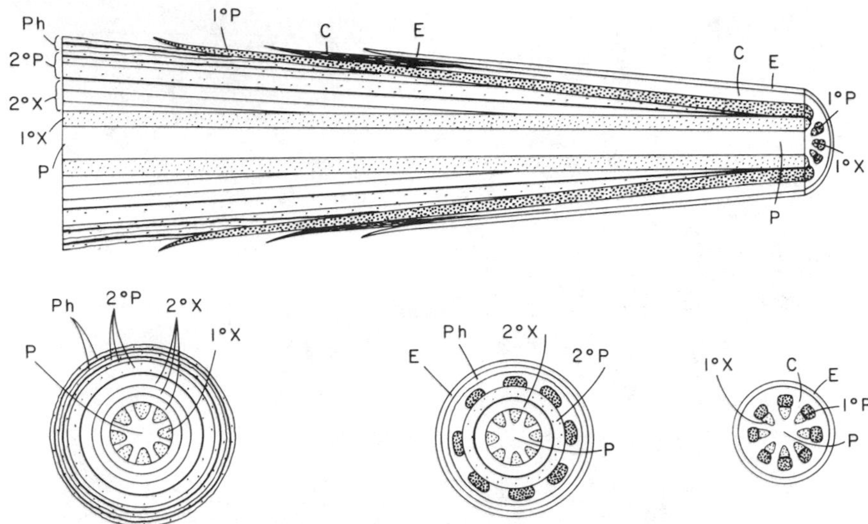

FIGURE 7.2 Idealized longitudinal and transverse sections through a stem (branch) exhibiting secondary growth. The base of the branch is to the left of the figure and contains the greatest volume fraction of secondary tissues. The distal portion of the branch lacks secondary growth and consists entirely of primary tissues. The idealized patterns of primary and secondary tissue distribution shown here are essentially the same for the trunk of a tree. C = cortex, E = epidermis, P = pith, Ph = phellem, 1° P = primary phloem, 2° P = secondary phloem, 1° X = primary xylem, 2°X = secondary xylem.

This chapter begins with a review of some preliminary topics essential to our main focus: criteria used to distinguish among plant organ types; the relation between growth and the magnitude of static loading; growth allometry; and thigmomorphogenesis (plant growth responses to mechanical stimuli). Readers already familiar with these topics can begin with the section "Mechanical Analogues to Plant Organs."

PLANT ORGANS

The vascular plant body consists of a shoot system and a root system (fig. 7.1). The former is divided or decomposed into stems, leaves, buds, and flowers. The primary shoot derives developmentally from the shoot apical meristem of the embryo. The leaf is the principal lateral appendage of the stem. One leaf or more is attached to a stem at a node that differs anatomically and morphologically from the portion of the stem spanning two successive nodes, called an internode. The shoot branches by means of axillary buds, typically found in the axils of leaves above nodes. By contrast the root system, which initially is derived from the root apical meristem of the embryo, is typically augmented by adventitious roots that may develop from root apical meristems produced anywhere on the plant body.

Typically, the three vegetative organs are delineated based on criteria that rely on developmental and positional information ultimately reflecting the meristematic origins and the locations of the organ types within the ground plan of the vascular plant body. The shoot produces appendicular organs—leaves—as a consequence of exogenous development (leaves develop from superficial meristematic tissue), whereas roots do not bear appendicular organs and have endogenous development (they develop from deep-seated meristematic tissues within stems, other roots, or in some cases leaves). Stems typically have an endarch xylem maturation (fig. 7.2); that is, primary xylem maturation is centrifugal (the first primary xylem cells to form, called protoxylem, are closest to the center of the stem, while subsequently formed primary xylem cells, called metaxylem, are closer to the epidermis. By contrast, roots are characterized by exarch xylem maturation (fig. 7.2); that is, xylem maturation progresses centripetally (protoxylem cells are farther from the center of the root axis than subsequently formed metaxylem cells, which occupy a more internal position).

Developmental criteria are not infallible, even within the particular context of vascular plant groups; for example, the xylem maturation in the stems of lycopods is typically exarch, not endarch as it is in most seed plants (gymnosperms and angiosperms). Also, as paleobotanical information accumulates concerning the evolutionary relations among the various vascular plant groups, it is becoming increasingly obvious that plant organs like leaves cannot be viewed as evolutionarily homologous structures. Finally, all attempts to extend the concepts of stem, leaf, and root to nonvascular plant groups (the algae and the bryophytes) based on the criteria established for vascular plants must be viewed with skepticism.

Functional distinctions among the three vegetative organs are not possible. Each vegetative organ type has the potential to fulfill the same functions as any other. Accordingly, the leaf cannot be functionally defined as the principal photosynthetic organ of the plant body. The leaves of many plant species are devoid of photosynthetic tissues and may function as part of the plant's defenses, as they do in many species of cacti. The photosynthetic leaves of the Venus's flytrap (*Dionaea*) serve equally well as digestive organs, while the leaves of the common garden pea and the much less common fern *Lygodium* function much like stems by providing mechanical support. By the same token, the stems of the cactus *Opuntia* and the roots of the orchid *Chilorhiza lucifera* are the principal, if not exclusive, photosynthetic organs of these plants. Thus it is much more meaningful to identify the functional role or roles of a particular plant organ and to quantitatively examine the extent of performance. This form-function approach to the organography of plants recognizes the anatomical, morphological, and developmental differences among organs while tactfully sidestepping definitions for stem, leaf, and root that remain conceptually elusive if not vacuous.

The Fallacy of Static Self-Loading

As discussed in chapter 3, most terrestrial plant organs experience self-loading, in that they must sustain their own biomass against gravitational force as well as resisting the relatively long-term loadings that may be imposed by the accumulation of organic debris, snow, ice, epiphytes, and the like. Unlike those in fabricated artifacts, however, the conditions of self-loading in plants are not static. A suspended apple fruit can increase over a thousandfold in volume and weight during its growth. Similarly, over the course of even a single year, a stem may increase in length by many orders of magnitude. Since shape influences the distribution of mechanical stresses within a structure, and since the size of a structure influences the magnitude of self-imposed mechanical stresses, the conditions of self-loading are continuously modified as organs grow.

Additionally, the material properties of an organ change as a result of development. Most plant organs begin their development as essentially undifferentiated parenchymatous (meristematic) tissue. Subsequent differentiation and maturation of cell types leads to the appearance of different types of tissue with distinctive mechanical properties. Changes in the material properties of organs can be seen by drawing on only a few examples. Fruits ripen and may become softer or harder depending on the types of tissues that develop within them. Leaves become stiffer and more resilient as the vascular tissues within them mature. Early in the year, the new shoots of beech and hemlock (*Fagus* and *Tsuga*) typically deflect under their own weight, yet they assume a more vertical orientation and increase in rigidity during the summer months. Likewise, the secondary growth layers of vascular tissues in tree branches and trunks result in a longitudinal gradient of material properties as well as the characteristic tapering in the geometry owing to the amortization of growth layers (fig. 7.2). True, in many cases organs exhibit determinant growth, as do the leaves of most dicot species. Stems may grow vertically to a certain point and then cease elongating. Nonetheless, even in these instances the mechanical fabric of organs may undergo developmental modifications, such as those attending senescence, altering the capacity of tissues to sustain the static loading of the tissues above them. Thus, while it may be legitimate to speak of the instantaneous static equilibrium of a plant organ, ontogeny invariably alters the conditions of static equilibrium from day to day or, as in the rapid growth of structures like tendrils, even from hour to hour.

The mechanical behavior of plant organs under a particular condition of dynamic loading depends on their state of development and maturation. Young stems are typically less elastic than their older counterparts. Accordingly, much smaller dynamic loadings are typically required to displace them from their static equilibrium conditions. Also, the way young and old plant organs fail mechanically under comparable magnitudes of dynamic loading may differ appreciably. For instance, in storms older tree trunks may uproot, while younger saplings typically snap along their length. Old leaves may shear from their twigs, while the lamina of young leaves may shred. These and other examples illustrate the importance of plant development in influencing mechanical behavior. Since each plant is a collection of many organs and each organ may be in a different developmental state than its counterparts, we rarely have the luxury of dealing with a single mechanical configuration for an organ type.

THE ALLOMETRY OF GROWTH

Morphological and anatomical data must be superimposed on the mechanical performance of each organ within the context of its particular developmental state. Accordingly, the allometry of changes in size, shape, and anatomy must be related to the allometry of mechanical parameters. Of particular interest is how far allometric changes in the geometry and material properties of plant organs are compensatory in terms of mechanical performance. For example, is flexural stiffness maintained at a relatively constant level throughout ontogeny, or does it change, making an organ more susceptible to bending and deformation at one point in organ development? If flexural stiffness changes ontogenetically, are these changes scaled to the self-loadings the organ experiences throughout its functional lifetime? To answer questions like these we must ask how organogenesis can scale mechanical performance to absolute size and shape. Since organogenesis is a feature of plant development, the inquiry must extend to the merits of different plant ontogenies. That is, have species with mechanically safe ontogenetic programs achieved prominence over species with more vulnerable mechanical configurations in certain habitats or environmental conditions? Some of the ecological literature still retains a conceptual distinction between K-selected and r-selected species—species that occupy relatively stable habitats and that are relatively large, slow-growing, and produce relatively few but large propagules per season (K-selected species), in contrast to species that occupy relatively unstable habitats and that are relatively small, fast-growing, and produce relatively more but smaller propagules per season (r-selected species). If the distinction between K-selected and r-selected species is meaningful, then their ontogenetic differences may reflect mechanical adaptations to stable and unstable habitats.

These issues define a biomechanical research agenda that must acknowledge a spectrum of biological attributes. Ultimately, every plant organ undergoes some sort of mechanical instability or failure. Indeed, some organs are *designed* to do so. Deciduous leaves are shed from their twigs, as are many types of fruits, because they form abscission layers of tissue that shear easily. Some stems undergo large deflections from the vertical because of bending under self-loading, ultimately make contact with the substrate, and subsequently produce roots and stems at their tips, thereby vegetatively propagating the individual plant (e.g., blackberries, or *Ribes* spp.). Other plant organs ontogenetically maintain mechanical stability, but only for a limited time. For example, secondary growth in the trunks and branches of trees alters their girth and material properties so that flexural stiffness increases from year to year. Recall that flexural stiffness is the product of the elastic modulus and the second moment of area (which, for a cylindrical column, is a function of the radius raised to the fourth power). The sequential deposition of secondary xylem and cork layers in trunks and branches increases their girth, thereby raising the second moment of area, while the deposition of secondary tissues also changes the elastic modulus (as the volume fraction of primary tissues is progressively reduced and the volume of secondary tissues is amortized). Nonetheless, the cantilevered branches of many tree species frequently undergo mechanical failure through shearing under the static loadings resulting from their own weight, while the branches of other species undergo large horizontal deflections until they ultimately come to rest on the ground, where their growing tips can reaffirm a vertical posture.

In the case of the tree trunk, the ratio of girth to length increases as secondary growth layers are added yearly. At first glance this mode of development appears ideally suited to allometrically maintaining compensatory changes in the flexural stiffness of tree trunks as the compressive stresses produced by the accumulation of biomass in the crown of a tree increase yearly. But is it? The answer in terms of static equilibrium depends on the allometric relation between the changes in the flexural stiffness or strength of the trunk and the changes in the weight of the crown. For some species, allometry may be slightly biased to the detriment of mechanical stability, so that progressively lower magnitudes of dynamic force (wind pressure) will induce mechanical instability or failure.

For plant organs lacking secondary growth, the allometry of primary growth and its influence on mechanical behavior are in large part influenced by whether an organ exhibits determinate or indeterminate growth—whether organogenesis stops after some developmentally prescribed period (determinate growth) or continues indefinitely (indeterminate growth). In both determinate and indeterminate growth, the elastic parameters of stems and leaves lacking secondary growth may reach some maximum value once histogenesis beneath meristematic regions is complete. Additionally, the second moment of area may reach some maximum and final value once the organ achieves its final girth and cross-sectional shape. Hence the flexural stiffness or strength of these organs achieves some final, absolute value. Provided the static loading condition of these organs involves mechanical stresses below the elastic limits of mechanical behavior, their functional life span would not be jeopardized by self-loading.

In the case of organs that are indeterminate in vertical growth, however, organogenesis in length continues indefinitely. From beam theory, we recognize that as the length of a vertical beam with a given flexural stiffness increases beyond a certain value, mechanical instabilities occur, manifesting themselves as a deflection from the vertical. As the length of the beam increases further, the magnitude of self-imposed bending stresses, and consequent bending deformation, increases. In some cases large bending deflections are followed by sudden buckling. Accordingly, we can appreciate the mechanical significance of determinate growth in organs lacking secondary growth or where secondary growth is poorly expressed developmentally. Determinate growth can confer a design factor or margin of safety whereby growth in height is truncated before large bending deformation or buckling occurs. By contrast, indeterminate growth in an organ's length can result in bending or buckling. When plant stems bend under their own or externally applied weight, the deformation in geometry may not be recoverable when bending stresses are removed. For example, the shoots of *Oenothera* can grow vertically as much as six feet, but they often bend significantly under their own weight. If the shoots of this plant are cut into segments, we can see that they retain as much as 90% of their curvature, indicating that the bending stresses have produced unrecoverable deformations. In fabricated beams these would be considered plastic deformations resulting from bending stresses that had exceeded the elastic limits of loading of the materials used in construction. In *Oenothera,* however, the unrecoverable deformations result from the way cells have grown under the influence of bending stresses. The walls of these cells have not undergone plastic deformation in a strict sense. Rather, histogenesis has elastically responded to the mechanical stresses imposed on tissues by the weight of the organism.

As mentioned previously, bending may have advantages in terms of vegetative propagation (when stems root at the point of contact with a suitable substrate, as in many blackberry species). In other cases, once in contact with a secure substrate, the growing tip of an organ can reassume a vertical growth posture. Indeed, many plants normally grow so that the older portions of stems lie horizontally on the ground while their growing tips are somewhat elevated. Continued growth reapportions vertical elements into a horizontal orientation. Species with this growth habit essentially snake along their substrates. One advantage to this design is that the individual plant can climb over obstructions. Species of *Senecio* and *Lycopodium* typically manifest this type of growth, while some of the earliest vascular land plants may have developed a pseudorhizomatous habit based on a similar growth pattern.

An intriguing aspect of this snaking mode of growth is that in some extant species vertically growing stems (those exhibiting orthotropic growth) become horizontal because of mechanically induced stresses (e.g., the lycopod *Lycopodium lucidulum*), whereas in other species physiologically controlled growth results in stems that normally develop horizontally (plagiotropic growth) but have slightly upturned growing tips that turn downward as they mature (e.g., prostrate species of *Juniperus*). The obvious question is whether plant stems that normally exhibit plagiotropic growth capitalize on bending stresses as a mechanical cue to physiologically adjust their orientation and, if so, when this was achieved in their evolution.

Thigmomorphogenesis

When plants experience mechanical perturbation (dynamic loadings), their growth and development may be altered. This effect, called thigmomorphogenesis (see chap. 3), involves the motion-induced inhibition of growth.

Thigmomorphogenesis can be expressed in plant organs either exhibiting or lacking secondary growth, as well as in organs that have indeterminate or determinate growth. For example, the swaying of branches or tree trunks or the stalks of inflorescences induced by wind typically results in a reduction in organ length compared with plants growing in more protected environments. Additionally, when the capacity for secondary growth is present, growth in girth may be disproportionately greater for organs that are mechanically perturbed. Changes in the elastic parameters of organs are not atypical. The elastic modulus of dynamically loaded stems and leaves is frequently higher than that of stems and leaves whose movements are naturally or artificially restrained. When handled or otherwise mechanically disturbed, the resulting changes in the ratio of length to width and in the elastic parameters, either individually or collectively, tend to increase the flexural stiffness of organs. The mechanical advantages are immediately apparent from the perspective of beam theory. An increase in the flexural stiffness of a beam (or plant organ) confers a greater capacity to resist bending moments, hence a greater capacity to maintain some preferred orientation under dynamic loading. Thus thigmomorphogenesis provides an adaptive, developmental growth response to the mechanical stresses experienced.

Perhaps much less obvious than its role as an adaptive feature of plants growing naturally, thigmomorphogenesis demonstrates the need for great care in the design and interpretation of experiments that involve handling plants under artificial conditions. For example, Turgeon and Webb (1971) examined the effect of handling on the ratio of length to volume of the petioles of glasshouse-grown squash, *Cucurbita melopepo*. The handling involved shaking petioles gently and stroking leaf lamina once across the upper surface each day. After twenty days of treatment, handled and control plants were compared. Turgeon and Webb (1971) found that the lengths and fresh weights of the petioles and stems of the handled plants were significantly less than those of the control plants, but the increase in volume per unit length of handled petioles was significantly greater than for control plants. The greatest inhibition of petiole growth occurred among the youngest leaves (those just about to unfold). The effects of handling were manifest along the entire growing portion of the stem, however, even in areas not directly handled. This suggests that the results of physiological experiments attempting to resolve the effects of growth substances or herbicides or any other substance that must be externally applied can be misinterpreted if the treated and control plants are mechanically handled in different ways—for example, if the control plants are not sprayed with a placebo.

Mechanical Analogues to Plant Organs

The column and the cantilever are the two principal structural analogues we can use to predict the geometric attributes of plant organs. These members are typically used when engineers need load-supporting members in construction, and therefore, understandably, the mechanical attributes of columns and cantilevered beams have been given considerable attention, both in engineering theory and in practice. This is fortunate for botanists, since virtually every plant organ can be viewed as a load-supporting member, and many closed-form solutions have been found to describe how columns and cantilevers behave under various conditions of loading.

Yet how far traditional engineering theory can be applied to biological cases depends on the robustness of the physical analogy that can be drawn between organic and inorganic geometries. Any analogy of this type relies on assumptions about function as well as shape (morphology) and structure (anatomy). A columnar plant organ that does not mechanically function as a load-supporting member cannot be legitimately analyzed as a column, regardless of physical appearances. Also, every plant organ has functions besides mechanical support, and how well the morphology and anatomy of an organ conform to a particular mechanical analogue is often compromised by these additional functions; for example, a cylindrical leaf or stem combines the functional roles of a load-supporting and a photosynthetic device. If the organ is viewed as being composed of two or more materials (tissues) differing in stiffness, then the ability to resist bending or torsion will be maximized by preferentially placing the stiffest material toward the periphery of the cross section. But this location is also the best for photosynthetic tissues, owing to the metabolic requirements for gas exchange and light interception. The anatomical conflict between support and photosynthesis can be resolved by a compromise—structural elements within a cross section are placed some distance from the neutral axis. Similarly, from solid mechanics, the bending moment is anticipated to increase with distance from the tip of a slightly deflected column. Thus the largest bending moment is typically at the base of stems and columnar leaves deflected from the vertical, which explains why the volume fractions of mechanically stiffer, nonphotosynthetic tissues typically increase toward the base of such organs, whereas the volume fractions of photosynthetic tissues increase toward the tips of plant organs lacking secondary growth. This anatomical compromise is evinced for most vertical stems of herbaceous species, whose green, photosynthetic cells are typically seen just beneath the epidermis, below shoot apical meristems, and whose thick-walled, nonphotosynthetic cells occupy similar locations farther from the growing tips. This design compromise is also evident in the color gradation of some stems that possess secondary tissues—green toward their tips and brown or gray (owing to corky external layers of tissue) toward their bases.

Compromise is also evident for the other biological functions of organs, such as hydraulics. The primary vascular tissues may be decomposed into cell types that conduct liquids and those that protect conducting cell types, as well as other cells and tissues, from mechanical damage. The principal load-bearing members of the vascular tissues are the vascular fibers—thick-walled cells that are dead at maturity and whose lengths run parallel to the longitudinal axis of organs. The greatest proportion of these fibers are typically found farthest from the neutral axis of vertical or nearly vertical stems. Mechanically, they operate as elastic beams that restrain neighboring, thinner-walled cells from bending and possibly buckling. From the treatment of the tensile and compressive stress distributions in beams (chap. 3), the mechanical function of fibers is enhanced by their location with respect to bending stresses, which are highest at the perimeter of each transection. In stems and cylindrical leaves (like those of onion, garlic, chive, or many species of sedges and rushes), phloem fibers (which occur toward the outside of vascular bundles, where tensile and compressive stresses occur) serve this role, whereas in spatulate, more or less horizontally oriented leaves, xylem fibers (found along the upper surfaces of vascular bundles running through the leaf) resist the tensile stresses induced by bending. Thus, even within the vascular tissue system there exists a certain compromise between cells that serve a hydraulic and a mechanical function.

We must also consider the mechanical theory appropriate to our task. For example, in chapter 3, the bending of a vertical beam was discussed in terms of the Euler column formula and the analogous equation from the theory of the Elastica. Both equations could be applied to the same physical system (a bending vertical beam), but the relative merits of using the Euler or the Elastica equation depend largely on whether the physical system fulfills certain assumptions underlying the formulation of these equations. The elementary theory of bending reveals that the stresses and deformations of cantilevers are directly proportional to the applied loads only when deformations in shape caused by bending do not influence the action of the applied loads; that is, the elementary theory is appropriate only for small deflections or deformations and only when Hooke's law holds (see chap. 3). The Euler column formula assumes the conditions specifying an ideal column—one that is perfectly

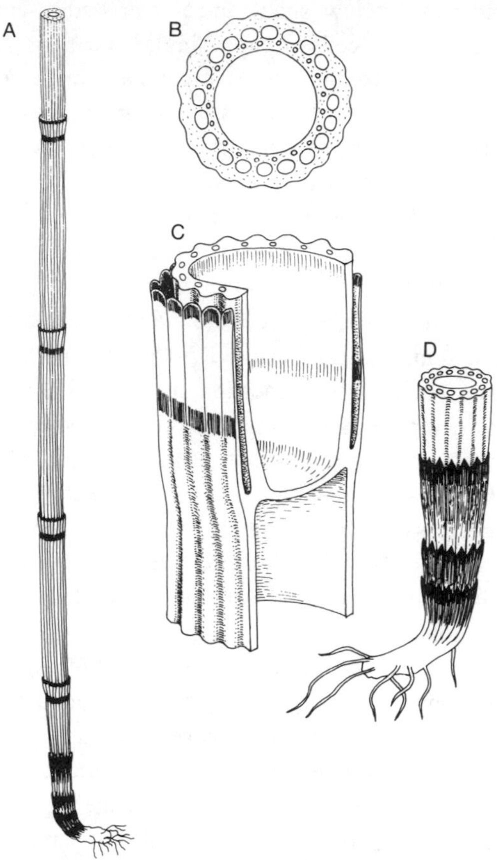

FIGURE 7.3 Morphology and some anatomical details of a single aerial shoot of the horsetail *Equisetum hyemale*. (A) Gross morphology of the lowest portion of an aerial shoot (shoot apex is not shown) illustrating the whorls of fused, highly reduced leaves at each node and the reduction in internodal length toward the base of the shoot. The rhizome the shoot is attached to is not shown. (B) Transverse section through an internode illustrating the large central (pith) canal and the smaller arrays of canals found in the cortex (vallecular canals) and associated with the vascular bundles (carinal canals). (C) Three-dimensional section through a single node with clasping whorl of fused leaves. The hollow internodes flanking the node are separated by a nodal diaphragm or septum. (D) Detail of the basal portion of an aerial shoot illustrating the shortening in internodal length and roots.

straight, dimensionally uniform, homogeneous in composition, and never stressed beyond its proportional limit. Whenever these boundary conditions are violated, or whenever axial and lateral forces operate together, mechanical features such as stresses, deflections, bending moments, and shear forces are

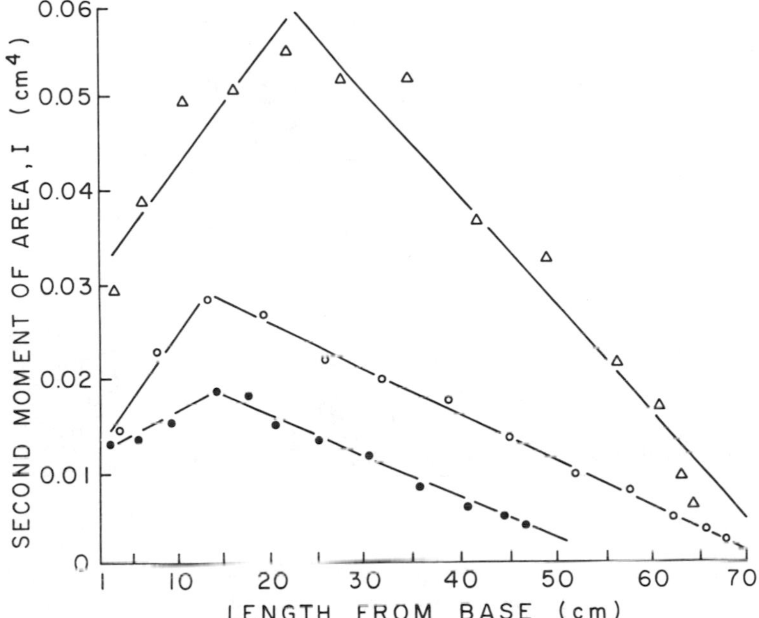

FIGURE 7.4 Changes in the second moment of area (given in units of cm^4 for convenience) plotted as a function of the length of the three vertical shoots of the horsetail *Equisetum hyemale*. The shoots of this plant consist of hollow internodes with septate nodes (see fig. 7.3). They are tapered at their base and toward their free ends; as a consequence, the second moments of area increase to a maximum and then decrease toward the tip of each shoot.

not proportional to the applied loads, nor is the Euler column formula appropriate. True, the Euler formula can be used for pedagogical simplicity—to show students relationships and to provide crude first-order approximations of behavior. But in meaningful analysis the expediencies of pedagogy must give way to often tedious, more sophisticated approaches. Indeed, understanding plant biomechanics requires exploring how and when plants violate the assumptions of elementary mechanical theory—these violations are at the very heart of biology because they set organic beings apart from the inorganic world.

The violation of the assumptions of the elementary theory of bending is illustrated by the horsetail *Equisetum hyemale* (Niklas 1989a). Plants of this species produce an underground, much-branched plagiotropic stem (rhizome) from which arise orthotropic, essentially unbranched photosynthetic shoots that are determinate in their growth (fig. 7.3). The leaves on the shoots are

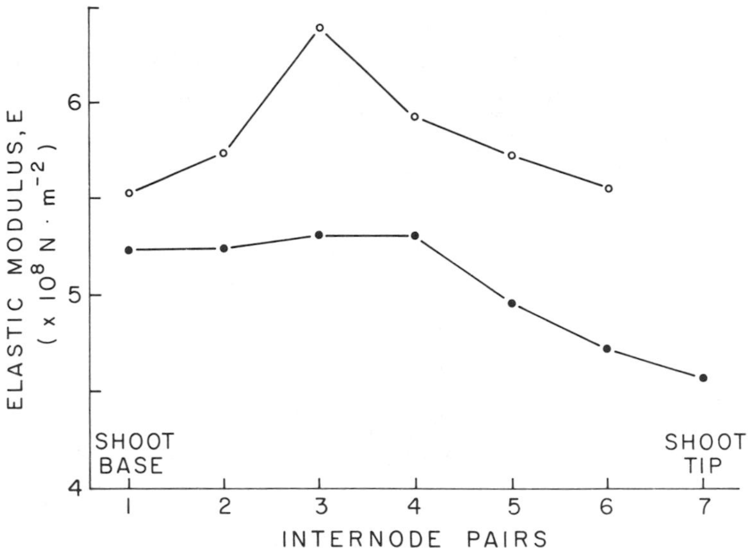

FIGURE 7.5 Changes in the elastic modulus measured for pairs of internodes (two internodes separated by a node) of two vertical shoots of *Equisetum hyemale*. The elastic (dynamic) modulus was measured from the natural frequencies of vibrations of segments of the shoot. Since single internodes were too short for measuring the dynamic elastic modulus accurately, pairs of internodes were used. The data indicate that the dynamic elastic modulus is not constant along the length of a shoot.

arranged in whorls at nodes. Anatomically, each whorl is associated with a transverse diaphragm of tissue that runs through an otherwise hollow stem. Leaves are essentially vestigial, since they contribute little by way of photosynthesis when mature. The bulk of photosynthesis is carried out by the internodes of aerial shoots. Running the length of each internode are two parallel arrays of canals. One array (the vallecular canals) is external to the vascular tissue; another (the carinal canals) is part of the vascular tissue system.

Each aerial shoot can be likened to a cylindrical column that at first glance seems an ideal candidate for the application of the Euler column formula. But this assumes that the column is homogeneous in its material properties and in its geometry—it assumes that flexural stiffness is uniform. This is not true, however. Aerial shoots are double tapered—they expand in diameter from the base and then diminish toward the tip (fig. 7.4). Also, the wall thickness of internodes varies as a function of the internodal distance from the base of the shoot; that is, the second moments of area vary. The modulus of elasticity also varies along the length of shoots, tending to reach a maximum value below or near the midspan (fig. 7.5). Finally, the shoots are hollow and septated rather

TABLE 7.1 Conversions of Complete Elliptic Integrals of the First Kind $K(p)$ and Tip Deflection Angle α

α (degrees)	$K(p)$	α (degrees)	$K(p)$
0	1.57079	46	1.63651
2	1.57091	48	1.64260
4	1.57127	50	1.64899
6	1.57187	52	1.65569
8	1.57271	54	1.66271
10	1.57379	56	1.67005
12	1.57511	58	1.67773
14	1.57667	60	1.68575
16	1.57848	62	1.69411
18	1.58054	64	1.70283
20	1.58284	66	1.71192
22	1.58539	68	1.72139
24	1.58819	70	1.73124
26	1.59125	72	1.74149
28	1.59456	74	1.75216
30	1.59814	76	1.76325
32	1.60197	78	1.77478
34	1.60608	80	1.78676
36	1.61045	82	1.79992
38	1.61510	84	1.81915
40	1.62002	86	1.82560
42	1.62523	88	1.83856
44	1.63072	90	1.85407

than solid in construction. This has many consequences for the mechanical behavior of shoots in bending. Transverse diaphragms restrain the cylindrical walls of internodes from crimping (Brazier buckling), by virtue of end-wall effects (see chap. 2). Surgical destruction of these diaphragms lowers the measured elastic modulus of shoots by as much as 20%. Accordingly, if the Euler column formula had been applied directly to estimating the mechanical behavior of the vertical shoots, then the analysis would have been misleading and would have entirely neglected the functional role of nodal septa. Perhaps incidental to all of this, but nonetheless interesting, nodal septa act as physical barriers to the accumulation of water (which freezes) in each internode during the winter. Extracellular freezing appears to be a physical mechanism that increases osmotic concentration in internodal tissues (see chap. 4), reducing their potential to freeze and undergo damage (see Niklas 1989b).

A RETURN TO THE ELASTICA

Mechanical theory has been developed for noncurved and curved, tapered and untapered columns and cantilevers that are undergoing large deformations and are not geometrically or materially homogeneous. Large deformations of columns and cantilevers are treated by the Elastica (see chap. 3), which will occupy our attention in this and the next few sections. Homogeneity in material properties is a more difficult matter and will be treated later in this chapter.

The Elastica assumes that loadings do not exceed the proportional limits of a structure's materials. Thus the elastic range of behavior can be assessed. Within this range of behavior, however, large deflections can be treated and, in addition, the loadings on the structure may be eccentric in their distribution. This frees us from many of the limits imposed by the proper use of the Euler column formula.

The most useful equation from the Elastica is

$$K(p) = \left(\frac{Pl^2}{EI}\right)^{1/2}, \tag{7.1}$$

where $K(p)$ is known as the complete elliptic integral of the first kind, which is a function of the deflection angle α measured at the tip of a bending beam, P is the load at the free end (tip) of the beam, l is beam length, and EI is flexural stiffness. The relation between the complete elliptic integral and the deflection angle can be quickly determined by measuring α and looking up its corresponding $K(p)$ value from a table such as table 7.1.

Equation (7.1) can be applied to many biological cases, but one—the flower stalk (peduncle) of the common garlic plant (*Allium sativum*)—will suffice as illustration. The flower stalks of this plant grow by means of a region of meristematic tissue at or very near their base (this region, because of its position, is called an intercalary meristem, in contrast to the apical meristem commonly found on the stems of other species), as well as diffuse growth in older, more distal portions of the organ. A cluster of flowers, produced by floral primordia at the tip of the stalk flowers, eventually develops and ripens to produce plantlets that grow precociously while still attached to parental tissue. Thus, as flowers and plantlets develop, the flower stalk is subjected to an increasing static load at its free end. Flower stalks lack the capacity for secondary growth, and their vascular bundles are scattered throughout their cross sections. Thus, in many respects the morphology, anatomy, and mode of development of these organs fulfill some of the basic assumptions made in the derivation of eq. (7.1).

For a perfectly vertical stalk, the deflection angle α equals zero, and the value of the complete elliptic integral of the first kind can be taken to equal 1.57079 (see table 7.1). Substituting this value of $K(p)$ into eq. (7.1) yields

$$l = \left\{[K(p)]^2 \frac{EI}{P}\right\}^{1/2} = \left(2.46738 \frac{EI}{P}\right)^{1/2}, \tag{7.2}$$

where l is the theoretical length to which a flower stalk with a flexural stiffness EI can growth before it will begin to deflect from the vertical under the load P of its flowers or plantlets (or both). Provided we can measure E, I, and P, we can calculate l and compare this value with the average height of flower stalks or the height of each flower stalk in relation to its individual flexural stiffness and loading condition. Fortunately, the second moment of area of flower stalks is easy to calculate, since they are terete and taper little in geometry. Hence $I = \pi R^4/4$, where R is the radius of the stalk. The weight of the flowers or plantlets is also easily determined by means of dissection and a balance. The modulus of elasticity can be measured by multiple resonance frequency spectra (Niklas and Moon 1988). For plants grown in the field, the average elastic modulus of flower stalks is $3.55 \times 10^8\ \mathrm{N \cdot m^{-2}}$, while the average second moment of area (measured at the base of stalks, where the maximum bending moment will occur) is $6.36 \times 10^{-11}\ \mathrm{m^4}$. Thus the average flexural stiffness is $2.26 \times 10^{-2}\ \mathrm{N \cdot m^2}$. If we take the average weight of the terminal cluster of flowers (not plantlets) as 6.57×10^{-2} N, then from eq. (7.2) we calculate that the average stalk length should be 0.92 m. The average length of flower stalks (for which E, I, and P were measured) is 0.507 m, or 55% of the calculated value. Remember that our calculations are applicable only to static loading conditions and that we have neglected the weight of the flower stalk tissues, as well as dynamic wind loadings, which would reduce the theoretical height stalks could grow to before they would be predicted to deflect from the vertical.

Equations (7.1) and (7.2) can also be used to predict the maximum static load P that can be sustained at the tip of an average flower stalk. Substituting the average values of E, I, and l into eq. (7.2) and solving for P yields

$$P = 2.46738 \frac{EI}{l^2} \approx 2.47 \frac{(2.26 \times 10^{-2}\ N \cdot m^2)}{(0.507\ m)^2} = 0.217\ \mathrm{N}. \tag{7.3}$$

This value is roughly 3.3 times the weight of the average flower cluster, but the average weight of a mature inflorescence (with plantlets) is 0.186 N, which is 86% of the estimated value. Once again, these calculations do not take into account the effect of wind loadings (or the casual visitation of a bee).

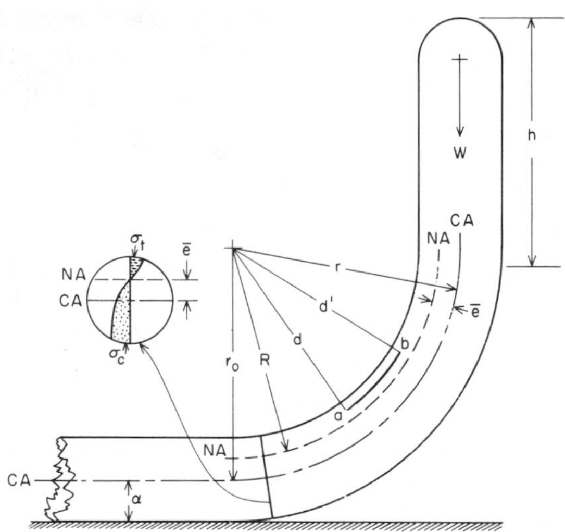

FIGURE 7.6 Relations among mechanical parameters for a curved beamlike plant axis resting on a substrate and bearing a weight at its tip. The transection to the left of the figure illustrates the distribution of tensile σ_t and compressive σ_c bending stresses relative to the centroid axis (CA) and neutral axis (NA). The tensile stresses develop within the upper, concave surface of the plant axis because the upward bending of the beamlike axis is the result of growth and gravity exerts a force on the vertical biomass. Thus compressive stresses develop on the convex surface of the beamlike organ. This spatial distribution of stresses is the reverse of the stress distribution that would develop in a beam forcibly bent into the shape achieved here by growth. The NA lies a distance *e* from the CA. The bending stresses developed within the curved beam depend on the radius of curvature *R*, the weight supported by the beam *W*, the uncurved, vertical height of the beam *h*, and the way the radius of curvature changes over the curved span of the beam. For further details, see text. (From Niklas and O'Rourke, 1982.)

If they had, then the predicted value would be even lower and closer to the actual value. Significantly, many mature inflorescences are observed to buckle in the field. Thus, it appears that the Elastica equation is useful for predicting the behavior of columnar or beamlike plant organs that have a fairly uniform (homogeneous) anatomy and second moment of area and that grow vertically with little or no curvature.

CURVED PLANT ORGANS

Unlike aerial organs that are relatively straight and can be modeled as columns or cantilevers (e.g., aerial shoots of *Equisetum hyemale* and the flower stalks of *Allium sativum*), many vertically growing plant organs have a natural curvature and bend in one fashion or another as they grow. At first glance this appears to present computational difficulties in terms of calculating some fundamental mechanical features, such as bending moments. But recall from chapter 3 that if we can measure the curvature K, then we can use the Elastica and even the elementary theory of bending to calculate bending moments, and from these bending moments we can calculate the maximum bending stress under the conditions of static equilibrium. Thus our fundamental task is to find some mathematical expression for the curvature of bending.

Typically, the form of the equation that approximates the geometry of curved organs is the logarithmic spiral (Niklas and O'Rourke 1982), which takes the form

$$r = r_o \, e^{k\theta}, \tag{7.4}$$

where r is the radius of curvature from the point of origin of the spiral to any locus on the curve, r_o is the original radius of curvature, k is a proportionality constant, and θ is the angle subtended from r to r_o. Some of these relations are summarized in figure 7.6. Notice that when $k - 0$, $r - r_o$, and the curve is some arc of a circle. Since the change in the arc length ds along the curve is given by the formula

$$ds = (1 + k^2)^{1/2} \, r_o \, e^{k\theta} \, \theta \cdot d\theta \tag{7.5}$$

and the curvature K is the reciprocal of the radius of curvature ($K = d\theta/ds$), the precise equation for K is

$$K = [r_o \, e^{k\theta} \, (1 + k^2)^{1/2}]^{-1}, \tag{7.6}$$

from which analytical equations defining the bending moment can be derived. For example, the maximum bending moment M_{max} consists of two components: a bending moment M_v produced by the vertical portion of the organ above the region of flexure, and a bending moment M_f referable to the region of flexure. (If the flexure or bending extends along the entire length of the organ, then $M_v = 0$.) The formula for M_v is

$$M_v = m \, \pi \, R^2 h \, r_o \, e^{k\phi} \, (\sin \phi + e^{-k(\pi/2)} \cos \phi), \tag{7.7}$$

where h is the height (length) of the organ, while the formula for M_f is

$$M_f = \frac{m \, \pi \, R^2 r_o^2}{(1 + 4k^2)} \left\{ e^{2k\phi} \left[4k^2 \, e^{-k(\pi/2)} + e^{-k(\pi/2)} + 3ke^{-k\pi} + 2k^2 \right] - 2 \sec \phi \right\}, \tag{7.8}$$

where ϕ is the angle subtended between a horizontal plane and the radius of curvature at the point where the plant organ becomes vertical ($k = \tan \phi$), m is the unit mass of the organ, and R is the radius of the organ (see fig. 7.6).

True, eqs. (7.7) and (7.8) are cumbersome, but they need be applied only if the geometry of flexure cannot be approximated as an arc of a circle with radius r_o. When an arc of a circle is a reasonable approximation for the curvature of flexure, the equation for the maximum bending moment reduces to a simple form:

$$M_{max} = m\ \pi\ R^2\ r_o^2 \left(\frac{h}{r_o} + 1\right). \tag{7.9}$$

The parameters in eq. (7.9) are easy to measure empirically, although the most difficult to measure is r_o, since this requires an educated guess as to the radius of the circle whose arc best approximates the region of flexure.

The maximum bending stress σ_{max} occurring in the plant organ as a result of static loading is given by the formula

$$\sigma_{max} = \frac{M_{max}\ (4r_c^2 - R\ r_c + R^2)}{\pi\ R^2\ (R\ r_c^2 - R^2\ r_c)}. \tag{7.10}$$

This is the general formula, where r_c is the radius of curvature for any curve approximated by a logarithmic spiral. In the case of an arc of a circle, $r_o = r_c$ and

$$\sigma_{max} = \frac{m\ \pi\ R^2(hr_o + r_o^2)\ (4r_o^2 - R\ r_o + R^2)}{\pi\ R^2\ (R\ r_o^2 - R^2\ r_o)}. \tag{7.11}$$

Equation (7.11) provides considerable insight. For example, if the length of the organ does not extend beyond the region of flexure, then $h = 0$ and $r_o = 1.6\ R$, where R is the radius of the organ. Conversely, if the organ extends vertically far above the region of flexure, then $h >> r_o$, and r_o equals the *diameter of the organ*. Thus, the optimum condition (where the radius of curvature equals the diameter) represents the upper, limiting case for organ flexure provided growth can accommodate this sharp bend and that growth is predicated on minimizing bending stresses. The latter assumption appears somewhat justified, regardless of the systematic affinity of a vascular plant. In a survey of genera in which plant stems had axial curvatures approximated by some form of a logarithmic spiral, none was found to have a radius of curvature equal to the diameter D of the stem (Niklas and O'Rourke 1982). However, two species, a lycopod (*Lycopodium complanatum*) and an angiosperm (*Plectranthus*) were found to have radii of curvature very close to the theoretical upper limit of bending (3.0 D and 2.7 D, respectively). Among the other plants examined, the radius of curvature was found to be as high as 8.5 D

(*Senecio*), indicating that some plants turn sharply upward as they grow (*Lycopodium* and *Plectranthus*), while others are extremely lax in their curvature of bending (*Senecio*). All the species examined appeared to have shoots characterized by orthotropic growth, but all redistribute originally vertical portions of their shoots into the horizontal, manifesting what appears to be plagiotropic growth. This appears to be a controlled form of mechanical failure, in the sense that at some point in their growth the vertical shoots of these plants can no longer sustain the bending stresses produced by their vertical static loadings and deform into a horizontal posture. *Lycopodium* is one of the evolutionarily oldest plant genera known, whereas most of the other genera examined were angiosperms—the most recent major group of plants to make its evolutionary appearance. Apparently, regardless of the evolutionary age of the plant group, the mechanical principles that dictate plant shape and form appear uniform and persistent.

Note that no matter how optimal the radius of curvature is in respect to minimizing the maximum bending stresses that develop at the base of a curved vertical plant organ, there is no curvature that prevents a vertical organ from bending to either side of its plane of flexure when subjected to lateral wind pressure. This form of deformation results from torsion along the length of the organ trailing behind the vertical portion of the organ. Adventitious roots provide a remedy to this design flaw, and many plants with trailing stems appear to produce extra adventitious roots near the base of vertical shoots. Another analogous solution to the torsion of vertical shoots is seen in the tropical lycopod *Lycopodium cernuum*. This weedy species is a common floristic element on Pacific islands, like those of the Hawaiian archipelago. Its vertical shoots can grow up to 3 m in height but almost invariably come to rest by leaning on the trunks of nearby trees. Nonetheless, wherever a vertical shoot develops, the horizontal stem it is attached to typically branches horizontally. The two derived branches (ahead of the vertical shoot), together with the horizontal stem that trails behind, form a tripod on which the vertical shoot rests. Most intriguing of all, the mutual angle among the three prongs averages 120°, precisely the angle one would use if designing a tripod. The tripod cannot prevent large elastic bendings of the vertical shoots, but it has a remarkable capacity to prevent the base of shoots from torquing.

TAPERING

Unless a 5% to 10% error margin is acceptable, eqs. (7.1) to (7.11) cannot be used when an organ is significantly tapered along its length or when transverse

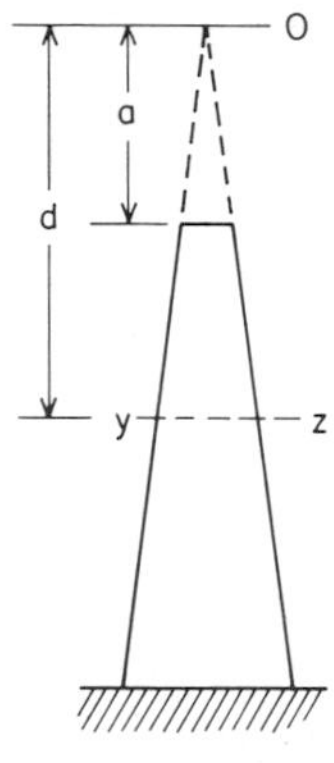

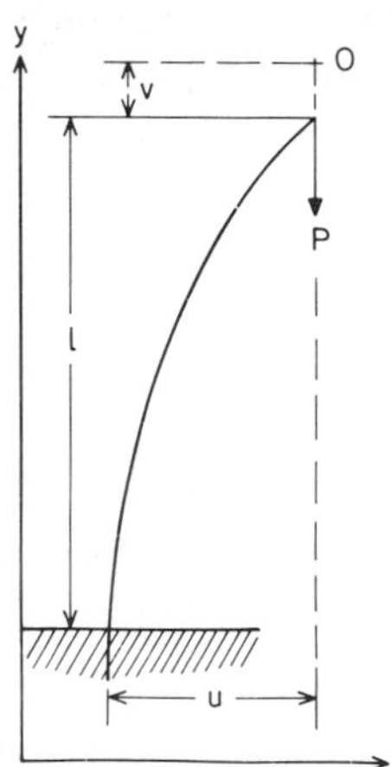

FIGURE 7.7 Linearly tapered beam for which changes in the second moment of area can be calculated. The second moment of area for any transverse plane y–z at a distance d from origin of taper O can be calculated based on a power function and the second moment of area at the free end of the beam (at distance a from the origin of taper). The less the degree of taper the less the tip load P required to cause the beam to deflect from the vertical (shown to the right of this figure).

or longitudinal variations exist in the material properties of tissues. Tapering of plant organs is very common, as is anatomical heterogeneity. Accordingly, we must treat tapered organs and anatomically heterogeneous organs with considerable discretion.

Tapering typically implies that the absolute size of cross sections varies while the cross-sectional geometry does not; that is, the second moment of area will have different absolute values depending on the absolute dimensions of each cross section, but the mathematical form of the equation for the second moment of area will be the same anywhere along the length of the organ. Accordingly, the magnitude of I of a tapered column can be taken to vary as a power of the distance d from the free end O of the column (fig. 7.7), where the second moment of area I_x measured for any section in the plane y–z is given by the ormula

$$I_x = I_a \left(\frac{d}{a}\right)^n, \tag{7.12}$$

where d is the distance between O and y–z, I_a is the second moment of area measured at a point a, and n relates to the column's degree of taper. For a solid truncated cone, $n = 4$, and the deflection curve of a tapered cone can be easily derived (given the assumptions of the Elastica) from the formula

TABLE 7.2 Values of the Factor m for a Truncated Conical Column

I_a:I_b =	0.1	0.2	0.3	0.4	0.5	0.6	0.7	0.8	0.9
m =	1.202	1.505	1.710	1.870	2.002	2.116	2.217	2.308	2.391

Note: $N = 4$. Various degrees of tapering are expressed as a ratio of the second moment of area at the top I_a to the base of a column I_b.

$$\frac{EI_a}{v^4} u^4 \frac{d^2y}{dx^2} = -Py, \tag{7.13}$$

where u and v are the deflection components in the horizontal and vertical directions, respectively. The critical buckling load is then given by the formula

$$P_{cr} = m \frac{EI_b}{l^2}, \tag{7.14}$$

where I_b is the second moment of area measured at the base of the tapered column and m is a factor that depends on the ratio of the vertical displacement v to column length l. (Some values for m are given in table 7.2.) From eq. (7.14) and table 7.2 it becomes obvious that tapering provides a mechanical advantage, since the flexural stiffness of a vertical tapered beam will increase toward its base, as will the weight that has to be sustained. Also notice that tapering has the advantage of requiring no longitudinal gradient in the material properties of the column. The modulus of elasticity can be uniform along the length of the structure, yet the flexural stiffness can increase basipetally.

Gere and Carter (1963) derived a very useful formula relating the critical buckling length l_{cr} of a top-loaded, linearly tapered cone to the basal and apical diameters D_b and D_a of the cone, the second moment of area measured at the top I_a, and the magnitude of the load P:

$$l_{cr} = 1.71 \left[\left(\frac{D_b}{D_a} \right)^{2.53} \frac{EI_a}{P} \right]^{1/2}. \tag{7.15}$$

This equation ignores the weight of the column (= tree trunk), which is only reasonable if the weight of the canopy is substantially greater than that of the trunk. Equation (7.15) reveals that the critical buckling length logarithmically declines as a function of the quotient of D_b and D_a. Holbrook and Putz (1989) have used this formula, in an elegant fashion, to study the allometry of trees. As expected, this allometry is shown to be affected by different conditions of shading and wind loading.

Tapering is expected to decrease the critical load, as well as the critical buckling length, as shown by deriving the closed-form solution for the critical load P_{cr} of a tapered column by treating it as if it were composed of a series of cylindrical elements, each with a uniform cross-sectional area. In the case of a tapered column consisting of two cylindrical elements, for which the differential equation of the deflection curve of each element is easily derived, let I_1 and I_2 denote the second moments of area for the upper and lower elements, where length and girth are measured along the x and y axes, respectively. The differential equations of the deflection curves of the two cylindrical elements are then given by $EI_1(d^2y_1/dx^2) = P(\delta - y_1)$ and $EI_2(d^2y_2/dx^2) = P(\delta - y_2)$. Using the notation $k_1^2 = P/EI_1$ and $k_2^2 = P/EI_2$, these differential equations can be solved for y_1 and y_2; that is, $y_1 = \delta + C\cos k_1x + D\sin k_1x$, and $y_2 = d(1 - \cos k_2x)$, where C and D are constants of integration. Since at the top of the column the deflection equals δ, while at the point where the two elements are jointed the deflections are equivalent, we find that $C = -D\tan k_1L$ and $D = (\delta \cos k_2l_2 \cos k\ l)/(\sin k_1l_1)$, where l_1 and l_2 are the lengths of the upper and the lower cylindrical elements and L is the total length of the tapered column. Since the two portions of the deflection curve have the same tangent at $x = l_2$, $\delta\ k_2 \sin k_2l_2 = -Ck_1 \sin k_1l_2 + D\ k_1 \cos k_1l_2$. Substituting the values of C and D into this relation, we find that the critical load is given by the condition $\tan k_1l_1 \tan k_2l_2 = k_1/k_2$. This formula is transcendental. When solved by trial and error, it shows that $P_{cr} = (\pi_2EI_2/4l^2)\ \{(l_2/L) + (l_1I_2/L\ I_1) - (1/\pi)[(I_2/I_1) - 1)] \sin(\pi l_2/L)\}^{-1}$, which reveals that taper decreases the critical load and results in a greater tip deflection of the column. For example, if $I_2/I_1 = 2$ and $f = l_1/l_2$, then critical load decreases as a function of $k\{(f+1)^{-1} + [2f/(f+1)] - (1/\pi) \sin[\pi/(f+1)]\}^{-1}$, where $k = (\pi_2EI_2/4L^2)$. For any column, the basal girth, length, and elastic modulus are constants. Thus, $P_{cr} = 1$ when $f = 1$ (when the ratio of the length of the upper, narrower cylindrical element to that of the lower, broader element equals one), and P_{cr} decreases as f increases, that is, as the length of the narrower element increases.

If tapering decreases both P_{cr} and l_{cr}, then what benefits does taper confer? Tapering can minimize the biomass of a columnar organ while maintaining a uniform working stress. The maximum stress σ_{max} for a self-loaded columnar organ, such as a tree trunk, equals $(P/A) + \rho L$, where ρ is the weight per unit volume, P is the compressive load of the crown, and L is the total length of the trunk. Substituting the working (safe) stress σ_w for σ_{max} and solving for the safe cross-sectional area A at the base of the trunk gives the formula $A = P/(\sigma_w - \rho L)$. Thus, for any load and working stress, the degree to which the trunk must taper increases as a function of the product of total length and tissue density. Consider a trunk 40 m in total height consisting of two cylindrical portions of equal length l (= 20 m), the uppermost sustaining a compressive load of 50,000 N (equal to the weight of the crown). If the bulk density of the trunk's tissue is 509.8 kg·m³ and the working stress of the tissue is 900,000 N·m², then the safe area for the top cylindrical element A_T equals 50,000 N/[900,000 N·m² – (509.8 kg·m³)(20 m)] = 0.0625 m², while the safe area for the bottom element A_B equals (50,000 N + 6.250 N)/[900,000 N·m² – (509.8 kg·m³)(20 m)] = 0.0703 m². Thus, the total volume of the trunk is (0.0625 m² + 0.0703 m²) 20 m = 2.66 m³. For a tree trunk that is uniform in girth (a prismatic cylinder), the uniform safe area A_U equals (50,000 N)/[900,000 N·m² – (509.8 kg·m³)(40 m)] = 0.0714 m². This prismatic trunk has a volume of 2.86 m³, which is 7.5% larger than that of the two-element (tapered) trunk.

Extending this analysis to determine the taper of a column such that σ_w is uniform throughout L gives a tapering law for a column of uniform strength. The condition for uniform strength is satisfied when $dA\sigma_w = \rho A dl_x$; the change in area times the working stress equals the product of density, area, and the change in unit length l_x. From this relation, we find that the safe area for the top of the column equals $(P/\sigma_w)e^{(\rho l_x/\sigma_w)}$, while the safe area at the bottom of the column equals $(P/\sigma_w)e^{(\rho L/\sigma_w)}$, where e is the base of natural logarithms. In addition to mathematically defining the taper of columns of uniform strength, these formulas reveal that tapering conserves the volume (biomass) of tree trunks whose mode of growth leads to a column of uniform strength. Since the volume of such a column equals (P/ρ) $[e^{(\rho L/\sigma_w)} - 1]$, given the previously specified values for the compressive load (P = 50,000 N), tissue density (ρ = 509.8 kg·m³), overall length (L = 40 m), and working stress (σ_W = 900,000 N·m²), the volume of our column of equal strength is 2.49 m³, roughly 13% less than that of the prismatic column (whose volume equals 2.86 m³).

Based on the same method of analysis for columns of uniform strength, Greenhill (1881) estimated the maximum height to which a tree trunk could grow based on its observed tapering. In the case of the first Kew "flagstaff," the trunk of the conifer given to the Royal Botanical Gardens at Kew, Richmond, Surrey, measured 221 feet in height and 21 inches in basal diameter. Based on these dimensions, Greenhill estimated that the maximum theoretical height of the original tree could not have exceeded 300 feet. As mentioned by D'Arcy Thompson (1942), Galileo suggested much the same theoretical limit (*ducente braccie alta*) for the maximum height of any tree. (In Florentine times length was often measured in *braccie,* whose magnitude varied according to context but was typically taken as the distance between the elbow and the wrist. Based on a measurement taken from my own arm, *ducente braccie alta*—or in modern Italian, *duecento bracci alto*—roughly translates to 183 feet. However, in the standard parlance of Italy, the magnitude of *braccie* was taken as equivalent to that of our modern yard. The ambiguity as to the precise value of Galileo's theoretical limit may have been intentional on the part of its author.) And note that Eiffel built his "great tree of steel," a thousand feet high, with a taper whose profile follows much the same logarithmic curve that provides equal strength throughout the trunks of many species of trees.

Is there, in fact, evidence that the tapering of tree trunks conforms to the pattern predicted for columns of uniform strength? Data from a limited number of studies suggest there is. Consider that the maximum bending stress at any point in a tapered trunk is given by the formula $\sigma_{max} = Mr/I$, where M is the moment, r is the radius measured along organ length at which the stress is being calculated, and I is the second moment of area for this radius. If an organ is linearly tapered, then I will vary as a function of length such that $I = I_o\,(ks + 1)^4$, where $k = (r - R)/RL$ and s is the arc length measured from the base of the trunk with radius R and second moment of area $I_o = \pi R^4/4$. Calculations reveal that the optimal taper (for which the maximum bending stress is minimized and the bending stresses are most uniformly distributed along organ length) is achieved when $kL \approx -0.60$, where kL is called the taper parameter—a dimensionless quantity. Thus, for an untapered column $kL = 0$, while the maximum taper is achieved when $kL = -1.0$; that is, when $r = 0$. (Note, however, that bending shear stresses have been neglected in this analysis. When these additional stresses are taken into account, r cannot equal 0.) The optimal taper is satisfied when $kL = -0.60$; bending stresses will be uniformly distributed in the region of a trunk or a branch where wood development is most advanced and will decrease rapidly toward the tip where wood development is least advanced. The taper parameter was measured for seventeen field- and container-grown saplings by Leiser and Kemper (1973), who report that kL for the former is -0.658 ± 0.094 ($n = 5$), while for container-grown saplings (staked and pruned according to horticultural practice) $kL = -0.325 \pm 0.214$ ($n = 12$). These data suggest that the taper of normally growing young trees optimizes stress distribution within trunks, while—though perhaps justified from the perspective of aesthetics—staking and pruning trees severely reduce trunk taper and produce an ugly mechanical design.

Tapering also benefits cantilevered branches by providing a cantilever of uniform strength—a beam in which the section modulus Z varies in the same proportion as the bending moment so that the working stresses σ_W along the length of the beam are equivalent. (The section modulus was treated in chap. 3; Z is the quotient of the second moment of area and half the depth of the cross section. The section modulus will be used here to compute the minimum amount of cross-sectional material that satisfies the condition of strength, since, as in the case of columns of uniform strength, the amount of material in each cross section through a cantilever of uniform strength must have the minimum area necessary to satisfy the conditions of uniform strength.) Consider a cantilever having length l and a rectangular cross section with height h and width b, subjected to an end load P. For such a cantilever, $I = bh^3/12$ and $Z = bh^2/6$. If x is the distance measured from the tip of the cantilever (where $x = 0$), then the bending moment M equals Px, while the maximum bending moment equals Pl. Since $\sigma_W = M/Z, \sigma_W = 6Px/bh^2 = 6Pl/bh_o^2$, where h_o is the height of the cross section measured at the base (fixed end) of the beam. Thus, $h^2 = x\ h_o^2/l$, which reveals that taper in height varies according to a parabolic law. Although the tip deflection of this cantilever equals $2Pl^3/3EI_o$, twice that of a prismatic cantilevered beam with equivalent flexural rigidity, a tapered cantilever of uniform strength can conserve 50% in volume compared with its prismatic counterpart while maintaining uniform strength along its length. One of the consequences of this conservation of volume is the effect on the elastic strain energy that can be stored in bending. Since the working stress in bending for a cantilever of uniform strength equals $6Pl/bh_o^2$, the maximum strain energy must equal $(1/9)bh_ol(\sigma_W^2/2E)$. This quantity of energy is 50% greater than what can be stored by a prismatic cantilever (with equivalent values of h_o and σ_W submitted to a comparable end load P). Since the cantilever of uniform strength has half the weight of its prismatic counterpart, it can store three times as much energy per unit weight of material. Thus taper provides a large elastic resilience.

ANATOMICAL HETEROGENEITY AND COMPOSITE MATERIALS

The consequences and benefits of taper (geometric heterogeneity in cross sections) can be treated, for the most part, with some very simple equations. Anatomical heterogeneity within an organ presents substantially greater analytical difficulty because variations can occur in either the relative volumes or spatial distributions of tissues differing in material properties, or in both, and because no current theory treating the mechanical properties of composite materials is entirely satisfactory.

The mechanical significance of anatomical heterogeneity was first recognized by Wilhelm Hofmeister (1859), who observed that when the outer and

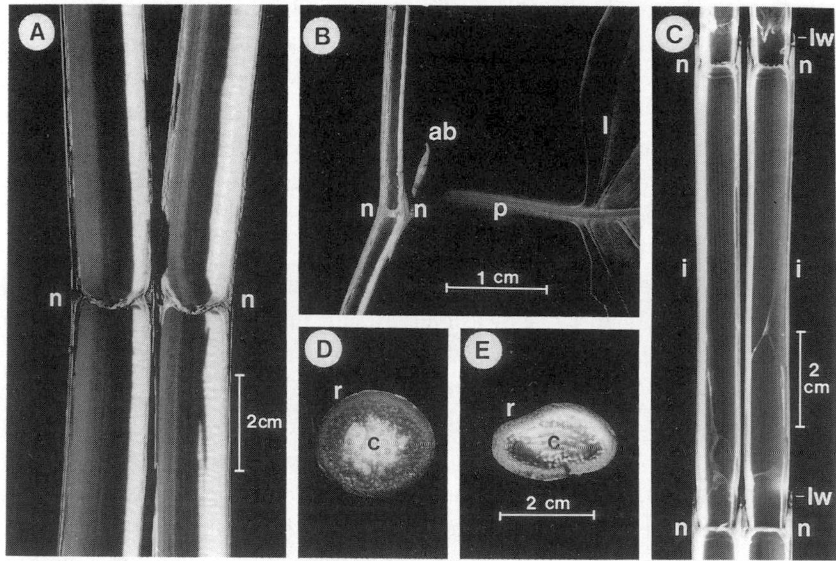

FIGURE 7.8 Examples of hollow stems (A–C) and stems that can be modeled as "core-rind" constructions (D–E): (A–B) Hollow stems (with septate nodes n to which leaves and axillary buds ab are attached) of the dicot *Polygonatum cuspitatum*. (C) Hollow internodes (i) of *Equisetum hyemale* with nodal septa bearing leaf whorls (lw). (D) Transection through the stem of corn (*Zea mays*) consisting of an epidermis and outer cortex with numerous vascular bundles (that can be modeled as a "rind" r) surrounding an inner cortex of scattered vascular bundles within a thin-walled parenchyma (that can be modeled as a "core" c). (E) Transection shown in D after it has been air dried for twenty-four hours. The "core" has decreased in volume, and the outline of the "rind" has deformed into an irregular ellipse.

inner tissues of the shoots of grape (*Vitis vinifera*) were separated from one another, they spontaneously changed their dimensions. Julius von Sachs (1865) extended Hofmeister's observations and preliminary conclusions and coined the term *Gewebespannung* ("tissue tension") to describe the way growth induces internal forces within tissues that differ in their capacity to sustain these forces (see Kutschera 1989 for an excellent historical review of this topic). From the observations of Hofmeister, Sachs, and many others, it is now generally appreciated that each plant organ, in its primary phase of development, may be considered to have a rind and a core. In most instances the rind may be treated as an anisotropic material that is placed in tension by an essentially incompressible, more or less isotropic core. This simple model conforms in principle to viewing the organ as a two-phase composite material in which each material phase behaves as a solid. The "core-rind model" can also be applied to hollow stems and leaves (fig. 7.8A–C), where, in contrast to an isotropic solid, the core is a gas-filled phase while the rind is a solid phase consisting of all the tissues in each cross section.

Regardless of the dubious merits of viewing the gas-filled chambers of hollow organs as a material component, however, the mechanical performance of hollow, tubular organs presents its own analytical format, which will be treated in the following section. This section will consider the majority of stems and leaves lacking secondary tissues that are not hollow in their construction (for example, that shown in fig. 7.8D–E). In the case of these organs, either a two-phase (core-rind) model or an n-phase model (in which each tissue is identified as a separate material component) can be used to predict mechanical behavior. Regardless of the number of phases identified by a researcher, the essential stipulation for any n-phase model is that the material properties of each phase (tissue) must show sufficient difference to warrant the notion that the tissue contributes to the mechanical performance of the composite material (the organ) as a whole.

Additionally, every n-phase model assumes relationships among the properties of the various phases. As previously noted, in the case of a core-rind (two-phase) model, the core might initially be considered as isotropic elastic material (if the core is a parenchymatous tissue, then this assumption is not a poor one), while the rind may be considered to have anisotropic material properties (as is typical for the epidermis and mechanical support tissues). Additional assumptions might be that shear stresses between the core and the rind are negligible and that axial compressive stresses within the core can likewise be neglected such that the core would be stress free in the longitudinal direction. With these assumptions, the soft core would be considered to behave like an elastic foundation that mechanically functions to resist localized bending stresses by distributing them throughout the immediate lengths of longitudinally aligned structural elements within the rind. A stem with a relatively large, parenchymatous pith would have this property, since parenchymatous tissue is a hydrostatic, essentially incompressible material against which phloem or sclerenchyma fibers could exert a mechanical influence. As the parenchymatous pith absorbs water and swells, it would hydrostatically place its anisotropic rind into ring or hoop stress, stiffening the organ as a whole and preventing the rind from crimping (Brazier buckling). The primary stems of many species of plants have this mechanical configuration (Kutschera 1989; Kutschera, Bergfeld, and Schofer 1987), whereas two-phase (core-rind) models have been successfully used by Anazado (1983), who modeled the behavior of corncobs, and Lu, Bartsch, and Ruina (1987), who similarly modeled the stalks of corn. Both studies report a remarkable correspondence between predicted and observed mechanical behavior.

The advantages of a two-phase composite are illustrated by considering a cantilevered stem consisting of two materials, one being stronger in tension than in compression—for example, phloem fibers and a parenchymatous hydrostatic cortex. When the stem is subjected to a bending moment, the tensile strength of the cortex on the convex side would be overcome before the fibers yielded under tension. For loadings that induce yielding in the fibers, the fibers would sustain all or most of the tension developed within the stem; that is, the fibers would take up all the tension, while the cortex would take up all the compression. A surprisingly small volume fraction of fibers can be shown to provide substantial mechanical benefits. Indeed, the minimum volume fraction of fibers can be calculated, based on the material properties of the fibers and the cortex, provided the location of the neutral axis is known. This latter stipulation introduces some complexity, however, because the neutral axis through the stem may not coincide with the centroid axis, particularly if the stem experiences a large bending moment and yielding of materials occurs. To determine the location of the neutral axis, we denote as kd the distance to the neutral axis measured from the top (convex surface) of the bent stem, where $k<1.0$ and h is the height of the representative cross section. To determine the minimum volume fraction of fibers, we must express kh in terms of the bending stresses, as well as specify the allowable stresses and relative elastic moduli of the fibers and the cortex. Since the unit longitudinal contraction of the cortex ε_c and the unit elongation of the fibers ε_f must be given by the formulas $\varepsilon_c = -kh/r$ and $\varepsilon_f = (1-k)h/r$, where r is the radius of curvature, the maximum compressive stress in the cortex and maximum tensile stress in the fibers equal $\sigma_c = -(kh/r)E_c$ and $\sigma_f = [(1-k)h/r]E_f$, where E_c is the elastic modulus of the cortex and E_f is the elastic modulus of the phloem fibers. If V_f denotes the volume fraction of the fibers within the stem and n denotes the ratio of E_f to E_c, then $V_f = k^2/[2(1-k)n]$ and $k = |\sigma_f/(\sigma_c + \sigma_f n)|$. Thus, if we know the allowable stresses for the fibers and the cortex (σ_f and σ_c, respectively), as well as the ratio of the elastic moduli of the two materials (n), we can calculate the location of the neutral axis kd, determine k, and finally calculate the V_f necessary to sustain loadings below the yield stress of the fibers.

For example, consider a stem, with a rectangular cross section of height h, composed of fibers with an allowable stress of 68 $\mathrm{MN \cdot m^{-2}}$ and a cortex with an allowable stress of 3.66 $\mathrm{MN \cdot m^{-2}}$. If the ratio n of the elastic modulus of the fibers to that of the cortex equals 15, then, from the previous formulas, k

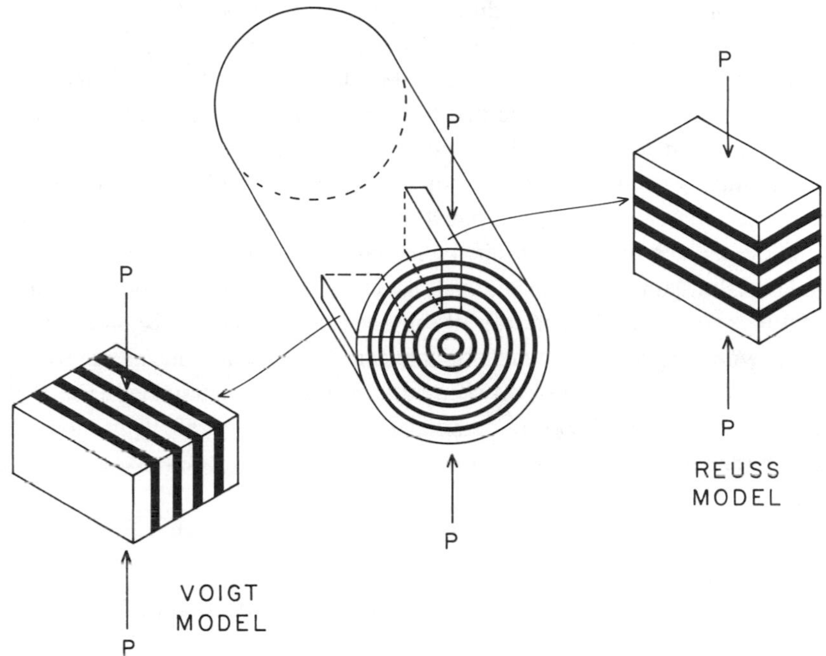

FIGURE 7.9 Limiting cases for the behavior of composite materials defined according to the equal strain (Voigt) model and the equal stress (Reuss) model. The Voigt model assumes that all the material elements (shown as alternating dark and light layers) are aligned parallel to the direction of the externally applied force. To maintain compatibility between adjacent layers, the deformations (strains) within all layers must have equal direction and magnitude. The Reuss model assumes that the material elements are aligned normal to the direction of the externally applied force and that the stresses within layers have equal magnitude. Depending on how materials are placed in a composite, both models may be operational (see middle drawing).

$= 68\ \mathrm{MN \cdot m^{-2}} / [68\ \mathrm{MN \cdot m^{-2}} + (3.66\ \mathrm{MN \cdot m^{-2}})15] = 0.553$ and $V_f = (0.553)^2 / [2(1 - 0.553)(15)] = 0.0228$, which indicates that only 2.28% of the stem's cross section need be composed of phloem fibers to meet the elastic requirements of this cantilevered stem. This simple example illustrates how adding a small amount of material with a relatively high elastic modulus confers a significant mechanical benefit. It is not surprising, therefore, that most organs manifest anatomical heterogeneity and possess elastic tissue components like phloem fibers at some considerable distance from the neutral axis. The volume fractions of these elastic components are typically much larger than those predicted based on the assumption that self-loading induces the maximum bending moment. The obvious explanation for this discrepancy is that these organs experience additional dynamic loadings.

An obvious question arises when we consider an organ composed of two or more materials differing in their material moduli: What is the elastic modulus of the composite material? The first basic theory for the mechanical behavior of composite materials was developed by James Maxwell, who in the nineteenth century considered two limiting models currently called the equal strain or Voigt model, and the equal stress or Reuss model (fig. 7.9). The Voigt model sets the upper limit for the magnitude of the composite elastic modulus. It assumes that all the material elements within the composite material lie parallel to the direction of the externally applied force and that the strains within each element are equivalent in direction and magnitude so that deformational compatibility between adjacent layers is maintained. The elastic modulus of the composite material E_{cm} according to the Voigt model is given by the formula

(7.16a) $$E_{cm} = \sum_{i=1}^{n} E_i V_i \qquad (i = 1, 2, \ldots n),$$

where E_i is the elastic modulus and V_i is the volume fraction of the ith component; that is,

(7.16b) $$E_{cm} = E_1 V_1 + E_2 V_2 + \ldots + E_n V_n.$$

The Reuss model sets the lower limit for the magnitude of the composite elastic modulus. It assumes that the material elements within the composite are aligned normal to the direction of the externally applied force and that the stresses developed within each element are equivalent in magnitude, whereas the strains may differ depending on the elastic modulus of each element. The elastic modulus of the composite material according to the Reuss model is given by the formula

(7.17a) $$\frac{1}{E_{cm}} = \sum_{i=1}^{n} \left(\frac{V_i}{E_i}\right) \qquad (i = 1, 2, \ldots n)$$

or

(7.17b) $$\frac{1}{E_{cm}} = \frac{V_1}{E_1} + \frac{V_2}{E_2} + \ldots + \frac{V_n}{E_n}.$$

For a simple two-phase composite material, the total volume of the composite must equal the sum of the volume fractions of the two material elements ($1 = V_1 + V_2$). Since the ratio of the two elastic moduli is already specified ($n = E_2/E_1$), eqs. (7.16) and (7.17) are reduced to the formulas

(7.18) $$E_{cm} = E_1[V_1(1-n)+n] \quad \text{(Voigt model)}$$

and

(7.17b) $$\frac{1}{E_{cm}} = \frac{1}{E_1}\left[V_1 + \frac{(1-V_1)}{n}\right] \quad \text{(Reuss model).}$$

In the case of the previous computations for the minimum volume fraction of phloem fibers within a stem, V_f was found to equal 0.0228 (thus $V_c = 0.9772$) when $n = E_f/E_c = 15$. By substituting V_c for V_1 in eqs. (7.18) to (7.19), the upper and lower limits for the magnitude of the composite elastic modulus of the stem are $E_{cm} = E_c[V_c(1-n)+n] = E_c[0.9772(1-15)+15] \approx 1.319E_c$ (for the Voigt model) and $(1/E_{cm}) = (1/E_c)[V_c + (1 - V_c)/n] = (1/E_c)[0.9772 + (1 - 0.9772)/15] \approx 0.9787/E_c$ or $E_{cm} \approx 1.023E_c$ (for the Reuss model). These limiting values reveal that the elastic modulus of a composite can be substantially increased by adding a relatively small amount of material with a high elastic modulus.

Clearly, however, the composite elastic modulus theoretically falls somewhere between these limits. A reasonable approximation for the middle ground magnitude of the composite elastic modulus is given by combining eqs. (7.18) and (7.19); that is, the material elements within a composite are assumed to behave according to both models (see fig. 7.9):

(7.20) $$\frac{1}{E_{cm}} = \frac{1}{E_1}\left\{\frac{x}{[V_1(1-n)+n]} + (1-x)\left[V_1 + \frac{(1-V_1)}{n}\right]\right\},$$

where x represents the proportion of the material that obeys the Voigt model and $(1-x)$ the proportion that obeys the Reuss model. Inserting $V_c = 0.9772$ for V_1, $n = E_f/E_c = 15$, and $x = 0.5$ (half the material obeys the Voigt model) into eq. (7.20), we find that $E_{cm} \approx 1.152E_c$, which, as expected (given the stipulation that $x = 0.5$), is roughly the mean of the upper and lower limits for E_c calculated separately by means of the two models. Note, however, that eq. (7.20) yields mathematically absurd results when the elastic modulus of either of the two components approaches or equals zero, indicating that the elastic moduli of composite materials can be determined by this method only when the elastic moduli of the components do not differ from one another in the extreme.

In many practical situations, we cannot neglect the effects of shearing between component phases (bonding shear stresses) within a composite material. These bonding shear stresses increase in magnitude as the total lateral surface area of all the stiffer elements per unit length of the composite decreases. By increasing the number of individual stiff fibers and decreasing their individual diameters, the total lateral area of the stiffer of the two materials can be increased while keeping the total cross-sectional area of the stiffer elements constant. An extreme expression of this strategy is seen in fiber-reinforced composites, such as plant cell walls (see chap. 5). The behavior of fiber-reinforced composites whose fiber component is finely distributed within an isotropic matrix has been variously modeled and can be extended to treat plant organs only very crudely. Based on one modeling approach, the composite elastic modulus of fiber-reinforced materials can be estimated by the formula

$$E_{cm} = E_f V_f(z) + E_m(1 - V_f), \tag{7.21}$$

where E_f and E_m are the elastic moduli of the fiber and the matrix components, respectively, and z is a quantity that depends on the stiffness, cross-sectional area, spacing, radius, and length of the fibers within the matrix (see Vincent 1990a, 128–29). For most practical situations, $E_f >> E_m$, and a reasonable approximation is given by $E_{cm} = E_f V_f(z)$ (see Wainwright et al. 1976, 147–49).

Hollow Organs

Without a central hydrostatic core or transverse diaphragms or some form of internal strutting, stems and columnar leaves would be susceptible to Brazier buckling, and indeed many do mechanically fail in this mode of deformation. (Note that in hollow plant organs the core is a gas phase that can be pressurized and therefore still operates as a hydrostat under some conditions. For example, if a hollow stem or leaf petiole is submerged in water and if the internal pressure can be increased physiologically, then the gas-filled core can be very effective in resisting localized bending moments as well as providing buoyancy. Examples are seen in the petioles of some species of water lilies that have an aerenchymatous core with numerous pressurized chambers.) Truly hollow plant organs represent an interesting mechanical case because the relation between an applied force and the resulting deformation is not linear even within the proportional limit of loading. This was discussed in chapter 3 when we considered Brazier buckling.

Two basic types of nonlinear collapse can occur when a hollow organ is subjected to a bending moment: Brazier buckling, or ovalization, and bifurcation buckling, known as shortwave axial buckling. Some of the salient features of both are shown in figure 7.10. Recall from chapter 3 that Brazier (1927) observed that when a thin-walled tube bends, its originally circular

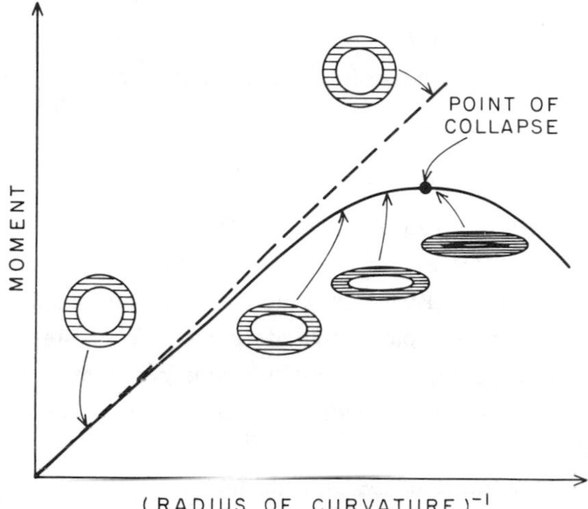

FIGURE 7.10 Bending moment plotted as a function of the reciprocal of the radius of curvature (l/R) for two hollow cylindrical beams made of different materials. A cylinder composed of a relatively stiff material will resist ovalization (dashed line) when subjected to the same bending moments as a cylinder made of a less stiff material, which will undergo Brazier buckling (solid line) owing to the progressive ovalization of its originally circular cross section. A beam that resists ovalization has a linear moment-curvature relation; a beam that undergoes Brazier buckling has a nonlinear relation between moment and curvature. At some bending moment (the critical bending moment) ovalization leads to Brazier buckling, at which point the beam collapses.

cross-sectional geometry progressively deforms into an ellipse. As this ovalization progresses, the force-deformation curve, which is initially linear, becomes progressively nonlinear (fig. 7.10). As the bending moment increases, the slope of the plot of the bending moment versus the radius of curvature decreases, and the tube collapses at the point of maximum moment. Brazier derived a formula for the critical bending moment M_{cr} relating the geometry of the tube and its modulus of elasticity. This formula was given in chapter 3 and is repeated here (note the caveats concerning E and υ mentioned in chapter 2):

$$M_{cr} = 0.31426 \frac{\pi E t^2 R}{(1 - \upsilon^2)^{1/2}}, \tag{7.22}$$

where t is the tube's wall thickness, R is the external radius of the tube, and υ is the Poisson's ratio of the tube's material. We can readily see that the critical bending moment increases as the elastic modulus, wall thickness, Poisson's ratio, or external radius increases. Equation (7.22) can be further simplified if we assign a value to the Poisson's ratio. A reasonable value, based on a number of studies treating a variety of primary tissue types, is 0.3. When this value is incorporated and end-wall effects are taken into account, eq. (7.22) takes the form

$$M_{cr} = 1.035\, E\, t^2\, R. \tag{7.23}$$

Lu, Bartsch, and Ruina (1987) found that the critical bending moments of cornstalks averaged 0.991 $E\, t^2\, R$, remarkably close to the predicted relation. True, when healthy and actively growing cornstalks are examined, they are not hollow but have a solid parenchymatous core. When stalks become diseased or when they age and the parenchymatous core deteriorates, however, wind pressure or just the static loading of the stalks themselves causes these organs to buckle.

Equation (7.23) shows that the wall thickness of a tubular organ is much morc influential in terms of the critical bending moment than the external radius of the organ; that the critical bending moment depends on the ratio of t to R; and that if a soft elastic core is present, the magnitude of the contribution of the core to the total structural stability of the organ increases as the ratio of t to R increases. Indeed, some very simple calculations show that the critical ratio of t to R is 0.20. If the ratio is less than 20%, then a tubular organ becomes very susceptible to Brazier buckling. Significantly, very few hollow plant organs have a ratio of t to R less than 0.20, and most are higher than this value (see chap. 3).

Although the second type of nonlinear collapse, bifurcation buckling, has been empirically studied (Seide and Weingarten 1961), there are no known analytical solutions to this problem. Bifurcation buckling is sudden and catastrophic and occurs without a preliminary (observable) deformation. In general, if the ratio of a tubular organ's length to radius equals or exceeds 100, then Brazier buckling is more likely to occur than bifurcation buckling. This is an interesting cutoff ratio, since the nodal septa that span the hollow stems of many plants reduce the ratio of length to radius of these hollow tubes well below the ratio at which Brazier buckling is likely to occur. Accordingly, these organs ought to undergo sudden, catastrophic bifurcation buckling when very large bending moments are applied to them. Indeed, they do. If the fully mature hollow stem of a typical grass species or of a horsetail is bent slowly by hand, little geometric deformation will be noticed before the critical bending moment is reached. When these organs buckle they do so suddenly, with little visible warning—and usually just above a node.

The apparent convergence in the mechanical design of the hollow, septated stems of some grass and dicot species and all known species of horsetail (fig. 7.8), which are two phylogenetically very distantly related plant groups, is even more remarkable in that their mechanical design is also used by engineers in the construction of some bridges, such as the Forth Bridge in Scotland. The main diagonal struts of this bridge, measuring 12 feet in diameter, were reinforced by perforated diaphragms set 12 feet apart. Within each wall of the strut, six T-shaped stiffeners add further support. The placement of these stiffeners corresponds almost precisely to the placement of the vascular fibers seen in the cross sections of grass stems. Simon Schwendener, a botanist and engineer, showed that the resistance to bending is at least twenty-five times greater than it would be if these vascular stiffeners were brought together into a solid core within the plant stem (see Schwendener 1874, 45, fig. 5).

Observations lead to the conclusion that plants' hollow organs in general undergo bending failure primarily as a result of compressive buckling rather than tensile failure of tissues. This is explicable in terms of the very high tensile modulus of most plant cell walls and tissues, in particular the phloem fiber stiffeners. Geometrically, a cylindrical, hollow plant organ fails by a compressive crimping that propagates circumferentially to either side of the initial site of failure. Crimping is resisted by the elastic fibers in the rind of stems and leaves, but these stiffeners eventually give way, and compressional failure propagates on the opposite side of the organ. This propagation is accompanied by lateral shear stresses, causing tissues to undergo a bending moment normal to their longitudinal axis. Under extreme bending loads, phloem fibers undergo tensile failure.

DYNAMIC LOADINGS ON PLANT ORGANS

Thus far we have treated plant organs primarily in the context of their static equilibrium. The importance of dynamic loadings cannot be overestimated, however. Indeed, it is very likely that all plant organs are overbuilt in the sense that they can sustain far greater loadings than they typically experience in terms of static loadings. In the terrestrial environment, wind pressure is the most ubiquitous and important dynamic loading, while in the aquatic environment the movement of water has similar effects. Dynamic loadings can cause instability when dynamic resonance between an organ and the frequency of wind gusting occurs. Additionally, repeated bending in response to wind or water pressure can fatigue the structure of an organ whose organogenesis is complete and whose capacity to repair damage is limited. Dynamic resonance

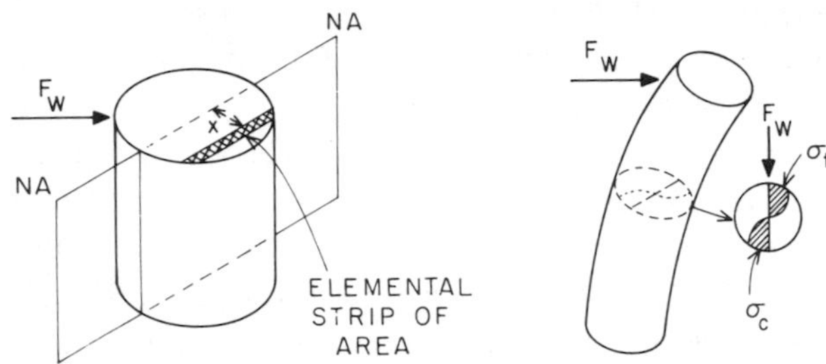

FIGURE 7.11 Bending of a beamlike plant axis owing to the lateral force exerted on it by wind (F_W). Lateral wind loading results in a bending moment, placing tissues in either tensile or compressive stress that increases in magnitude as a function of the distance x from the neutral axis NA. Within each transection, the intensity of both the tensile and the compressive stresses that develop in each elemental strip aligned parallel to the plane of bending increases as a function of the distance x of the elemental strip from the neutral axis. For further details, see text.

is an increase in vibration that results when a periodic motion imposed on a structure has approximately the same frequency as one of the natural modes of vibration of the structure. Air or water flow patterns generated around and by a plant organ can form eddies that are periodically shed downstream. As successively formed eddies detach, the drag on an organ is temporarily reduced, and the organ elastically rebounds as strain energy is converted into potential energy. If the period of eddy detachment coincides with one of the natural frequencies of an organ's vibration, then the resulting mechanical stress may equal or exceed that breaking or yield stress, and the organ will mechanically fail. Even if the dynamic loading is below the breaking or yield stress, oscillation can result in repetitive shear stresses that induce structural fatigue. Shear stresses and structural fatigue can be beneficial, as when they remove fruits and old leaves from stems, or they can be destructive, as when branches are broken from trunks in storms.

We can explore dynamic wind loading by considering the limiting condition for an ideal elastic behavior of a cantilevered column. The dominant type of stress within the limits of elasticity is the bending stress caused by airflow. Recall from the elementary theory of bending that the maximum bending stresses along the length of such a member will occur at the base. Also, within each transection through the member, bending stresses will have a distribution such that the highest tensile and compressive stresses will occur just at the surface. If a cantilevered column is to remain mechanically stable, the force of wind F_w must not exceed the maximum bending stress σ_{max} for elastic behavior:

$$F_w \leq \frac{\sigma_{max}\, I}{x\, l} \approx \frac{\sigma_{max}\, I}{R\, l}, \quad (7.24)$$

where I is the second moment of area measured about the neutral axis of bending (perpendicular to the direction of the application of F_w), x is the maximum distance of an elemental strip of area from the neutral axis measured parallel to the direction of F_w (therefore, to calculate the maximum bending stress in a transection, x must be set equal to the radius R of the column), and l is the length of the member (fig. 7.11). For a cylindrical stem or leaf with a solid cross-sectional area, $I = \pi R^4/4$. Therefore

$$F_w \approx \frac{\pi\, \sigma_{max}\, R^3}{4\, l}, \quad (7.25)$$

since R equals the unit radius. Notice that the maximum force that can be sustained increases as the cube of the radius and decreases as a function of the length. This provides some insight into the thigmomorphogenetic response of trees to dynamic wind loadings—the trunks and branches of wind-loaded trees typically grow disproportionately more in girth than in length. Thus their ratio of radius to length is larger than for trees growing in sheltered habitats. Indeed, when trees are cleared away from a dense population of plants, those that remain, which tend to be tall and spindly, typically fall over under wind loadings that are remarkably small.

As mentioned previously, eq. (7.24) follows directly from the elementary theory of bending, which predicts that the maximum bending stress of a member with a circular cross section is given by the formula

$$\sigma_{max} = \frac{M\, R}{I}, \quad (7.26)$$

where M is the bending moment, which equals the product of the critical load P_{cr} and length. Thus

$$\sigma_{max} = \frac{P_{cr} l\, R}{I} = \frac{4\, P_{cr}\, l}{\pi\, R^3}. \quad (7.27)$$

By substituting $P_{cr}l$ for M in eq. (7.26), we get eq. (7.27). (Note that the term R/I in eq. 7.26 is the reciprocal of the section modulus Z; that is, $R/I = 1/Z$.)

Speck and Vogellehner (1988b, 264) provide a useful formula for the percentage contribution of a single tissue (ith tissue) toward the flexural rigidity of an organ:

$$(EI)_i\% = \sum_{i=1}^{n}(E_i I_i \times 100\%), \quad (7.28)$$

where $(EI)_i$ is the percentage contribution of the *i*th tissue to the organ's flexural stiffness. This equation is based on the parallel-axis theorem, which assumes that the structural components within a beam are configured in parallel arrays running the length of the structure; that is, if the second moment of area I_z of an element with respect to the axis z through the centroid is known, then the second moment of area with respect to any parallel axis z' can be calculated from the formula $I_{z'} = I_z + Ad^2$, where A is the area of the element and d is the distance between z and z' (see Timoshenko 1976a, 422). Equation (7.28) also assumes that the percentage contributions of tissues within the organ to flexural stiffness are additive. These assumptions may be valid, but they must be tested against empirical data before eq. (7.28) is used.

As mentioned before, dynamic wind loadings can induce harmonic motion in plant organs provided resonance frequency is achieved. It is reasonable for us to assume that a relation exists between the natural frequency of vibration of a structure and the static deflection produced under self-loading, because a structure's natural frequency of vibration is the result of cyclic exchanges of kinetic and potential energy (see chap. 3). The potential energy is associated with the strain energy stored within a structure as a consequence of elastic deformations. One measure of the strain energy is the static deflection under the acceleration of gravity. Thus, if the static deflection is known, we can estimate the natural frequencies of vibration even of complex structures.

Recall from chapter 3 that an organ's static deflection δ due to bending equals the product of unit mass and the acceleration due to gravity divided by some constant ($\delta = mg/k$) and that the natural frequency of vibration f_i is given by the formula

(7.29) $$f_i = \frac{1}{2\pi}\left(\frac{k}{m}\right)^{1/2}.$$

Thus

(7.30) $$f_i = \frac{1}{2\pi}\left(\frac{mg}{\delta m}\right)^{1/2} = \frac{1}{2\pi}\left(\frac{g}{\delta}\right)^{1/2}.$$

For an untapered cantilever sustaining a load at its tip, $k = 3EI/l^3$. With the appropriate substitutions, we can relate the natural frequency of vibration to the dynamic loading of wind pressure, such that

(7.31) $$f_i = \frac{1}{2\pi}\left(\frac{3g\ EI}{F_w\ l^3}\right)^{1/2}.$$

Thus, as the force of wind pressure increases, the natural frequency of vibration of a cantilevered organ will decrease. For a cantilevered organ subjected to some wind force, the frequency of vibration will increase as *EI* increases. This explains why older stalks of grass oscillate with a higher frequency of vibration than younger stalks (*EI* increases to some limit with the aging of tissues). It also explains why leaves with longer petioles have a lower frequency of vibration in the wind than leaves on the same tree with shorter petioles. Notice too that for plant organs that exhibit motion-induced inhibition in length (a common symptom of thigmomorphogenesis) *I* will increase and *l* will decrease; therefore there will be a reduction in the frequency of vibration under dynamic loading conditions.

An interesting example of wind-induced natural frequencies of vibration is seen for the petioles of the poplar *Populus deltoides*. As leaves mature their petioles increase in length, but they also change their cross-sectional geometry, becoming more elliptical in transection. The major axis of the elliptical cross section is oriented vertically. From eq. (7.31), we would expect that the natural frequency of vibration would decrease as the length of petioles increases. As petioles mature, however, their flexural stiffness increases disproportionately in such a fashion that the natural frequency of vibration remains relatively constant regardless of petiole length. Also, since the bulk of the mass of each petiole is oriented vertically, the second moment of area measured in the plane of bending due to self-loading is very high, and young petioles can maintain a slightly cantilevered orientation in the horizontal plane. But because older petioles have a low second moment of area measured in the other direction, they torque around their longitudinal axis and droop downward. Ontogenetic changes in petiolar length, flexural stiffness (as measured along the major and minor axes of petioles), and the weight of the leaf lamina collectively result in the leaves' trembling even in a slight breeze, regardless of their developmental age. This wind-induced harmonic behavior promotes gas exchange between the leaf and the atmosphere and cools leaves heated by the sun. Similar growth patterns and consequent mechanical responses to wind movements are seen for species of birch (*Betula*).

LODGING

All of the preceding explanation has been based on the implicit assumption that a statically or dynamically loaded organ is firmly anchored at its base. Indeed, most organs are clamped at their base and are more or less free to deflect along their length. When this is so, lateral displacements are largest at the free end and approach zero toward the base. In some instances, however,

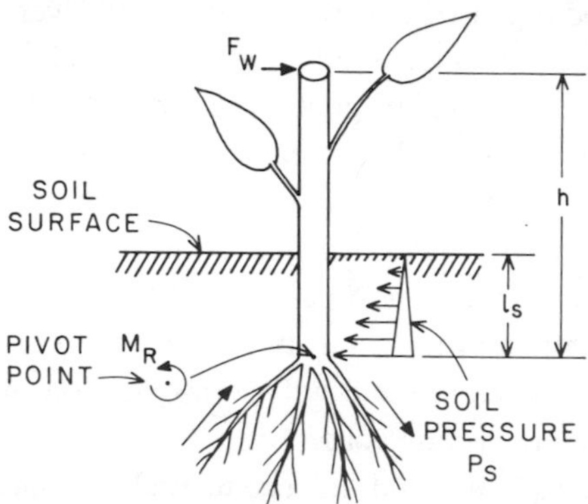

FIGURE 7.12 Pivoting of a plant owing to wind pressure that may lead to lodging. When a lateral wind loading F_W is applied, the plant pivots about the root crown, placing roots on the leeward and windward sides in tension. In addition to the small resisting moment attributable to the roots, the soil acting against the stalk of the plant embedded below ground level provides a second (and dominant) source of resistance. The parameters that influence the mechanics of pivoting are the moment arm (the product of the lateral loading F_W and the distance through which it operates h), the resistance to bending that results from the length of the stem underground (l_S), and the soil pressure ($\mathcal{P}_S$), which increases in intensity as a function of depth.

plant organs anchored in the soil by their roots behave as if their subterranean portions pivot; the base of the organ has a rotational moment. Pivoting is a mode of potential mechanical failure. It is frequently, but not exclusively, the mode of failure when plant stems lodge, a term agronomists use to describe plants uprooting, breaking, or otherwise mechanically deforming to the ground owing to the effects of wind or rain on their stems and leaves. We have already examined the mechanics of breakage and large lateral deformations. In this section we will examine pivoting and how this can cause a plant to lodge.

When plants lodge as a result of dynamic loadings from wind or rain, it is a consequence of a rotational moment about the point where roots are attached, along with soil failure. The stems of most plants resist deflections at their base as a result of tensile stresses that are resisted by the root system and by the pressure exerted on roots by the soil overburden. Thus the geometry of the root crown and the depth to which it is buried within the soil influence how great a dynamic force will cause the plant to lodge as a result of pivoting. Clearly, if the root crown is superficial or level with the ground surface, then soil pressure will be minimal or virtually zero. If the root crown is submerged well below the ground surface, then soil pressure becomes an important aspect of resisting wind loadings. At extreme depths of burial, pivoting may not even be an important feature, since the mode of mechanical failure may be breakage of stems well above the ground surface.

The root crown and the submerged portion of the stem subjected to wind can be compared to a pier foundation under lateral loading. Walker and Cox (1966) and Walker and Haan (1974) have developed the pertinent relations among the resistance of a pier to the magnitude of a lateral force, the depth of burial beneath the soil, and the diameter of the pier. Casada, Walton, and Swetnam (1980) used these relations to examine the wind resistance of tobacco plants as influenced by the depth of burial of the pivot point—roughly the root crown. These authors found that the roots of tobacco plants provide only a small percentage of the total resisting moment, while soil pressure against the buried portion of stems provides most of the resistance to wind loadings (fig. 7.12). Casada, Walton, and Swetnam (1980) provide an empirically determined formula that relates the force of wind F_w, the distance h of the application of the wind force above the pivot point, the resisting moment M_r, soil pressure $\mathcal{P}_s$, the projected area A of the stem acting against the soil, and the distance l_s of the pivot point below the surface of the soil:

$$F_w h = M_r + \frac{\mathcal{P}_s A l_s}{2}. \tag{7.32}$$

When $F_w h$ is plotted as the ordinate and l_s is plotted as the abscissa, eq. (7.32) has the form of a straight line, where the resisting moment is the y-intercept and where soil pressure and the projected area divided by two are constants that define the slope of the line. Notice that $F_w h$ is a bending moment (it has units of force times distance) and that it will increase as either the wind pressure or the lever arm h, or both, increases. Thus, long stems with a crown of leaves aggregated at the top exert considerably more leverage at the pivot point than do shorter stems with leaves attached along their lengths. This equation can be used only if the soil pressure is considered not to vary significantly as a function of the depth of burial of the root crown. Up to about 0.25 m in depth, this assumption provides little error ($\leq$ 5%), but for loose soil that can be compacted under compression, oscillatory motion of the stem can cause $\mathcal{P}_s$ to decrease as soil is removed from the vicinity of the stem owing to the movement of the stem about the pivot point.

For very shallow root systems, root suction must be considered the principal force resisting deflections at the base of a stem or tree trunk. Root suction refers to the combined capacity of roots to resist tensile stresses owing to their tissue tensile modulus and how well the root-soil interface resists shearing.

The capacity of roots to resist shearing is a complex function of the geometry of root crown and root hairs and the amount of soil particles that adhere to the surfaces of roots. All other things being equal, root suction increases as the water content of the soil increases, to the limiting condition where the soil begins to behave as a non-Newtonian fluid, at which point mechanical stresses cause the soil to flow like a liquid and the shear modulus of the soil is lowered.

Plants lodge for a variety of mechanical reasons other than pivoting at the base of their stems, but it is fair to say that all modes of lodging in one way or another result from failure of a plant's growth to provide an adequate margin for safety with respect to dynamic loadings. For example, Metzger and Steucek (1974) report differences in the thigmomorphogenetic response of two varieties of barley (*Hordeum vulgare*) that differ in their propensity for lodging. Barsoy, a variety of barley resistant to lodging, responded to mechanical perturbation by a significant reduction in shoot length, while Penrad, a variety that is much more susceptible to lodging, showed little variation in shoot length in response to shaking or handling. Since shoot length provides the lever arm through which wind pressure operates, a thigmomorphogenetic inhibition of the shoot growth in length can be seen to provide a mechanical advantage. Nonetheless, this advantage comes at a cost—a reduction in plant height that may reduce the capacity of a plant to gather light and avoid being shaded by its neighbors.

Some plants can recover from lodging by diffuse meristematic growth that can reorient plant stems. I have seen milkweed plants (*Asclepias*) reaffirm their vertical posture over the few days following a violent wind and rain storm. The plants retained some of their initial curvature of bending, particularly toward the base of their stems, where the capacity for meristematic growth is less than toward the midspan or tip. The curvature of bending was approximated by an arc of a circle (see eqs. 7.9 to 7.11); thus it was possible to calculate the maximum bending moment and maximum tensile stresses each curved stem experienced under its own static self-loading.

SAFETY FACTORS

Engineers plan for accidents by designing structures that can sustain stresses and strains many times those that are likely to occur under normal conditions. This mechanical design factor or safety factor, which can be taken as the ratio of the maximum loadings likely to be experienced to the operational (normal) loadings on a structure, is typically based on the statistical probability that certain types of loadings will occur. When this probability drops below some minimum level, economic considerations take precedence over considerations of legal or ethical liability. Clearly, structural reinforcement beyond certain limits becomes disproportionately expensive unless the likelihood of mechanical failure is high enough to warrant the extra expense. Thus the magnitude of a safety factor is an operational decision based on a number of factors.

It is not uncommon to hear biologists discussing plants in terms of cost and profit. Within this context, cost typically refers to the metabolic resources that must be invested to achieve some level of performance, while profit refers to the dividends that result in terms of vegetative or reproductive growth. Biologists have debated how far a cost-performance perspective can be taken in terms of plant biology. The principal issue in this debate is not whether biological functions involve a metabolic investment—clearly, they must—but whether metabolic investments are limited, retroactively through the operation of natural selection, by the dividends they accrue. The literature dealing with photosynthetic adaptation is full of the notion that natural selection will favor genotypes that are more efficient in allocating their limited resources over competing genotypes that are more wasteful. This notion can be transferred with little intellectual effort into the area of plant biomechanics. Mechanical stability is in part defined by the material properties of tissues and the geometry of plant organs that in turn are bought at the expense of metabolic energy. Mechanical stability is also defined by the environment and by the loadings applied to plants. Thus the safety or design factor of a plant should be contingent on the probability of its experiencing various types of loadings, as well as the frequency distributions of their duration and magnitude. Thigmomorphogenesis appears to reflect a possible adaptation in plant growth regarding the types, magnitude, and duration of loadings. As in the case of the literature treating photosynthetic adaptation, however, evidence for the efficiency or economy of mechanical design is sparse and debatable, because efficiency, economy, and safety factors are extremely difficult to quantify in unambiguous ways, particularly since plants must perform a variety of biological functions other than sustaining their own weight and the dynamic loadings imposed by the environment. In this section we will focus only on the design factors associated with mechanical stability. Clearly this is a narrow perspective, since design factors enter into the hydraulic architecture of plants, just as they affect the interception of sunlight and the exchange of gases between the plant body and the external atmosphere. The design factors and requirements for these biological functions are very important, but their treatment will be delayed until chapter 10, where we will view this topic in terms of plant evolution and biomechanics.

The literature contains only a few examples of attempts to calculate the design or safety factors of plants. One of these examples is particularly noteworthy, however. Tateno and Bae (1990) calculated the lodging safety factor for mulberry trees (*Morus bombycis*) treated with succinic acid 2,2-dimethylhydrazide (SADH), which is known to retard (dwarf) growth by stimulating the production of ethylene and reducing auxin levels within plants. The lodging safety factor was calculated on the basis of the ratio of the critical lodging load (the empirically determined minimum leaf fresh weight required for lodging) to the leaf fresh weight observed on the plant. In the untreated trees, the lodging safety factor averaged 3.2 (Tateno and Bae 1990). When treated with SADH, stem elongation was inhibited by about 80%, but the percentage of shoot dry matter partitioned into the leaves was found to be always larger than that of the untreated plants. Therefore dwarfing increased the critical lodging load but also increased the leaf fresh weight. As a result, the lodging safety factor of treated and untreated trees was roughly comparable. Tateno and Bae (1990) concluded from these experiments that shoot formation in the mulberry tree is developmentally controlled so as to maintain a relatively constant lodging safety factor. This and other examples from the literature suggest that plants have an intrinsic capacity to grow in a manner that provides a design factor against unusual externally applied loadings.

The issue of economy in mechanical design is complex and will be debated for years to come. Yet a critical part of the debate must revolve around the philosophy underlying the way design factors are calculated. In this regard it is extremely useful to examine how engineers determine and measure the design factor in constructing a structure. True, the standards engineers use are often determined by entirely human factors, such as legal liability, and have a limited bearing on organisms. Also, a biologist cannot set an arbitrary standard for an organism but will be interested in measuring the design factor of an organism's construction. The fundamental question, however, is whether the biologist and the engineer can use parallel methods to calculate design factors.

Two considerations figure prominently in determining the magnitude of a design factor. First, it is essential to discover what type of stress (dynamic versus static) is likely to determine the conditions under which a structure may mechanically fail. Since dynamic stresses and static stresses are additive (thus, structural failure is likely to occur under dynamic loading conditions), a very simple design factor can be calculated from the stress ratio, in which the numerator represents the dynamic loading condition and the denominator is the static loading condition. And second, it is necessary to determine the

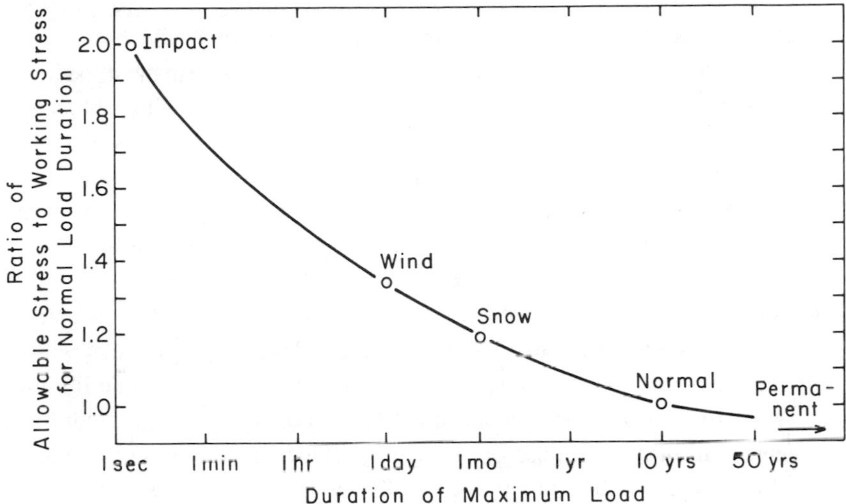

FIGURE 7.13 Design factor ("safety factor") for a vertical wooden column plotted as a function of the duration of maximum load. The design factor is taken as the ratio of the allowable stress to working stress for the normal load duration. The allowable stress refers to the maximum stress the column is designed to withstand. Working stress refers to the stress calculated to result from the static loading of the structure (based on the "normal" load duration). As the duration of the maximum load experienced by a column decreases, the design factor increases. Wooden columns that are expected to experience impact load durations (2 sec or less) have the highest design factor (which equals 2); wooden columns that sustain permanent load durations (fifty years or more) are designed with the lowest design factor (which equals 0.9). See text for further details.

magnitude and duration of all the types of stresses that a structure will probably experience. In this regard, one of the simplest design factors is the dimensionless ratio between the maximum load the structure is designed to bear and everyday working stresses. An example of this kind of design factor for a column is the length ratio, in which the numerator is the critical length at which buckling or compressive failure will occur and the denominator is the actual length of the column. This ratio is convenient because it can be applied to a variety of vertical support members that differ in size and shape or material properties. Ideally, a design factor ought to incorporate the magnitude and duration of application of loadings as well as the types of loading conditions a structure will experience in its functional lifetime. At first glance this appears difficult to accomplish with a single dimensionless number. But in the case of wooden support members, like tree trunks, the type of loading and its duration can be incorporated into a design factor by a very simple scheme, as shown in figure 7.13. In this figure, the type of loading is actually defined based on its duration of application (given as the abscissa), which can range from a few seconds (impact loadings) to scores of years (permanent loading). The design factor (given as the ordinate) is the ratio of the allowable stress (always taken as less than the working stress) to the working stress (the loads that will actually be experienced) for the normal, everyday load duration. As the duration of a loading that is likely to be experienced decreases, the design factor increases to a maximum value of two, which is designated for support members that are likely to experience impact loadings. This mode of calculating a design factor is comparable to the one engineers use for wooden structures that will experience dead and live loads. A dead load is one that will be sustained for a long time and that is accommodated under static equilibrium. A live load is one that will be sustained for a short time and that induces a dynamic equilibrium response in a structure. Office furniture is an example of a dead load in a building; the traffic of office workers constitutes a live load. If loads of different durations are applied simultaneously, then the design factor reflects the total of all the applied loadings at their allowable stress limits adjusted by the factor for the shortest load duration among them. This ensures that the maximum design factor will be used in constructing a wooden support member.

Three features of how the design factor shown in figure 7.13 is calculated are noteworthy if a similar procedure is applied to an organism. First, the ratio of the working stress to the allowable stress is normalized based on load durations of one to ten years. If a structure experiences impact load durations, then the highest design factor must be used to ensure safety. From a biological perspective the normal load duration on an organ or individual could vary and depends on the lifetime of the structure. For example, the load duration used to normalize the design factor could be one to ten years (normal load duration) for an annual or biennial plant species, or it could be ten to fifty years (permanent load duration) for a woody perennial species.

Second, a lower design factor can be accommodated by an organ than by an individual. For example, in designing a leaf with a functional life expectancy of one year or less (as in most deciduous trees), a lower design factor may be involved than in designing a leaf that will functionally persist for two years or more (as in many tropical plants or conifers). A woody stem, by contrast, may have a very high design factor, since its potential life expectancy would be high.

Third, whether the absolute design factors engineers use in constructing wooden structures are higher or lower than those that occur during plant growth may be less significant than the relative differences among the design factors for plant organs subjected to different types of stress. Design factors used in industry are subjective and reflect a litigious society.

The slenderness ratio is another example of how a dimensionless ratio can be constructed and used as a design factor. It is the result of two dimensional ratios—the ratio of the length of a column l to the least radius of gyration r, which in turn is the square root of the ratio of the member's second moment of area I to its cross-sectional area A:

$$\frac{l}{r} = l\left(\frac{I}{A}\right)^{-1/2}. \tag{7.33}$$

For a column with a solid circular cross section,

$$\frac{l}{r} = l\left(\frac{\pi R^4}{4 \pi R^2}\right)^{-1/2} = \frac{2l}{R}, \tag{7.34}$$

where R is the radius of the transection. (Provided the second moment of area and cross-sectional area of a member are known, the slenderness ratio can be calculated with ease from eq. 7.33, as shown for a solid terete column in eq. 7.34.) A low slenderness ratio means very high compressive stresses will be required to induce lateral elastic buckling. A high slenderness ratio means relatively low compressive stress may induce failure through global buckling. The lowest permissible slenderness ratio—the ratio at which elastic buckling will occur—is given by the formula

$$K\frac{l}{r} = \left(\frac{\pi^2 E A}{P}\right)^{1/2}, \tag{7.35}$$

where K is the effective length parameter and P is the critical load. When the slenderness ratio is greater than 120, K equals 1. For slender columns, global elastic buckling usually occurs when the critical stress σ_{cr} is equal to or less than 0.5 of the yield strength F_y. Thus, for elastic buckling $\sigma_{cr} \leq 0.5\ F_y$, and the lowest value of slenderness ratio at which elastic buckling will occur is

$$K\frac{l}{r} = \left(\frac{\pi^2 E}{\sigma_{cr}}\right)^{1/2} = \left(\frac{2\pi^2 E}{F_y}\right)^{1/2}. \tag{7.36}$$

The expression to the right of eq. (7.36) is the ratio of the elastic modulus of a material to its yield strength. It reflects the relative likelihood of global elastic versus inelastic buckling and is sometimes referred to as the elastic-inelastic buckling ratio, symbolized by C_c. This ratio can be used to calculate the design factor of any columnar plant organ.

Engineers have standardized the design factor for columns subjected to axial compressive loads by means of both the slenderness ratio and the elastic-inelastic buckling ratio. For columns subjected primarily to axial loadings, with effective slenderness ratios equal to or greater than C_c, the allowable stress σ_a is obtained by dividing the critical stress σ_{cr} by a safety factor equal to 1.92; that is, $\sigma_a = \sigma_{cr}/1.92$. Accordingly, a design factor roughly equal to 2 is used in constructing axially compressed columns.

When similar computations are done for the primary stems of annual plants, a design factor of 2 is frequently found, while for woody perennials a design factor of 3 to 4 is not uncommon. By contrast, the design factor for most tree trunks is 4 or more. This is not too surprising, since the stems of an annual plant can fail mechanically at the end of the reproductive season with relative impunity, whereas the stem holding the crown of a tree must last for many decades. An intriguing feature of the herbaceous stems of annuals is that their design factor may increase throughout the growing season. As tissues mature and subsequently dehydrate as a result of senescence, the modulus of elasticity of these stems increases. In many cases the second moment of area does not change, owing to a stiff outer rind of thick-walled tissue that resists geometric deformation attending the hydrostatic collapse of an inner core of parenchymatous thin-walled tissue (see fig. 7.8D–E). As water is lost from tissues, the weight of the entire vegetative organ diminishes. Therefore the load that must be sustained decreases. The phenology of changes in the flexural stiffness and loading of primary stems and leaves provides insight into why biennials bolt in their second year of life and not in their first. Snow loadings and the limited developmental capacity to repair tissue damage make primary organs in the North Temperate latitudes mechanically short-lived. This will be taken up in chapter 8, when we treat the mechanical architecture of the individual plant.

As mentioned earlier, a similar method for measuring the design factor when constructing a column is to use length ratio; that is, l_{cr}/l, where l_{cr} is the critical length of a column at which elastic deflections are predicted to occur and l is the actual length of the column. The critical buckling length can be calculated from either the Euler column formula for small deflections or the analogous equation from the Elastica for large deflections. When l_{cr}/l is applied as a design factor, it takes the same value as the design factor based on σ_{cr}/σ_a; that is, $l_{cr}/l \geq 1.92$ for axially compressed columns.

From the foregoing, we can readily appreciate that in proportioning columns subjected only to axial compressive loads, the principal concern is to define an average allowable stress. The allowable stress is a remarkably convenient parameter in that, once it is specified, the transverse area A for a given

TABLE 7.3 Experimental Data from *Allium sativum* Flower Stalks Removed from Plants Growing under Three Experimental Conditions

Stem Radius (cm)	Second Moment of Area (10^{-11} m^4)	Elastic Modulus (10^8 N·m^{-2})	Flexural Stiffness (10^6 N·m^2)	Weight of Flower (N)	Stem Length (m)
		Field-Grown Plants (Unprotected)			
0.30±0.02	6.34±1.56	3.55±0.09	2.27±0.1	6.57±0.5	0.507±0.12
		Field-Grown Plants (Protected)			
0.30±0.03	6.27±1.93	3.53±0.08	2.23±7.3	6.57±0.7	0.713±0.15
		Glasshouse-Grown Plants			
0.31±0.02	7.35±1.53	3.56±0.06	2.63±5.7	7.06±1.9	0.86±0.05

design load P is defined; since stress is force per unit area, $A = F/P$.

In the design of a compression member, the maximum loads and the effective length of the member are specified by the engineer, whereas the shape and dimensions of the cross section of the member are unknown and must be determined analytically. In biological systems, however, the situation is reversed: the shape and the dimensions of the structure are known (they can be empirically measured and determined), while the loadings, other than self-loading, are largely unknown. In these circumstances we may wish to standardize the design factors of organs based on the magnitudes of the dynamic loadings experienced against the dynamic loadings that would incur mechanical failure, or based on the loadings that would cause mechanical failure under some applied static load against the loading of self-weight. In either case, we would want to know both the slenderness ratio and the critical buckling length ratio. This is illustrated in the following section.

AN EXAMPLE

Design factor analysis is illustrated for the flower stalks of the garlic plant *Allium sativum*. I selected these stems as an example because their mechanical performance is vital to reproduction and the fitness of the species—they sustain the weight of flowers and new plant propagules (seeds and vegetatively produced plantlets), and the length of flower stalks influences the potential for the long-distance dispersal of these reproductive structures. Another reason for selecting this plant is that genetically identical populations of plants can be produced by the vegetative propagation of bulbs tracing their genetic ancestry to a single plant.

Table 7.3 presents data for flower stalks harvested from three populations of plants grown under three different conditions: plants grown in the field and left unprotected from the force of wind; plants also grown in the field but whose flower stalks were protected from the wind by plastic barriers that were elevated as the stalks grew in length; and plants grown under glasshouse conditions that were watered from below and protected from any mechanical stimulation. The objective of this experiment was to determine the design factor of flower stalks that resulted from growth under these different conditions of wind loading. After the flower stalks were mature, they were harvested and examined morphologically and mechanically to determine their flexural stiffness, stem length, and critical buckling length (by means of the equations of the Elastica).

The data indicate that the lengths of flower stalks varied significantly among the three populations of genetically identical plants, presumably as a result of different degrees of wind loadings and their effect on plant growth. Other parameters, such as the radius of stalks and flexural stiffness, did not appear to vary significantly. From the weight of the terminal flower clusters, stem length and radius, and elastic modulus, we can calculate the average critical buckling length for each of the three populations. From these calculations we can see that the average critical buckling lengths were 91.3 $\pm$ 10.3 cm for the unprotected plants grown in the field and 90.0 $\pm$ 12.7 cm for the protected plants. By contrast, the average critical buckling length for the flower stalks of the glasshouse-grown plants was 95.6 $\pm$ 5.2 cm. Thus we can easily calculate the length ratios for the three populations: for the unprotected, protected, and glasshouse-grown stalks, the length ratios are 1.85 $\pm$ 0.29, 1.29 $\pm$ 0.15, and 1.11 $\pm$ 0.07. These ratios are entirely consistent with a thigmomorphogenetic response. Mechanical perturbation by wind resulted in a motion-induced inhibition in growth in stem length: unprotected stems grew less in length than stems protected from mechanical stimulation. Accordingly, the highest length ratio was found for flower stalks growing in the field that were not protected from wind movements, while the lowest ratio was found for stalks produced by plants grown in the glasshouse and protected from handling.

If we examine these data in terms of the design factor relationships shown in figure 7.13, some striking parallels are evident. For example, figure 7.13 indicates that a design factor of 2 (the maximum design factor) should be used when a columnar (woody) structure is subjected to impact load durations (1 sec), while the design factor can drop below 1.1 for columns that need only sustain their own self-loadings (normal load duration). Although perhaps fortuitous in this respect, the design factor for unprotected, field-grown flower stalks is very near 1.9 based on the length ratio, while the design factor for glasshouse-grown plants is precisely 1.1. Obviously the flower stalks of the garlic plant are not made of wood (secondary xylem); indeed, the bulk of the tissue is parenchyma. Nonetheless, the logic underlying the design factor specifications for columns made of wood appears to correspond to the biological design factor seen in garlic.

Accordingly, based on the length ratio as a criterion for the design factor, we can conclude that thigmomorphogenesis and the determinate growth of flower stalks operate in tandem to produce a very reasonable correspondence between the expected and observed design factors for these organs. But other design factor criteria lead to the opposite conclusion. For example, the slenderness ratios for columns made of steel or wood are not permitted by engineers to exceed 200. True, flower stalks are not made of steel or wood, but these materials are substantially stronger and stiffer than the tissues composing the flower stalks of garlic plants. Thus the slenderness ratios of flower stalks ought to be even lower than those of steel or wood columns. They are not, however. The lowest slenderness ratio that can be calculated from the experimental data set is 338. This value is for unprotected plants. The highest value, calculated for glasshouse-grown plants, is 555, over 2.5 times the minimum safe value. Although one might argue that the consequence of a building's collapsing as a result of the compressive failure of its columns are dramatically greater and require a disproportionately greater investment in reinforcement than the consequences of the buckling of stems bearing flower clusters at their summits, we must acknowledge that the biological data from the garlic experiment fail to reflect the design factors used by engineers. Nonetheless, garlic flower stalks are well built, and even under substantial wind loadings they rarely fail. Perhaps the hydrostatic cellular solids of their tissues hold a lesson for the engineer as well as the botanist.

NONCOLUMNAR PLANT ORGANS

Thus far I have focused on the mechanical behavior of columnar and beamlike plant organs. The reasons should be fairly obvious—we have solid and simple physical analogues that let us derive closed-form solutions from the elementary theory of bending. Also, many important plant organs manifest a morphology and an orientation that are entirely compatible with the mechanical behavior of columns and beams. Many other organs, however, particularly

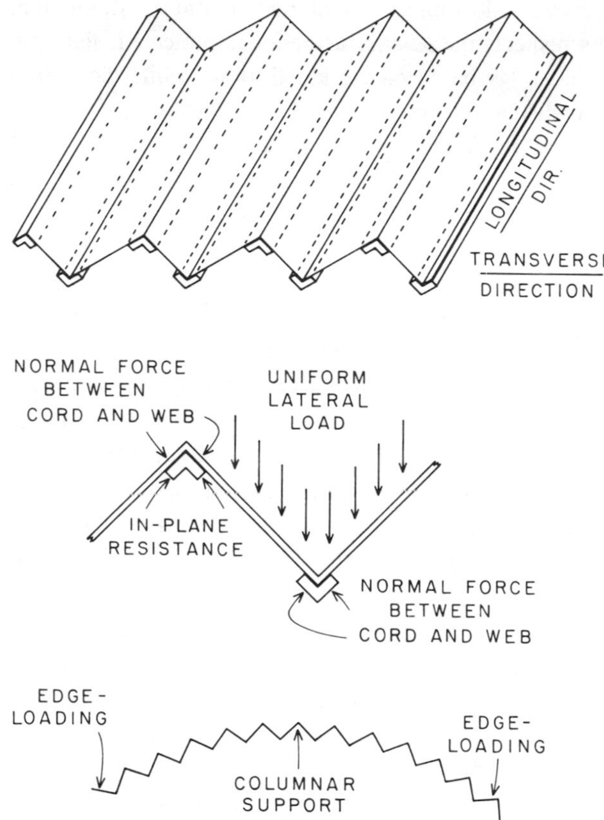

FIGURE 7.14 Cord-web model for some leaves. The leaf can be considered to operate mechanically as a series of longitudinally aligned beamlike cords interconnected by a relatively thin webbing. The uniform lateral loading of the webbing generates normal forces between the cord and web elements. When such a structure is supported by a column, it undergoes edge loadings that cause its unsupported span to deflect. The "column" could be a more robust "midrib" running the length of the corrugated cord-web shell; the edge loadings could simply be the weight of the cord-web structure on either side of the midrib.

photosynthetic leaves and stems, are flattened in one plane. This dorsiventrality in large part reflects their photosynthetic function, which typically requires a large surface area for light interception and gas exchange. Indeed, when these biological functions of photosynthesis and mechanical support tasks are examined in terms of their design requirements, some very simple calculations reveal that there are only two design compromises—cylindrical columns and flattened spheroids (see chap. 10). Since we have treated many of the mechanical aspects of the former, we are obliged to examine the latter here.

From the perspective of material properties, flattened stems and leaves can be modeled as *n*-phase composite materials. We saw how this can be done for cylindrical organs when core-rind and *n*-phase models were treated earlier in this chapter. In principle, a core-rind model can also be used to study flattened leaves or stems. The epidermis, however it is spatially configured, is a rind, while the mesophyll (the parenchymatous, often spongy ground tissue of the leaf) or the hydrostatic tissues within a stem can be modeled as the core. Just as in cylindrical organs, if the parallel or otherwise oriented vascular bundles in a dorsiventral organ are mechanically important, then the core-rind model can be modified or a three-phase model can be used to examine mechanical attributes. When spatulate organs are considered, however, it is their geometry that presents problems, since we must seek some physical analogue in terms of shape that provides us with closed-form solutions.

Although a number of physical analogues should always be explored, two are immediately obvious—folded plates, and membranes with heterogeneous compositions (fig. 7.14). If these are selected as the physical analogue, then flattened leaves and stems can be treated as stress-skin panels, where the thin facings (made of epidermis) are rigidly bonded to stringers (thick-walled tissues like sclerenchyma, or thick-walled cellular components of tissue systems like the fibers found in vascular bundles) that interconnect the external surfaces. A stress-skin panel construction is a particularly apt model when the organ has numerous gas-filled chambers, as in the spongy mesophyll.

An exceptionally interesting example of how a dorsiventral leaf can be mechanically modeled is given by Gibson, Ashby, and Easterling (1988), who treated the leaf of the common iris as a sandwich-laminate beam with fiber composite faces separated by a low-density foam core. Their model essentially operates along the same lines as a core-rind model, but the rind is a stressed skin and the overall geometry is not columnar. Fabricated sandwich laminates have a very high specific stiffness because the faces, separated by a lightweight core, confer a large second moment of area with little increase in overall weight. Modern skis have this construction—they are very light and resilient for their overall size. Gibson, Ashby, and Easterling (1988) were able to predict the bending stiffness of iris leaves with substantial success; the average stiffness measured for four leaves was $0.465\ \mathrm{N \cdot mm^{-2}}$, and they had calculated an average stiffness of $0.603\ \mathrm{N \cdot mm^{-2}}$. Thus it is worth examining their modeling procedure to see how it could be applied elsewhere.

The bending stiffness of a sandwich beam is measured from load-deflection experiments. A load P is placed at the end of the plant organ and the deflection

δ at the tip is recorded. The slope of the plot of load versus deflection is the bending stiffness P/δ. The bending stiffness is dependent upon the beam length l and beam width b, the thickness t of the face on either side of the beam, and the thickness of the inner core c, as well as the flexural stiffness EI of the sandwich beam and the longitudinal shear modulus G of the core:

(7.37) $$\frac{P}{\delta} = \left(\frac{l^3}{3EI} + \frac{1}{bcG}\right)^{-1}.$$

If the sandwich beam is composed of relatively thin faces and has a compliant core, then

(7.38a) $$EI = \frac{E_f\, b\, t^2\, c^2}{2},$$

where E_f is the elastic modulus of the material composing the faces. If the faces are relatively thick, however, then

(7.38b) $$EI = \frac{E_f\, b\, t^3}{6} + \frac{E_f\, b\, c(c+t)^2}{2} + \frac{E_c\, b\, c^2}{12},$$

where E_c is the elastic modulus of the core. Gibson, Ashby, and Easterling (1988) assumed that the value of the shear modulus of the core was half the value of the elastic modulus of the core. Recall from chapter 3 that for an isotropic material like parenchyma (of which the mesophyll of the iris leaf is composed), the relationship between the Poisson's ratio and the elastic and shear moduli permits this assumption. The exact values of the elastic moduli of the core and the rind (the faces) must be known before the bending stiffness can be estimated theoretically from eq. (7.38). This is very difficult to determine, however, particularly since surgical manipulation of the tissues within the leaf could affect empirical measurements and since elastic moduli for hydrostatic tissues like parenchyma depend on the tissue water content. Gibson and colleagues elected to use what is known as the rule of mixtures to estimate the elastic moduli of the core and the rind. The rule of mixtures assumes that the elastic modulus of a composite material equals the sum of the product of the elastic modulus of each component material and its volume fraction V; that is, it assumes that the Voigt model holds true (see eq. 7.16). In the particular case of the iris leaf, they reasoned that the elastic modulus of the cell wall was of paramount importance. Thus, if the elastic modulus of cell walls is known, then this value times the volume fractions of the cell walls in the core and in the rind equals the elastic modulus of the core and the rind. With the aid of computers and digitization software, the volume fractions of cell walls in a given tissue type are easily measured, but these workers approximated the volume fractions of cell walls in the iris leaf core and rind. (It is likely that their estimates are responsible for the difference in expected and observed bending stiffnesses.)

The foregoing approach to modeling flattened leaves or stems has great potential for broad application. Although monocot leaves like those of the iris have a different vascular configuration than dicot leaves (most monocot leaves have parallel arrays of vascular bundles that run the length of the leaf, whereas dicot leaves have a robust midvein of vascular tissue and secondary venation that diverges into a reticulated vascularization throughout the rest of the leaf lamina), the flattened leaves and stems of many plant species can be treated as sandwich-laminate beams.

There are other approaches that differ in degree and kind from that of Gibson, Ashby, and Easterling (1988). Many leaves can be viewed as a curved shell or folded plate consisting of cords spanned by webs. The cords provide the principal means of resisting bending moments, while the webs would resist in-plane shearing (fig. 7.14). This model would be most appropriate for leaves with parallel arrays of vascular bundles, where the sclerenchymatous or phloem fiber components act as cords spanning the length of the leaf.

Fabricated folded plate structures are typically curved constructions that are relatively thin in transection. They resist low-intensity forces applied over their relatively large surface areas, while their principal, if not distinctive, feature is that lateral loadings are resisted primarily by in-plane membrane forces (which develop within the webbing) rather than by the usual bending forces typically seen in structures made up of columns or cantilevered beams. Curved folded plates have the mechanical advantage of supporting lateral loads by arch compression—a curvilinear deformation spanning the width and breadth of the structure in which all components share the distribution of strains. In-plane shear resistance either supports the loadings directly or redistributes stresses as the plate distorts. In terms of biological analogues, the palmately compound leaves of many palm species and the large, simple leaves of some dicots (particularly those in the water lily family, Nymphaeaceae) have a mechanical construction analogous to a large folded plate. A particularly interesting example of a leaf constructed as a folded plate is that of *Corypha umbellata,* a species of fan palm that produces the world's largest leaf. The leaves of this species sustain remarkably large self-loadings and rarely buckle unless dynamically stressed by wind pressure. Even when they mechanically fail by means of buckling, they do so with minimum loss of function.

A folded plate can be arched in one or two directions. When it is arched in only one direction, the suspension action can be used to support loadings in only one direction and provides little support along its free edges (those perpendicular to the direction of arching, much like a grass leaf lamina far from its curved sheath). Plates that are curved in two directions are much less sensitive to the presence of holes or discontinuities or loadings placed toward their edges. The difference in the mechanical behavior of folded plates arched in one or two directions is significant in both qualitative and quantitative ways. In both cases, however, engineered folded plates are typically constructed of cords and webs, giving them a corrugated appearance. The mechanical interactions among the cords and webs can give insight into how leaf lamina operate mechanically.

For example, the corrugations in a folded plate provide one-way bending strength and rigidity that precludes overall buckling provided the length of the folded plate, parallcl to thc longitudinal axis of the cords, is not great. This mechanical morphology is very like that of grass leaves. The anatomical elements within these leaves that confer the one-way bending rigidity are sclerenchyma fibers beneath the upper and lower epidermis. These cords are aligned parallel to the longitudinal axis of the leaf and flank more centrally positioned, parallel aligned vascular bundles. The mesophyll in these leaves is fairly homogeneous in anatomy and probably isotropic in behavior. Together with the leaf epidermis, the mesophyll provides the webbing in the corrugated architectural design.

In an examination of the perennial English rye grass *Lolium perenne,* Vincent (1982) found that 90% to 95% of the stiffness of leaves measured along leaf length could be attributed to the sclerenchymatous fibers (the cords) even though this tissue has a volume fraction of only 4.24%. The volume fraction of vascular bundles is comparable to that of the sclerenchyma fibers (4.12%), yet Vincent reports that the bundles contribute very little to the rigidity and strength of leaves. Vincent (1982) modeled the leaves as a composite material with a longitudinal modulus E_L that equals the sum of the transverse modulus E_T and the product of the elastic modulus and the volume fraction of each tissue, according to the rule of mixtures:

$$E_L = E_T + (EV)_{vb} + (EV)_{sf}. \tag{7.39}$$

In this equation, which is a modification of the Voigt model (eq. 7.16), the transverse elastic modulus reflects the material properties of the webbing within the leaf (epidermis and mesophyll), while $(EV)_{vb}$ and $(EV)_{sf}$ are the contributions made by the vascular bundles and by the sclerenchyma fibers (the cords). The transverse modulus, as reported by Vincent, is 14.1 $MN \cdot m^{-2}$, while the longitudinal modulus is 554 $MN \cdot m^{-2}$. Since the volume fractions of the cords in this type of leaf are very low, the elastic moduli of these materials must be very high. Indeed, the elastic moduli of the sclerenchyma fibers and vascular bundles are 22.6 $GN \cdot m^{-2}$ and 0.838 $GN \cdot m^{-2}$, respectively. In fact, the mature leaves of many species of grasses are very tough and difficult to fracture. The average stress-intensity factor for *Lolium perenne* is on the order of 10^5 $N \cdot m^{-3/2}$, while the specific work of fracture is on the order of 10^1 $J \cdot m^{-2}$ (see Vincent 1982, 860, table 2).

A more recent study of the same grass species has added further insight. Greenberg et al. (1989) have shown that the leaves have different tensile properties depending on the test strain rate used and the location of the tissue sample with respect to the length of the leaf. As the strain rate is increased, the measured stiffness, toughness, and strength increase, while the ductility of leaves decreases. These findings are typical for polymeric materials (Hertzberg 1983) and reflect the behavior of viscoelastic materials in general. Also, the distal portions of leaves were shown by Greenberg and colleagues to be stiffer, stronger, and tougher than more proximal portions. This should come as no surprise, since most grass leaves grow by means of an intercalary meristem at the base. Thus older tissues are found at the tips of leaves, while younger, less mature (and thinner-walled) tissues are found toward the base. The result is that grasses expose the tougher and stronger portions of their leaves to grazing animals and lawn mowers.

Greenberg et al. (1989) also demonstrated that the mechanical properties of grass leaves could be correlated with the combined volume fractions of the epidermis and vascular bundles that increase toward the tips of leaves. Thus the epidermis (which was neglected in Vincent's analysis) appears to contribute significantly—presumably it could act as a load-bearing membrane. Some care must be exercised in interpreting correlative relationships, however; statistical correlations need not reflect a cause-and-effect relationship. Because of the way grass leaves grow, the volume fraction of the epidermis will be highly correlated with how fully vascular bundles and sclerenchyma fibers mature. Thus some correlations among anatomy, mechanical properties, and the location of tissues along the length of leaves must exist a priori. Nonetheless, when taken together, the work of Greenberg et al. (1989) and Vincent (1982) has demonstrated a strain-rate dependency of the mechanical properties of grass leaves and shown that these properties correlate with the volume fractions of different tissues within a plant organ.

As I mentioned earlier, the mechanical behavior of folded plates is influ-

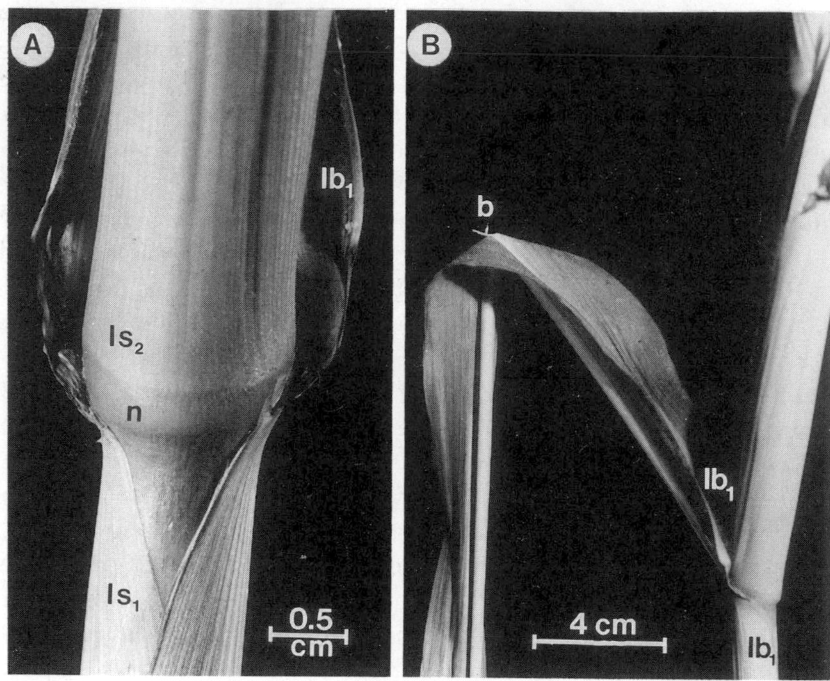

FIGURE 7.15 Morphology (A) and mode of mechanical failure (B) for corn leaves (*Zea mays*). (A) Leaf blade (lb_1) and clasping leaf sheath (ls_1) around internode of stem; the leaf sheath of the next higher leaf (ls_2) completely envelops the next higher internode of the stem. (B) Buckling of the leaf blade shown in A. Tearing of the leaf blade exposes ruptured vascular bundles (b).

enced by the way they are curved. For any given transverse curvature, the critical bending moment decreases as the length of the plate increases. Also, for any given ratio of length to width, the bending moment increases as the transverse curvature increases. These relationships are a direct consequence of the second moment of area, which is larger for a curved transverse section than for a flattened plate with equivalent thickness. Thus a shell rolled into a hollow tube or with its lateral margins bent toward the longitudinal axis is much more rigid than a flattened plate made of the same material. These tendencies are biologically expressed by a variety of monocot leaves, such as those of grasses. The basalmost portions of grass leaves are recurved in the direction of the stems they are attached to and tend to unroll along their length toward the tip. At some point the transverse curvature drops below a critical lower limit, and the weight of the more distal portions of the leaf is sufficient to induce a bending moment that can buckle the leaf (fig. 7.15). The moment can be a simple bending moment, or it can also involve torsion. Corrugated leaves can resist larger bending and torsional moments than their uncorrugated counterparts because of their capacity to resist in-plane shearing effects. These properties can be easily appreciated by cutting long strips of paper with equivalent ratios of length to width. A flat strip of paper bends easily when it is cantilevered into the horizontal direction. If its base is pinched to form a localized V-shaped region, the same strip of paper becomes stiffer and bends less easily under its own weight. If the paper is corrugated into a fan, it can sustain its own weight plus some substantial end loading before it buckles. The corrugated leaves and leaflets of palm species have discovered these mechanical principles. Similarly, the V-shaped bases of petioles of many dicots essentially capitalize on regionally high second moments of area for structural reinforcement. By contrast, the more distal portions of petioles can have a more or less elliptic cross-sectional geometry, which is particularly good at sustaining torsion when leaves flutter under dynamic wind loadings.

Exact mathematical solutions for the mechanical behavior of folded or flattened plates are very complex, though with the advent of high-speed computers numerical solutions can easily be obtained. Crude approximate solutions can be achieved with some simple, well-known equations of statics. Thus the conditions of static equilibrium for a cord-web folded plate are relatively easy to describe. The web of a folded plate transmits the uniform lateral loadings of the plate from one cord to another. The net reactions of the webbing on the cords are resisted by the entire plate acting on the parallel array of cords as if they were elastic beams. The cords resist bending moments by longitudinal thrustings into the webbing, and the webs resist in-plane shearing. Yet even though we can model a leaf as a folded plate, the mechanical behavior of this system still involves the simple relations reviewed in chapter 3 in terms of beams.

The following equations are useful in the analysis of simply supported folded plates:

$$EI = \frac{E\ t^3}{12\ (1 - \upsilon^2)}, \tag{7.40a}$$

where t is the thickness of the plate, and

$$\delta_m = \frac{5\ g\ l^4}{384\ EI} + u + \frac{2(l\ e)^{1/2}}{\pi}, \tag{7.40b}$$

where δ_m is the midspan deflection, g is the uniform lateral loading per unit area, l is the span length (the shortest dimension of the plate), u is the mid-span lateral deflection before buckling, and e is the longitudinal deflection (the distance from one edge to another). Also,

$$T = \frac{\pi^2 \alpha EI}{l^2}, \tag{7.40c}$$

where T is the membrane tension developed within the webbing and α can be obtained from the numerical solution of the following relation:

$$\alpha (1+\alpha)^2 + \frac{12l}{\pi^2 t^2}\left(e + \frac{\pi^2 u^2}{4l}\right) = \frac{3\delta_m{}^2}{t^2}. \tag{7.40d}$$

The bending moment M at the midspan is given by the formula

$$M = \frac{g\, l^2}{16}\left[\frac{1 - \sec h(\pi^2 \alpha/4)}{\pi^2 \alpha}\right], \tag{7.40e}$$

while the maximum combined unit stress σ_u is given by the formula

$$\sigma_u = \frac{S}{t} + \frac{6M}{t^2}. \tag{7.40f}$$

Equations (7.40a–f) provide a crude method for analyzing static equilibrium in simple folded (corrugated) plates. Nonetheless, one can readily appreciate that the conditions of equilibrium are complex and depend on a number of parameters. One interesting feature emerging from these equations is worth noting here. Collectively, the equations for static equilibrium show that the optimal number of cords is dictated by the ratio of cord diameter to the distance between neighboring cords. The optimal ratio is 10. Thus the distance between neighboring cords is predicted to be thirty times the average width of the cords. Fewer cords would reduce the stiffness of the plate; more cords would disproportionately increase the weight of the plate compared with the gains in stiffness. If cords are the physical analogues to either sclerenchymatous fibers or vascular bundles or both, then the extent to which leaf laminae are optimally reinforced could be assessed by means of some very simple morphological and anatomical investigations. Clearly, however, the vascular bundles in leaves have other functional roles, such as translocation of liquids (water and sap). These roles must have their own design requirements, which may or may not be antagonistic to structural optimality. Both structural support and hydraulics require some level of redundancy in case a load-supporting member or hydraulic conduit fails, particularly from additional dynamic loadings imposed on a leaf by the external environment.

Unfortunately, the mechanical behavior of flattened leaves and stems under dynamic loadings is very difficult to predict based on theory. In part this is because the material properties of plant tissues depend on the strain rate and because the shape of organs, particularly very flat ones, can deform when they are subjected to wind or water movements. With low strain rates, plant tissues remain within the proportional limits of their loadings and behave elastically. With high strain rates, strength tends to increase but the tensile stresses experienced may exceed the breaking stress of the organ's materials. Also, under conditions of high dynamic loading by wind, tissues may dehydrate before stomatal closure occurs. Thus the elastic modulus of wind-stressed leaves may be temporarily reduced, leading to mechanical failure under loadings that would otherwise be easily resisted. The dominant mode of mechanical failure, however, is likely to be shearing, particularly when strain rates are very high, as when wind speeds and direction vary over short time intervals. Geometric deformations can help in this regard. Simple leaves and leaflets on compound leaves tend to reorient their longitudinal axes in the direction of water and wind flow, as well as folding upon themselves. Taken together or separately, flagging and lateral collapse reduce the projected surface areas of leaves, thereby reducing drag and the loadings that must be sustained. Harmonic motion can also be helpful. If the leaflets of a compound leaf oscillate at different natural frequencies, then their collective movements can dampen the system as a whole. That is, the kinetic energy within the system can be quickly dissipated. Dampened systems are generally more dynamically stable and can resist greater loadings than can otherwise equivalent undampened systems.

VINES AND TENDRILS

Many plant species grow vertically by anchoring themselves to a mechanically supportive substrate. These species are said to possess the vine growth habit. The biomechanical properties of vines illustrate a number of form-function relations that are much less evident when we examine the biomechanics of herbs, shrubs, or trees, although the basic engineering principles that have been reviewed thus far apply equally well to all plant species. Among these properties are the growth dynamics and biomechanics of making contact with a mechanically supportive surface, called the climbing phase, and the biomechanical properties of the organs that anchor the vine to the surface it ascends. This section will examine these two aspects of the vine growth habit.

A vine in the climbing phase of its life history was described by Julius von

Sachs (1875; see Goebel's 1887 translation of this work) as possessing a normally etiolated morphology. Etiolation occurs in plants that have been deprived of adequate light for normal growth. It is usually characterized by a marked increase in the lengths of internodes, a reduction in how far leaf lamina expand, and a temporary suppression of the growth of lateral branches. Referring to the climbing growth habit of vines, however, normal etiolation means that lack of light reaching a vine seedling or mature individual is not the proximate cause for these morphological features. Rather, the internodes of vines that are not in contact with a support surface are usually much longer than those that develop after contact has been made, while there is a marked if temporary suppression of lamina expansion and of the primary growth of lateral branches. Intuitively, we can see that the probability of making contact with a suitable substrate increases if the actively growing distal internodes of a vine are extended farther. Also, from first principles, engineering theory indicates that reducing the mass of lateral appendages such as leaves and branches means the unsupported portion of the vine will bend less under its own weight. Indeed, the leader (the distal portions of a plant shoot) of many vine species is extended into a whiplike organ. Studies by Troll (1937) and Courtet (1966) reveal that the anatomy of the unsupported leaders produced by many vine species also helps maintain vertical support. Leaders typically have a large centrally positioned mass of thin-walled ground tissue (the pith) surrounded by densely packed relatively thick-walled tissue, while the older portion of the shoot in contact with a support is typically more woody. The hydrostatic mechanical properties of the core-rind anatomical configuration of the distal leader can be inferred from our previous discussions, but it is worth noting that the core-rind configuration of the leaders of many vines may play a role both in hydrostatically sustaining unsupported weight and in the circumnutation of leaders. In circumnutation the leaders of vines typically sweep through the air and grow vertically by some endogenously driven growth rhythm until they make contact with a support. The ascending spiral resulting from circumnutation can have a radius of search ranging from a few centimeters to as much as a meter, while the average period of sweep can be as long as one to two hours, although environmental and interspecific differences are pronounced. The mechanism of circumnutation may vary from one species to another, but the periodic elastic changes in the volume and shape of thin-walled cells within the cross-sectional anatomy of the leader theoretically are the most efficient, since these changes are recoverable over relatively short periods, permitting cells to deflate when the cells on the opposite side of the shoot expand. Inflation of thin-walled cells on one side of the leader would cause the shoot to bend in the opposite direction. Ideally, the hydrostatic cells responsible for circumnutation should be farthest from the centroid axis of the shoot. It is worth noting that circumnutation attends axial elongation. Therefore periodic but asymmetric elongation of cells could account for the phenomenology of "sweeping." Millet, Melin, and Badot (1986) have reported differences in the osmotic potential of epidermal cells on the concave and convex surfaces of circumnutating leaders. Nonetheless, the physiological mechanism responsible for propagating the differential osmotic potentials in epidermal cells remains unknown.

Once contact has been made between a leader and a support, vines anchor to the support in a variety of ways. Among the most common devices are twining of the entire shoot, production of adhesive adventitious roots, and twining of lateral appendages (lateral branches and leaves), collectively called tendrils. Although the mechanisms of climbing and twining around a support attracted the attention of botanists well before the time of Charles Darwin, who was himself interested in this topic, comparatively recent studies by Silk (1989a,b) and Putz and Holbrook (1991) have contributed much to our current understanding. Emerging from these studies is the notion that climbing stems and tendrils must generate a force operating perpendicular to the support surface to maintain the anchorage of a vine or clasping organ by friction. Evidence of this perpendicular force is the capacity of some twining vines to crush hollow paper tubes and to coil tighter when removed from their vertical supports. Putz and Holbrook (1991) investigated the importance of friction in climbing by modeling a shoot coiled around a cylindrical support as a loosely coiled, frictionless spring. How far this model's predictions deviated from observation could be used to estimate the importance of friction to climbing vines. In their model, the cylindrical support prevented the lateral buckling of the vinelike spring, which could slide downward under self-loading only by a collapse of its helical gyres. Clearly, the mechanical properties of a vertically oriented spring depend on its elastic and shear modulus, the unit mass of the spring, the vertical ascent angle, and the radius of curvature—the last being a function of the diameter of the cylindrical support. The elastic and torsion moduli are influential because they define the resilience of the spring and the strain energy stored within it as a result of vertical compression on self-loading. The unit mass of the spring defines the magnitude of the self-loading, while the vertical ascent angle and the radius of curvature influence the bending moment imposed upon the spring. Putz and Holbrook report that frictionless springs with ascent angles less than 70° should significantly condense in length, whereas *Dioscorea bulbifera* can grow with vertical ascent angles as small as 33°. The discrepancy between the model and reality indicates that friction is important. As with the potential role of cellular turgor pressure in circumnutation, the anisotropic inflation of a climbing shoot could account for the inward force generating friction. Since there is an intrinsic relation between the radius of curvature of a spring and its mechanical stability, there must be a relation between the mechanical stability of a climbing vine and the diameter of its supporting member. Putz and Holbrook (1991) report that vines of *Dioscorea bulbifera* grown on small-diameter poles bend sideways when they are removed but retain their original helical geometry, whereas vines removed from large-diameter poles collapse vertically under their own self-loading.

Similar mechanical relations are envisioned for tendrils that wrap around supporting members and generate inward frictional forces. The work that must be performed to pull a climbing stem or a tendril from a supporting member must be proportional to the change in curvature of the springlike shoot or tendril. Since the curvature of the springlike structure is inversely proportional to the radius of curvature, the work expended in dislodging a coiled shoot or tendril increases as the diameter of the supporting member decreases. This can be seen from a formula (devised by Hibbeler 1983) and applied by Putz and Holbrook (1991) to the mechanics of tendrils:

$$F = f^* e^{\mu\phi}, \tag{7.41}$$

where F is the tensile force necessary to pull a limp rope over a support surface, f^* is the opposing tensile force resisting the pull, μ is the coefficient of friction between the rope and the support surface, and ϕ is the total angle of contact (in radians). The shoots and tendrils of vines are not limp, of course, as are ropes and pulley belts, but eq. (7.41) shows that there is an intrinsic relation between the degree of curvature and the tensile force required to pull a clasping, springlike structure from its supporting surface. For any given length of the spring, the total contact angle ϕ (= the total angular wrap) is a function of the radius of curvature; ϕ = spring length/radius of curvature. Thus, as the radius of curvature decreases, ϕ increases and more work must be expended. Equation (7.41) also reveals the importance of the magnitude of the inward friction generated by a clasping, springlike organ—as the coefficient of friction increases, the tensile force F must increase. When tendrils and climbing shoots make contact with the surface of a support member, they often expand laterally and become elliptical in cross section. This increases the contact between the surface area and the supporting member, thereby increasing the friction that develops when tensile forces are applied along the length of the clasping organ. Also, when tendrils coil around a supporting member, the lateral surfaces of their gyres make contact, further increasing the friction developed when they are placed in tension. Additional refinements in clasping biomechanics are seen when tendrils are examined anatomically. Clasping shoots and tendrils become lignified, and their breaking strength increases compared with that of unsupported shoots and tendrils. Indeed, tendrils become more rigid when they are placed in tension (Brush 1912).

One further interesting feature of tendrils relates to their possible role as shock absorbers when vines are placed in dynamic loading. Portions of tendrils not appressed to a supporting member are frequently coiled. These free coils can contract, bringing the subtending stem of a vine closer to its supporting member. But free coils can also absorb and dissipate energy when the vine is subjected to wind pressure. The extension of free coils requires strain energy, while the resumption of their original geometry, once the dynamic load is removed, elastically dissipates this energy. We are only beginning to appreciate the remarkable biomechanical attributes of vines and tendrils. Further research in this fascinating area (along the paths already laid out by Silk, Holbrook, Putz, and others) is expected to yield many additional findings.

Eight
The Plant Body

> A biologist, regardless of his line of specialization, cannot afford to lose sight of the whole organism if his goal is the understanding of the organic world.
>
> Katherine Esau, *Plant Anatomy*

Implicit throughout the previous discussion is the notion that cells, tissues, and organs function within a larger biological context—the individual plant—whose manner of growth makes it convenient to define various levels of organization but whose biology reflects a single functional entity. From this perspective, the previous chapters are thus nothing more than a pretext, a pedagogical format, for approaching the individual plant body. In this chapter my task is to redress the prior reductionist perspective and achieve some synthesis among most of the topics covered earlier.

The task is by no means simple. Analyzing an organism requires a precise and accurate description of internal structure and organic form, which, as we have seen, is remarkably intractable even at the level of the fabric of a single tissue or the shape of a single organ. Indeed, the relatively few cell cycles required to construct a mature organism from its zygote generate remarkably large numbers of cells whose arrangements and interrelationships are extraordinarily complex. For example, notwithstanding the need to replace cells over its lifetime, the human body is fabricated from the products of not more than fifty cell cycles that result in over 10^{15} cells, while the transition from a fertilized egg to a mature oak tree may involve only sixty cell cycles with a resultant 10^{18} cells. As a consequence of this progressive increase in size and cell number, both the individual plant and its immediate environment change. A seed germinates underground and experiences compressive and shear stresses as it grows through the soil; a sapling exerts itself through gradients of light, humidity, and wind pressure; a mature tree deals with the historical legacy of mechanical fatigue in its support tissue and with the physiological burden of a potential for reduction in the volume fraction of photosynthetic tissue. No single physical analogue or metabolic model can be used as a proxy for this type of organism, nor can any single environmental factor be given priority throughout the lifetime of an individual plant.

Our task can be simplified somewhat if we focus on a few of the critical aspects of organic form while dispensing with all but the most essential environmental factors. In terms of organic form, we have at our disposal a few mathematical tricks, among the most useful being a dimensionless ratio, which can treat some aspects of shape as independent of size. Dimensional analyses of plants is also a goal, since several important mechanical parameters, like stress and strength, depend on size. As biologists we recognize that changes in shape typically attend changes in the absolute size of an organism, and we are obliged to examine how mechanical and physiological features scale to changes in both shape and size. Indeed, if a similar or identical level of mechanical performance is to be maintained, it becomes evident that the shape or material properties of a structure, or both, must change as size increases. This principle, called the principle of similitude, was originally articulated in 1638 by Galileo in his *Discorsi e dimostrazioni mathematiche* and is as applicable to biomechanical devices as to machines. Indeed, the logic of Galileo provides us with a rebuttal to the notion that all plants can grow indefinitely in size without incurring mechanical liabilities.

The literature treating the mechanical design of the entire plant body is sadly limited and evinces a bias toward mature, arborescent plants. True, trees are of great economic and ecological importance, and in many habitats arborescence is the dominant growth form. But trees do not spring fully formed from the brow of Zeus or the pen of Joyce Kilmer. They grow from seeds and undergo numerous morphological, anatomical, and physiological modifications as they mature. Nor are all terrestrial plants arborescent. Shrubs, vines, and herbs are ecologically and economically important as well. Indeed, the grasses are perhaps the single most important group of plants, yet comparatively little is known about their biomechanics. Also, if our objective is to uncover the underlying principles of plant mechanical design, then a narrow view stemming from the notion that the plant body is a tree cannot be tolerated. Accordingly, this chapter will treat a variety of plant growth forms. Although the paradigm remains the vascular sporophyte, the growth of seedlings and herbaceous plants that lack secondary growth, as well as the more obvious arborescent sporophytes that dominate the vegetative landscape and the biomechanical literature, will be given equal attention. In our efforts to be as comprehensive as possible, we will also compare and contrast the vascular sporophytes of monocots, dicots, gymnosperms, and pteridophytes.

As in other chapters, we will start with what appear to be first principles and ask the question, Can the form of a vascular plant body be deduced from theoretical considerations of engineering principles?

The Form of the Plant Body

It is of no little consequence that most large and sedentary plants, regardless of their evolutionary and systematic affiliations, have converged on a modular organographic construction involving more or less cylindrical axes used for mechanical support and flattened surfaces used to intercept sunlight. The plant bodies of macroscopic algae and of the nonvascular and vascular plants have a modular construction. That is, their gross morphology appears to be constructed by reiterating only a few different kinds of parts, which we refer to as organs. Among the algae these parts are called stipes, holdfasts, and fronds; among nonvascular land plants, like the mosses, they are axes, rhizoids, and phyllids; and among the vascular plants they are called stems, roots, and leaves. The modularity conferred by meristematic growth provides a redundancy of organs, a superb safety factor against the loss of parts owing to inclement weather, disease, or herbivory. The redundancy in plant parts can be truly impressive. A large American elm typically bears several million leaves at a time, and it is not uncommon for a single leaf of *Acacia* to produce over five thousand leaflets.

Regardless of their developmental origins, the organs of all manner of plants converge morphologically and anatomically when they fulfill the same function. Likewise, the gross morphology and internal structure of all large plants are convergent. True, there are numerous exceptions. The spherical plant body of *Volvox,* the filamentous strands of *Ulothrix,* and the subterranean gametophytes of *Lycopodium* and *Psilotum* are but a few examples of plant constructions that lack evident analogues to leaf, stem, and root. Many more could be named, but when the morphologies and anatomies of most plant species are cataloged and sorted, it becomes evident that there exists a predominant organographic ground plan for the plant body, regardless of whether we speak of aquatic or terrestrial, vascular or nonvascular, extinct or extant plants. It is not surprising, therefore, that many have attempted to formulate the design principles dictating organic form and structure in general and those of plants in particular.

One of the earliest workers to treat the general design principles of organisms was Nicolas Rashevsky (1899–1972). Rashevsky formulated what has become known as the principle of adequate design: "The design of an organism is such that the organism performs its necessary functions adequately and with a minimum expenditure of energy and material both in the performance of the functions and in the construction of the organism" (Rashevsky 1973, 146). To be sure, this definition contains an element of circular reasoning ("performs its necessary functions adequately"). But Rashevsky's principle emphasizes two very prevalent notions in the current ecological and physiological literature: that every organism must perform a number of biological functions simultaneously, each of which must be performed "adequately" for growth, survival, and reproduction; and that the compromises that must be reached among these functions should involve a "minimum expenditure of energy." From an engineering perspective, how well a system like a plant achieves a balance among all its various design components, such as mechanical support, hydraulics, gas exchange, and light interception, can be treated by optimization theory, a branch of mathematics that examines how interconnected functions can be collectively maximized. With the aid of optimization theory, Rashevsky's view that an organism must "perform its necessary functions adequately" and that organisms should perform their various functions "with a minimum expenditure of energy and material" escapes its logical circularity. Unfortunately, however, the completion of such an optimization analysis has escaped even the most enthusiastic researchers. Nonetheless, Rashevsky provided a preliminary draft in terms of a simple exploration of the principle of adequate design for trees. Since his derivation is relevant to some of the topics treated later in this chapter, we will reconstruct some of the broad features of Rashevsky's reasoning here. (We will return to the issue of the general design principles of the plant body in chapter 10, where the topic of plant evolution will be treated in greater detail.)

Rashevsky argued that the branching nature of the vascular sporophyte is a result of the metabolic constraints imposed on a terrestrial plant by its sedentary existence. Although the entire plant body cannot move as a unit, the branching of stems provides for an increasing number of photosynthetic organs (leaves) that fill in available aerial space, while the branching of roots permits exploration of an ever larger subterranean space containing minerals and water. Thus, branching relieves a plant of some of its sedentary restrictions. He reasoned that the mass of the whole plant body M is given by the formula

$$M = \pi \rho \, (l_o \, r_o^2 + n \, l \, r^2) \, , \tag{8.1}$$

where ρ is the average tissue density, l_o and r_o are the length and radius of the trunk, and l and r are the average length and radius of n number of branches. As Rashevsky cautions, however, eq. (8.1) is a very crude approximation and provides little geometric information about the plant. With a few more assumptions, however, his formulas take on more definite form. For example, if q denotes the average rate of metabolism per unit mass, then the total metabolic rate qM of the plant can be expressed by the formula

$$qM = k\,n\,l\,r\,, \tag{8.2}$$

where k is a proportionality factor that contains the average tissue density and has units of $g^2 \cdot cm^{-5} \cdot s^{-1}$. Notice that r is not squared, since Rashevsky assumes that metabolism is related to total leaf area, which in turn is proportional to the area of all n branches. In terms of mechanics, Rashevsky argues that branch length is determined by branch radius, lest branches break under their own weight. Since the weight of a branch depends on the average tissue density, branch length must be some function of branch radius and ρ:

$$l = f(r, \rho) \tag{8.3a}$$

and

$$l_o = f_o(r_o, \rho, M)\,. \tag{8.3b}$$

The total mass of the plant figures in eq. (8.3b) because the trunk has to support the entire plant. Similar considerations provide comparable formulas for branch and trunk radii in terms of plant metabolism:

$$r = f_1\left(\frac{qM}{n}\,,\ r\right) \tag{8.4a}$$

and

$$r_o = f_1(qM, r)\,. \tag{8.4b}$$

That is, the flow of metabolites through the average branch and the trunk is a function of branch and trunk radius. (Leonardo da Vinci had much the same idea, as kindly pointed out to me by Professor Enzo Macagno. Apparently da Vinci used the word *grosseza* ("size") to mean cross-sectional area when discussing the volume of water flow through pipes and trees; see Macagno 1989.)

Rashevsky's derivation, thus far, consists of six equations and six quantities, five of which determine the shape of the plant body (l, l_o, r, r_o, and n). A small trunk length and large trunk radius together with a small number of branches and large branch length yield a tree very much like a crab apple, whereas large values for all five parameters yield trees that look like oaks, maples, or ashes. The fundamental question, however, is What dictates the magnitudes of these parameters? From chapters 3 and 7, we know that the deflections of beams, columns, and stems are dictated by flexural stiffness EI, whereas the second moment of area I of the trunk and the average branch is a function of the trunk radius r_o and the average branch radius r. Likewise, for the present we can assume that the average elastic modulus of a plant is proportional to the average tissue density ρ. These relations did not escape Rashevsky's attention, and from them he derived a mathematical relationship to define average branch length:

(8.5) $$l = k_1 \rho^{1/3} r^{2/3} .$$

Since the trunk of the plant is axially compressed by a force equal to $M/\pi r_o^2$ per square centimeter of transection, M must equal $k_2 \rho r_o^2$, where k_2 is yet another proportionality factor. Finally, since the metabolic flow through the trunk is proportional to r_o^2 and inversely proportional to density, the metabolic rates of the trunk and branches are given by the formulas

(8.6a) $$qM = k_3 \left(\frac{r_o}{\rho}\right)$$

and

(8.6b) $$qM = k_3 \left(\frac{r}{\rho}\right) .$$

From these relations, Rashevsky provides the dimensionless ratio of the average branch length to trunk length exprcssed as a function of the average rate of metabolism per unit mass of tissue,

(8.7) $$\frac{l}{l_o} = \frac{q^{2/3} M^{1/3}}{k^4 - q^{5/3} M^{1/2}} ,$$

which essentially states that for any given total mass M, the ratio of the branch length to trunk length is dictated by the flow of nutrients q. As the value of q decreases, the value of n will increase (more branching) and l/l_o will decrease (a longer trunk with shorter branches). An additional feature of Rashevsky's equations that he did not comment on is that at some point, no matter how the plant is constructed, it will begin to starve after it reaches a certain stage in its morphological development. That is, the allometry of its shape and size is not compensatory in terms of the amount of nutrients the plant will produce versus the amount it will consume. At some point in its development, the ratio of production to consumption will drop below unity, and the plant will become progressively depauperate in metabolic resources. One of the symptoms of the reduction in metabolic resources would be a general reduction in overall growth. Some of my colleagues tell me that the rate of growth of many trees is inversely proportional to age and that some trees do appear to starve to death when extremely old.

The logic behind Rashevsky's derivation of plant shape rests on many arguable assumptions, and it is much to his credit that he pointed out many of these in his last paper, published posthumously (Rashevsky 1973). Yet the issue here is not whether eqs. (8.1) to (8.7) are *true*, but rather whether the general approach to plant design is reasonable. Of particular interest is that Rashevsky's global view of plant form was based on a delicate (if not logically precarious) balance between hydraulics (flow rates of metabolites) and structural mechanics (whose design is controlled by the strength of the trunk). In terms of hydraulics, no assumption was made about the direction of flow (photosynthates from leaves, water from roots); flow was simply assumed to exist. As a counterpart to flow, the mechanical design of the plant was assumed to be based on the strength of the structure, and with considerable insight, Rashevsky made the assumption that the elastic modulus of plant tissues is proportional to tissue density. Research since Rashevsky's death has revealed that the density of wood correlates very nicely with E measured in compression along the grain. Thus, for a broad range of wood species, E/ρ can be taken as a constant.

There have been many other attempts to formulate the design principles of plants with a tree growth habit (Murray 1927; Horn 1971; Jankiewicz and Stecki 1976; McMahon and Kronauer 1976; Honda and Fisher 1978; King and Loucks 1978; Niklas 1986a). But one of the earliest attempts was made by Leonardo da Vinci, who postulated that the diameter measured anywhere along the length of a tree is roughly equal to the sum of the diameters of branches above the level where diameter is measured. One of Leonardo's assumptions (concerning the relation between the diameter and the mass of a branch) appears to be very nearly correct. Murray (1927) demonstrated empirically that the mass M_b of a branch measured anywhere above where the branch is cut is highly correlated with the circumference C_b of the branch measured at the cut cross section:

$$C_b^{2.49} \approx M_b \,. \tag{8.8}$$

This relation can be very useful, particularly since the circumference (πR) of a branch with a terete cross section is related to the second moment of area (π

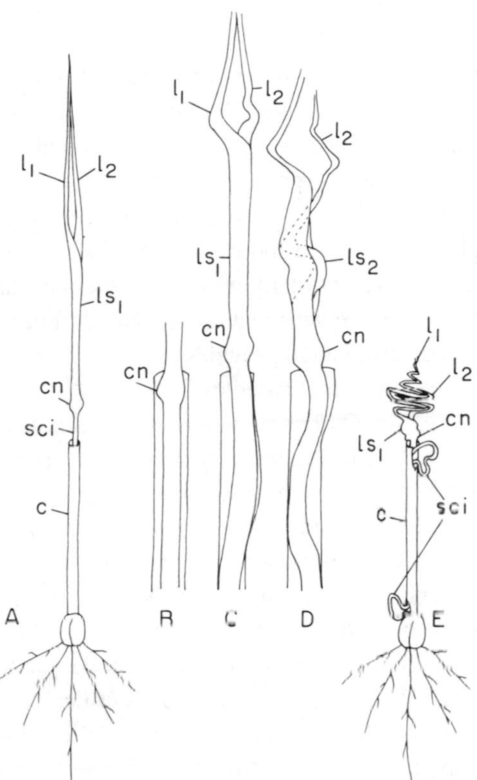

FIGURE 8.1 Diagrams of seed germination and modes of mechanical failure of wheat (*Triticum aestivum*): (A) Mechanically undeformed wheat seedling with shoot (consisting of two visible leaves, one visible node, and one internode) extending beyond the confines of the coleoptile. (B–D) Elongation and gradual bending of subcrown internode (as the crown node emerges from the coleoptile) and deformations of leaf laminae and sheaths owing to the resisting soil overburden. (E) Extreme deformation of seedling with subcrown internode emerging from sheared coleoptile and folded leaf laminae: c = coleoptile, cn = crown node, l = leaf lamina, ls = leaf sheath, sci = subcrown internode.

$R^4/4$) and since the mass of the branch measured above the cut is the static load that must be supported.

As we will see in the following sections, more recent attempts to deal with the mechanical design of branching make many of the same assumptions Rashevsky made but frequently neglect the yin and yang of mechanics and hydraulics, emphasizing the former while all but ignoring the fact that plants are living, metabolic-transporting structures.

SEEDLINGS

There are three primary reasons to begin a treatment of the plant body with a review of the mechanics of seedlings: (1) Germination affords an opportunity to view the vascular sporophyte as an integrated mechanical structure while it is still relatively small and morphologically simple. (2) Subsequent ontogenetic modifications in shape and increases in size make an integrated view of the sporophyte plant body much more difficult. (3) The mechanics of penetrating the soil are different in many ways from the mechanics of sustaining loads for aerial organs. Since germination involves soil penetration by organs other than roots, seed germination provides us with a glimpse of the versatility with which organs are functionally deployed.

The germination of wheat, *Triticum aestivum,* illustrates the mechanics of seed establishment typical for a relatively large number of plant species, the grasses. Wheat has hypogeal germination; that is, the cotyledon, the first embryonic leaf of the plant, remains underground (fig. 8.1). Upon germination, the wheat embryo vertically extends a tubular embryonic structure, called the coleoptile, which penetrates the soil above the seed and provides a conduit through which the shoot axis, the epicotyl, grows. The coleoptile provides mechanical support and protection to the juvenile photosynthetic shoot. Therefore the mechanical role of the coleoptile is critical to the survival of the wheat plant as a whole. The coleoptile is a thick-walled cylindrical shaft that sustains axial compressive loadings from the soil overburden. Solutes within actively growing cells provide the driving force for water flux, and the resulting turgor pressure within the coleoptile provides the hydraulic force necessary to push through the soil. Although the coleoptile is hollow, its mechanical behavior is that of a solid shaft or column owing to the presence of the epicotyl (fig. 8.1). Thus we can look upon the coleoptile-epicotyl as a two-phase composite column, consisting of an outer elastic rind (the coleoptile) and an inner core (the epicotyl) that operates as an inner elastic foundation, preventing short and long wave buckling of the coleoptile.

This mechanical configuration changes dramatically when the shoot extends beyond the length of the coleoptile, which has determinate growth in length and girth. When the shoot emerges from its shaft, the inner elastic foundation it provided is essentially eliminated, and the coleoptile is free to buckle. Likewise, if still underground, the shoot is directly subjected to the compressive stresses resulting from its growth against the resisting soil overburden. The mechanical deformations that can result as the seedling continues to grow are due to self-imposed compressive stresses. The soil overburden

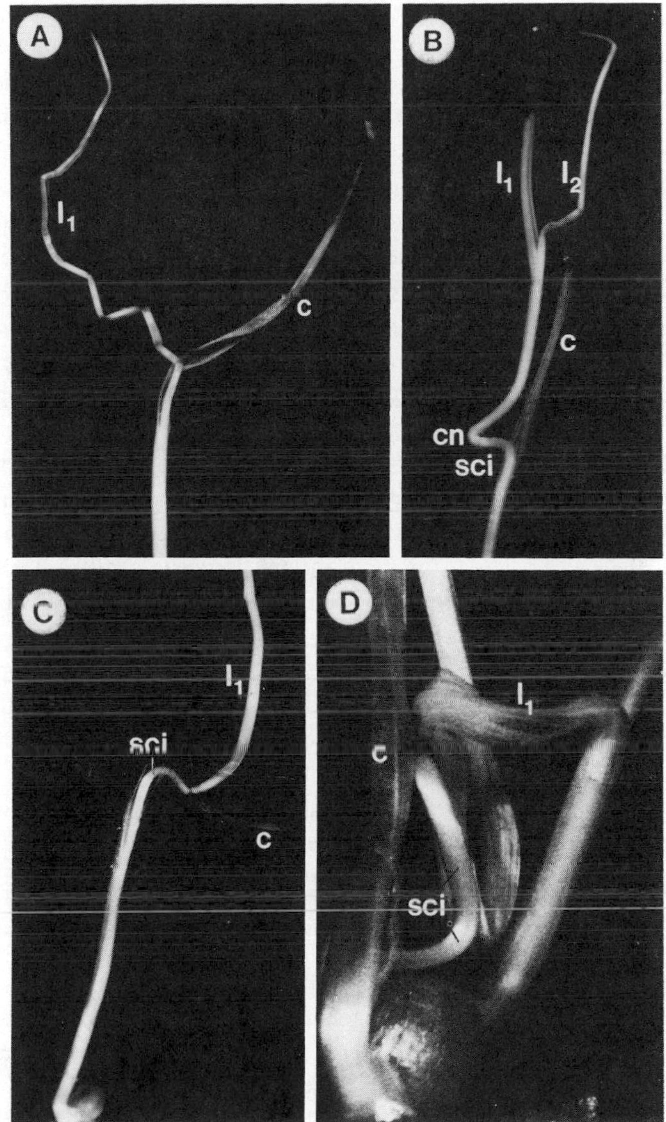

FIGURE 8.2 Seedling deformations of wheat (*Triticum aestivum*) resulting from compressional deformation from soil overburden. (A) Folding of leaf lamina and shearing of coleoptile. (B) Bending of the lamina of the second-formed leaf and of the subcrown internode beneath the crown node, with resulting shearing of the coleoptile. (C) Bending of leaf sheath and subcrown internode. (D) Extreme bending of subcrown internode and the leaf lamina of the first-formed leaf: c = coleoptile, cn = crown node, l = leaf lamina, sci = subcrown internode (see fig. 8.1A).

exerts a constant compressive stress on the seedling, but the magnitude of the compressive stresses the plant experiences increases as a result of the hydraulic force the seedling exerts against the constant weight of the soil above it. The deeper within the soil the seed is buried, the greater the initial compressive stress exerted on the seedling by the soil overburden. The types of mechanical deformations that can result are summarized in figure 8.2. These deformations are easily understood in terms of the second moments of area and the material properties of the various organs growing through the soil. For example, the tips of the first few leaves to form on the shoot are the first portions of the epicotyl to meet resistance as they grow through the soil. Since the blades of leaves are flattened in cross section, they have a preferred axis for bending. By contrast, the basal portion of the leaves forms a tubular leaf sheath with a terete cross section. Much like the tubular coleoptile, the leaf sheath has no preferred axis for bending. However, the leaves of wheat, like many other grasses, grow from their base by means of an intercalary meristem. Thus the base of each leaf consists of tissues that behave very much like a plastic material, whereas the more mature tissues toward the tips of leaves behave like an elastic material. When leaf blades mechanically fail under compression, they do so by folding like a concertina, primarily but not exclusively toward the basal portions where juvenile tissues occur (see fig. 8.2A). The leaf sheaths can also deform. When they do, they typically shear along their length, where their margins make contact.

The mechanics of vertical growth influences organs other than the leaves of the juvenile shoot. Young leaves are pushed upward by the elongation of the basalmost internode of the shoot, called the subcrown internode. Before the leaves emerge from the coleoptile, the subcrown internode is physically restrained from bending by the coleoptile. The extent of bending is further reduced by the presence of a pistonlike node at the tip of the subcrown internode, called the crown node, to which the oldest, first-formed leaf sheath is attached. When the crown node eventually emerges beyond the confines of the coleoptile, the subcrown internode is no longer laterally restrained by the coleoptile and can undergo Euler buckling owing to the axial compressive loading caused by its elongation through the soil above it. As elongation progresses, the magnitude of the bending load increases, and the deformation geometry of the subcrown internode changes. Under an axial load roughly equal to the critical bending load, the subcrown internode undergoes simple Euler buckling. This is seen as a C-shaped bending geometry (fig. 8.1C). When the axial load approaches nine times the critical buckling load, an S-shaped deformation occurs (fig. 8.1D, E). Also, the bending of the subcrown

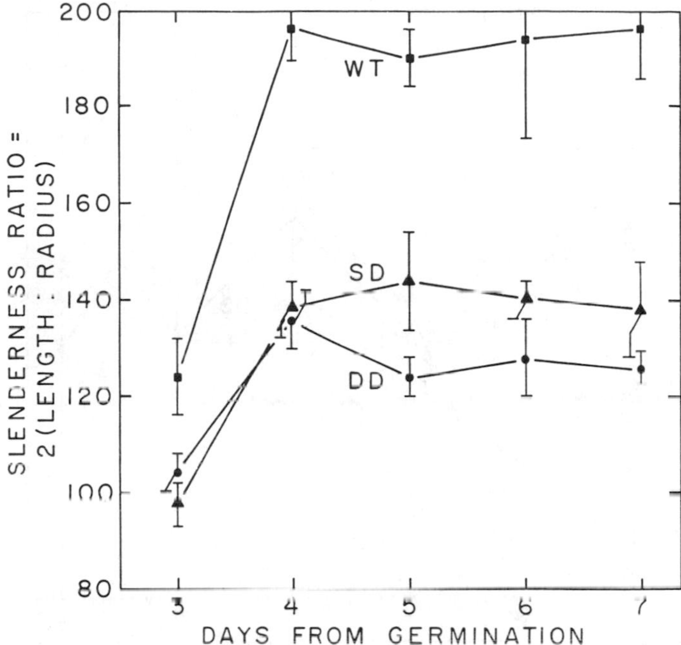

FIGURE 8.3 Slenderness ratio (twice the length:radius; $n = 5$) of coleoptiles from wheat seedlings (*Triticum aestivum*) differing in *Rht* gene dosage plotted as a function of days from germination: WT = wild type, SD = single dwarf, DD = double dwarf. (From Niklas and Paolillo 1990.)

internode can shear the coleoptile along its length (fig. 8.2B–D).

From the foregoing, we can readily appreciate that the mechanics of wheat germination is not simple. Virtually every mechanical principle thus far reviewed is required to understand seedling deformation. The complexity of seedling biomechanics is furthered by genetic differences among wheat cultivars, since they influence the geometry and the material properties of organs like the coleoptile (Niklas and Paolillo 1990). For example, the *Rht* gene increases the girth of wheat coleoptiles, hence this gene increases the second moment of area of this organ. The *Rht* gene also decreases the length of coleoptiles; hence, depending on the dosage of the *Rht* gene, the slenderness ratio of the coleoptile can be dramatically affected. From chapter 7, we know that an increase in the second moment of area increases the capacity of a column to sustain axial compressive loadings. Also, we learned that a decrease in the slenderness ratio decreases the potential for long and short wave deformations like Euler and Brazier buckling. Remarkably, the slenderness ratios

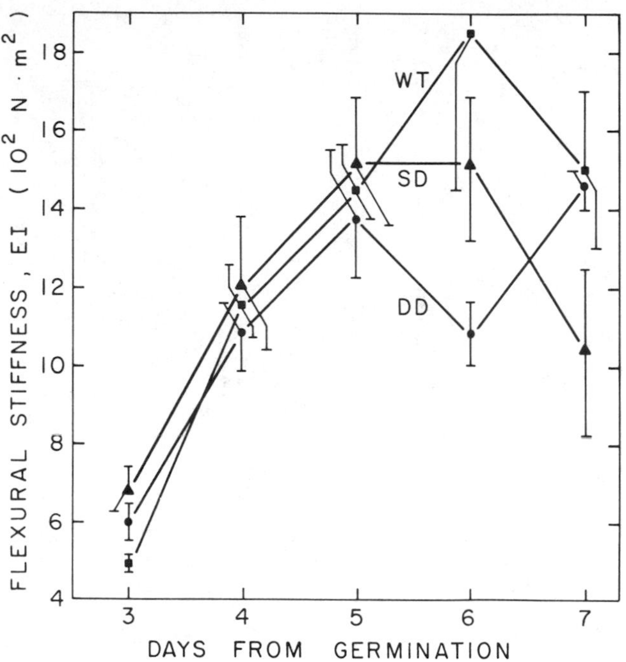

FIGURE 8.4 Flexural stiffness (*EI*) of coleoptiles from wheat (*Triticum aestivum*) differing in *Rht* gene dosage plotted as a function of days from germination: WT = wild type, SD = single dwarf, DD = double dwarf. (From Niklas and Paolillo 1990.)

of *Rht* gene bearing seedlings are almost always equal to or less than 120 (fig. 8.3)—the precise limit for the slenderness ratio of a safe column. True, the coleoptile is a hollow tube and the limiting case of 120 for the slenderness ratio is true only for a column with a solid cross section, but for most of their functional lives the coleoptile and the subcrown internode operate mechanically in tandem to produce a more or less solid column.

Unfortunately, the *Rht* gene also reduces the elastic modulus of seedling tissues (Niklas and Paolillo 1990). This is disadvantageous since, all other things being equal, a decrease in *E* reduces the load a structure can sustain. The decrease in *E* and the increase in *I* produced by the *Rht* gene are compensatory, however. That is, the flexural stiffness of coleoptiles containing the juvenile shoot from plants differing in their *Rht* gene dosage is virtually identical to that of the wild type (fig. 8.4). From this we may be tempted to think that the effects of the *Rht* gene on seedling establishment and biomechanics are inconsequential, but this is not so. The reduction in the length of the coleoptile forces the shoots of plants with the *Rht* gene to push through a greater

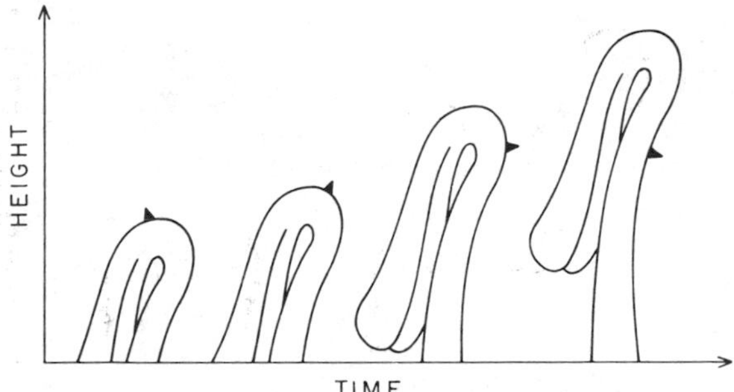

FIGURE 8.5 Diagram of hypocotyledonary hook of the bean (*Phaseolus vulgaris*). The position of an epidermal cell (marked with a black triangle) is seen to "flow" over the curved region of the hook as the hypocotyl grows and elongates.

soil overburden than do the shoots of wild type seedlings planted at comparable depths. Field studies indicate that the seedling establishment of plants with the *Rht* gene is less than that of wild type seedlings. If they survive beyond the stage of a seedling, however, plants with the *Rht* gene, which are shorter and stouter than their wild type counterparts, are extremely resistant to wind and rain lodging. Although the effects of the *Rht* gene on plant growth and survival are complex, we can learn a very important lesson—biomechanical analyses at the whole plant level are absolutely essential. The *Rht* gene may be disadvantageous at the stage of seedling growth but very advantageous when plants reach adolescence or maturity.

Unlike wheat, with its hypogeal germination, many plants have epigeal germination. That is, the first-formed embryonic leaves, called cotyledons, are elevated above the ground by the extension of the region of the embryo below the cotyledonary node. This region is called the hypocotyl, and as it grows it typically forms a hook so that the cotyledons are reflexed against the hypocotyledonary axis with their tips directed downward toward the soil (fig. 8.5). Although the radius of curvature of the hypocotyledonary hook is relatively constant, individual tissue elements formed at the junction between the cotyledons and the hypocotyl undergo progressive elongation and limited expansion. Thus portions of the hypocotyl are essentially displaced from a meristematic growth zone and gradually recurve upward so that the basal portions of the hypocotyl assume a vertical orientation. By crude analogy, the hypocotyl grows much like a fountain of water that maintains a constant shape, although the fluid within the stream changes position. The extension of the hypocotyl provides a mechanical force that can be applied against the soil pressure; newly produced tissue elements can displace adhering soil particles, and me-

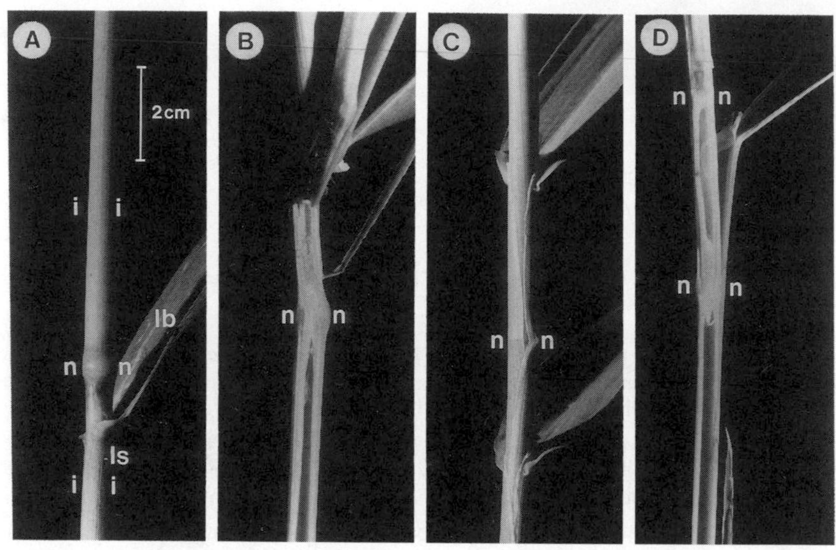

FIGURE 8.6 Morphology of oat (*Avena sativa*) showing relations between clasping leaf sheaths and internodes: (A) External morphology of the basal portion of a mature stalk of oat. (B) Longitudinally bisected specimen (shown in A) revealing hollow internodes below and above a nodal septum. (C) External morphology of the distal portion of a mature stalk showing a leaf sheath enveloping the node above it. (D) Longitudinally bisected specimen (shown in C) revealing two nodes and hollow internodes: i = internode, lb = leaf blade, ls = leaf sheath, n = node.

chanical damage is minimized. The kinematics of the growth of the hypocotyledonary hook has been elegantly studied by Silk (1980, 1984) and provides a splendid example of an organ whose geometry appears to be steady yet whose constituents are capable of flowing, much like a fluid. Recent workers have recognized the analogy between plant growth and fluid dynamics, leading to a deeper appreciation of plant morphogenesis and developmental physiology (Silk and Erickson 1979; Skalak et al. 1982; Gandar 1983). Of particular interest is the growing recognition that as tissue elements flow through and past growth fields, their material properties undergo changes. The formulation of growth in terms of continuum mechanics represents a starting point for integrating mechanical principles into existing models of plant growth (Silk and Wagner 1980; Plant 1983).

MATURE SHOOT SYSTEMS OF MONOCOTS

The establishment of seedlings as self-sufficient, photosynthetic plants involves a series of ontogenetic changes leading to a structurally more complex plant axis bearing appendages, the aerial shoot system. For many plant species a mechanical intimacy exists between leaf and stem; the mechanical behavior of the shoot as a whole largely depends on the mechanical operation of leaves. This is seen in monocots like the grasses, whose leaf sheaths typically clasp internodes so that they structurally reinforce growing and mature stem internodes (fig. 8.6), as well as in a variety of orchid species with equitate leaves. Indeed, abutting leaves and clasping leaf sheaths may be the principal supporting elements early in the ontogeny of the monocot shoot, and in arborescent monocots, such as the palms, the vascular system of clasping leaf bases may continue to provide structural reinforcement as the shoot system matures and ages. Since monocots lack secondary growth, the growth and development of herbaceous and arborescent monocot species provide us with a mechanical design alternative to plants that produce wood. In this section we will consider the biomechanics of monocots by examining a typical herbaceous monocot, oat (*Avena sativa*), and by reviewing what is currently known about the mechanics of arborescent monocots, the palms.

The common cultivar of oat has a structure and mode of growth fairly typical for grass species. Early in its development, the shoot consists of a series of closely spaced nodes, each bearing a single leaf. Each leaf consists of a distal, more or less flattened leaf lamina and a basal, tubular leaf sheath that envelops the portion of the shoot above it. In longitudinal cross section, the unextended shoot consists of a relatively short plug made up of stacked nodes and short internodes surrounded by a cylinder of concentrically packed, clasping leaf sheaths. By intercalary meristematic growth at the base of each internode, the shoot extends in overall length, so that nodes and their leaves are drawn farther apart (fig. 8.6). Older, more extended internodes occur at the base of the shoot, while younger, less extended internodes are found progressively toward its tip. As a consequence, internodes are more visible progressively toward the base of the shoot. During the elongation phase of growth, internodes become hollow (fig. 8.6). The distalmost internode, called the peduncle, is the subtending axis of the inflorescence. It is the last to emerge from the confines of the enveloping leaf sheath borne on a proximal node. The way oat shoots elongate can be crudely compared to the extension in length of the cylindrical elements making up a collapsible telescope. If the eyepiece is held firmly and the outermost, largest cylindrical element of the telescope is gradually pulled, each successively smaller cylindrical element is progressively exposed to view, and the eyepiece cylinder (analogous to the peduncle) is the last to be exposed.

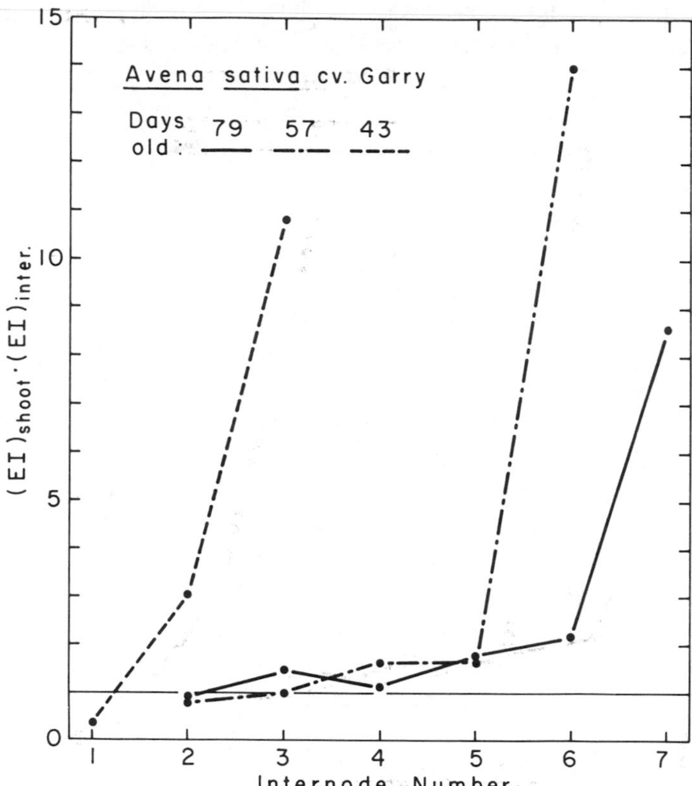

FIGURE 8.7 Relation between the flexural stiffness (*EI*) of the shoots (clasping leaf sheaths plus internodes) and the stems (internodes without their leaf sheaths) for a cultivar of oat (*Avena sativa* cv. Garry; see fig. 8.6). The dimensionless ratio of the flexural stiffness of the shoot to stem is plotted against internode number (ascending in number toward the tip of the shoot) for plants differing in age (see inset legend). Data points falling above the thin horizontal line indicate that *EI* of the leaf sheath contributes significantly to the stiffness of the shoot; points falling on or near the line indicate that internodes are as stiff as leaf sheaths. As the age of the plant increases, internodes stiffen acropetally until only the distalmost internode is less stiff than its clasping leaf sheath. (From Niklas 1990b.)

The critical feature in understanding the mechanical consequences of the growth of plants like oat is that cellular differentiation in each internode proceeds in a basipetal direction. That is, older tissues occur toward the upper portion of each internode. Thus the tissues at the base of each internode are younger and typically exhibit plastic or viscoelastic behavior as opposed to the elastic behavior of older tissues. The mechanical significance of clasping

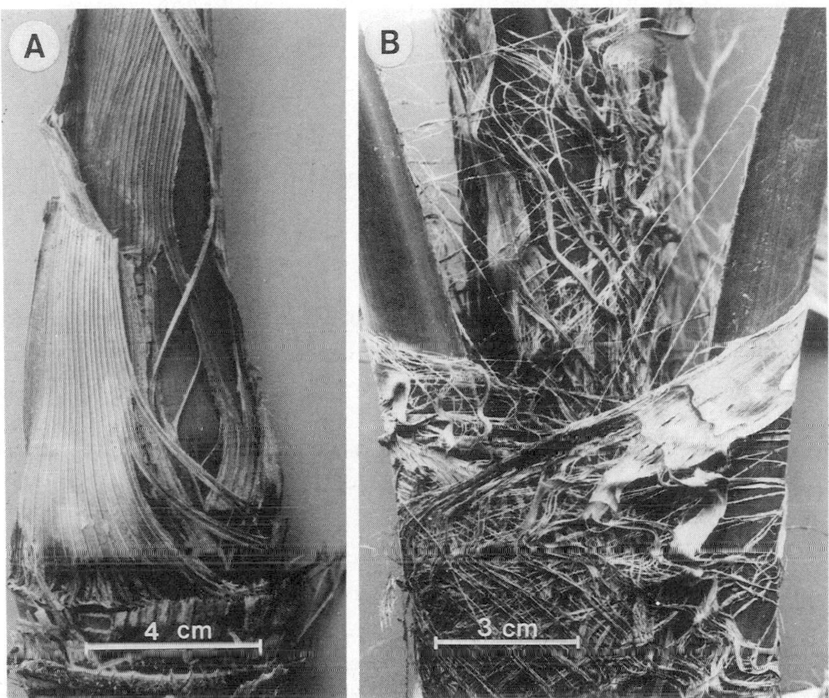

FIGURE 8.8 Leaf bases surrounding younger portions of palm shoots. (A) Shearing of leaf bases (owing to circumferential expansion of shoot) revealing fibrous leaf construction. (B) Leaf base fibers girdling successively younger leaves of palm shoot. See text for further details.

leaf sheaths is that these structures are stronger and more elastic than the juvenile internodal tissues they envelop. Data on the elastic moduli of leaf sheaths and internodes from young and old shoots of a cultivar of oat called Garry are presented in figure 8.7. These data reveal that the elastic modulus of the leaf sheath tissue can be two orders of magnitude greater than that of the enveloped internode (Niklas 1990b). By contrast, the elastic moduli of the leaf sheath and the internode are roughly equivalent for fully extended and mature shoots, presumably because cellular differentiation and maturation are completed.

The geometric arrangement of the clasping leaf sheaths ought not to be neglected, however. From first principles, we have learned that the farther supporting tissues are placed from the centroid axis, the greater the benefits to the second moment of area, conferring flexural stiffness. Leaf sheath tissues are mostly older, hence stronger, than the elongating internodes they surround, and they also clasp the internodes above them. In a mechanical context, this design is very similar to the coleoptile-epicotyl system discussed for wheat seedlings. And as in this system, we see another example of a core-rind (leaf sheath–internode) model in which a composite material maximizes flexural stiffness. The ontogeny of the oat shoot essentially involves a bootstrapping mode of growth in which the stem grows vertically by leaning on and pushing through its stronger leaves until internodes are developmentally rigidified.

By retaining portions of encircling leaf bases even after leaves die and leaf lamina shred off, arborescent palms mechanically behave very much like their herbaceous relatives the grasses. The adhering tissue of the leaf base has a fibrous composition that is reoriented as the stem it is attached to expands in girth (fig. 8.8). The mechanical role of the leaf base of palms was alluded to by Schwendener (1874), but the first detailed treatment of this feature was that of Schoute (1915), who described the morphology, anatomy, and mechanical behavior of the leaf base of the palm *Hyphaene*. Later Tomlinson (1962) provided additional insights in his review of the taxonomic distribution of different leaf-base morphologies among the various palm genera. Typically, the leaf base begins its development as a hollow cylinder encircling the palm stem. After maturation, the portion of the leaf base on the side of the stem opposite the leaf lamina dies, leaving behind the fabric of the fibrovascular bundles when the ground tissue degrades. These bundles are organized into two layers that are arranged into antiparallel chevrons (fig. 8.8B). As the stem increases in girth, the fibers within the outermost of the two layers are progressively reoriented circumferentially. (The passive reorientation of the fibrovascular bundles in the leaf base of the palm is not unlike that envisaged by the multinet hypothesis describing the reorientation of the microfibrillar cellulosic network in expanding and elongating cell walls; see chap. 5.) Although shearing occurs among the fibers toward the lower portions of the outer layer of fibrovascular bundles, the leaf base can remain attached to its stem a considerable distance from the growing shoot apex and continues to provide hoop reinforcement. The tensile modulus of isolated dehydrated fibrovascular bundles is truly astonishing ($\approx$ 100 $GN \cdot m^{-2}$). The significance of this modulus is diminished within the fabric of the leaf base, however, since bundles shear and since the leaf base has a comparatively low shear modulus. Nonetheless, the shear modulus of an entire leaf base measured along the length of parallel arrays of fibers is from thirty to fifty times higher than the shear modulus measured in the plane bisecting the two antiparallel chevron layers of fibers.

Growth in arborescent palms does not permit maintaining geometric similarity; palms grow in height but cannot maintain constant slenderness ratios because they lack a vascular cambium that can sequentially add growth layers of secondary vascular tissues. Palms get proportionally thinner as they grow taller. This is often assumed to be a mechanical disadvantage because, as we have repeatedly seen, a vertical column becomes increasingly more susceptible to Euler buckling as its slenderness ratio increases. But Euler buckling is actually a safe mechanical response to growth and represents a minimum weight solution for growing in height provided tissues are flexible. (Indeed, the stems of many plants typically manifest the primary mode of Euler buckling after they reach their critical buckling height.) Nonetheless, arborescent palms have evolved a number of ontogenetic and structural design factors that compensate for the lack of a vascular cambium and that reduce (or delay) Euler buckling. An excellent review of these design factors is provided by Rich (1987; see also Rich 1986; Rich et al. 1986), who has done much work in this area of plant biomechanics. For example: (1) During early ontogeny, the shoot apex of some palms increases and produces a stem diameter sufficient for future support requirements before growth in height is initiated. Essentially, the maximum girth is achieved before the stem elongates significantly—subsequent growth in height reduces the slenderness ratio, but not below a critical margin of safety. As in most treelike plants, the rate at which palms grow in height decreases with the age of the plant. Thus the palm approaches its critical buckling height with progressively less speed. An interesting but unanswered question is whether palms grow in the manner envisioned by Zeno's dichotomy (see chap. 3). (2) For some species, stem diameter increases by sustained cellular expansion at a considerable distance from the shoot apex. Although new cells are not added in significant numbers, as they are in arborescent dicots, cells grow bigger, increasing the girth of shoots. (3) Stem tissues undergo sustained lignification and become stiffer as they age. Lignification of cells preferentially occurs more toward the perimeter of the shoot cross section than toward its centroid axis. This maximizes the mechanical dividends of lignification because it places the stiffest materials in the locations within the stem that will experience the highest tensile and compressive bending and torsional shear stresses. The extent of lignification increases with the age of the tissue. Thus the base of the palm stem will have the most lignified tissue. Rich (1987) found that increases in stem tissue stiffness and strength and the allometry of shoot growth collectively confer an adequate design factor against mechanical failure as shoots elongate and grow in height. Two aspects of this work bear further comment.

First, it is intriguing that shoot diameter is apparently maximized before

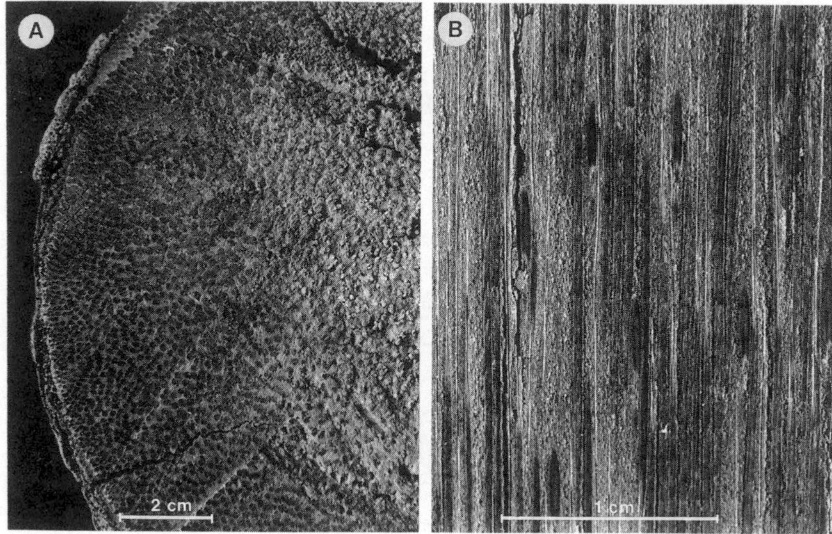

FIGURE 8.9 Transverse (A) and longitudinal (B) sections through the stem of a palm (?*Chrysalidocarpus lutescens*). Vascular bundles are seen scattered throughout the planes of the section but tend to be more concentrated toward the perimeter of the transection (A).

shoot elongation occurs in an effort to attain an adequate design factor, implying that shoot growth in height is limited below some critical length. If no developmental mechanism exists to truncate the elongation, then regardless of how fully the diameter of the mature shoot is prefigured before elongation begins, the palm's design factor progressively erodes as it continues to elongate. Thus the growth of some species of palms may be determinate in terms of mechanical susceptibility to failure.

Second, modifications in the dry weight, elastic modulus, and modulus of rupture of older, more peripheral tissues in the trunks of palms are compatible with the theory of cellular solids presented in chapter 6. A single mature palm trunk can encompass the entire range of tissue density and elastic moduli measured in tension for dicot wood. In older, peripheral tissues, dry density ranges from roughly 0.1×10^3 kg·m^{-3} to 1.0×10^3 kg·m^{-3}, while the maximum values of E reported for the palms *Welfia georgii* and *Iriartea gigantea* (whose specific epithet says it all) exceed the maximum value of E reported for the hardwood of the dicot *Tabebuia serratifolia* (21 GN·m^{-2}), which has one of the highest tensile elastic moduli known. Significantly, Rich (1987) shows that E increases with the 2.46 power of tissue dry density, while the modulus of rupture varies as a 2.05 power of overall tissue dry density and as a 1.60 power for tissues with low plastic deformation. Exponents greater than unity for any of these relations are expected for materials that behave as three-dimensional cellular solids (Gibson and Ashby 1982).

The heterogeneity in the distribution of vascular bundles and the stress-interactions between bundles running the length of the stem and those diverging into leaf bases also have significant mechanical consequences on vertical stability, particularly since the frequency, diameter, and curvature of vascular bundles running the length of palm trunks change as a function of their distance from the centroid axis. Vascular bundles tend to become more numerous and smaller in diameter toward the periphery of cross sections (fig. 8.9), whereas peripheral bundles tend to be more vertically aligned than their more central counterparts, which tend to run along shallow spiral courses (Zimmermann and Tomlinson 1972; Esau 1977, 265). As noted previously, the vascular bundles of the leaf bases, which are extensions of the peripheral vascular bundles of the stem, follow a circumferential course beginning 180° opposite each leaf lamina. Thus the peripheral vascular bundles of the stem eventually come to wrap around the external portions of the palm trunk. A consequence of this leaf-stem interconnectedness is that the vascular fibers toward the perimeter of the stem cross sections, together with the leaf bases, operate collectively as a series of intertwining guy wires that can transmit compressive stresses into tensile stress throughout the entire shoot. This can be verified by means of a small wire model fitted with strain gauges. Compressive loadings are found to be transmitted throughout the model in the form of tensile stresses. Since these elements are just beneath the exterior of the stem, they are optimally positioned for their mechanical task and reduce the magnitude of compressive stresses occurring within the parenchymatous matrix (which has a low shear modulus) as stems bend. This mechanical design is remarkable because any stress on any part of the stem is dissipated throughout a fairly large region, thereby reducing the local stresses experienced by any given region of the matrix and its associated vascular bundles. Thus the mechanical design of palms is as impressive as these plants are staggeringly beautiful.

Pteridophytic Trees

Taxonomically minded readers may be somewhat surprised by a treatment of the seedless vascular plants, collectively referred to as pteridophytes, following a discussion of the much more evolutionarily recent monocots. True, from an evolutionary perspective, the pteridophytes are treated out of sequence.

From a biomechanical perspective, however, the archaic pteridophytic lineages are by no means primitive. Indeed, as I will show for *Psilotum,* some extant pteridophytes have achieved mechanical designs at least as sophisticated as the arborescent monocots and dicots. In this regard it is worth noting that many extinct pteridophytes grew as tall as or taller than the trees produced by some flowering plants. For example, the arborescent lycopods and horsetails, such as *Lepidodendron* and *Calamites* of the Carboniferous period, were very tall, rivaling many modern-day plants. (Reasonable estimates suggest that *Lepidodendron* grew to a height of 54 m, with 12 m rootlike extensions into a swamp substrate, and even *Calamites* reached the impressive height of 12 m.) These plants achieved their stature in part by an evolutionary convergence—the production of secondary tissues. In *Lepidodendron* an external cone of secondary cortex (periderm) was produced, often in such abundance that the bulk of the fossil plant remains in the Pennsylvanian Coal Measures are peridermal tissue. By the same token, the mechanical principles underlying the construction of the tree ferns is not too unlike that discussed for the arborescent monocots—the deployment of the stiffer of two or more materials toward the periphery of cross sections. The structural and mechanical analogies that can be drawn among the pteridophytic trees, gymnosperms, and some angiosperms testifies that terrestrial plants have faced many of the same physical limits on vertical growth over their 500 million year history and that, though convergence is a common biomechanical theme throughout the evolutionary history of plants, different plant lineages have capitalized on different structural components (wood, root mantles, periderm, clasping leaf bases, etc.) to achieve a vertical posture.

There is also a pedagogical reason for discussing pteridophyte mechanics before treating gymnosperm and dicot trees. The previous treatment of arborescent monocots does not prepare us for dealing with branching, a characteristic of the trees produced by gymnosperms and dicots. Extensive vertical branching creates a potentially very complicated mechanical system, most unlike the columnar trees of palms. True, each of the large pinnately compound leaves of many species of palms is mechanically analogous to a single shoot on a dicot or gymnosperm tree, but the dendritic branching so typical of dicots and gymnosperms presents its own mechanical subtleties.

A nonwoody but copiously branched pteridophyte is illustrated by *Psilotum nudum* (see fig. 1.6G for its typical vertical growth habit). The sporophyte of this species grows horizontally and vertically by the repeated division of shoot apical meristems. Each vertical portion of the plant can grow to produce a three-dimensionally branched truss, frequently consisting of well over a hundred branch elements. The vertical trusses of *P. nudum* are the principal photosynthetic organ of the sporophyte. Trusses also produce and bear sporangia, elevating them above the ground, thereby promoting spore dispersal, sometimes for considerable distances. The three-dimensionally branched trusses of *P. nudum* together with the rootless horizontal portions compose a minimalist vascular land plant that has captured the imagination of the developmentalist as well as of those interested in the early evolutionary history of vascular plants. Nonetheless, *P. nudum* is a remarkably successful plant not infrequently found growing in the otherwise vegetatively barren volcanic islands of the Pacific. My field experience with this genus is limited to a population of plants growing on the steep slopes of an old caldera (Mount Kilauca on the island of Hawaii). Many individuals had their densely aggregated horizontal axes embedded within clumps of lichen, which appeared to serve as water traps.

A vital clue to the mechanical design of *Psilotum nudum* is the change in the external coloring of branch elements along the length of the vertical truss. The distalmost branches look bright green, owing to chlorenchymatous tissues just beneath the epidermis, while the basalmost branch element of each vertical truss produced even by healthy plants is often pale yellow or brown. When dissected, the region occupied by chlorenchyma in the upper portions of the truss is seen to be sclerenchymatous in the lowest branch elements. Since the sclerenchyma has lignified, thickened cell walls, how far this tissue occupies any cross section can easily be seen by staining transverse sections with phloroglucinol, which stains lignified cell walls bright red and stains cell walls loaded with phenolic compounds yellow (plate 2).

For our purposes it is important to remember that, in addition to their hydraulic function, which is accomplished by means of a slender vascular strand running the length of each branch element, the branched trusses of *Psilotum nudum* fulfill the functions of both leaf and stem—both photosynthetic and mechanical functions. The relative doses of chlorenchyma and sclerenchyma in any particular branch element reflect the proportions of these two functions—a large volume fraction of chlorenchyma means photosynthesis is the primary functional role of the branch element, while a large volume fraction of sclerenchyma means its primary function is mechanical support. Clearly, both can be accomplished in a single branch element, but the optimal location for photosynthetic tissue in a cylindrical stem or leaf is just beneath the epidermis, since this reduces the attenuation and change in spectral quality as light passes through the organ, as well as permitting direct access to the external atmosphere (via stomata) for gas exchange. Likewise, the optimal location for sclerenchyma—a thick-walled, mechanically stiff material—is just beneath the epidermis, since this places the stiffest material where bending and torsional shear stresses will always reach their highest intensity. However, two tissues cannot occupy the same location in the same branch element. *P. nudum* provides an example of the compromise between the dual functions of mechanical support and photosynthesis. The solution is elegantly simple—establish two countergradients in the relative volume fractions of sclerenchyma and chlorenchyma, such that the former increases in abundance basipetally (toward the base of each truss) while the latter increases acropetally (toward the branch tips of each truss). Thus both tissues occupy the same location, but not in the same branch element. This design compromise is clearly evident in the external color of each truss, since the relative dose of chlorenchyma in each successive level of branching in a truss can be gauged by an element's relative greenness, which in healthy plants always increases toward the tip branch elements.

Another feature of the mechanical design of *Psilotum nudum* is the tapering of the trusses. The girth of branch elements increases basipetally. As we have seen, tapering provides for a gradient in the second moment of area along the length of an organ. For a cylindrical organ like the trusses of *P. nudum,* even a relatively small increase in cross-sectional radius can effect a substantial increase in the second moment of area, since radius is raised to the fourth power. The basipetal rise in the second moment of area combined with the basipetal rise in the volume fraction of sclerenchyma provides for a significant increase in the flexural stiffness of branch elements toward the base of each truss.

We can begin to appreciate the full significance of the mechanical design of *Psilotum nudum* when we look at the collective consequence of the changes in the elastic modulus, second moment of area, and branch length and at the distribution of loadings along the longitudinal course of trusses. Unlike large dicot trees, which by their sheer size and weight make detailed mechanical analysis difficult, the morphology and elastic parameters of *P. nudum* are relatively easy to measure by carefully labeling each of the hundred or so branch elements in each truss and then determining E, I, l, and P by gradually dissecting the truss from the tip branch elements downward (Niklas 1990c). Recall that flexural stiffness EI, branch length l, and the self-loading P of a plant (above the point at which EI is determined) can be used to construct the loading parameter Pl^2/EI, a dimensionless parameter that can estimate the bending deformation (deflections from the vertical) of a beamlike organ (chap. 3). A low value of the loading parameter means a branch element will not bend

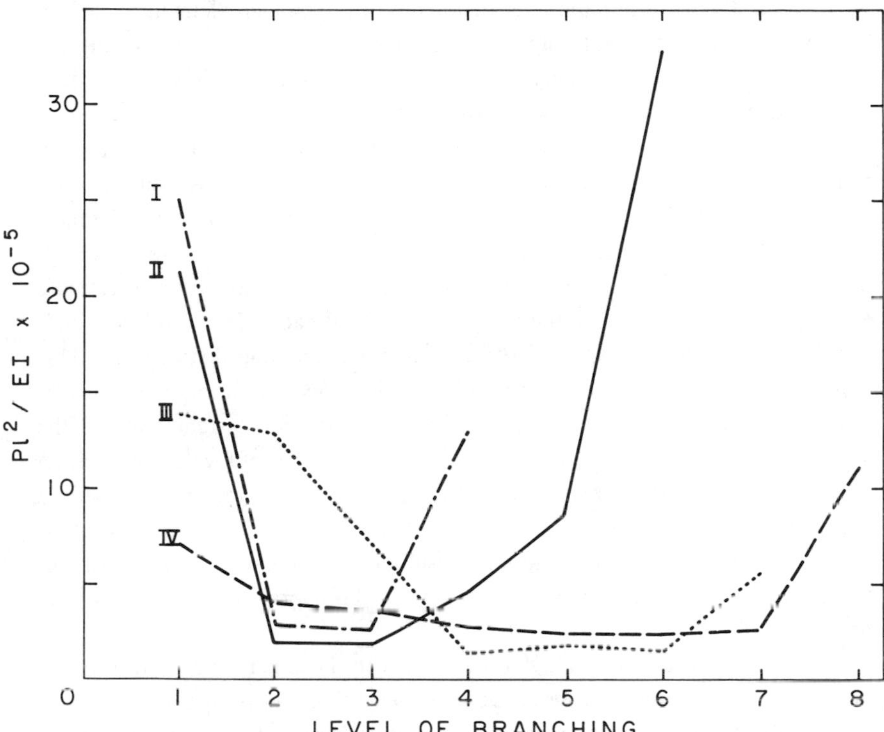

FIGURE 8.10 Loading parameter (Pl^2/EI) of *Psilotum nudum* plotted as a function of the level of branching in four (I–IV) vertical trusses of the sporophyte generation. The basalmost branch element (level 1) and highest level of branching in each truss have higher loading parameters than intermediate branching levels. However, as the size of the truss increases (measured by the number of the levels of branching) from truss I to truss IV, the loading parameter of level 1 decreases, indicating that the flexural stiffness of the subtending branch element has increased relative to the load (biomass) it must support. (From Niklas 1990c.)

significantly as a result of the loadings it experiences, while a high loading parameter means large deflections are likely. Figure 8.10 is a plot of the loading parameters versus the level of branching for four trusses of *P. nudum*. The basalmost branch element in each truss is numbered 1, while higher levels of branching are shown in ascending numerical order. The data in figure 8.10 reveal two important features. First, the distalmost branch elements appear to have the poorest design—their loading parameters are extremely high compared with those of the subtending branch elements. This is not particularly surprising, however, since the most distal elements must support only their own weight. Much more significantly, the second feature revealed in figure

8.10 is that the loading parameter of the basalmost branch element in each of the four trusses is typically higher (hence less efficient) than the loading parameters of intermediate branch elements. From a purely engineering point of view, the high loading parameters of the basalmost, subtending branch elements constitute a very poor design, since each of these elements must support the entire weight of the branch truss it subtends. Two observations mitigate this conclusion. First, each of the four basalmost branch elements operates as a more or less point-loaded column during most of the development of a vertical truss, because the geometry of branching within each truss is fairly symmetrical around the vertical axis. Thus each of the single subtending branch elements experiences a concentric loading distribution. This branching configuration tends to reduce the development of large vertical deflections in response to self-loading. Perhaps much more significant, as the size of the truss (measured in terms of either the number of levels of branching or total truss weight) increases among the four trusses examined, the loading parameter of the basalmost branch element in each truss decreases. That is, the mechanical attributes of the single subtending element in each truss appears to be mechanically scaled according to the overall size (weight) of the truss.

A fundamental question is, Are the trends in these data simply fortuitous, or does *Psilotum nudum* grow in such a manner that it developmentally scales its elastic parameters to the intensity of its self-loading? The answer appears to be that *P. nudum* exhibits an intrinsic mechanical allometry. The scaling of elastic parameters to the loadings on branch elements is achieved in three ways. First, each vertical truss is determinate in its growth. After a developmentally prescribed period of growth, the truss no longer branches or increases in height or weight. Thus the moment arms on each branch element achieve their final and maximum intensity once growth ceases. Second, the cortical tissues of *P. nudum* have the capacity for sustained lignification after a truss completes its branching. Thus the elastic modulus of the cortical tissues, particularly in the basalmost branch element in each truss, can increase as a function of tissue age. (The bifurcational symmetry in the branching of *P. nudum* trusses allows us to anatomically examine positionally analogous branch elements at different stages in their ontogeny. When this is done, we find that the intensity of staining with phloroglucinol, which reacts with lignin, increases as branch elements mature. Also, the intensity of staining continues to increase even when a truss ceases branching and growing in length.) And third, the basalmost branch elements in each truss can increase in diameter as they get older. The way this expansion is achieved—whether by cellular division or expansion or both—still remains unclear, but the consequence on the second moment of area is very straightforward: expansion in girth can dramatically increase the second moment of area, and hence flexural stiffness.

Surprisingly, the way *Psilotum nudum* mechanically scales its elastic parameters to its overall size is very like the way some arborescent palms achieve their design factor (Rich 1986). Recall that sustained lignification and lateral expansion by means of meristematic activity in the ground tissue increase the flexural stiffness of palm trunks in the basipetal direction. Both *P. nudum* and palms lack a vascular cambium, yet both have developmental systems operating at the level of primary growth that are remarkably convergent and strikingly effective in mechanically sustaining vertical growth. That this developmental convergence has occurred in two taxonomically very dissimilar plant groups should come as no surprise, since lignification and diffuse meristematic growth are plesiomorphic features of all known vascular plant lineages.

Before leaving the genus *Psilotum,* it is instructive to note that within this genus there is another species, *Psilotum complanatum,* that has a wholly different mechanical design. *Psilotum complanatum* produces aerial trusses that branch two-dimensionally and assume a morphology much like that of a leaf. The trusses of this species mechanically operate as cantilevered beams. The basalmost branch elements of each truss are more or less terete in cross section, and cross sections become more elliptical as branching levels are ascended acropetally, much like the petioles of many dicots. The single subtending branch element has the highest elastic modulus of any branch element in each truss, and just as in *Psilotum nudum,* the elastic modulus and second moment of area of *P. complanatum* are scaled to the overall size and weight of each truss. When we compare the flexural stiffness of these two species, however, *P. complanatum* has much the weaker mechanical configuration, which may account for the pendulous growth habit of this epiphytic species. Comparisons between the two species reveal the mechanical design of *P. complanatum* is convergent with the mechanical design of fern leaves, whereas the mechanical design of *P. nudum* is much more like that of a branch. This may account for the suggestion that the aerial trusses of *Psilotum* are homologous with the leaves of ferns (Bierhorst 1971).

Psilotum nudum and *P. complanatum* are only two species among thousands of pteridophytic plants whose collective biomechanical repertoire is little understood and, unfortunately, is given scant attention. The material properties and mechanical geometry of the ferns in general and the tree ferns (Cyatheaceae and Dicksoniaceae) in particular provide fertile ground for research. Some species of tree ferns can grow over 20 m in height and produce leaves that measure 5 m from tip to base. The trunks of these plants are rarely branched and are covered with a dense mat of adventitious roots typically associated with the bases of leaves. By contrast, the climbing fern *Lygodium japonicum* (Schizaeaceae) has leaves that continue to grow at their tips and can wrap around branches and twigs of larger plants much like the tendrils of the garden pea. Of general interest is that the leaves of many fern species have a hypodermal layer of sclerenchyma, which appears to act as a rigid rind providing very high stiffness (comparable to dicot twigs with secondary xylem and cork tissues). Indeed, I have measured the dynamic elastic modulus of the petioles of the maidenhair fern, *Adiantum pedatum,* at $5.12 \times 10^{10}\ \mathrm{N \cdot m^{-2}}$, which is very near the tensile elastic modulus of dry cellulose.

Dicot and Gymnosperm Trees

The vertical stems of all plants simultaneously provide mechanical support and transport and store fluids. Arborescent dicots and gymnosperms can compensate for increased structural and hydraulic demands during vertical growth by augmenting their stem diameter through the activity of lateral cambium. Each new layer of secondary xylem functions for a year or more primarily as a water transport tissue, but this function is gradually left behind as the vascular cambium expands in girth to accommodate new layers of secondary xylem. Over the years, each layer of wood is progressively internalized as the vascular cambium advances outward from the centroid axis within the trunk and branches of a tree. Thus the mechanical role of each layer of wood is amortized over decades, centuries, or in extreme cases millennia.

Unlike arborescent palms, which tend not to branch, and unlike *Psilotum,* which branches but has determinate growth in the length and weight of branch elements, gymnosperm and dicot trees typically branch and exhibit indeterminate growth in the length and weight of branches. Thus the trees that we are about to discuss have mechanical attributes in static and dynamic equilibrium that differ from those of most plants that have occupied our attention thus far. Studies focusing on static or dynamic equilibrium or both have produced two mechanical models that attempt to predict how allometric growth in the length and girth of branches maintains mechanical stability. Both models are predicated on the mechanical principles treated by beam and buckling theory reviewed in chapter 3. The two models are called the elastic-stability model and the constant-stress model. The elastic-stability model emphasizes static loadings (self-loading as well as snow loadings; King and Loucks 1978), whereas the constant-stress model emphasizes dynamic wind loadings (Metzger 1893). The elastic-stability model argues that stems and branches taper so as to maintain a constant elasticity throughout the tree, whereas the constant-stress model argues that branches taper so that the maximum bending stress in any transection is independent of the length of branches or of the trunk.

The elastic-stability model considers the stem a vertical column that resists buckling from its own self-loading. The principal features of this model can be derived from the Euler column formula, discussed and criticized somewhat in chapter 3. And we have already discussed the modified formula for the buckling of a column owing to its own weight (see chap. 3). Returning to these previous discussions, we see that the critical buckling length l for such a column is given by the formula (see Timoshenko and Gere 1961, 101–3)

$$l = \left(\frac{2.788\ EI}{q}\right)^{1/3}, \quad (8.9)$$

where q is the uniformly distributed load intensity and EI is flexural rigidity. For a cylindrical column, I equals $\pi R^4/4$ and the uniformly distributed load intensity equals $\rho\pi R^4$, where ρ is the average tissue density of the columnar tree trunk. Inserting these expressions for I and q into eq. (8.9) gives the following formula:

$$l \approx 0.887 \left(\frac{E}{\rho}\right)^{1/3} R^{2/3}. \quad (8.10)$$

The ratio E/ρ is approximately constant across a wide spectrum of woods (McMahon and Kronauer 1976); therefore $l \propto R^{2/3}$. McMahon (1973) and McMahon and Kronauer (1976) present a similar derivation predicting the relation between the length and the radius of a tapered stem. (It is worth noting that Rashevsky 1973 also derived the same relation as that given by eq. 8.10 [see eq. 8.5, where r is branch radius], as did Greenhill 1881.) From eq. (8.10), we see that if the height and diameter of trees are allometrically maintained during growth so that their tapered trunks are elastically similar throughout their length, then a log-log plot of trunk length and radius ought to have a slope equal to two-thirds, or roughly 0.666. When data from 576 trees from nearly every North American species are plotted in this fashion, a slope of 0.666 is actually calculated (McMahon 1973; McMahon and Kronauer 1976). The data also reveal that trees are generally limited in overall height to about one-fourth their predicted critical buckling height. Accordingly, most species are mechanically overbuilt and have a design factor of roughly 4. The magnitude of this design factor can be evaluated in another way. We need simply cut wedges out of the vertical trunks of trees and measure their cross-sectional area just before they begin to topple over from their own weight. Needless to say, this is sample destructive and can be dangerous, but in general about three-fourths of the trunk's cross section can be removed from most trees before they begin to irreversibly fail under their own weight. Hence a design factor of 4 appears to be a reasonable estimate whether we use a pristine equation or a sharp ax. Also, tree trunks can sustain the weight of their canopies even when their heartwood is almost completely eroded by fungal or insect damage, particularly in dense stands of trees where neighboring plants buffer the effects of wind. Nonetheless, the mechanism by which a design factor of 4 is achieved is not clear. It is possible that the maximum height trees reach is limited by how far water can be transported vertically in the xylem conduits of the secondary vascular tissue system. Or it may be that as trees grow they progressively starve owing to a shifting in the ratio of productive photosynthetic tissues to consumer nonphotosynthetic but living tissues, or perhaps a stochastic phenomenon involving the death of trees results from dynamic mechanical instabilities caused by storms. The design factor of any plant involves more than the static loadings brought about by growth and must logically reflect the magnitudes of dynamic loadings. Thus, regardless of the ultimate reason or reasons for a design factor of 4, we must assume that a proximate reason why trees are overbuilt in some manner relates to wind-pressure loadings.

The two-thirds-power rule predicting the relation between stem length and radius can be derived from the bending theory for cantilevered beams as well. Assuming that the deflections δ from a nearly vertical cylindrical cantilever subjected to a uniformly distributed load intensity q are relatively small, bending theory gives us the formula

$$\delta = \frac{ql^4}{8EI} = \frac{\rho l^4}{2ER^2} . \tag{8.11}$$

And if the deflection per unit length of the cantilever is assumed to be constant, that is, $\delta / l = k$, then

$$l = \left(2k \frac{E}{\rho}\right)^{1/3} R^{2/3} \propto R^{2/3} . \tag{8.12}$$

Thus, when uniformly distributed static loads are considered, the optimal taper that ensures a uniform elastic stiffness is given by a two-thirds-power rule.

The constant-stress model is predicated on the physical analogy of flexure under static loading and the deflections that could result from wind-pressure dynamic loadings against the crown of a tree. The model identifies the beam taper that ensures that the maximum bending stress is uniform along the length of a beam. Recall from chapter 3 that the maximum bending stress σ_{max} is given by the formula

(8.13) $$\sigma_{max} = \frac{MR}{I},$$

where the bending moment M equals the product of the load and beam length. For a cylindrical beam submitted to a uniformly distributed load intensity q, eq. (8.13) can be rewritten to yield the formula

(8.14) $$\sigma_{max} = \frac{2ql^2}{\pi R^3} = \frac{2\rho\pi R^2\, l^2}{\pi R^3} = \frac{2\rho l^2}{R}.$$

Rearranging eq. (8.14) to solve for l, we see that $l = (\sigma_{max}/2\rho)^{1/2}\, R^{1/2}$. Thus $l \propto R^{1/2}$, and assuming that σ_{max} and the average tissue density are constants, a log-log plot of beam length versus beam radius ought to yield a regression line with a slope equal to 0.5. That is, the taper of the beam should conform to a one-half-power rule. McMahon and Kronauer (1976) provide extremely elegant derivations for the one half and the two-thirds-power rules. However, eqs. (8.10) to (8.14) provide the gist of these derivations without undue mathematical fuss.

Dean and Long (1986) reconsidered the constant-stress model in terms of the dynamic loadings of wind pressure. In their analysis, the self loading is replaced by the force F of wind on the tree canopy. This force was taken to be the product of the total leaf area A of the plant and the wind pressure $\mathcal{P}$. Although wind exerts a pressure everywhere on the projected area of a tree, including the tree's trunk, the bulk of the wind pressure is applied to the center of the crown. Dean and Long, citing unpublished data, argued that the total leaf area of a tree is proportional to the silhouette area of the crown. Thus, if we substitute $A\mathcal{P}$ for the load and $D/2$ for the radius in eq. (8.14), where D is the diameter of the trunk, then we can derive the formula

(8.15) $$D \propto \left(\frac{A\mathcal{P}\, l}{\sigma_{max}}\right)^{1/3}.$$

If the wind pressure $\mathcal{P}$ and maximum stress σ_{max} are taken as constants, then we can see that trunk diameter must be proportional to the 0.33 power of the product of the silhouette area of the crown and the length (height) of the trunk:

(8.16) $$D \propto (A\, l)^{0.33}.$$

TABLE 8.1 Statistics for the Exponents of Eqs. (8.16 and 8.17) Based on Nonlinear Regression Analysis of Data from Mature and Sapling *Pinus contorta*

Sample	Exponent ± SE (n)
Regression according to Eq. (8.16)	
Mature	0.31 ± 0.0037 (306)
Destructive	
Nondestructive	0.32 ± 0.0067 (532)
Sapling	0.28 ± 0.0005 (284)
Destructive	
Regression according to Eq. (8.17)	
Mature (nondestructive)	1.36 ± 0.056 (532)
Sapling (nondestructive)	0.91 ± 0.032 (258)

Source: Data from Dean and Long (1986, tables 2 and 3).

Actually, the exponent in eq. (8.16) can vary between 0.33 and 0.50 depending on how free the base of the tree is to move. This range in the exponent somewhat diminishes the ability of eq. (8.16) to discriminate between the one-third-power rule and the one-half-power rule, which is unfortunate, since we would like to use empirical data to test how trees are designed.

However, the elastic-stability model (which by rights should be called the elastic-stiffness model, since it says nothing about the allometry of strength) predicts that trunk diameter should be proportional to the 1.5 power of trunk length (height):

$$D \propto l^{1.5} . \quad (8.17)$$

Thus the constant-stress model adapted for dealing with wind pressure and the elastic-stability (or elastic-stiffness) model provide two different predictions for the allometry of growth in trunk diameter and length that can be empirically tested by regression analysis of data from trees. In an exhaustive study of over 11,000 *Pinus contorta* trees, Dean and Long (1986) found that both mature trees and saplings grew in a manner consistent with the predictions of the constant-stress model for trunk taper. The allometry of mature trees simultaneously followed the predictions of the elastic-stability model, but that of saplings did not. From their data, Dean and Long concluded that the tapering of *Pinus contorta,* which equalizes the bending stress due to wind loadings, appears to be essential for all stages of growth, while mature trees tend to limit their height to below the theoretical critical buckling length (table 8.1).

Although none of the models thus far discussed has been disproved, it appears that the most generally applicable is the constant-stress model, provided the local environment inflicts significant wind loadings. Recall that over 80% of all the plant species examined exhibit thigmomorphogenesis such that the allometry of growth is responsive to how much an organ is mechanically perturbed. The shoots of woody species show different taperings if they are restrained by guy wires or unrestrained. In a study of *Pinus radiata,* Jacobs (1954) showed that restrained trees grew less in diameter than unrestrained trees. When released from their restraining wires, trees were no longer mechanically capable of sustaining their own weight and underwent large deflections from the vertical or simply snapped. Also, Larson (1965) showed that the growth in the upper portions of the stems of *Larix radiata* tended to equalize bending stresses when lower branches of trees were pruned to change the distribution of stresses. Collectively, these studies indicate that plant growth can accommodate both static and dynamic loadings simultaneously. The tapering of branches and tree trunks most likely reflects a trade-off between the optimal geometries predicted by the constant-stress and elastic-stability models. The reconciliation of design factors predicted by these two models should not come as a surprise, nor should the fact that at some stages in the development of a plant one or the other design factor may predominate.

In an extremely interesting and thoughtful paper, Holbrook and Putz (1989) compared predicted critical dimensions (diameter and stem length) based on a number of theoretical treatments of critical buckling dimensions with the empirically measured dimensions of sweet gum (*Liquidambar styraciflua*) saplings that were open grown, prevented from swaying in the wind by means of guy wires, and both guy wired and artificially shaded. Holbrook and Putz considered five formulas:

Greenhill (1881):

(8.18a) $$L_{cr} = 1.26\left(\frac{E}{\rho}\right)^{1/3} D_b^{2/3} \text{ (uniform column)}$$

and

(8.18b) $$L_{cr} = 1.97\left(\frac{E}{\rho}\right)^{1/3} D_b^{2/3} \text{ (tapered column),}$$

where D_b is the diameter at the base of the column (tree trunk).

King and Loucks (1978) and King (1981, 1987):

(8.19) $$D_{\min} = 2\left(\frac{\rho}{CE}\right)^{1/2} H^{3/2}\ ,$$

where $C = (5.33 + 60.6k + 23k^2)/(1.0 + 20.4k + 119k^2 + 429k^3)$, k is the ratio of crown to trunk mass and H is 10/9 of the height to the center of the crown.

Gere and Carter (1963):

$$L_{cr} = 1.71 \left[\left(\frac{D_b}{D_a} \right)^{2.53} \frac{EI_a}{P} \right]^{1/2} , \tag{8.20}$$

where D_a is the diameter at the apex (the crown center of mass) and I_a is the second moment of area at the apex.

Holbrook and Putz (1989, 1743):

$$\int_0^L EI(Y)[X(Y)'']^2 dy = \int_0^L [(W_s(Y) + W_c(Y)] \{\int_0^Y [X(Y)']^2 dY\}_r dY \tag{8.21}$$

(based on the principle of virtual work, which states that for a body to be in elastic equilibrium the total work done on it by external forces must equal the increase in the elastic energy stored within the body), where Y is the axis along which length is measured, X denotes the horizontal displacement, $I(Y)$ is the second moment of area, given by $\pi[r(Y)]^4/4$, $r(Y)$ is the linear dependence of stem radius on height, $X(Y)$ is the horizontal displacement, $X(Y)'$ and $X(Y)''$ are the first and second derivatives taken with respect to Y, $W_s(Y)$ is stem weight as a function of Y, and $W_c(Y)$ is the crown weight as a function of Y.

Holbrook and Putz found that three formulas (Greenhill's first formula, Gere and Carter's, and their own) approximated the experimental results fairly well. By contrast, Greenhill's second formula (for a tapered column) overestimates the critical buckling height because it neglects the weight of the crown and assumes that the trunk is tapered to a point (so that the weight at the top of the tree vanishes to zero and therefore underestimates the moment arm). The formula of King and Loucks holds the height and the ratio of crown to stem mass constant while calculating the critical buckling diameter. Therefore the critical dimensions of trees are overestimated (they are larger than the actual measured critical diameter).

Although the predictions of the Gere and Carter formula were found to agree with the experimental data, this formula should also have overestimated critical buckling heights because the weight of the tree trunk is neglected and the weight of the crown is assumed to be applied to the trunk as a single apical load (at the crown's center of gravity) rather than as a distributed load; that is, the formula's assumption of load distribution results in a lower bending moment than actually occurs. That formulas, like Gere and Carter's, based on assumptions that clearly do not comply with botanical realities nonetheless yield predictions that appear to agree with experimental data ought to generate considerable skepticism about how well we truly understand tree biomechanics in terms of the scaling of critical dimensions. The findings of Holbrook and Putz (which ought to be read in detail by anyone interested in the mechanical design of trees) illustrate a point raised in chapter 1—the only good model is one that can be shown to be wrong.

One major assumption of all the models used to investigate the mechanics of trees is that the allometry defining the relative proportions of a tree is uniform throughout. Contrary to this assumption, a careful analysis of a silver maple (*Acer saccharinum*), measuring 13 m in height with a wet mass of 370 kg, revealed that branches below a certain critical size (equal to or less than 3 m in length) are allometrically more slender as size decreases, while those branches above the critical size are more robust as size increases (Bertram 1989). That is, distal branches, which incidentally bear most leaves on the tree, were disproportionately more slender than the branches that bore them in turn. The allometry of this silver maple appears not to be idiosyncratic, since that of a white oak (*Quercus alba*) reported by McMahon and Kronauer (1976; compare their fig. 6 with Bertram's fig. 3) is directly comparable. Why the radial growth of peripheral branches is less robust than the overall allometry of the tree remains something of a mystery, although several hypotheses can be advanced: for example, the radial growth (thigmomorphogenetic) response to mechanical stimulation of peripheral branches may be repressed in some manner, perhaps because leaves may be the source of some diffusible photosynthate that limits radial growth. Regardless of how differences in the scaling of branches are effected, the mechanical benefits appear much clearer. Peripheral branches are far more flexible than their larger, more basal counterparts, and so critical strain levels induced by dynamic loadings may be avoided. Thus deflections typically do not induce fracture, and leaves remain attached to trees.

Does a general model exist for the scaling of mechanical parameters among all woody plants? Circumstantial evidence suggests the answer may be yes. Norberg (1988) concluded that a geometric similarity model (shape remains constant as size increases) adequately describes the allometry of small growth forms like mosses, ferns, grasses, herbs, and very small trees, while the elastic similarity model more adequately describes the scaling of large arborescent growth forms. Data provided by Whittaker and Woodwell (1968) reveal that the allometry of shrub species conforms well to that of the peripheral branches of the silver maple Bertram examined, whereas the scaling exponent of tree species was only slightly less than that found for the larger, nonperipheral branches of the same tree. These findings suggest that the small peripheral branches of trees can be considered, in a structural sense, shrubs that are simply held aloft by the larger stems and branches of trees (Bertram 1989, 252).

MECHANICAL FAILURE OF MATURE SHOOTS

Wind, rain, and other loadings on trees can uproot or snap trunks. A variety of factors influence a plant's susceptibility to these types of mechanical failure. Principal among these is the strength of wood. Uprooting of trees tends to occur among species with dense, stiff, and strong wood. Trees with weak wood tend to snap rather than uproot. In predicting whether a tree will uproot or snap, however, a number of other factors are as important as the mechanical properties of wood. In uprooting, external loadings do not exceed stem strength but do exceed the root-holding capacity of the soil. Accordingly, the strength of root systems is critical in determining a particular tree's susceptibility to uprooting. The overall strength of the root system is difficult to determine (Fraser 1962; Fraser and Gardiner 1967) and depends on soil type and moisture content (Sutton 1969) as well as on the age of the roots (since the relative tensile properties of roots change as a function of age). In general, older roots are less tensile and operate as compression-supporting members. (The tensile portions of roots have a central cable of primary vascular tissue and occur toward actively growing tips that grow away from the foundation of the root crown and trunk.) In younger root systems, all portions of the root crown experience tension when the shoot is placed in dynamic bending, and the junction between the root crown and the trunk pivots. In older root systems, roots on one side of the root system are placed in tension while those on the opposing side are placed in compression, and the pivoting of the root crown that precedes uprooting is skewed toward the side experiencing compression.

Crown size and shape and how well trees are buttressed influence the mode of mechanical failure. Crown size and shape influence wind resistance; artificial pruning or natural abscission or breakage of branches lessens wind-induced root movements (Hutte 1968). Since the relative flexibility of branches also affects their ability to deform and reduce drag, the stiffness of wood and leaves influences a tree's ability to resist wind pressures. Buttressing increases the resistance of trees to mechanical loadings (Richards 1952; Smith 1972), and Henwood (1973) concluded that by lengthening the moment arm of the base of a tree, buttresses may reduce the tensile stresses developing within the root system. Thus buttresses may have a complex influence on the

TABLE 8.2 Examples of Forest Type and Percentage of Tree Mortality Due to Uprooting

Forest Type	Percentage
Tilia-Carpinus-Quercus forest (Poland)	48
Sirena forest (Costa Rica)	37
Llorona forest (Costa Rica)	34
Abies forest (California)	25
Tropical moist forest (Panama)	25
Mesic gap forest (eastern USA)	19
Fagus-Magnolia forest (Texas)	0

Source: Data from Putz et al. (1983, 1017).

mode of failure of trees. Mergen (1954) reports that unbuttressed trunks tend to snap near the ground. Thus buttressing might be expected to reduce the probability of uprooting and favor snapping above the portion of the trunk where buttressing occurs.

In a very interesting study, Putz et al. (1983) evaluated the importance of a number of factors that might affect the mode of mechanical failure for Panamian tree species. Among 310 fallen trees, these authors report that 70% snapped somewhere along the length of the trunk, 25% uprooted, and 5% broke at ground level. After performing a stepwise discriminant analysis between uprooted and snapped individuals, they found that wood properties principally correlated with the mode of failure. Uprooted individuals tended to have denser, stiffer, and stronger wood and to be shorter for a given trunk diameter. Buttressing appears to have had little effect on the mode of failure, nor did the depth of the soil in which trees grew. Uprooted trees had wood with higher elastic and rupture moduli than the wood of snapped trees. However, Putz et al. (1983) reported that the material properties of wood could not be used to predict the mechanical behavior of trees. They found no significant correlation between wood flexibility and tree flexibility ($r = 0.078$, $n = 16$, $p = 0.4$). These experimental data are significant because they show that the mechanical behavior of a structure cannot be invariably or accurately estimated from the material properties of the shoot alone.

In an excellent review of the literature, Putz et al. (1983) point out that the dominance of uprooting or snapping varies from one type of community composition to another (table 8.2). Obviously, several biological and environmental factors influence this variation, but the relative age of the community and of individual plants may be most significant. The relative susceptibility to uprooting or snapping may change with floristic composition (species with different woods), the age of individual plants (older trees approaching their critical buckling lengths), and a variety of environmental factors such as the quantity of rainfall, soil conditions, and the direction, magnitude, and duration of wind.

Ways Plants Recoup Their Losses (Reaction Wood)

Woody angiosperms and gymnosperms are capable of reorienting branches and tree trunks when plants are mechanically displaced from their original growth position (Archer and Wilson 1970; Wilson and Archer 1977). This orientation involves producing reaction wood. In leaning stems, the vascular cambium produces secondary xylem (wood) that mechanically acts to bend the stem upward toward the vertical by either contraction or expansion. When the wood contracts it is called tension wood (TW); when it expands it is called compression wood (CW). This terminology does not reflect that TW is produced in the regions of the stem that experience tensile stresses or that CW is produced in regions of the leaning stem that experience compressive stresses. In fact, when a gymnosperm is tied into a vertical loop, CW forms on the bottom of each part of the loop. From first principles (chap. 3), we know that the concave surface of the upper portion of loop is in compression while the convex surface of the lower portion of the loop is in tension. Thus the locations where CW is formed do not conform to the nature of the stresses experienced by the stem. Rather, the terminology CW and TW refers to the way the reaction wood mechanically operates to restore the orientation of the bent shoot.

All the available evidence concerning reaction wood appears to support the hypothesis that the stimulus that largely induces its formation is gravity (Wardrop 1964; Westing 1965; see Wilson and Archer 1977). If gravity were the only factor influencing the distribution of reaction wood in branches, however, then all branches would be oriented vertically. From our everyday experience we know this is not so. A more reasonable view of the factors that stimulate the formation of reaction wood sees each growing shoot as having an equilibrium position, EP (a term suggested by Little 1967). The EP is dictated by as yet unknown physiological and mechanical factors whose interactions and consequences differ among shoots programmed to grow vertically or horizontally or somewhere in between. When a shoot is reoriented, reaction wood is produced to restore the shoot's EP. How this is accomplished is still unknown. Thus we must view the equilibrium position hypothesis as a heuristic device to explain complex growth responses in wood plants.

The complexity of these growth responses is illustrated by reviewing a few experiments in which reaction wood formation can be induced, as well as the information available on where reaction wood is produced in naturally growing plants. To do this we must first consider the anatomical characteristics of reaction wood.

Plant anatomists have long noticed that horizontal branches of conifers are eccentric in cross section, with the pith displaced toward the upper surface. Below the pith there is a wedgelike region of reddish wood that is still referred to as rotholz (literally, "red wood") in the lumber industry of today. The rotholz is the reaction wood of conifers, and from the description of where it is found we can understand why it was referred to as compression wood. As we saw earlier, however, CW is formed in regions of a bent shoot that experience compressive or tensile stresses. Thus the earlier notion that compressive stresses alone induce CW is incorrect. But note that conifers produce CW only, while angiosperms produce CW and TW.

Compression wood in conifers always results from a differential growth response in the vascular cambium and is identified by its rounded, relatively short tracheids. The S2 layer within secondary cell walls is thick, and the secondary wall layers have a high lignin content and a high microfibrillar angle. In the initial stages of the formation of tension wood, the xylem fibers that are produced appear normal until an inner gelatinous layer (S_g) within the secondary walls is produced. This layer has a low lignin content and a low microfibrillar angle (see Wilson and Archer 1977, 24–25, for an excellent review). The increase in gelatinous fiber formation is typically accompanied by a significant reduction in the size and number of tracheary elements produced in the tension wood. Kennedy (1970) has described opposite wood, which is found on the opposite side of a shoot in which CW if formed. Opposite wood differs from normal wood (found on either lateral side of a shoot) in that it has thicker cell walls and so looks very much like the secondary xylem produced toward the end of each seasonal cycle of wood deposition; that is, opposite wood looks like the latewood in each growth layer of normal wood. Münch (1938) pointed out that there is a continuous gradient of wood characteristics among normal, opposite, and reaction wood.

Sinnott (1952; see also Sinnott 1960, 355–58) showed that when CW is induced in an artificially bent branch of white pine (*Pinus strobus*), CW is also induced for a short distance below the point of attachment within the vertical trunk. Jankiewicz (1966) also showed that CW can form above the branch attachment site in the trunk. Thus the stimulus to induce CW appears to be transmitted over a relatively short distance along the vertical axis of the trunk. Experiments indicate that exogenously applied auxin can induce CW formation in vertical shoots, while the antiauxin TIBA reduces the amount of CW formed. Gibberellin and cytokinins appear to have little direct effect but can stimulate cambial activity in general.

The mechanical action of reaction wood appears to be the capacity of TW to contract and of CW to expand, thereby generating growth strains that deform the branch so that its original EP is reestablished. As we saw earlier (chap. 6), normally developing wood generates growth strains that must interact with those produced by reaction wood, but the relative magnitudes of normal and reaction wood growth differ substantially. For example, Nicholson, Hillis, and Ditchburne (1975) showed that in severely bent *Eucalyptus* stems, TW produces compressive strains ten times greater than those measured in normal wood.

Münch (1938) appears to have been the first to suggest a mechanism for the mechanics of CW—the swelling of lignin among the cellulose microfibrils within the differentiating secondary cell wall. The microfibrillar angle appears to be the major factor determining the magnitude and direction of growth strains produced when the secondary wall swells (Boyd 1973, 1974).

Attempts have been made to predict the internal strain distribution within an artificially bent shoot and to calculate the strains that can restore the shoot's original orientation. These analyses are made very difficult because the wood present in the shoot before it is bent is subjected to external bending strains associated with the new, deformed shape and will retain the prior growth strains produced by normal growth, whereas reaction wood formed in response to the imposed bending stresses will not be under the same bending stresses as the older wood but will produce new growth strains. Archer and Wilson (1970, 1973) showed that counterintuitive shifts in tensile and compressive stresses can occur within a bent shoot. Their computations indicate that the strains in the younger wood can be reversed from those induced by the initial bending. These shifts are actually not counterintuitive at all when we consider that the growth of the shoot superimposes a new field of growth strains over the artificially induced field of stresses caused by bending. Wilson and Archer (1977) reviewed much of their prior work on reaction wood and empirically demonstrated that the consequences of growth strains responsible for restoring a shoot to its original position can be predicted by continuum mechanical models.

Before leaving the topic of reaction wood, an interesting case of tension wood formation is worth noting. Zimmermann, Wardrop, and Tomlinson (1968) reported that the aerial roots of the subtropical fig tree (*Ficus benja-*

mina), which extend downward to the ground and develop secondary growth once anchored in the substrate, produce tension wood until they are between 10 and 15 mm in diameter. The capacity of the reaction wood in these aerial roots to contract organ length was amply demonstrated by these workers. Roots were planted in pots during their free-hanging stage and could lift the pots from the ground during their subsequent development. By inference, we can assume that the aerial roots of *Ficus benjamina* place the branches they are suspended from in tension once they anchor to the ground and mechanically operate as guy wires. The mechanics of tension wood in this species and its consequences on the mechanical stability of branches would provide an extremely interesting subject for future investigations.

Another interesting manifestation of reaction wood is seen in plants that normally produce pendulous branches, such as the weeping willow and weeping birch. As some branches elongate, they assume a vertical posture some distance from their decumbent apical meristem. Since the intensity of the tensile and compressive stresses developing within each branch is highest in the region of bending that has the largest radius of curvature, and since this is where reaction wood develops, the branches of weeping species of trees would make an interesting experimental system in which the ever-changing static loading of branches could be used to evaluate the mechanics and physiology of reaction wood.

In summary, although we now have a fairly firm conceptual view of how reaction wood operates mechanically, we are woefully ignorant of the physiological mechanisms responsible for its formation. Additionally, we still do not have a firm grasp of how CW and TW achieve their growth strains at the cellular level. The mechanics of reaction wood has obvious consequences and benefits to mature plants and remains fertile ground for continued research.

Nine

Fluid Mechanics

> He learnt to swim and to row, and entered into the joy of running water; and with his ear to the reed-stems he caught, at intervals, something of what the wind went whispering so constantly among them.
>
> Kenneth Grahame, *The Wind in the Willows*

All organisms operate physiologically and mechanically within a fluid that is either gaseous or liquid. Consequently, an understanding of the physical properties and behavior of fluids is requisite to virtually every level of biological inquiry. Although some of the generic physical properties of fluids were reviewed in chapter 2 and the biological importance of water to plants was specifically treated in chapter 4, we have largely neglected the kinetic properties of fluids—the physics of fluid motion. The science treating this subject is called fluid mechanics, and a review of this science and the insights it provides the botanist will occupy us throughout this chapter.

Virtually every body of fluid is in motion, because all fluids have little or no capacity to resist externally applied forces; that is, fluids shear easily. (Even glass, which is a non-Newtonian liquid, flows under the influence of gravity, and as a consequence an old pane of glass is often much thicker at its base than at its top.) As a consequence of this propensity for movement and rapid deformation, the characteristic problems addressed by fluid dynamics are those where viscous forces are largely neglected. This intrinsic capacity for deformation has two biologically relevant consequences. First, fluids mix and transport suspended or dissolved materials. The capacity for fluids to mix and transport materials is critical to the mixing of the atmosphere and oceans and to a variety of physiological processes such as respiration and photosynthesis. Vertical mixing of the oceans and atmosphere results in the mass transport of carbon dioxide and oxygen, preventing most habitats from becoming physiologically stagnant. Monteith (1973) calculated that the photosynthetic activity of an average crop would consume all the available CO_2 within a 30 m thick layer of air above the plants in a single day. Obviously this does not happen, but Monteith's calculations indicate that individual plants would die if the air did not behave as it normally does. Also, since most plants are sedentary, many species require the movement of air or water to propagate. Wind-pollinated plants rely on mass transport of pollen, and many biotically and abiotically pollinated species use air currents to disperse seeds and fruits. Perhaps less understood but equally important is the way marine plants use water flow for pollination (e.g., the sea grasses) or gamete/zygote transport (the algae).

The second consequence of the movement of fluids is that they exert a direct mechanical influence on objects immersed in them—momentum is transferred from the moving fluid to any object that obstructs its flow (see eq. 1.4). Hence the movement of fluids displaces objects from their static equilibrium. As already noted, these displacements influence physiological processes, such as photosynthesis, but under extreme conditions the motion of air and water can result in the mechanical instability and failure of plants.

Let me begin the discussion of fluid dynamics with a brief recapitulation of the physical properties of fluids, followed by a treatment of some of the general equations of fluid flow and the simplifying concepts that make these equations less mathematically formidable. The bulk of this chapter, however, is devoted to plant aerodynamics and treats the subjects of wind pollination and wind dispersal of airborne seeds and fruits.

Physical Properties of Fluids

In chapter 2 we saw that the elastic modulus could be used to describe the material properties of an ideal (elastic) solid, whereas Newtonian fluids do not have an elastic modulus. However, we also learned that for Newtonian fluids ("other things being equal") the ratio of stress to the rate of shear is constant, and that the proportionality factor is called dynamic viscosity (μ). The dynamic viscosity of a Newtonian fluid reflects the fluid's material properties in much the same fashion as the elastic modulus describes the material properties of an elastic solid. Highly viscous non-Newtonian fluids like tar, molasses, and wet soils, however, can be extended in the form of beams or columns, and they have a property that is equivalent to an elastic modulus, called the coefficient of viscous traction. Since the Poisson's ratio of all known liquids, Newtonian or non-Newtonian, equals 0.5, it follows that the coefficient of viscous traction must always be three times as great as the viscosity.

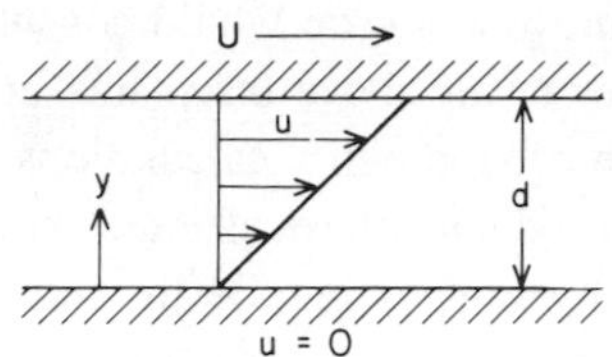

FIGURE 9.1 Couette flow between a stationary (bottom) plate and an upper plate (moving with a velocity U) separated by distance d. The velocity of the trapped fluid u (measured at distance y) increases as y approaches d. This creates a gradient in velocity between the two plates such that u increases from zero (at the surface of the stationary plate) to U (at the surface of the moving plate).

Also, all fluids, whether gaseous or liquid, have a bulk modulus K, which equals the shear modulus of the fluid.

Returning specifically to Newtonian fluids, dynamic viscosity can be formalized mathematically by considering Couette flow, created within a thin layer of a fluid that is trapped between two flat plates, one stationary and the other moving at a speed U (fig. 9.1). The fluid has zero speed at the interface between it and the surface of the stationary plate and has a speed equal to U at its interface with the moving plate. A gradient of speed exists between these two extremes such that

$$u = \frac{y\,U}{d}, \tag{9.1}$$

where u is the speed of fluid flow at any distance y from the stationary plate and d is the maximum distance between the two plates. It is easy to see that eq. (9.1) is reasonable, since when $y = d$ the fluid has a speed of U, and when $y = 0$ the fluid is stationary. Dynamic viscosity can be conceptualized from the physics of Couette flow by recognizing that a tangential force is required to keep the moving plate in motion and that this force must be equal to the force required to keep the stationary plate at rest. The tangential force divided by the surface area of the plate gives us a shear stress that must be in balance with the viscous stresses within the fluid. Since all the layers within the fluid are subjected to the same shearing stress τ, the shearing stress must be proportional to the change in the speed u with respect to the changes in the distance y, symbolized in terms of the calculus as du/dy, which is the velocity gradient. The change in u with respect to the change in distance y, or du/dy, is the rate of shear. Since we know that the dynamic viscosity μ is the proportionality factor between the shear stress and the rate of shear strain, we can derive the following mathematical relation between the shear stress and the dynamic vis-

TABLE 9.1 Physical Features of Dry Air and Fresh Water at Atmospheric Pressure

Temperature (0°C)	Density, ρ ($kg{\cdot}m^{-3}$)	Dynamic Viscosity, μ ($kg{\cdot}m^{-1}{\cdot}s^{-1}$)	Kinematic Viscosity, υ ($m^2{\cdot}s^{-1}$)
	Dry Air		
0	1.293	17.08×10^{-6}	13.21×10^{-6}
10	1.247	17.62[a]	14.13[a]
15	1.226	17.84[a]	14.55[a]
20	1.205	18.08	15.00×10^{-6}
25	1.185	18.36[a]	15.49[a]
30	1.165	18.59[a]	15.96[a]
40	1.128	19.04×10^{-6}	16.88×10^{-6}
	Fresh Water		
0	0.99984×10^3	1.785×10^{-3}	1.785×10^{-6}
5	0.99997	1.519	1.519
10	0.99970	1.307	1.307
15	0.99910	1.139	1.140
20	0.99820	1.002	1.004
25	0.99704	0.890	0.893
30	0.99565	0.798	0.801

[a]Values extrapolated from measurements at 0°C, 20°C, and 40°C.

cosity for any Newtonian fluid:

$$\tau = \mu \frac{du}{dy}. \tag{9.2}$$

For a Newtonian fluid, μ is constant only for a given temperature. Thus the value of μ is temperature dependent. The dynamic viscosity increases for air (and decreases for water) as temperature increases (table 9.1).

Up to this point we have considered a fluid's viscous characteristics. Like all forms of matter, fluids have inertial characteristics as well—forces must be applied to either slow down or speed up a fluid or to change its direction of flow. Density is a measure of the inertial characteristics of a fluid, since it gives us the mass of a fluid particle. The ratio of μ to density ρ is a very convenient parameter with which to relate a fluid's viscous and inertial characteristics. This ratio is called the kinematic viscosity (υ). All other things being equal, a fluid with a high kinematic viscosity has a flow dominated by viscous forces, while a fluid with a low kinematic viscosity has a flow dominated by inertial forces. Perhaps counterintuitively, air is kinematically more viscous than water (table 9.1). Since density is temperature dependent, υ varies as a function of the temperature of the fluid.

Static Equilibrium, Steady Flow, and Bernoulli's Theorem

At some levels of comprehension we can draw a fundamental distinction between a liquid and a gas. For example, liquids are largely incompressible materials, while gases tend to be easily compressed. All fluids tend to be isotropic in their material properties, however, and at some levels of understanding we need not draw a real distinction between these two states of matter. Indeed, if their velocities are equivalent, a gas and a liquid will flow in much the same way. More precisely, the density of a fluid is generally irrelevant in understanding the characteristics of flow, because the forces and mass acceleration within any fluid are proportional to density. But density does enter into our understanding of the forces a moving fluid produces on an object obstructing flow. Thus we can use gas or liquid interchangeably if our intent is to characterize flow patterns, but we must distinguish between liquids and gases when we are concerned with the transfer of momentum.

A fluid at rest has no shear stress components, and the only stress within it is the hydrostatic pressure $\mathcal{P}$; that is, $\sigma_x = \sigma_y = \sigma_z = -\mathcal{P}$. From the equations of equilibrium (see eq. 3.1 and the attending discussion in chap. 3), we can quickly deduce that the equilibrium of force components F_i along any axis x_i through a body of fluid with density ρ is given by the formula $(\partial\mathcal{P}/\partial x_i) - F_i\rho = 0$. Thus, in the absence of gravity or any other body force, the hydrostatic pressure is constant throughout the body of fluid. When the force of gravity operating in the negative direction z of depth is considered, however, the force component in the z direction is $F_z = -g$, and the equations of equilibrium become $d\mathcal{P}/dz = -\rho g$, which gives us the familiar equation of hydrostatics, $\mathcal{P} = \mathcal{P}_o - \rho gz$, and Archimedes' principle, $F = \rho gV$; that is, the former states that the hydrostatic pressure will decrease as a function of decreasing depth (see chap. 2), while the latter states that the upward buoyant force F on an element of fluid of volume V is equal to the weight of the displaced fluid (see chap. 1).

When addressing the steady flow of fluids, we find that the instantaneous flow of a fluid can be expressed by a field of arrows, each indicating the velocity of an element of fluid at a particular instant in time. The lines that can be drawn from sequences of arrows in the direction of flow are called streamlines. When a fluid is in steady flow (and only then), streamlines precisely represent the lines of motion of fluid elements. (A streamline indicates the motion of many fluid elements; a line of motion represents the path followed by a particular element of fluid over time.) In steady flow, elements of fluid never cross from one streamline to another. This constraint is very useful in treating the behavior of fluids, since streamlines can be conceptually arranged into a tube (a stream tube) that circumscribes a limited portion of fluid; for example, each stream tube operates as a frictionless pipe through which the fluid moves. If the cross-sectional areas of the stream tube through which the fluid enters and leaves are denoted as A_1 and A_2, respectively, and if the stream velocity is similarly denoted as u_1 and u_2, then it must be the case that $A_1 u_1 \rho_1 = A_2 u_2 \rho_2$. Since for an incompressible fluid like water $\rho_1 = \rho_2$, it also must be the case that $A_1 u_1 = A_2 u_2$. This mathematical identity translates into the old proverb "still waters run deep"—streamlines must tend to converge wherever the fluid flow is rapid and must diverge wherever the flow decreases. The same conclusion is reached when we deal with air because, if velocities are less than roughly one-fourth the speed of sound, then gases can be treated as essentially incompressible fluids.

What happens when a volume of fluid decreases in speed? The answer is given by Bernoulli's theorem, which expresses one of the fundamental conclusions of fluid dynamics—that when a fluid decreases in speed, its pressure increases. Conversely, when a fluid speeds up, its pressure decreases. If this attribute of fluids did not exist, then airplanes and seagulls could not fly. When fluids move they do work, and if we assume that the viscous forces operating within the moving fluid are negligible (a fair assumption for ideal gases and very rapidly moving liquids), then all the work done by a fluid's motion must equal the sum of the kinetic and potential energies within the fluid. The work W done by a fluid moving through a stream tube with a diminishing cross-sectional area (a volume of fluid that is increasing in speed) equals the product of the pressure difference between the points of entry and exit, $\mathcal{P}_1 - \mathcal{P}_2$, and the change in volume dV; that is, $W = (\mathcal{P}_1 - \mathcal{P}_2)\, dV$, which must equal the kinetic and potential energies. Since the volume of the fluid element at A_1 has a kinetic energy equal to $(1/2)\rho u_1^2\, dV$, and since this volume has been replaced by an equivalent volume with kinetic energy $(1/2)\rho u_2^2\, dV$ at A_2, the kinetic energy has been increased by $(1/2)\rho(u_2^2 - u_1^2)dV$. By the same token, the potential energy U of a volume of fluid subjected to an external force will increase by $\rho(U_2 - U_1)dV$. If the external force is gravity, then $U_2 - U_1 = (z_2 - z_1)g$, where z is the vertical dimension. Since the total energy within the system must be conserved, $\mathcal{P}_1 + (1/2)\rho u_1^2 + \rho z_1 = \mathcal{P}_2 + (1/2)\rho u_2^2 + \rho z_2$. This identity resolves itself into Bernoulli's theorem; $\mathcal{P} + (1/2)\rho u^2 + \rho z =$ a constant along any streamline.

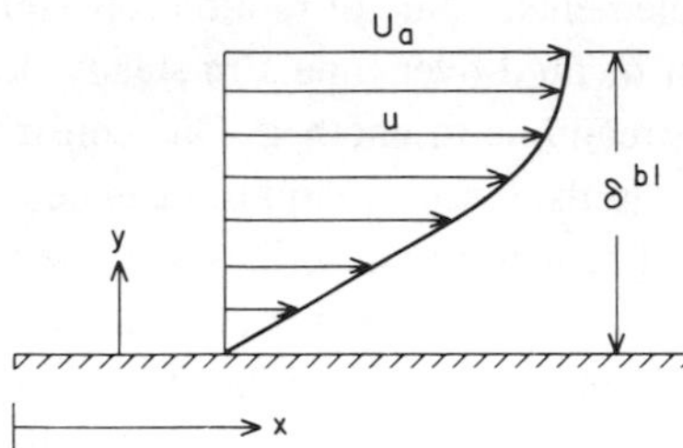

FIGURE 9.2 Boundary layer created at the surface of a stationary plate over which a fluid is moving with an ambient velocity U_a. At some distance x from the edge of the plate and at some vertical distance y from the surface of the plate, the velocity of the fluid u equals U_a. The distance y at which the local velocity equals 90% of the ambient velocity is the thickness of the boundary layer.

Bernoulli's theorem reveals that a fluid exerts a force (caused by an increase in pressure) when it strikes the leeward surface of any obstruction to its movement. The pressure $\mathcal{P}_1$ exerted by the center streamline (the streamline that collides with the leewardmost point on the obstruction) must exceed the ambient pressure $\mathcal{P}$ in the undisturbed streamline such that $\mathcal{P}_1 = (1/2)\rho U_a^2 + \mathcal{P}$, where U_a is the ambient velocity of the undisturbed streamline. By the same token, as streamlines pass around the obstruction they converge, their local speed increases, and their local pressure drops. Accordingly, a fluid moving around an object exerts a force on the object, while the object in turn influences the flow characteristics of the fluid moving around it. Although this all appears intuitively obvious, the effects of fluids on objects and those of objects on fluids can be very complex as well as biologically important. The following section addresses some of these effects and their relevance to plants.

BOUNDARY LAYERS

In a system where a fluid flows around a stationary solid, the velocity of the fluid is zero at the interface between the two. At some distance from this interface, the velocity of a fluid equals that of the fluid's mass flow. These two statements provide us with the logic to construct a mathematical relation between the velocity gradient and the shear stresses when considering Couette flow (see fig. 9.1). They also allow us to understand a very important concept in fluid dynamics called the boundary layer.

The boundary layer is a blanket of fluid surrounding a solid. Within this layer, viscous forces are dominant. Thus, regardless of a fluid's kinematic viscosity or ambient flow speed, there exists a layer of fluid around each object submerged within it for which we cannot ignore viscous forces. As we shall see, the geometry of the boundary layer is influenced by an object's shape and orientation to fluid flow and by the ambient characteristics of flow. For the time being, however, we can consider the boundary layer by examining a simple case. All we have to do is remove the moving plate in figure 9.1 and replace it with a layer of fluid extending indefinitely from the stationary plate (fig. 9.2). At some distance from the leading edge of the plate x the velocity of a moving layer of the fluid equals the ambient velocity U_a. At the plate-fluid interface, the velocity of the fluid is zero. As in Couette flow, the velocity of each layer of fluid at a distance from the plate's surface y is given by u. The distance y at which u equals 90% of the ambient velocity of flow U_a is the thickness of the boundary layer, symbolized as δ^{bl}. Provided the flow of the fluid is laminar—each layer of the fluid moves parallel to the plate's surface—the thickness of the boundary layer is given by the formula

$$\delta^{bl} = k\left(\frac{x\mu}{\rho U_a}\right)^{1/2}, \tag{9.3}$$

where k is some constant. Since the thickness of the boundary layer increases in proportion to the square root of x, the outer limit of the boundary layer has a parabolic geometry; the profile of the velocity gradient as u approaches U_a is curved as shown in figure 9.2.

Note that the stationary plate shown in figure 9.2 has to be extremely thin (much thinner than drawn in the figure), otherwise its leading edge will obstruct flow and disturb the geometry of the boundary layer. In nature we rarely deal with very thin plates and more commonly deal with the relatively thick and complex geometries of stems, leaves, and pinecones. Thus eq. (9.3) is limited in its applicability, though it does describe the general features of a boundary layer's velocity gradient around very flat objects. Pearman, Weaver, and Tanner (1972) considered the boundary layer around platelike leaves and derived an approximate expression for the average (boundary) layer thickness $\bar{\delta}^{bl}$:

$$\bar{\delta}^{bl} = 4.0\left(\frac{l}{U_a}\right) \quad \text{(platelike)}, \tag{9.4a}$$

where l is the mean length of the leaf (given in m) in the downwind direction, U_a is the ambient wind speed (given in $m \cdot s^{-1}$), and $\bar{\delta}^{bl}$ is given in mm. Similarly, Nobel (1974, 1975) shows that the thicknesses of the boundary layers that form around cylinders and spheres in air are expressed by the formulas

(9.4b) $$\bar{\delta}^{bl} = 5.8 \left(\frac{d}{U_a}\right) \quad \text{(cylinder)}$$

and

(9.4c) $$\bar{\delta}^{bl} = 0.28 \left(\frac{d}{U_a}\right) + \frac{2.5}{U_a} \quad \text{(sphere)},$$

where d is diameter (given in m). Once again, to be dimensionally correct, U_a must be given in $m \cdot s^{-1}$ such that $\bar{\delta}^{bl}$ is in mm. Equations (9.4a) to (9.4c) are useful under normal airflow conditions but are limited when airflow is very slow or very fast. They are presented here to illustrate that the ambient flow speed U_a and the dimensions of an object (either l or d) influence $\bar{\delta}^{bl}$. Implicitly (because there are different equations for plates, cylinders, and spheres), these equations also indicate that the boundary layer thickness depends on the shape of the object as well as its size.

We shall return to boundary layers later in this chapter, since they are important to a variety of phenomena such as heat transfer and pollen capture. But the full significance of boundary layers and the equations used to approximate their geometry will not be seen until we review some other features of fluid motion, principally the concept of Reynolds number, which provides us with a measure of the relative importance of viscous to inertial forces in moving fluids as well as a convenient context in which to discuss the characteristics of flow in general.

Reynolds Numbers and Flow around Cylinders

The pattern of fluid flow around an object and the resistance to motion that the fluid experiences depend on the relative magnitudes of the inertial and viscous forces produced within the flow. Although kinematic viscosity (hence density and dynamic viscosity) is an important physical property of a fluid, the velocity gradient of a fluid depends on the geometry of the obstructing solid and the fluid's ambient flow. The English engineer Osborne Reynolds recognized that a single dimensionless ratio could combine all these parameters and hence could be used to characterize flow patterns. This ratio has been named in his honor and is called the Reynolds number, symbolized by *Re,* which is given by the formula

(9.5) $$Re = \frac{\rho\, l\, U_a}{\mu} = \frac{l\, U_a}{\upsilon},$$

where l is the characteristic dimension. Since *Re* is computed based on the ratio of density to dynamic viscosity, the magnitude of *Re* reflects the ratio of inertial to viscous forces. Thus a high *Re* means that flow is dominated by inertial forces. From eq. (9.5) we can see that l and U influence the proportion of inertial to viscous forces in addition to υ. The characteristic dimension l is usually taken as the largest dimension of the obstructing solid that is parallel to the direction of fluid flow. Thus a cylinder with a length:diameter ratio of 10 will have an *Re* ten times higher if it is aligned parallel to flow than if flow is normal to its diameter. For many biological structures, selecting a characteristic dimension may present difficulties. Not only are they often irregular in shape, but many important biological structures change their orientation to the direction of fluid flow as a function of ambient flow speed or as a consequence of periodic (harmonic) motion caused by the transfer of momentum between the fluid and the structure. Therefore our example of the cylinder is somewhat trivial in terms of biological structures and their complex patterns of movement. *Re* is a powerful tool as well as an elegant concept. Provided *Re* is equivalent for any two geometrically identical systems, such as a scaled model and the real prototype the model is patterned on, the geometry of fluid flow around the model and its real prototype will be identical. Aerodynamics engineers have capitalized on this principle by constructing scale models of aircraft and examining their airflow characteristics before constructing expensive full-sized aircraft.

Wind tunnel experiments also exploit the principle of relative motion. The velocity of a moving object relative to a (more or less) stationary fluid is equal and opposite to the velocity of the fluid relative to a stationary object provided the velocity of the fluid relative to the object is kept constant. A researcher is free to choose whichever frame of reference is most convenient. If the velocity of flow is not constant, however, then the distinction between an accelerating object and an accelerating fluid is important, and the principle of relative motion does not apply.

The nondimensionality of *Re* provides a basis for describing flow characteristics around objects based on Reynolds numbers. In practice this is accomplished by examining a simple geometry and varying the ambient velocity, the magnitude of the characteristic dimension, or both to achieve a wide range of *Re*. Typically, two-dimensional flow characteristics are described. For example, if a very long cylinder is aligned normal to the direction of ambient flow, the behavior of fluid flow will be identical in every plane normal to the longitudinal axis of the cylinder. Thus the flow patterns around all transections can be described by considering flow around a single circular transection. By

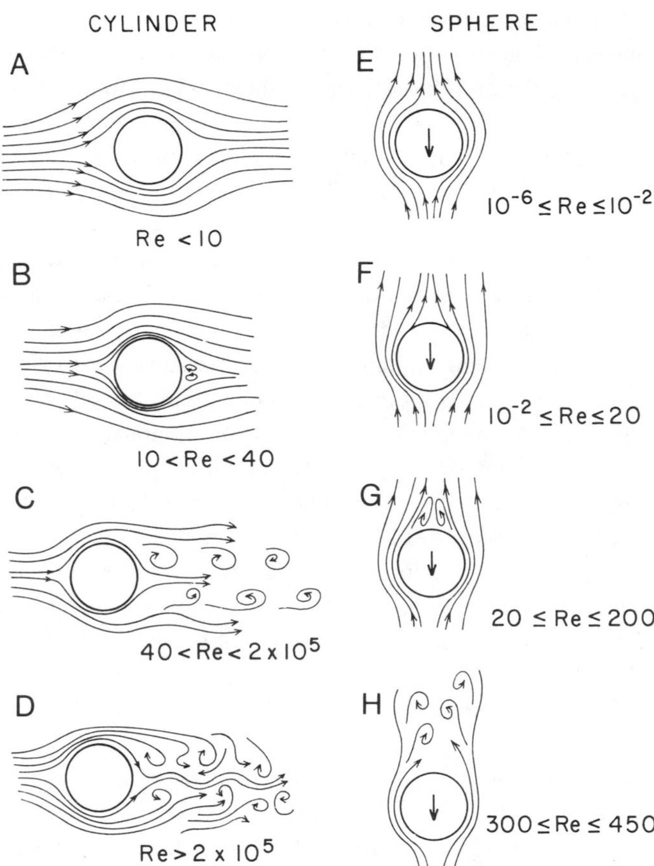

FIGURE 9.3 Flow characteristics around a transverse section of a circular cylinder (left) and a descending sphere (right) as a function of different Reynolds numbers (Re). See text for further details.

contrast, three-dimensional flow can be very complex, even around an infinitely long cylinder. This is because, even though flow patterns are the same for each transection through the cylinder, they may not be in synchrony along its entire length. Transfer of momentum can make solids vibrate and cause subtle deflections along the length of an object, with profound consequences for three-dimensional flow. Therefore it is wise to characterize flow around an object by referring to orthogonal planes (transverse and median-longitudinal) unless an object is essentially radial in symmetry.

We will consider first the changes in airflow patterns around a cylinder, because this geometry is very common among plants. Herbaceous plant stems, twigs, branches, and tree trunks are more or less cylindrical, as are many types of leaves. Figure 9.3A–D shows the flow characteristics around a single plane normal to a cylinder's longitudinal axis. (For the time being we will neglect the flow patterns that occur along the length of the cylinder.) When *Re* is much less than unity, flow is dominated by viscous forces and is symmetrical around the transection upstream and downstream of flow. The physical presence of the cylinder exerts a fluid dynamic effect over a distance measured in terms of many diameters to each side of the cylinder. The symmetry and range of influence change as *Re* increases beyond 4. The fluid is displaced by the cylinder more on the downstream side than on the upstream side, and in addition to this asymmetry, the influence of the cylinder is reduced laterally. Also, the fluid that comes very close to the sides of the cylinder never touches its downstream surface. Thus portions of the fluid become trapped downstream and rotate as attached eddies because of shearing forces. The size of these eddies increases as *Re* rises within the range $4 \leq Re \leq 40$. Momentum is transferred from the fluid to the cylinder. The decrease in the fluid's momentum means a reduction in fluid velocity. Since the rate of momentum transport is smaller behind than in front of the cylinder, a wake region develops downstream. At $Re \approx 40$, the flow in the wake becomes unsteady, and the instability takes on a form of flow called a Von Karman vortex street. That is, eddies detach in an alternating pattern from the downstream surface of the cylinder's cross section (fig. 9.3C). This instability continues up to $Re \approx 2 \times 10^5$. The Von Karman vortex street results from portions of the downstream flow assuming high vorticity (rapid rotation). At $Re > 100$ these portions of rotating fluid detach from the cylinder's sides and are shed downstream in a periodic manner with a frequency *f* related to the diameter of the cylinder *d* and ambient flow speed U_a by the Strouhal number, symbolized by *St:*

$$St = \frac{fd}{U_a}. \tag{9.6}$$

As *Re* increases to about 200, highly irregular and rapid velocity fluctuations occur downstream, and the Von Karman vortex street becomes turbulent at some distance from the cylinder. Tritton (1977) points out that periodic vortex shedding continues even as *Re* exceeds 10^7, despite the turbulence downstream of the cylinder. Fourier analysis of oscillograms (see Tritton 1977, 25, fig. 3.10) reveals the presence of a harmonic signal with a frequency *f* for $200 \leq Re \leq 10^7$, except at $200 \leq Re \leq 400$ and $3 \times 10^5 \leq Re \leq 3 \times 10^6$. This periodicity is largely ignored by the literature; flow at $Re > 2 \times 10^5$ is generally de-

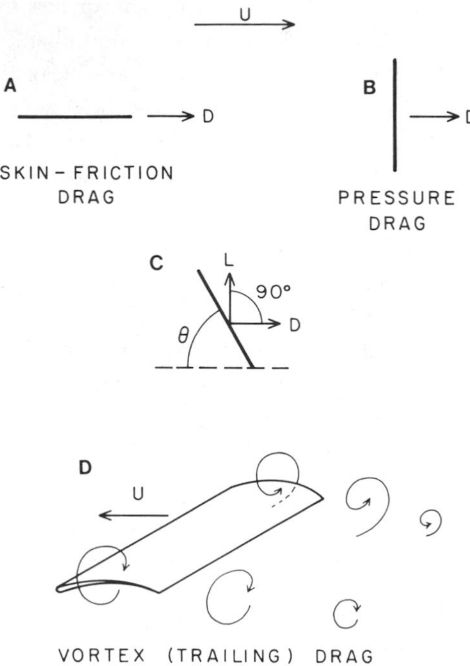

FIGURE 9.4 Drag and lift created around a plate held at some angle to a moving fluid. The movement of the fluid and the resistance of the plate to the fluid's movement create two orthogonal forces called drag and lift (top and middle). The drag force has three components: skin-friction drag, pressure drag, and vortex (trailing) drag. The last is seen as eddies of fluid circulating over the upper surfaces of a moving airfoil (bottom).

scribed as turbulent (see Vogel 1981, 74, fig. 5.5)—a term that inadvertently suggests there is little structure to the flow pattern.

The transition to turbulent flow at *Re* above 2×10^5 is a consequence of changes in the boundary layer around the cylinder. Below this value, flow within the boundary layer is laminar. Above this *Re* value, however, flow within the boundary layer undergoes a transition to turbulent flow, where fluid within the boundary layer moves into the downstream wake. We can now begin to appreciate why the formulas given for boundary layer thickness, eqs. (9.3) and (9.4), must be considered with caution. As I noted before, these formulas are useful only when flow within boundary layers is laminar. When *Re* exceeds roughly 2×10^5, the flow within boundary layers around cylinders (and other objects) takes on complex features. Vogel (1981, 129–34) provides some excellent advice on when (and when not) to use standard equations to calculate boundary layer thicknesses based on ranges of *Re*.

As we can see (fig. 9.3A–D), the flow patterns generated around a circular cylinder change as a function of *Re*. Likewise, flow around other geometries will show transitions dependent on *Re*. For biological structures with complex geometries, such as flowers, pinecones, and leaves, these transitions may be much less abrupt than those seen around simple, smooth-surfaced structures. The description of flow patterns for these complex objects must be based almost entirely on experimental observations, since analytical (and even numerical) solutions are typically unavailable. Nonetheless, these observations must be couched in terms of the dependency of flow patterns on *Re*.

DRAG AND LIFT

Plant structures come in a variety of shapes other than cylinders. Spores, pollen, seeds, and fruits tend to be spherical or elliptical or highly ornamented. But before we can discuss the flow around these plant structures in terms of *Re*, we must first understand the concepts of drag and lift.

All objects in moving fluids experience a force that is resolvable into two orthogonal components, one parallel to and the other normal to the direction of fluid flow (fig. 9.4). These two forces are called drag and lift, respectively. Generally, lift is important only at relatively high Reynolds numbers, since efficient (high) ratios of lift to drag occur only at high flow speeds. (However, in principle lift can be generated at any *Re*.) By contrast, drag is always important.

Drag always consists of at least two components, called pressure drag and skin-friction drag. The physical meaning of these two components is illustrated in figure 9.4 for a simple flat plate. As we can see, when the plate is aligned parallel to the direction of flow, drag is entirely skin-friction drag (the plate is considered infinitely thin and therefore projects no frontal area to flow). This drag component is the result of the net force produced by the resolvable components of all the tangential forces operating along the surface area of the plate. These forces are due to viscous shear stresses produced by two layers of moving fluid (one moving above the plate, the other moving under it) with equivalent velocity vectors (a vector has both speed and direction). When the plate is aligned normal to the direction of flow, drag is entirely due to the pressure drag component, which arises from the resolved components of the pressure acting normal to the surfaces of the plate. Because all real objects project some surface area toward the direction of fluid flow, all real objects experience both skin-friction drag and pressure drag.

Experience tells us that some objects are streamlined while others are not. Streamlined objects are those that smoothly separate airflow upstream and generate little downstream disturbance in fluid flow patterns. In general, streamlined objects tend to project little surface area toward the direction of fluid flow, thereby minimizing pressure drag, and have smooth surfaces even at relatively high Reynolds numbers. Objects that are not streamlined have comparatively high pressure drag components. Bluff-bodied objects generally project a large surface area in the direction of fluid flow and generate substantial downstream disturbance in fluid flow patterns, even at relatively low Reynolds numbers. The same object can be considered either streamlined or nonstreamlined depending on its orientation to the direction of fluid flow, and many biological structures, such as leaves and very flexible stems, reorient (flag) when subjected to relatively high fluid flows, thereby reducing their projected area and lowering pressure drag. The dependency of drag on orientation alerts us to suspect that it depends on *Re*. Thus we may intuitively expect drag to depend on flow speed, the size of the object, and the nature of the fluid. Indeed, the general formula for drag *D* is

$$D = \left(\frac{\rho\, U_a^2}{2}\right) (Re)^a A, \tag{9.7}$$

where *A* is surface area. The first parenthetical factor in eq. (9.7) is called dynamic pressure. It occurs because any portion of a moving fluid must conserve energy (as we saw earlier, there must be a balance between its rate of increase in kinetic energy and the rate at which pressure does work). The second factor in eq. (9.7), which is *Re* raised to some exponent *a*, is called the drag coefficient, symbolized by C_D. Thus, eq. (9.7) can be rewritten as

$$D = \left(\frac{\rho\, A\, U_a^2}{2}\right) C_D. \tag{9.8}$$

Clearly, C_D is not a constant. It varies as a function of *Re* and depends on the geometry of the object. For smooth objects such as circular cylinders and spheres, C_D changes dramatically as *Re* increases (see Tritton 1977, fig. 3.14). At low *Re*, however, C_D is roughly proportional to Re^{-1}. Thus, for a given geometry in a prescribed fluid, eq. (9.8) can be used to calculate drag because a direct proportionality between *D* and U_a exists for very low speeds. From $10^2 < Re < 3 \times 10^5$, C_D varies very little. This means that for a given object in a prescribed fluid, *D* is proportional to the square of the ambient flow speed. These generalities hold until $Re = 3 \times 10^5$, at which point the boundary layer around cylinders and spheres undergoes a transition from laminar to turbulent flow. C_D actually decreases (until about $Re = 6 \times 10^5$), despite an increase in the ambient velocity, because of the narrower wake resulting from the delayed separation of the boundary layer. The momentum extracted from the moving fluid is decreased as a consequence, and the drag is lowered.

Returning to figure 9.4, we see that the second force generated by fluids moving over objects is lift. This force occurs perpendicular to the drag vector. Since lift and drag operate orthogonally and drag depends on the orientation of the object, we must expect lift to be orientation dependent as well. Indeed, under constant flow speeds, lift increases linearly as the incidence angle θ rises from 0° to 10°–15° (depending on the object's geometry; fig. 9.4). Drag increases slowly at first and then much more rapidly as θ rises.

Thus far we have talked about only two of drag's components (skin-friction drag and pressure drag). Drag has a third component called vortex (commonly referred to as induced or trailing vortex) drag, which is a direct result of the generation of lift. A physical manifestation of vortex drag is seen every time an aircraft lands on or takes off from a dusty landing field—swirls of sand and dust (carried on vortices) are seen over the upper portions of the aircraft's wings. This vorticity is generated outside the airfoil's boundary layer, near its tips, and is carried into the wake (fig. 9.4). Farther downstream, the action of the fluid's viscosity dissipates the momentum of the (tracking) vortices.

Lift results from the acceleration of flow over the surfaces of obstructing objects, because high velocities produce low pressures. This is formalized by Bernoulli's theorem:

$$\mathcal{P} + \frac{\rho\, U_a^2}{2} + \rho z g = \text{constant}, \tag{9.9}$$

where $\rho z g$ is referred to as the elevation factor (see Static Equilibrium, Steady Flow, and Bernoulli's Theorem, above). Equation (9.9) contains the now familiar term of dynamic pressure (see eq. 9.7). Under most conditions of flow, liquids (and even gases) can be considered incompressible; therefore density ρ remains relatively unchanged. Thus, according to Bernoulli's theorem, at any given elevation or depth, as U_a increases $\mathcal{P}$ must decrease. (This is beautifully illustrated by slow-motion pictures of flying birds that show feathers on the upper surface of their wings popping upward owing to localized low pressure. As discussed earlier, eq. (9.9) tells us why.) Lift L is given by the formula

$$L = \left(\rho \frac{A\, U_a^2}{2}\right) C_L, \tag{9.10}$$

where C_L is the lift coefficient.

Terminal Settling Velocities and Flow around Spheres

Here we treat the aerodynamics of very small spheres. This is a prelude to discussing the behavior of plant spores and pollen in moving fluids, followed in turn by considering the aerodynamic behavior of wind-dispersed seeds and fruits. All the discussions will require some geometric simplification, because spores, pollen, fruits, and seeds can be morphologically complex and because analytical solutions to their exact behavior in fluids are lacking. From an aerodynamic perspective, however, all these plant structures are simply objects differing in mass and shape. Thus we can ignore taxonomic differences and categorize them according to morphological and physical features relevant to predicting their aerodynamic behavior.

The simplest geometry to begin with is the sphere, since it has a complete rotational symmetry about all three principal axes. The assumptions that it is not spinning and that it has a completely smooth surface greatly simplify the investigation of the aerodynamics of the sphere. Indeed, we must make these assumptions, because when they are violated even a simple geometry like a slightly roughened sphere becomes very complex aerodynamically (Mason 1978). The palynologist or microbiologist instantly recognizes that our assumption of a smooth-walled sphere eliminates the vast majority of spores or pollen, but we can derive formulas for the behavior of real spores and pollen only if we understand how simpler objects (like spheres) operate as fluid-borne particulates.

The aerodynamics of small, relatively low-density spheres is dominated by viscous forces because these shapes descend slowly in a column of air or water and thereby produce very low Reynolds numbers. Since most spores and pollen grains have diameters less than 2×10^{-4} m, we can assume their aerodynamic behavior is influenced by viscous forces. Under low *Re*, flow is essentially laminar, which greatly simplifies the analytical solutions for the behavior of small spheres suspended in a fluid.

Some of the airflow characteristics of spheres with different Reynolds numbers are shown in figure 9.3E–H. The values given for *Re* are approximate, since transitions from one flow regime to another can be very gradual. For very low *Re* (10^{-6} to 10^{-2}), flow around the sphere is laminar and steady. As in the case of *Re* for a cylinder, the flow patterns upstream and downstream are symmetrical. This symmetry deteriorates at about $Re \approx 20$, where a pressure difference between the front and rear of the sphere is generated—lower pressures occur at the rear. At $20 \leq Re \leq 200$ flow remains laminar and steady, but attached eddies form. The size of these eddies increases as *Re* rises to 200. At *Re* greater than 300, lateral asymmetry in flow occurs, and vortex shedding can produce regular oscillatory lateral movements. Above $Re \approx 450$, the frequency with which vortices are shed increases and the location where shedding occurs moves closer to the rear of the sphere. Also, lateral displacements of the falling sphere increase as *Re* rises and descent paths become progressively nonvertical.

Comparing the flow patterns generated around a sphere and around a circular cylinder reveals many similarities, but vortex shedding occurs at lower *Re* for spheres than for cylinders.

From eq. (9.8), we can calculate the drag force generated by a sphere with a relatively slow descent in a fluid, that is, at lower Reynolds numbers at which laminar flow occurs. This calculation requires that we specify the area *A* in eq. (9.8). If we select the total surface area of a sphere with radius *r*, then eq. (9.8) takes the form

$$D = \left(\frac{\rho U^2}{2}\right)(4\pi r^2)\left(\frac{6}{Re}\right), \tag{9.11}$$

where *U* is the velocity of the sphere relative to the fluid. As previously noted, at very low *Re* (under conditions of laminar flow), C_D is proportional to the reciprocal of *Re*. Thus, if we select $2r$ (the diameter of the sphere) as the characteristic length in the equation for *Re*, then eq. (9.11) becomes

$$D = \left(\frac{\rho U^2}{2}\right)(4\pi r^2)\left(\frac{6\upsilon}{2rU}\right). \tag{9.12}$$

Simplifying eq. (9.12) yields Stokes' law (named in honor of the great nineteenth-century physicist George Stokes, who determined the laws of motion for spheres at low *Re*):

$$D = 6\pi\mu rU. \tag{9.13}$$

(Notice that if we select the projected surface area of the sphere πr^2, then the proportionality factor for the relation between C_D and *Re* becomes 24. This yields the same formula as eq. 9.13.) Stokes' law was derived by mathematically considering the total force per unit area in the direction of flow acting at every point on the surface of the sphere. The sum of the pressures operating on this system is equal to $3\mu U/2r$. This value multiplied by the total surface area of the sphere once again equals $6\pi\mu rU$.

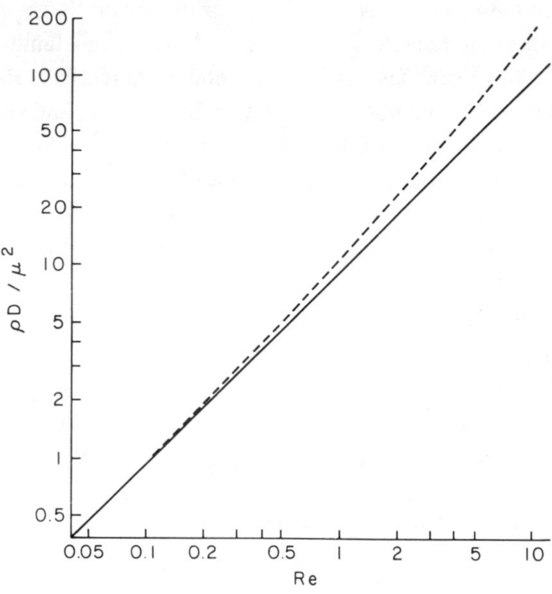

FIGURE 9.5 Drag on a sphere (expressed as the ratio of the product of density and drag to the square of dynamic viscosity: $\rho D/\mu^2$) plotted as a function of low Reynolds number (Re). Dashed line indicates empirical observations; solid line calculated from Stokes's law (see text, eq. 9.13).

The general form, but not the numerical factor 6π, of Stokes' law can be derived from a relatively simple dimensional argument. We suppose that the drag force D depends on the radius r and the velocity U of a sphere and on the density ρ and dynamic viscosity μ of the fluid it falls through. If we symbolize mass, length, and time as m, l, and t, then the drag force, which is the product of m and acceleration, has the units mlt^{-2}. Since density has the units ml^{-3} and viscosity has the units $ml^{-1}t^{-1}$ (force per unit area divided by the velocity gradient), then D equals $kr^aU^b\rho^c\eta^d$, where k is a real number. In terms of dimensions, $mlt^{-2} = l^al^bt^{-b}m^cl^{-3c}m^dl^{-d}t^{-d}$. By equating the powers of m, l, and t, we can see that $a = b$, $c = b - 1$, and $d = 2 - b$. Thus, D must equal $k(rU\rho/\mu)^b(\mu^2/\rho)$. If we suppose that D is proportional to U when U is very small, then $b = 1$ and $D = k\mu rU$, which is equivalent to eq. (9.13) when $k = 6\pi$.

Figure 9.5 plots the drag on a sphere (as a function of the product of fluid density and drag divided by the square of the dynamic viscosity) over a range of low *Re*. The dotted line reflects the distribution of experimental observations, while the solid line is based on the predictions from Stokes' law. As we can see, the analytical solution is very good for $Re<1.0$ but deviates significantly from observations made at $Re>1.0$. White (1974) has provided a formula to fit experimental observations made on spheres with $Re \leq 2 \times 10^5$:

$$C_D = \frac{24}{Re} + \frac{6}{1 + Re^{1/2}} + 0.4. \tag{9.14}$$

Equation (9.14) is based on computations in which the *projected* surface area was used to calculate drag.

If we can calculate the drag on an object, we can predict the object's terminal velocity, symbolized by U_T. When an object like a sphere falls through a column of fluid, it accelerates until the drag force acting on it precisely equals the object's net body force (weight minus buoyancy), at which point no further acceleration occurs and a terminal (constant) velocity is achieved. The weight of any object is its mass times the gravitational constant g (which is 9.81 $m \cdot s^{-1}$). Mass is volume times density. Since the volume of a sphere is $4\pi r^3/3$ or $4.189 r^3$, if we know the density of the sphere, we can easily calculate its weight. The buoyancy of an object is the difference between its density ρ and the density of the fluid ρ_o in which it is immersed. Accordingly, the net body force F_r on a sphere is given by the formula

$$F_r = g \left(\frac{4\pi r^3}{3} \right) (\rho - \rho_o). \tag{9.15}$$

Setting eq. (9.15) equal to the drag force given by Stokes' law (eq. 9.13) gives the formula

$$U_T = \frac{2r^2 g}{9\mu} (\rho - \rho_o), \tag{9.16}$$

where U_T is the terminal velocity. Equation (9.16) indicates that the terminal velocity increases in proportion to the square of the radius of the sphere. Larger spheres have a higher terminal velocity than small ones provided their densities are equivalent, but for any given mass, the larger the value of r the smaller the U_T. Spores and pollen from many plant species have numerous cavities or empty volumes that increase their size with little or no investment in mass. An example of this is seen in the air bladders or sacci found on pollen grains from most species of pine. For wind-dispersed pollen a smaller U_T is beneficial, since the grains take longer to settle out of the air column and thus will travel farther horizontally. Also, given the principle of relative motion, a smaller U_T means that a smaller airflow speed is required to elevate a spore or pollen grain. Remember that the size of palynomorphs (pollen grains, patho-

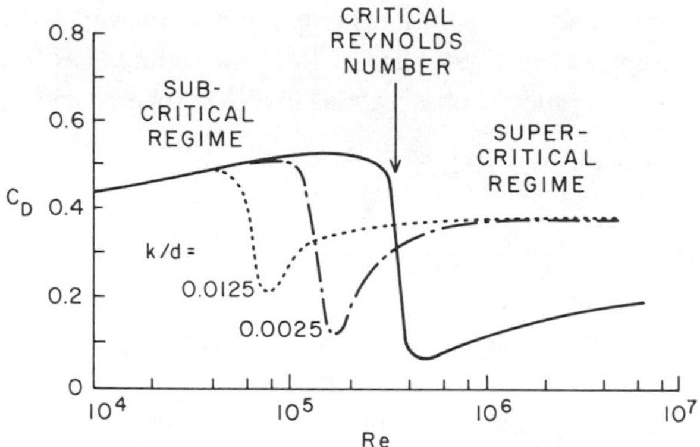

FIGURE 9.6 Relation between the drag coefficient C_D and the Reynolds number for a sphere with a smooth surface (solid line) of diameter d and for two spheres with roughness elements (dotted and dashed lines) with two different heights k. The critical Re is seen as a sharp decrease in C_D.

gens, fungal spores, etc.) that have a preferred site of arrival cannot be reduced indefinitely because very small palynomorphs have so little inertia that they rarely collide with surfaces or settle out of airflow patterns.

Equation (9.16), however, can be used only for $Re \leq 1.0$. At higher Reynolds numbers, Stokes' law is not applicable. One of the major obstacles in calculating drag is determining the drag coefficient for irregularly shaped objects. Even very small projections on the surface of a sphere can result in dramatic departures from the relations between C_D and Re found for smooth-surfaced spheres (see fig. 9.6).

The densities of real spores and pollen are hard to determine because of their often complex morphologies (hence the inappropriateness of $4\pi r^3/3$ to approximate their volumes). However, their terminal velocities can be empirically determined by stroboscopically photographing pollen or spores as they descend through a long column of unmoving air (formed by a material that carries no electrical charge). At some point along their downward trajectory, the spores or pollen achieve terminal velocity. By photographing their descent with a specified frequency ($400 \cdot s^{-1}$ generally works, though other settings need to be explored), one can calculate the velocity of the particulate from the successive images on a negative or print (fig. 9.7). (The terminal velocity of a sky diver is 120 miles per hour, or roughly 54 $m \cdot s^{-1}$. Thus, without a parachute, a sky diver's *terminal* velocity takes on a new meaning—jumping to a conclusion.)

One of the assumptions we made at the very beginning of this section is

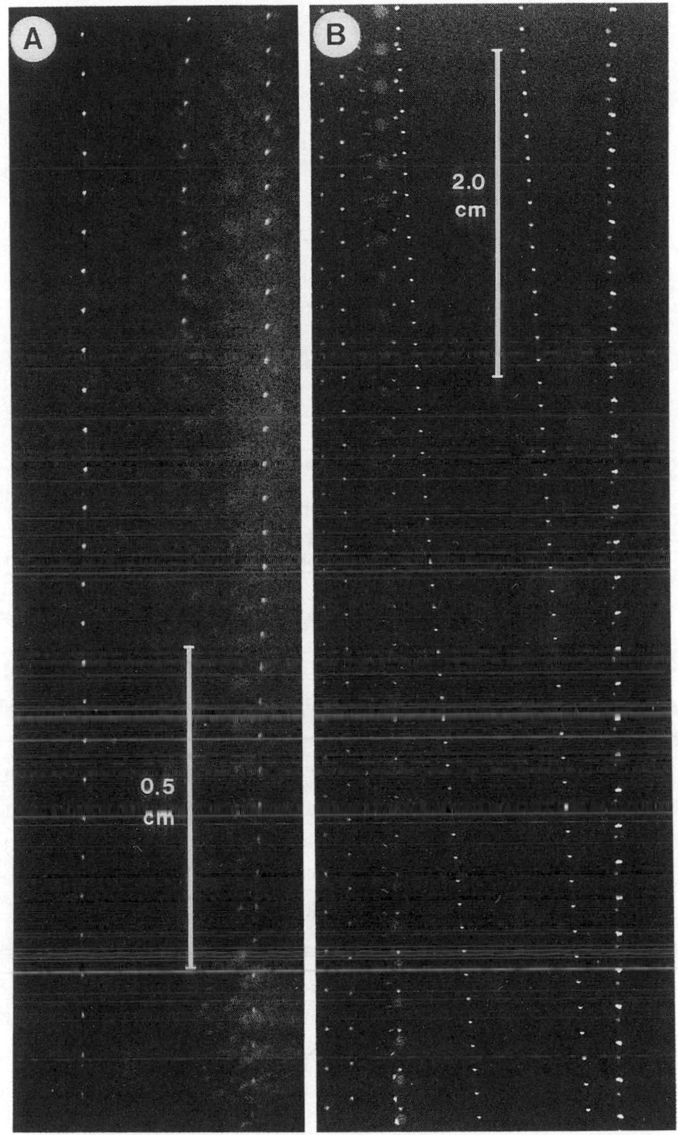

FIGURE 9.7 Stroboscopic photographs (40 cycles per sec) of pine (*Pinus strobus*) pollen (A) and the megaspores of a lycopod (*Selaginella*) (B) descending through a column of air. Terminal descent velocities can be calculated by measuring the distance between successive stroboscopic images and dividing the distance by the time interval between successive flashes of light. The descent velocities of the pollen grains to the left and right of the scale in A are 2.56 and 2.19 $cm \cdot s^{-1}$, respectively. The descent velocities of the two megaspores to the right of the scale in B are 5.3 and 8.5 $cm \cdot s^{-1}$.

TABLE 9.2 Summary of Motion Characteristics of Platy and Columnar Objects

	Plates (thickness:width < 0.2)	
$w < 0.2$ mm	0.2 mm $< w < 4.0$ mm	$w > 4$ mm
$U_T < 0.3$ m·s^{-1}	0.3 m·s$^{-1} < U_T < 1.5$ m·s^{-1}	$U_T > 1.5$ m·s^{-1}
$Re < 1.0$	$1.0 < Re < 100$	—
$U_T \propto wt$	$U_T \propto (t/C_D)^{1/2}$	—
Descent: stable, regardless of orientation	Descent: stable, oriented for maximum drag	Descent: tumbling motion
	Columns (1 < length:diameter < 10)	
$w < 0.1$ mm	0.21 mm $< w < 1.0$ mm	$w > 1$ mm
$U_T < 0.7$ m·s^{-1}	0.7 m·s$^{-1} < U_T < 6.0$ m·s^{-1}	$U_T > 6.0$ m·s^{-1}
$Re < 1.0$	$1.0 < Re < 100$	—
$U_T \propto$ diameter:length	$U_T \propto$ (diameter$/C_D)^{1/2}$	—
Descent: stable, regardless of orientation	Descent: stable, oriented for maximum drag	Descent: unstable

Source: Data on ice crystals from Ward-Smith (1984, 37, table 3.1).

that the sphere is not rotating about an axis. When rotation at significant speeds does occur, spheres develop transverse forces owing to the Magnus effect in the flow around them. This rotation generates lift force, with a resulting curvature in the flight trajectory (Prandtl 1952; Hoerner 1965; Burrows 1987; see also Niklas 1988b), a feature of particular interest to a baseball pitcher. Very high axial rotation speeds are not normally encountered for objects with small diameters, in part because of the dominance of viscous forces at low *Re*. But large objects lacking spherical symmetry are likely to generate aerodynamic moments that can cause flight paths to deviate significantly from those predicted based on an approximately spherical geometry. Large is a relative term that requires a dimensional perspective: I have seen tumbling descent trajectories for flattened pollen grains less than 6×10^{-5} m in diameter.

Many species of plants produce spores or pollen that are not spherical. Hence, when released from a resting position they are unlikely to achieve attitude stability shortly after they attain free flight. The aerodynamic moment resulting from suction forces causes these nonaxisymmetric shapes to rotate. For real fluid flows, friction occurs and a portion of the total mechanical energy available to the object is invested in rotational movement, decreasing the net forward movement of the object and thus its terminal velocity. Perhaps more significant, in free-fall considerable lateral displacements from vertical descent can occur without lateral air movements. Geometric irregularities are commonplace among spores and pollen, and the trajectories of these objects even in free-fall are anything but straight lines. Attempts to model the behavior of spherical and nonspherical particulates moving relative to a fluid have had limited success owing to the extreme complexity in calculating boundary layer characteristics, aerodynamic moments, and even such fundamental physical characteristics as the center of mass of irregularly shaped particles (see Burrows 1987; Niklas 1988b). Table 9.2 provides some insights into how geometry can influence the characteristics of the motion of free fall. For example, in flattened plates (with a thickness/width <0.2) tumbling occurs when the Reynolds number exceeds 100, whereas a stable descent (with an orientation that maximizes drag) is usual when the *Re* ranges between 1 and 100. Similar features are observed for columnar shapes with a ratio of length to diameter greater than 1 and less than 10.

WIND POLLINATION

Among abiotically pollinated species, wind pollination (anemophily) is the dominant mode of pollen transfer and is found in plant families with widely varying evolutionary histories. Faegri and van der Pijl (1979, 34) have estimated that perhaps 98% of all known abiotically pollinated species are anemophilous. It is the dominant mode of pollination in the monocot families Poaceae, Cyperaceae, and Juncaceae as well as within the dicot families Fagaceae and Juglandaceae. Wind pollination dominates among extant gymnosperms, owing to the taxonomic prominence of the Coniferales, and has been reported by species within the Cycadales and Gnetales (Niklas and Norstog 1984; Niklas and Buchmann 1987), although other species within these two orders are evidently insect pollinated (entomophilous). Paleobotanical data indicate that anemophily was most likely the ancestral condition among early gymnosperms, although some extinct nonflowering seed plant groups were entomophilous, such as species of Cycadeoidales and possibly Pteridospermales. By contrast, the earliest angiosperms have traditionally been viewed as biotically pollinated. Evidence that wind pollination has been derived secondarily among some flowering plants comes in part from the appearance of anemophilous species in lineages that are thought to have been ancestrally entomophilous (Compositae, Plantaginaceae). Nonetheless, both anemophilous and entomophilous flowering plant species seem to be present early in the history of angiosperms (Crepet and Friis 1987 and references cited there). Considerably more data, particularly from the fossil record of Cretaceous angiosperms, are needed to resolve the evolutionary origins of anemophily among angiosperms.

Traditionally, wind pollination has been viewed as a wasteful and inefficient mode of dispersal because pollen transfer is passive and much of what is produced fails to be captured by pollen-receptive surfaces (stigmas, pollination droplets, etc.). Hence proportionally more pollen is required per ovule than in biotically pollinated species. The number of pollen grains per ovule produced by anemophilous and entomophilous species bears out this assertion. Pohl (1937) reports that the pollen:ovule ratio of anemophilous species can be orders of magnitude more than that of entomophilous species. For example, the hazelnut *Corylus avellana* and the beech *Fagus sylvatica* (both anemophilous species) have pollen:ovule ratios on the order of 10^6:1 (Pohl 1937; see Faegri and van der Pijl 1979, 367, table 1), whereas the entomophilous species of maple (e.g., *Acer pseudoplatanus*) and basswood (*Tilia cordata*) have ratios on the order of $\approx 10^5$:1 and 10^4:1, respectively. The number of pollen grains per ovule is not always this clear-cut, however. For example, the birch *Betula verrucosa,* an anemophilous species, and *Polygonum bistorta,* an entomophilous species, have pollen:ovule ratios of comparable magnitudes (10^3:1). The number of pollen grains per ovule a plant produces may have less significance than its metabolic investment in producing pollen grains and ovules. Unfortunately, data on this issue are poor or lacking. Similarly, the use of pollen as food by insect pollinators may significantly diminish the efficiency of entomophilous species. Once again, data on pollen harvesting by insects are scanty.

Wind-pollinated species have evolved a number of features that can potentially increase pollination efficiency. Among them are featherlike stigmas that can collect airborne pollen; brush inflorescences with aggregated flowers that operate as aerodynamic units in trapping pollen; the exertion of pollen-receptive organs above the leaf canopy of a plant or the production of flowers before leaves are produced and expand; the release of pollen predominantly under warm, dry atmospheric conditions; and the production of small, light, nonclumping pollen grains, enhancing the separation of adjoining grains and long-distance dispersal of microgametophytes. These and other features are beautifully summarized by Faegri and van der Pijl (1979, 34–40).

More recently, empirical studies have demonstrated that anemophilous species use the aerodynamic properties of their ovulate organs to capture pollen. The ovulate cones of *Pinus* can generate complex airflow patterns that direct airborne pollen grains toward micropyles. There is evidence that some species of pine can selectively filter conspecific pollen from the airflow patterns around their ovulate cones. This implies that the physical properties of pollen grains (size, shape, density) and the geometry of airflow generated around ovulate cones (hence the morphology of cones) operate in aerodynamic reciprocity for some species (Niklas 1985a). A similar aerodynamic effect has been reported for two sympatric species of the gymnosperm *Ephedra* (Niklas and Buchmann 1987), as well as for the dicot *Simmondsia chinensis* (Niklas and Buchmann 1985). Much like aircraft carriers that are designed to permit only a certain type of aircraft to land, the airflow patterns generated around the seed-producing structures of these species aerodynamically generate idiosyncratic airflow patterns through which only certain types of particulates—pollen—can navigate and successively reach their reproductive destination. These observations show that anemophilous species have evolved so that the airborne pollen within the aerodynamic influence of their conspecific ovulate structures behaves less randomly, while the close-proximity trajectories of grains can be mathematically predicted and have some deterministic, nonrandom properties. Nonetheless, in terms of the mass transport of pollen by large-scale airflow currents, there is little doubt that anemophily is dominated by essentially unpredictable atmospheric factors.

Wind pollination involves three processes, each dominated by subtly different aerodynamic factors: pollen release, long-distance pollen dispersal, and close-proximity pollen capture by ovulate organs. Each has been investigated, and numerous reviews are available. However, some features of pollen release and capture merit comment here.

Pollen release from a microsporangium (the reproductive structure in which meiospores develop that will give rise to pollen) or from some substrate to which pollen grains have secondarily adhered can occur either passively or dynamically. Passive release is accomplished primarily by gravity, while dynamic release involves an exchange of momentum between moving air and a plant structure. Dynamic release is by far the more common mode, and it can involve vortex shedding, siphoning, or forced separation. These three modes of dynamic release require airflow of some kind to disengage pollen grains from the surfaces they adhere to. This is no simple matter because, by virtue of their small size, pollen grains are tenaciously held within boundary layers generated over surfaces.

In passive release, pollen simply separates from a substrate because of gravity. The microsporangia of *Pinus* are on the abaxial (lower) surfaces of microsporophylls (the modified leaflike structures on which microsporangia are produced). When microsporangia dehisce, pollen can fall from the chambers in which they are produced. Without airflow, grains come to rest on the adaxial (upper) surfaces of the microsporangia below. With even modest (0.5 $cm \cdot s^{-1}$) airspeeds, however, grains are swept away into ambient, large-scale airflow currents. Pollen may or may not be at rest relative to the surface it is attached to. If the surface is in motion, owing to wind-induced movements, then the pollen can have an initial velocity of projection. Since most pollen grains are small ($<< 1.5 \times 10^{-4}$ m) and their motion is dominated by drag coefficients on the order of 24 Re^{-1}, grains tend to settle into steady vertical fall when they reach terminal velocities. But even though pollen grains settle with respect to the portion of air they move in, large-scale air currents can move upward, carrying grains farther from the ground. The speed of these air currents need not be large. I have seen *Pinus* pollen that has come to rest on black asphalt rise owing to airflow generated by convective heat.

Although much less dramatic, some species of plants appear to use gravity as their pollination vector. That is, either the pollen grains of these species are shed in environments that normally receive little wind movement, or the pollen grains are so massive that their capacity for substantial long-distance dispersal, even in fairly windy habitats, is minimal and they simply fall from their microsporangia and land on pollen-receptive sites that just happen to be directly underneath (or nearly so). One example where gravity may have been the principal pollination vector in the past is in the extinct pteridosperm family Medullosaceae, some genera of which produced some of the largest pollen grains known (over 2×10^{-4} m in diameter, e.g., *Medullosa*). Taylor and Millay (1979) found it "difficult to envision the large grains of this group being dispersed any distance by air currents" and speculated that "the principal dispersal mechanism for these pollen grains involved some form of insect vector" (1979, 337). Although this speculation is entirely reasonable, we must note that living plant species can use gravity for pollen release and recognize that the density of a pollen grain is just as important to long-distance dispersal as its size. Provided their density was low, the pollen grains of *Medullosa* could have moved through the air with the greatest of ease. Indeed, the pollen of some wind-pollinated extant plants can clump into aggregates with diameters greater than the single pollen grains of species of *Medullosa* and still fall 200 m away from the parent plant.

Before leaving the topic of passive release of meiospores from sporangia, it is worth noting that some nonseed plants have evolved relatively elaborate mechanisms to actively push spores out of the confining sporangial walls that envelop them. For example, the spores of the horsetails (*Equisetum*) are furnished with an outer spore wall layer that peels back to form spatulate armlike extensions called elators. The spore wall material of elators expands or contracts in response to changes in the relative humidity of the environment. When the spores are fully mature and the sporangium containing them splits, the closely packed spores soon become separated as their elators shift their positions in response to changes in external atmospheric humidity. The initially densely crowded spores thus become disaggregated and are literally pushed out of their confining sporangia. (If the reproductive conelike structure of *Equisetum* is placed under a bell jar while sporangia are still closed and allowed to dehisce in unmoving air, then in a few days thousands of spores can be found directly under cones, and most of the sporangia will be emptied of their contents.) In this regard it is interesting to speculate whether inflation of the sacci found on gymnosperm pollen is a potential mechanism to rupture and passively release pollen from the microsporangia of these plants.

Although passive release is not infrequent, dynamic release is by far the most common mode of pollen dispersal. All plants experience dynamic loadings owing to wind. These loadings cause plant organs to move, by generating lift and drag or through the shedding of vortices (Von Karman vortex streets), which produces transverse forces. In either case plant organs are set into oscillatory movement with natural frequencies of vibration depending on the geometry and flexural stiffness of their tissues (see chaps. 3 and 7). Pollen grains, however small, have mass and hence inertial properties. Thus pollen grains tend to resist sudden changes in their motion relative to the motion of the plant parts they are attached to. (Beating the dust out of a carpet capitalizes on dynamic loadings and on the inertia of dirt.) By the same token, the fluttering of anthers attached to delicate stamen filaments (e.g., the anthers of many monocots like the grasses) or the movement of leaves and petals can release adhering pollen grains.

Vortex shedding was discussed in the context of airflow around cylinders and spheres (see fig. 9.3). At the time I neglected to talk about the transverse forces this effect can generate. Before leaving the lateral surfaces of a cylinder, each vortex induces lateral forces whose magnitude F equals the product of the vortical circulation Γ, ambient airflow speed U_a, and air density: $F = \rho U_a \Gamma$. These forces alternate from one side of the cylinder to the other, since vortices are shed in an alternating periodic manner. Thus the cylinder is caused to move from side to side with a natural frequency of vibration. (From the equation for the Strouhal number, we know that the natural frequency of vortex shedding is related to the characteristic dimension of a structure like a cylindrical stem or anther filament and to the ambient airflow speed.) The same effect occurs for flat, platelike structures like leaves. Significantly (you will recall) vortex shedding occurs only at certain Reynolds numbers. Low

Reynolds numbers do not generate vortex shedding. Thus pollen tends not be released at low *Re* (which translates into low wind speeds, since the characteristic dimension of the pollen grain and the kinematic viscosity of air can be considered constants in the equation for *Re;* see eq. 9.5). This is particularly advantageous, since pollen tends to be released only when wind speeds are above some critical threshold, ensuring that some long-distance dispersal will occur. Periodic oscillatory motion can occur under steady or unsteady winds. A pulse of air (wind gust) can cause a plant organ to vibrate through impact loading. The oscillatory motion of the organ gradually decays, but the period of vibration remains constant. Impact (wind) loading is probably the dominant mode of pollen release.

Siphoning describes the often substantial suction pressures produced when air flows around a structure. Negative pressures are generated along the leeward surfaces of cylinders and plates; as we have seen, these pressure differentials are described by Bernoulli's theorem. Ingold (1953, 1965) speculated that siphoning may play an important role in the spore dispersal of some fungi, while Burrows (1987) suggested a similar role in fruit and seed dispersal. Siphoning may be important in pollen release from the microsporangiate cones of gymnosperms, since airflow around cones generates a leeward pattern of air currents with a negative pressure. Siphoning, together with vortex shedding, may be very important in the release of spores from the capsules of moss sporophytes. The spore-bearing capsules of mosses are generally well within the boundary layers produced by the substrates where the moss gametophyte grows (damp soil, tree bark, etc.), but the seta (a very thin and elongated stalk that the capsule rests on) elongates and its tissues tend to dry as the sporophyte matures and spores ripen. From our treatment of the equations governing the relations among the natural frequencies of vibrations of beams, beam length, and flexural stiffness (chap. 3), we know that these developmental shifts are conducive to an increase in the natural frequency of vibration as the seta matures. Higher frequencies of vibration, together with even a slightly negative pressure induced at the top of the moss capsule, can promote spore release at remarkably low ground-level wind speeds ($< 0.05\ m{\cdot}s^{-1}$).

Other aerodynamic mechanisms, such as forced separation, may play a role in pollen release, but little research has been done to examine the prevalence of these mechanisms under natural conditions. Forced separation occurs when loadings on structures exceed the structural strength of attachment points between two plant structures. Pollen masses within microsporangia may have some adherence, and grains could be released in small clumps owing to forced separation. Clumping of pollen in anemophilous species is not advantageous, since clumps have higher terminal velocities than single grains (Niklas and Buchmann 1987). Nonetheless, forced separation of individual grains may occur in some circumstances.

Provided pollen is released into the air, mass transfer effects play important roles in bringing grains within the aerodynamic influence of ovulate organs. These effects are influenced by the structure of plant communities air passes through. The influence of atmospheric conditions on wind pollination efficiency can be determined in part by relating the dispersal of airborne pollen to the atmospheric vertical velocity gradient. (High turbulence in the atmosphere increases pollen dispersion and can reduce the probability of pollination by diluting the concentration of airborne pollen grains.) The variation in the average horizontal wind speed U_z with height z from ground level ($z = 0$) is a logarithmic function and is described in general terms by the equation

$$U_z = \left(\frac{U^*}{k}\right) \ln \left(\frac{z + z_m - d}{z_m}\right), \tag{9.17}$$

where U^* is the friction velocity, which equals $(\Gamma/\rho)^{1/2}$, k is Von Karman's constant (generally taken as 0.4), z_m is the roughness parameter for momentum, and d is the zero plane displacement. The values of z_m and d depend upon the spacings among plants and average plant height. For dense vegetation, d can be estimated from the average plant height h by the formula

$$d = 0.64\ h. \tag{9.18}$$

If plants are more sparsely distributed, then eq. (9.18) does not hold, and d must be empirically determined. The momentum roughness parameter z_m also depends upon the shape, height, and spacing of roughness elements within the landscape. For uniformly distributed elements of height h, z_m can be approximated from the formula

$$z_m = 0.13\ h. \tag{9.19}$$

Other formulas are available to accommodate roughness elements with more complex spacings and geometries (see Bussinger 1975).

If we know the ambient wind speed above the plant canopy (outside the canopy's boundary layer), then we can use eq. (9.17) to backtrack and solve for the value of the frictional velocity U^*. Using this value of U^* in eq. (9.17), we can then calculate U_z for different heights z within the canopy. For example, suppose the average wind speed at 3 m above a grass stand with an average height of 0.20 m is 3 $m \cdot s^{-1}$. From eqs. (9.18) and (9.19) we calculate that $d = 0.128$ m and $z_m = 0.226$ m. The value of $\ln[(z + z_m - d)/z_m]$ equals

4.71. Taking U_z as 3 $m \cdot s^{-1}$ and a value of 0.4 for k gives us a value of 0.255 $m \cdot s^{-1}$ for U^*. Thus the wind speed at a height of 0.20 m (just at the average height of the grass canopy) would equal 0.846 $m \cdot s^{-1}$, while at a height of 1 m, $U_z = 2.26\ m \cdot s^{-1}$. By reiterative numerical solution, eq. (9.17) can be used to construct the wind profiles expected over plant stands of various heights. Comparisons between predicted and observed velocity profiles are convincing as to the utility of the equations. Graphing z as a function of U_z yields the curved plot typical for the change of wind speed within a boundary layer, while graphing $\ln[(z + z_m - d)]$ as a function of wind speed U_z yields a straight-line plot whose y-intercept equals z_m. Provided the terminal velocity of a species of pollen is known, plots like these can be used to estimate the horizontal transport distance of pollen released at various levels within the wind profile for particular types of canopies and community types. (Incidentally, a quick inspection of eq. 9.17 ought to reveal why pollen-shedding organs are typically elevated well above the vegetative canopy layer of wind-pollinated species.)

Equation (9.17) cannot be used to calculate wind profiles *within* a plant canopy, because U^*, z_m, and d are very complex parameters once airflow enters the vegetated environment. However, the canopy flow regime can be viewed as consisting of three horizontal layers (see Bussinger 1975): a top layer whose aerodynamics is principally governed by foliage within the canopy; a middle layer whose characteristics are governed by the structure of the vegetation, as well as some ground effects; and a bottom layer that can be similar to the airflow above the canopy but whose maximum wind speed usually matches that of the bottom of the middle layer. The bottom layer is the airflow regime in the understory of a plant stand. The top layer is from the top of the canopy down to a height z equal to d. This layer exerts drag on the wind above the plant stand. The wind speed within this layer decreases exponentially with distance down from the top of the canopy but has an overall direction of flow similar to the ambient wind direction above the canopy. The second or intermediate layer extends from the zero plane ($z = d = 0.64\ h$) to from 5% to 10% of the canopy height. Here drag is dominated by stems, branches, or tree trunks. Finally, the bottom layer has a logarithmic profile whose shape (but not magnitude) is similar to that of the wind profile above the canopy. The bottom of this lowest layer has an airflow regime dominated by the boundary layers created in the understory or at the ground surface.

A formula that can be used to describe the wind speed U_c within the top layer of most plant canopies (and for the top and middle layers of plant stands with uniform canopies) is

$$U_c = U_t \exp\left[a\left(\frac{z}{h} - 1\right)\right], \tag{9.20}$$

where U_t is the wind speed at the top of the canopy, h is the canopy height, and a is the attenuation coefficient, which ranges from 0 (for very sparse canopies) to 4 (for very dense canopies); for example, for short grass canopies $a = 4$, for corn canopies $a = 2$, for isolated conifer stands $a = 0.02$. The wind speed at the top of the canopy U_t can be estimated from eq. (9.17). Bussinger (1975) provides graphs of z/h versus U_c/U_t (two very useful dimensionless parameters) for various types of crops and forest canopies based on eq. (9.20). These graphs reflect the complexities in airflow resulting from understories or their lack. For example, in plant stands without understories, such as old stands of sparsely planted pine trees, wind speeds among tree trunks can equal those at the very top of the canopy ($U_c/U_t = 1$). Thus, pollen transport distances can be considerable. (For detailed treatments of the modeling of pollen dispersion and deposit of pollen in a forest canopy, see Di-Giovanni, Beckett, and Flenley 1989; Di-Giovanni and Beckett 1990.)

Once pollen grains enter the aerodynamic environments created by and around their conspecific ovulate reproductive organs, their airborne trajectories become complex but manifest statistically predictable characteristics (plate 3). These close-proximity trajectories result from the inertial features of pollen grains (dictated by the mass and instantaneous velocity of each grain) and the airflow patterns generated by the geometry of the ovulate organ. Pollen grains are not neutrally buoyant with respect to air. They have inertial characteristics and, however small and light they may be, resist sudden changes in their speed and direction of motion. Hence their trajectories rarely, if ever, conform precisely to the motion of the airflow currents carrying them. Nonetheless, the pollen grains of a particular species have an *average* size, shape, and density. The statistical behavior of a large number of pollen grains can therefore be used to evaluate the aerodynamic properties of pollen from a particular species. By the same token, the morphology of ovulate organs is relatively circumspect for each species. Remembering that morphology translates into the geometry of surfaces that obstruct airflow, we quickly come to realize that for each specified direction and speed of ambient airflow there exists a fairly well defined airflow regime generated around a specific ovulate structure. Consequently the aerodynamic environment around an ovulate organ and the physical properties of conspecific pollen, and thus the trajectories of grains, have some statistically deterministic features. Given that many anemophilous species have evolved phenological characters and other adaptations that tend to maximize pollen dispersal and capture, it would be surprising to find that the deterministic features of fluid and pollen motion around ovulate structures are not used to aid anemophilous pollination. Indeed, empirical studies have shown that they are so used in taxonomically diverse plants (Niklas 1985a,b, 1987a,b, 1988b).

Some of these points can be illustrated by referring to a study I was involved with concerning the gymnosperm *Ephedra trifurca* (the data are reviewed in detail in Niklas and Buchmann 1987; see also Buchmann, O'Rourke, and Niklas 1989). Plants of this species are much branched, relatively short (< 2 m), and essentially leafless. Ovules (unfertilized seeds) are produced at nodes, sometimes singly but much more often in large numbers. Each ovule is more or less teardrop shaped, with an attenuated tip consisting of a hypodermic needle-like micropyle (a hollow tube through which pollen gains direct access to the female gametophyte, or more properly the megagametophyte, which produces the egg). The ovules' receptivity to pollen is shown externally when tissues beneath the internal base of the micropyle deliquesce to form a highly viscous fluid that exudes from the micropylar tip as a relatively obvious pollination droplet. (The droplet is very sweet and is sought after by scrounging ant and botanist alike.) Airborne pollen grains adhere to the pollination droplet, and when it is reabsorbed by ovular tissues, they are withdrawn into the inner ovule, where fertilization occurs. Plate 3 reveals some of aerodynamic complexity observed when the statistical behavior of over five hundred pollen grain trajectories (stroboscopically photographed) was assessed with the aid of a computer. (In each of the six computer-generated diagrams, airflow is from left to right. The mean direction and speed of pollen grains passing through each of many very small air spaces is indicated by an arrow and by color-coding—the legend is provided on the plate.) Careful analysis of these data reveals that the geometry of ovules focuses airflow toward the pollination droplets at the tips of micropyles. Regardless of their taxonomic affiliation, pollen grains moving in the airflow around ovules are essentially directed toward micropyles, but depending on their aerodynamic characteristics (size and density), some settle downward while others move upward—the aerodynamic environment generated by and around each ovule effectively operates as a centrifuge, sorting pollen grains according to the inertial attributes, which are in turn dictated by size, shape, and density.

We must recognize that the symmetry of airflow around any particular ovule will change as a function of the direction of ambient airflow, since internodes and other ovules are always upstream of the flow. Since ovules tend to be symmetrically positioned around each node, however, for any particular direction of airflow there is always at least one ovule in position to optimally filter the air for pollen. Because the direction of ambient airflow changes from hour to hour, each ovule has more than ample opportunity to exercise its capacity for aerodynamic filtration. Plate 3 illustrates only the statistical behavior of conspecific pollen (pollen from the same species of *Ephedra*), showing that pollen grain velocity diminishes along the leeward side of ovules and that pollen grain direction has the greatest variance near the pollination droplet. What is not illustrated (because it is a truly dynamic effect) is that each micropyle vibrates as a harmonic oscillator, causing the Reynolds numbers in the immediate vicinity of each pollination droplet to change over three orders of magnitude in a few seconds. Thus, around each pollination droplet, neither viscous nor inertial flow predominates—each pollination droplet essentially acts like flypaper, trapping anything that comes near it. Since conspecific pollen are the most likely to reach the vicinity of the pollination droplet, they have the highest probability of being captured. Indeed, computer simulations predict that *Ephedra* would most likely capture more conspecific pollen than pollen from any of the other species commonly found growing in the same habitat. After carefully counting pollen grains adhering to pollination droplets and determining their systematic affiliations, Buchmann, O'Rourke, and Niklas (1989) were able to show that the frequency distribution of pollen grains trapped on pollination droplets almost precisely conformed to computer-generated predictions based on an aerodynamic analysis of airflow around *Ephedra* ovules.

Let me emphasize, however, that the speed and direction of ambient airflow around ovulate organs are highly variable. Hence the aerodynamic environment around pollen-receptive organs will change as ambient conditions of airflow change. Since the physical attributes of conspecific pollen do not change with wind direction and speed, it must be acknowledged that the trajectories of pollen around ovulate organs will be altered depending on ambient airflow conditions. Currently little is known about the influence of variation in ambient airflow on the efficiency with which ovulate organs capture their conspecific pollen. As we have seen, however, the flow regimes created around simple geometries like cylinders remain relatively unchanged over fairly broad ranges of Reynolds numbers (hence, wind speeds). Similarly, irregular shapes undergo less abrupt transitions from one flow regime to another, so flow characteristics may be shared by two types of flow regimes over the transitional Reynolds numbers. Thus, future research into pollen-capture efficiency as a function of ambient airflow speed may reveal that changes in am-

FIGURE 9.8 Fruits and pappus of the oyster plant (*Tragopogon*): (A) Cluster of fruits produced by a composite inflorescence. (B) The pappus (modified calyx) from a single fruit (taken from the specimen shown in A). The delicate filaments of the pappus interdigitate to produce a parachute-like structure that creates drag, reducing the speed at which the fruit falls.

bient airflow conditions induce moderate changes in the airflow regimes generated around some ovulate organs.

Another research topic of considerable interest is the range of variation in pollen size, shape, and density typical for a species. If these parameters vary in some predictable manner for a species, then it is possible that some anemophilous plants produce different populations of pollen grains and that each population operates optimally under different aerodynamic conditions. Little is known about the variance in physical properties of pollen produced within a single microsporangium or about the variations in these properties from one sporangium to the next within microsporangiate cones or inflorescences. This topic may well be fertile ground for future research into wind pollination.

AIRBORNE SEEDS AND FRUITS

Plants have evolved a variety of methods to discharge and disperse their seeds and fruits, and many species have capitalized on the passive dispersal of their propagules by wind. The seeds of orchids are remarkably small (about 0.1 mm in diameter) and, because of their very small mass (about 5×10^{-7} g), achieve very slow terminal velocities (about 0.27 $m \cdot s^{-1}$, with Re roughly 1.8). The seeds of orchids are nearly equivalent aerodynamically to the pollen of some wind-pollinated species (the pollen grains of *Quercus robur* are 0.036 mm in diameter and have a mass of 2×10^{-8} g; $U_t = 0.03$ $m \cdot s^{-1}$, with $Re \approx 0.07$). By contrast, the seeds and fruits of most plant species are macroscopic compared with the dust seeds of orchids, and when wind dispersed these propagules often rely on plumage or wings to reduce their terminal velocities and so enhance dispersal. In this section I will review the basic aerodynamics of these airborne seeds and fruits. Rather than considering individual taxa, I will depend on the fact that seeds or fruits with plumage and with wings mechanically operate in ways dissimilar enough to provide us with two categories transcending taxonomic boundaries.

Plumed seeds and fruits are produced by a number of families (Onagraceae, *Epilobium;* Ranunculaceae, *Anemone;* Valerianaceae, *Centranthus*), but perhaps the most familiar are the fruits of Compositae (*Centaurea, Sonchus, Taraxacum, Tragopogon*) that have a parachute-like pappus derived from the calyx (the collective term for the sepals of a flower). The large pappus of these fruits has a relatively low porosity to airflow, and their aerial motion is such that they rapidly achieve terminal velocity. Indeed, in some fruits the delicate fibers of the pappus appear interwoven like a spiderweb (fig. 9.8). The pappus of these fruits primarily functions as a drag-producing structure below which

FIGURE 9.9 Stroboscopic photographs (20 cycles per sec) and morphology of the parachute-like fruits of the dandelion (*Taraxacum officinale*). The bulk of the fruit (an achene) is suspended beneath the pappus (a modified calyx). (A–B) Vertical descent of fruits (velocities are 22 $cm \cdot s^{-1}$ in A and 21 $cm \cdot s^{-1}$ in B). (C) Vertical descent of a pappus from which the fruit has been removed (velocity = 7.14 $cm \cdot s^{-1}$).

the fruit is suspended (fig. 9.9). When moving laterally, the longitudinal axis of the fruit is held at an angle to the vertical direction, with the pappus leading along the flight trajectory. Hoerner (1965) referred to this configuration as a drag parachute as opposed to a guide parachute, in which a small, very porous pappus operates primarily to provide orientation and guidance rather than drag. The differences between these two designs are a matter of degree rather than kind, however. Fruits with guide parachutes tend to have larger terminal velocities than those with a drag parachute pappus. Also, guide parachute fruits attain terminal velocities only after falling a considerable distance. The primary distinction between drag and guide parachutes is that the latter achieve much less lateral movement under comparable ambient airflow speeds.

Burrows (1987) has provided the equations for the motion of fruits with drag parachutes. These equations show that fruits of this kind move with the local horizontal flow vector virtually all the time, while the vertical component of their trajectories is almost entirely dependent on the vertical convection velocity. Accordingly, their flight trajectories are very complex compared with fruits of the guide parachute type. The basic conclusion we can draw is that they can be carried extremely large distances by even modest convection flows.

The functional significance of fruits with guide parachutes is unclear except that they have lower terminal velocities than if the pappus is removed. Burrows (1987, 36) comments that fruits of this type are "well equipped for utilizing projection, catapulting or ejection at the moment of release from the mature inflorescence." Indeed, the small amount of drag generated by the guide parachute could be useful in the torque it generates along the length of a fruit still attached to its inflorescence. This torque could provide the force necessary for fruit detachment. Clearly, the drag and guide parachute types of fruits represent conceptual extremes along what is in reality a biological continuum in morphology.

Examples of plants that produce plumed seeds are *Epilobium* and *Asclepias*. The aerodynamic behavior of these seeds is comparable to the guide and drag parachute fruit types, respectively. Winged seeds or fruits are produced by gymnosperms (*Pinus, Picea*) and a host of angiosperms (e.g., *Acer, Heteropterys, Tilia, Zanonia*). Typically these seeds or fruits can be placed into one of two categories—plane winged, in which the seed coat or fruit wall grows more or less symmetrically with respect to its longitudinal axis and generates a linear gliding flight, and autogyroscopic winged, in which lift is created by wings on the propagule that permit autorotation during free-fall.

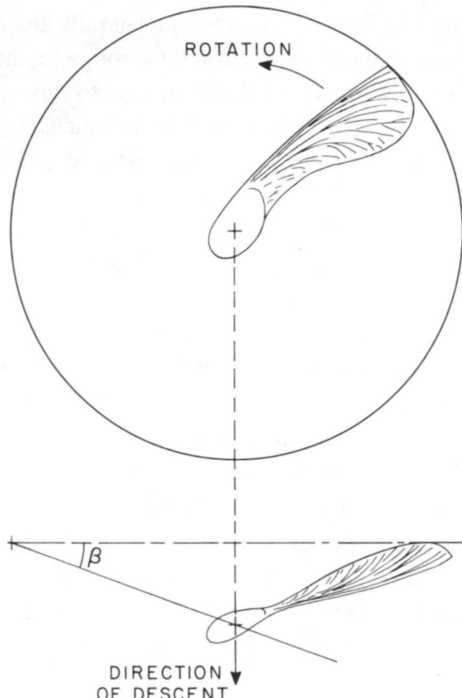

FIGURE 9.10 Top and lateral views of an autogyroscopic fruit (the samara of *Acer saccharum*). The wing of the samara rotates in free-fall creating a shallow conical section (with an area A_D) with a cone angle β, measured as the subtending angle between the longitudinal axis of the wind and a horizontal plane.

These two categories are not hard and fast distinctions, however. Small imperfections or bendings of plane winged seeds or fruits can result in irregular (rather than linear gliding) flight, and the single seed coat of *Pinus* or *Picea* can generate lift and autorotation of the seed body.

Ward-Smith (1984, 56–66) provides a beautifully concise review of the aerodynamics of plane winged and autogyroscopic fruits and seeds, while Burrows (1987) contributes many significant insights into the subject as well as an updated bibliography.

Plane winged seeds and fruits benefit aerodynamically by concentrating their centers of mass in a position relative to the chord of their wings, which stabilizes the location of their center of pressure (where resultant aerodynamic forces act on the wing). Equilibrium during gliding requires that the centers of mass and pressure coincide. This is achieved by weighting the wing with the bulk of the fruit, since a wing with a homogeneous weight distribution is highly unstable in flight. This is easily illustrated by observing the behavior of an ordinary sheet of paper in free-fall and comparing its behavior with that of an identical sheet with a paper clip attached to one end. The latter has a much more stable orientation during its free-fall. The flat piece of paper continually alters its incidence angle, and since incidence angle in large part dictates the ratio of lift to drag, the flat sheet goes through many flight reorientations (changes in incidence angle), with consequent effects on its lift:drag ratio. The precise position of the center of mass determines the incidence angle at which equilibrium occurs. This in turn determines the glide angle ϕ: $\tan \phi = (\text{lift:drag})^{-1}$. The smaller the glide angle, the larger the dispersal range of the seed or fruit.

The aerodynamic behavior of autogyroscopic seeds and fruits was commented on over 150 years ago by Sir George Cayley, one of the great aeronautical pioneers. Cayley observed the airborne motion of a samara from *Acer pseudoplatanus*. (A quotation from Cayley's notebook, dated 9 October 1908, is provided in Ward-Smith [1984, 60]. Readers will avoid confusion by noting that both Sir George and Ward-Smith confused the common name for *Acer pseudoplatanus*, which is the sycamore maple, with the proper sycamore tree, *Platanus occidentalis*. Maples have winged fruits, called samaras, whereas the sycamore produces an achene with tufts.) The behavior of autogyroscopic seeds and fruits is remarkably similar to that of a helicopter and results from the combination of angle of attack and sideslip generated by the slightly curved blade of fruit wing or seed wing. The local speeds developed at the tip of the wing are relatively high and, together with the corresponding moments, produce great dynamic stability. When viewed from above and below, the rotating wing sweeps out a shallow conical section with an area A_D (fig. 9.10). The angle subtending the longitudinal axis of the fruit or seed and the horizontal plane is the coning angle β, the equilibrium condition between the centrifugal forces acting on the rotating wing (which produce a turning moment tending to force the wing into a horizontal position) and aerodynamic forces (which tend to displace the wing into the vertical position). Thus the coning angle is the result of moments due to centrifugal and aerodynamic forces.

Ward-Smith (1984, 61–66) analyzes the rotational motion of autogyroscopic seeds and fruits. The following explanation is based on his treatment. The descent speed U_s of an autogyroscopic seed or fruit is given by the formula

$$U_s = \left[\frac{2}{k(2-k)}\right]^{1/2} \left(\frac{W}{\rho\, A_D}\right)^{1/2}, \tag{9.21}$$

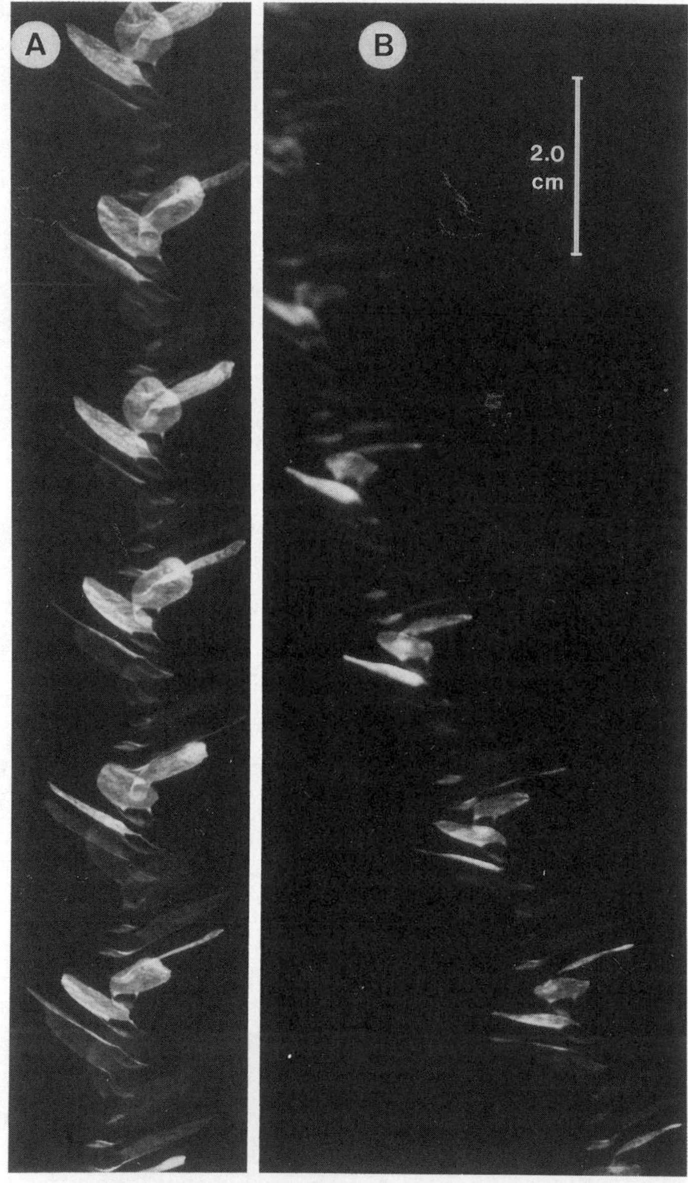

FIGURE 9.11 Stroboscopic photographs of the vertical descent (speed = 44.5 cm·s^{-1}) of an autogyroscopic seed of spruce (*Picea abies*) showing a simple helical trajectory (A) and a compound helical trajectory (B). The latter is characterized by a sideslipping of the seed to create a double helix (the seed rotates around its axis, as seen in A, and the overall descent trajectory is helical).

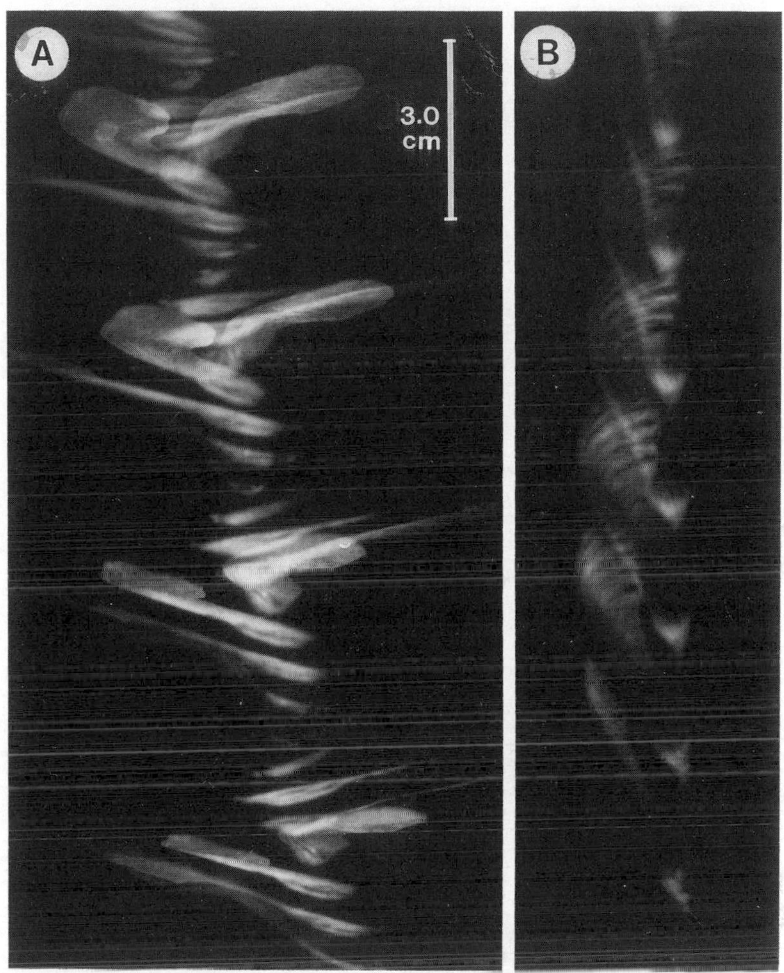

FIGURE 9.12 Stroboscopic photographs of an autogyroscopically rotating (A) and "stalled" (B) samara of *Acer saccharum*. The autogyroscopic motion of the fruit shown in B was temporarily interrupted by a strong gust of wind, resulting in a rapid descent (298 cm·s^{-1}) compared with that of the fruit shown in A (72.9 cm·s^{-1}).

where k is a constant of proportionality depending on the wake velocity generated downstream of a disk with area equivalent to the conical section A_D of the rotating propagule, W is the weight of the winged seed or fruit (given in mg), and ρ is the density of air. When $k = 1$, the descent speed is minimum, and eq. (9.21) becomes

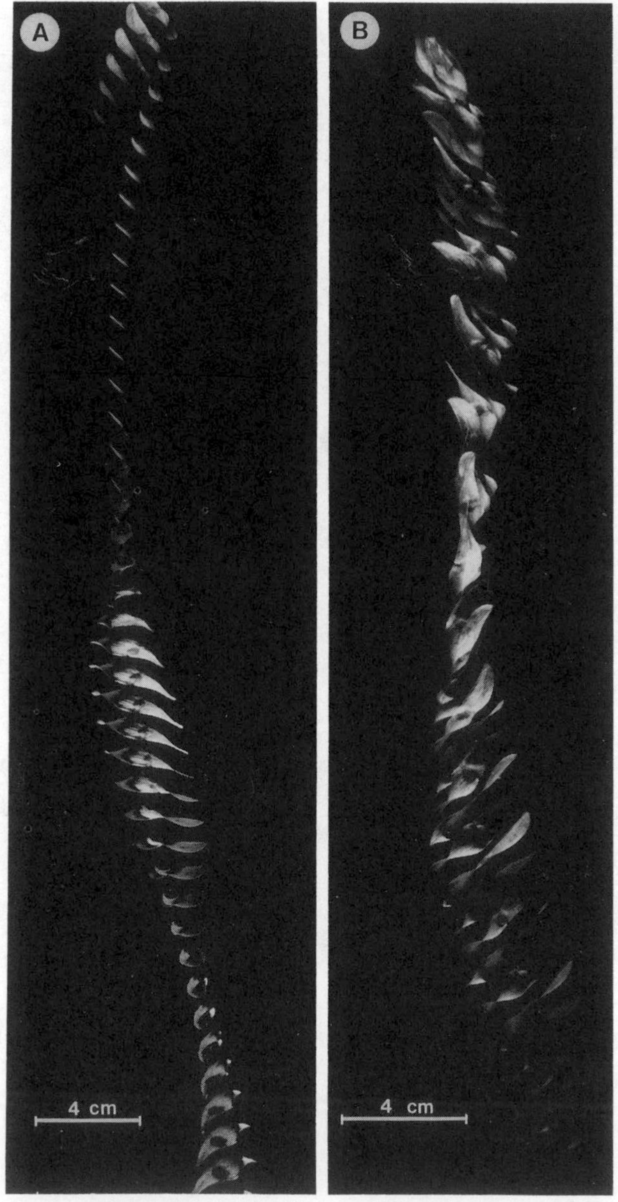

FIGURE 9.13 Stroboscopic photographs of the complex autogyroscopic motion of the fruits of the tree of heaven (*Ailanthus altissima*). The fruits rotate around the longitudinal axis of their wings and exhibit a compound helical trajectory (compare with fig. 9.11B).

$$U_s^{\min} = \sqrt{2}\left(\frac{W}{\rho\, A_D}\right)^{1/2} \approx 1.4\left(\frac{W}{\rho\, A_D}\right)^{1/2}. \quad (9.22)$$

Thus the descent speed varies as the square root of W/A_D, which is called the disk loading of the propagule (a term that refers to either a seed or a fruit). When $0.5 \leq k \leq 1.0$, calculations based on eq. (9.21) deviate no more than 15% from observed values of $U_s^{\min}$; therefore eq. (9.22) can be used with considerable success to predict the behavior of autogyroscopic propagules. Values of W and A_D are easy to come by for a particular fruit or seed: W is simply the weight of the propagule, where A_D can be measured from photographs taken above a descending seed or fruit or can be closely approximated as πl^2, where l is the length of the propagule. (Recall, however, that the actual area of the actuator disk will always be less than πl^2 because of the coning angle β.)

Three general kinds of descent trajectories are possible for an autogyroscopic propagule: (1) a simple rotating trajectory, in which the center of mass of the autogyroscopic propagule descends in a straight vertical line (fig. 9.11A); (2) a compound helical trajectory, in which the autogyroscopic propagule spirals downward and equilibrium in flight is established in a sideslipping motion, with stable autorotation (fig. 9.11B); and (3) a complex compound helical trajectory that is similar to a compound helical trajectory except that the propagule is displaced laterally by its intrinsic motion. Each of these three descent trajectories is stable but can be destabilized by gusting crosswinds of sufficient magnitude that stall the propagule. The normal simple rotating trajectory of a samara is compared with the same samara's stalled downward trajectory in figure 9.12 (A and B, respectively). Stable autogyration is reestablished only if steadier wind flows are reached. Since the descent trajectories of the compound helical type (1 and 2) involve additional types of relative motion of the propagule and since destabilization of trajectories can occur, eqs. (9.21) and (9.22) must be used with caution. They are legitimately applicable only in the case of still air and a simple helical trajectory.

As in the case of a free-falling sphere, which accelerates before achieving terminal velocity, acceleration precedes autorotation when an autogyroscopic propagule is initially released. The unsteady motion during the acceleration phase of descent is very complex (Norberg 1973). Measurements have shown that before achieving autogyration, propagules must accelerate by a factor of six times the final autorotation descent speed.

Before leaving this general topic, let me note that not all winged seeds or fruits conform to the simple treatments presented here. For example, the fruits of the tree of heaven (*Ailanthus*), the ash (*Fraxinus*), and the tulip tree (*Liriodendron*) rotate about the longitudinal axis of their wings in addition to undergoing autogyroscopic movements (fig. 9.13). Rotation about the longitudinal axis results from a twist along the length of the winged portion of the fruit. The actuator disks of these propagules produce complex patterns of overall movement that are poorly understood (McCutchen 1977). It is known, however, that as the wings of these fruits go through a full 360° angle of rotational movement, suction forces must be generated near the leading edge of the wing to supply the additional kinetic energy needed to maintain rotation in pitch. Thus, part of the potential energy of the propagule (owing to the height of the release point above the ground) is expended in this additional component of motion, and so the terminal velocities of these propagules are greater than if rotation in pitch did not occur (D. Green 1980). Propagules that autogyrate with this additional twisting component of motion are not always immediately apparent from a morphological examination, and individual fruits with this capacity may develop on plants that produce propagules not characteristically of this type. Statistical studies of populations of air-dispersed seeds and fruits from individual plants are required if a researcher wishes to discuss the fluid dynamics of propagule dispersal for a particular species. Unfortunately this is rarely done, and conclusions are typically drawn from empirical observations made on a few "representative" specimens.

The two general categories of air-dispersed seeds and fruits (plane winged and autogyroscopic propagules) have flight characteristics that differ markedly in their potential dispersal ranges and susceptibility to in-flight instability. Plane winged propagules tend to glide in long lines, and their range is highly susceptible to small morphological deviations. Also, small gusts of wind destabilize their trajectories. By contrast, autogyroscopic propagules have potentially smaller dispersal ranges than plane winged propagules but are substantially more stable in flight and become destabilized only when there is significant turbulent airflow. Research into the ecological preference (if any) of plant species with either type of dispersal mechanism would provide important insights.

Clearly there exist a variety of wind-dispersed propagules, among which plane winged and autogyroscopic seeds and fruits are only two general categories. Other categories of wind-dispersed propagules are the helicopter type, which rotates around the longitudinal axis (e.g., *Cordia alliodora,* Boraginaceae, and *Triplaris cumingiana,* Polygonaceae) and the tumbler type, which, as its names implies, has no single preferred axis of rotation (e.g., *Cavanillesia platanifolia,* Bombacaceae). For the entire range of morphologies of wind-dispersed propagules, Green (1980) suggests that the terminal descent velocity is highly correlated with the square root of the wing loading of propagules. (Wing loading is the ratio of the weight to the maximum cross-sectional area of the propagule. It is analogous to the disk loading discussed in the context of autogyroscopic seeds and fruits; see eq. 9.22.). Likewise, in a highly comprehensive and thoughtful study, Auspurger (1986) reported that a regression of the rates of descent on the square root of the wing loadings of the wind-dispersed propagules of thirty-four species (in sixteen families) of Neotropical trees yields a significant overall correlation; however, the slopes of the separate regressions differed significantly for each of the five aerodynamic types of wind-dispersed propagules considered in the analysis. Thus, within each category of wind-dispersed seed or fruit (within each type of propagule morphology), wing loading can provide a very reasonable estimate of the terminal descent velocity of propagules.

LONG-DISTANCE WIND DISPERSAL OF PROPAGULES

The dispersal of seeds and fruits by wind is of general interest to the plant ecologist and evolutionist (see, for example, Harper 1977), since the survival of the next generation of plants and the capacity of individuals to colonize new locations are influenced by the capacity for the long-distance dispersal of propagules. Accordingly, numerous workers have investigated the shape of the curve that relates the number of propagules to the distance from their source plant in an attempt to provide phenomenological and mechanistic insights into long-distance wind transport (Frampton, Linn, and Hansing 1942; Gregory 1968; Cremer 1977; Green 1980; Augspurger 1986). One of the most widely used mathematical expressions for the dispersal curve is known as the inverse power model (Gregory 1968), which is given by the formula

$$y = ax^{-b}, \tag{9.23}$$

where y is the probability density associated with the dispersal, x is the distance from the source, and a and b are constants. Another frequently used formula is known as the negative exponential model (Frampton, Linn, and Hansing 1942):

$$y = ae^{-bx}. \tag{9.24}$$

There are advantages and disadvantages to both these mathematical models, however. The inverse power model has the advantage that it transforms to a straight line when it is plotted on log-log paper. Thus the numerical values of

a and b can easily be estimated. By the same token, the negative exponential model transforms to a linear relation when plotted on semilog paper, and the probability density y remains a finite number as the distance from the source x converges on zero. Unfortunately, the disadvantage of both models is that neither provides any insight into the mechanistic attributes of dispersal and their effects on dispersal curves. Also, a comparison between predicted and observed dispersal curves indicates no clear preference for either model. For example, Gregory (1968) evaluated 124 empirical dispersal curves in terms of both models and found that 59 conformed better with the inverse power model, while 65 had a better fit with the negative exponential model.

From first principles, we may assume that the shape of any dispersal curve will be governed by three parameters, each of which should be treated as having a statistical distribution within a plant population or even within a population of propagules produced by a single plant: the heights at which propagules are released, their terminal descent velocities, and the vertical and horizontal wind velocity components attending the period when the propagules are released from their parent plant. For any given descent speed and wind velocity profile, the mean and mode of the dispersal curve (the frequency distribution of propagules as a function of the distance from the source plant) should be positively correlated with the height of release. For any given height of release and wind profile, the mode of the dispersal curve should be negatively correlated with the terminal descent velocity. And for any given height of release and terminal descent velocity, the mode of the dispersal curve should correlate positively with the ambient horizontal wind velocity component. As noted, however, although these expectations may be fulfilled for any one propagule, released from any given height and at any ambient wind speed and direction, the real dispersal curve will obviously reflect a statistical summary of the frequency distributions of terminal descent velocities, heights of release, and the vertical and horizontal velocity components of the ambient airflow. Clearly, variations in the weight, size, and shape of propagules, even from the same source plant, will affect the dispersal curve, as will the variations in the height at which propagules are borne on the plant and in the ambient wind speed and directional components attending propagule release.

Indeed, great caution must be exercised in evaluating how well the observed shapes of dispersal curves conform to predictions based on closed-form solutions. As seen from eq. (9.23) and (9.24), the dispersal curve is predicted to fall off with distance from the source plant, but it may achieve its mean and mode at some distance from a point source of propagules. It must be remembered that a two-dimensional normal distribution of propagules will have a mean or mode that is displaced toward the source plant when the distribution is expressed in polar coordinates (that is, in terms of the number of propagules per unit ground surface area) (see Augspurger 1986, 362; Okubo and Levin 1989, 329). This artifact results because the number of propagules per unit distance from the source plant is distributed over a greater surface area as the distance from the source plant increases linearly. This is easily visualized by imagining a hypothetical source plant situated at the very center of a polar coordinate system. If all the propagules borne by the plant are at a single point source and if the ambient wind speed and direction are constant, then all the propagules would be disseminated along a perfectly linear transect running through the polar coordinate system, and the number of propagules plotted as a function of distance from the point source would adequately reflect the reality of a one-dimensional dispersal curve. If the direction of lateral airflow varied over some angle, however, then propagules would be disseminated over a wedge-shaped region increasing in width outward from the source plant, and the number per unit area within this region would decrease as a function of the distance from the point source. Thus a sampling of the number of propagules along a linear transect away from the point source would be inadequate to the reality of a two-dimensional distribution of propagules on the ground surface. Obviously, variations in the height at which propagules are borne on the source plant and in the terminal settling velocity of propagules are as biologically real as variations in ambient wind speed and direction attending the release of wind dispersed seeds and fruits. Thus the application of relatively simple phenomenological equations to predict an ideal dispersal curve or to test an empirically determined dispersal curve can be very misleading. In general, the mode of the dispersal curve should be used rather than the mean, since the calculation of the mean dispersal distance depends on the tails of the spatial distribution of propagules and therefore is prone to more errors in measurement than is the mode.

More recent attempts to predict the dispersal curves of wind-disseminated seeds and fruits have generated mechanistic rather than descriptive equations. For example, Okubo and Levin (1989) applied a tilted Gaussian plume model to predict the mode of the dispersal curve x_m in terms of the height of propagule release h, the ambient horizontal wind speed U_a, the propagule descent speed U_s, and the vertical airflow mixing velocity W^* attending turbulence. Recall from our prior discussion that even though a propagule always descends with respect to the unit of air in which it is suspended, the unit of air itself may ascend because of updrafts in the mass airflow attending turbulent flow. Accordingly, the descent velocity of the propagule and the vertical mixing velocity must be considered simultaneously. Intuitively, we might expect W^* to be more influential in predicting the mode of the dispersal curve for propagules with a very low descent velocity, since their descent in the air column would be slow and updrafts would significantly delay their contact with the ground. (Notice that U_s and W^* reflect the horizontal and vertical velocity components of the general ambient airflow.) Indeed, Okubo and Levin (1989) mathematically demonstrated that the following relations should hold true:

(9.25a)
$$x_m \approx h\frac{U_a}{U_s} \quad \textit{(for heavy propagules)} \quad (U_s > 1\ m\cdot s^{-1})$$

and

(9.25b)
$$x_m \approx h\frac{U_a}{W^*} \quad \textit{(for light propagules).} \quad (U_s < 1\ m\cdot s^{-1})$$

(Equation 9.25a is similar to one derived by Cremer 1977 relating the horizontal transport distance D to the height of release and the ambient and descent velocities: $D = h\ U_a/U_s$.) Equations (9.25a) and (9.25b) were empirically examined based on data from fifteen independently published studies. The data from these publications were plotted in terms of two dimensionless ratios: the ratio of ambient wind speed to the descent velocity of the propagule U_a/U_s, and the ratio of the mode of the dispersal curve to the height of release x_m/h. The plot revealed that as the descent velocity of propagules increases, the normalized modal dispersal distance (x_m/h) decreases. For very heavy propagules ($U_s \geq 3\ m\cdot s^{-1}$), the normalized modal dispersal distances that were empirically determined coincided almost exactly with the predictions of the model proposed by Okubo and Levin. For very light particles (spores and pollen), however, their model predicted greater normalized modal dispersal distances. A more detailed analysis of the topic in terms of diffusion, settling, and advection yielded the formula

(9.26)
$$x_m = \frac{h\ U_a}{(U_s + W^*)},$$

which provided very reasonable predictions for the normalized modal dispersal distances of a broad range of airborne plant structures (Okubo and Levin 1989). One of the many advantages to the approach taken by Okubo and Levin is that data from a variety of plants growing in very different habitats can be placed within a single objective classification scheme for the aerodynamic properties of airborne spores, pollen, seeds, and fruits.

Finally, the sporophytes of most arborescent vascular land plants get taller as they grow, while some seed plants bear separate microsporangiate (pollen-producing) and megasporangiate (ovulate) organs on the same individual. Based on our previous discussion, it seems reasonable to speculate that considerable ecological benefits would be conferred on species that developmentally bias the vertical distribution of ovulate cones toward higher altitudes as the size of an individual increases. My experience with a grove of pine trees suggests that short trees tend to have microsporangiate and ovulate cones more or less uniformly distributed along their height. But as an individual tree gets taller, more of its ovulate cones are produced on higher branches than on lower ones, while the production of microsporangiate cones appears to be increasingly limited to lower and intermediate branches. Although I have not analyzed the data in detail, I also have the impression that smaller trees have a higher ratio of microsporangiate cones to ovulate cones and that this ratio shifts in favor of ovulate cones as the size of an individual tree increases. This is interesting, since smaller trees tend to be found in more open and windy habitats. The shift in the vertical distributions of the two types of reproductive organs may confer a benefit in terms of long-distance seed dispersal, while the shift in their relative numbers may reflect the possibility that larger trees can physiologically sustain the growth of more seeds than smaller trees. If we crudely extend these speculations to the arena of plant evolution, it seems reasonable to argue that shifts in the position and relative number of micro- and megasporangiate cones conferred the same benefits on the earliest seed plants. Indeed, heterosporous species within very separate evolutionary lineages (e.g., lycopods and horsetails) tend to have been much taller than their homosporous phyletic counterparts. In any event, ecologists and paleontologists should focus on shifts in the location and relative number of different types of sporangiate organs as they examine individuals from the same population that differ in overall size.

Ten

Biomechanics and Plant Evolution

The future isn't what it used to be.

Attributed to a dean of faculty

The agenda of biomechanics is not limited to the analysis of living organisms. Like the mythical Janus, its perspective can extend backward in time to examine the history of life. But this retrospective mode of inquiry has two requisites: organic form and structure must be well preserved, and their biological functions must be correctly identified. Once these requisites are satisfied, the task of quantifying how well form and structure fulfill their functional obligations is relatively easy to accomplish. Yet to some, excellent preservation and objective function-ascribing statements may sound difficult or nearly impossible, particularly since the "poor fossil record" and "the Panglossian view of the adaptationist" have received much notoriety. In point of fact, for many types of organisms the fossil record is remarkably complete, even for relatively transient, albeit influential episodes of evolutionary innovation. Indeed, paleontological findings over the past few decades should make us very skeptical of the statement, "You'll never find that preserved in the fossil record." Exquisitely preserved soft-bodied invertebrates have been found in strata hundreds of millions of years old, and even the delicate molecular fingerprints of DNA have been lifted from fossil leaves over twenty million years old (Golenberg et al. 1990). By the same token, objective hypotheses in the form of function-ascribing statements for fossil organisms need not be procured with the aid of a top hat. The dental formula of a fossil mammal or the venation of a fossil leaf reveals much to those already familiar with the workings of the jaws and leaves of living animals and plants. No magic is required to compute the compressive or shear stresses exerted on a tooth by the closing of a jaw or the volume of water that could have flowed through a vessel in a leaf. And for fossil organisms whose shapes and general organizations appear radically different from any seen before, it may still be possible to draw tentative hypotheses about metabolic and ecological requirements to be tested as more data are gleaned from the fossil record. Thus, paleobiomechanical inquiries into the form-function relations are much more accessible than one may initially think. And just as for all aspects of scientific inquiry, hypotheses can be proposed, tested, and either subsequently modified or rejected through experimental methods.

The claim that paleontology is an experimental science comes as no surprise to the paleontologist, but it often shocks others. True, the history of life cannot be repeated as can the various fates of an isogenic cell line cultured under differing but highly controlled conditions. In these and other ways, some of the experimental methods of paleontology differ substantially from those of other experimental sciences. However, we can contrive controls in paleontology against which to test the effects of varying one experimental parameter at a time. We can build physical models of animals and plants long dead and learn how they flew or radiated heat or intercepted sunlight. We can alter these models in accordance with the actual changes in the morphology of geologically younger fossils to mimic past evolutionary events and determine whether the performance levels of various biological functions remained constant or changed for the better or worse. From experimental formats like these, we can examine hypotheses and test them against subsequent paleontological findings. In a more global sense, the history of life in toto can be decomposed into the various components we call lineages, and their evolutionary responses to the same or similar challenges from the environment can be compared. As a backdrop to this deconstructive mode of analysis, we have a record spanning three billion years documenting the community composition of individuals, species, and higher taxa from which the relative success of morphological, anatomical, and reproductive changes may be gauged. These and many other approaches provide experimental formats that, though not canonical, are highly informative.

In science, observations are used to generate hypotheses that are subsequently tested. In the case of paleontology, our observations are the many thousands of fossil organisms that are chronologically listed in order of their appearance within the geologic column. These observations, when referenced against the geologic time line, show that organisms have changed, as has the physical environment. To account for these changes, a variety of hypotheses have been put forth, and one of the most widely subscribed to is natural selection. In its simplest form, the basic thesis of Darwinian evolution is that organisms predominantly exhibit adaptive responses (mediated by the process of natural selection) to the challenges imposed on them by the environment. Over geologic time, the environment operates much like a semipermeable membrane or filter, through which some genetic variations pass by virtue of their phenotypic manifestations, whereas others do not. Those that survive this rite of passage contain the genetic materials that will provide the basis for subsequent genetic changes, hence phenotypic variations. As a consequence of this process, the challenges the environment imposes alter the genetic composition of populations and so indirectly influence the appearance, structure, and reproductive potential of surviving phenotypes. It is important, therefore, to see that natural selection is not the fountainhead of the genetic changes within organisms; it operates to eliminate, not engender, genetic alterations. By the same token, those genetic alterations that are removed by natural selection are those that are phenotypically expressed (or indirectly linked to those that are expressed) and affect survival and reproduction. Those genetic properties that are hidden or suppressed may pass through the process of natural selection for many generations until their presence affects the phenotype and so becomes liable to the consequences of natural selection.

Unlike the diffusion of molecular species through a semipermeable membrane, in which their material properties remain unaltered, the passage of phenotypic variants through the barriers imposed by the environment does not ensure the long-term survival of a species. The nature of the environment changes over time, just as does the genetic composition of a species. The reason becomes readily apparent once we see that the environment itself is altered by the process of evolutionary diffusion.

The environment consists of two components. One of these is the biotic component comprising coexisting organisms and the various biological interactions among them. The other is the abiotic component that consists of purely physical features. Both the biotic and abiotic components influence the growth, survival, and potential for reproduction of every individual, and ultimately the fate of the human constructs we call species and "higher taxa." Both the biotic and abiotic components change over time. Indeed, the study of natural selection is really a study of how ecology has changed over time, since the biotic component is altered by the process of evolution (Wade and Kalisz 1990). As organisms pass through the environmental barrier, their appearance and their very nature may change, altering the biotic component of successive environmental barriers to survival and reproduction. At the same time, various aspects of the physical environment may change. The global climate has warmed and cooled over millions of years. Land masses have drifted over the earth's core. Their geologic schisms and collisions have resulted in the physical separation and juxtaposing of the organisms drifting along with these land masses, with consequent effects on climate and ecology.

An obvious question arises, given the hypothesis of natural selection. Which of the two components, abiotic or biotic, is more influential? Darwin felt that abiotic factors in the environment varied too randomly to provide directionality to the vectors he felt were evident in the history of life. Accordingly, if the history of life evinces adaptation, which Darwin believed it did, then the influence of biotic factors must predominate over that of the abiotic environmental component. Darwin was not alone in this opinion. And the supremacy of biotic over abiotic factors continues to be more often cited as an argument for biomechanical improvements in the history of animal lineages (Allmon and Ross 1990). Given the way most heterotrophs accumulate carbon and the fact that many animals have a limited capacity to alter their environment through behavioral responses such as locomotion, this consensus is not unexpected. Indeed, it may be correct. Nonetheless, any dichotomy in opinion is prone to the "black and white" fallacy of logic—the relative influence of abiotic versus biotic factors has likely varied over geological time as well as from one group of organisms to another. Indeed, if the debate over the importance of biotic versus abiotic factors in evolution is meaningful at all, then it can be resolved only by determining the relative frequency with which either has been the principal agency in evolutionary patterns.

In terms of this relative frequency of either biotic or abiotic factors as the principal agency influencing the course of evolution, it seems appropriate to repeat a point made in chapter 1. Plants constitute over 90% of the world's biomass. As such, plant biology and history surely must provide weight to any debate concerning evolution whose resolution rests on "relative frequency." And biomechanical analyses of plant evolution provide some of the most sophisticated instruments with which to assess the relative importance of abiotic and biotic factors. Certainly the principles of biophysics and physical properties of light and water are constants in the physical environment of plants. These principles and physical properties may not have overpowered the roles of the biotic component in the history of plant life, but they assuredly were important in the evolution of eukaryotic photoautotrophs.

To evaluate some of the issues raised here, we must turn our attention from adaptation as a state of being (organisms work; they appear suited to their environment) to adaptation as a process of becoming or staying adapted (organisms change in their biomechanical attributes over evolutionary time). In this regard, four aspects of adaptation as a process are important to remember: (1) Adaptation (whether defined as a state or a process) cannot be evaluated outside the context of the environment. (2) Adaptation is a *relative* attribute; it depends on an individual organism's capacity to function as well as or better than its contemporaries. (3) Although characters or attributes may be advantageous or disadvantageous, the individual organism operates as a montage of its character states; accordingly organisms, not individual characters or their states, are perpetuated through natural selection. It is the collective effect of this montage that determines survival and reproductive potential. (4) Adaptation as a state permits comparisons among organisms at three levels: among individuals within a single population; among individuals from different species that have similar character states; and among individuals from different species that have different character states. The first of these comparisons is relevant to an examination of intraspecific (more precisely, intrapopulational) competition, while the second and third relate to interspecific competition. A truly comprehensive examination of adaptation as a process requires the chronological assessment of all these aspects of adaptation as a state.

This chronological assessment requires understanding the properties and evolutionary fate of organisms on time scales of 10^6 years or longer. Thus the fossil record, which provides anatomical, morphological, and biochemical data for quantitative assessments of form-function relations, is indispensable. To a lesser degree of accuracy, the fossil record also provides paleoecological data that can place a fossil taxon in the context of its physical and biological environment. The ability to document the long-term fates and consequences of form-function relations distinguishes the geological from the ecological time scale. Only rarely can we view the evolutionary fate of an organism in ecological time, and even then we see only the finale of an evolutionary history that preceded our direct observations. The fossil record reflects an evolutionary experiment that has been in progress for billions of years. The analysis of this experiment is the task of every biologist.

The evaluation of plant adaptation is made easier in that virtually every plant, apart from a relatively few parasitic species, uses the same resources and, with a few exceptions, the same biosynthetic pathways to produce primary and secondary metabolites. Therefore, whenever intra- or interspecific competition occurs, it primarily involves antagonists with the same metabolic requirements (light, water, minerals, and space), even if acquisition of these requirements is not the direct cause of hostility. Significantly, however, this metabolic homogeneity contrasts sharply with the morphological, anatomical, and reproductive diversity seen among fossil and extant plant species. Indeed, biochemical and ultrastructural evidence is often required to prove common ancestry among plant groups. Accordingly, the biomechanicist faces what appears to be a dilemma—the growth and survival of plants appear predicated on similar metabolic imperatives, yet plant form and structure appear to be disproportionately diverse by comparison. Four hypotheses are conventionally provided to reconcile this dilemma:

1. *The structural heterogeneity among plants may be much less than it appears and masks essential similarities.* Convergence in shape and structure among organographically distinct plant parts is common among plant species. The phyllids of mosses, the thalli of liverworts, the microphylls and megaphylls of lycopods, horsetails, ferns, gymnosperms, and angiosperms, and the stems and roots of many plants frequently possess a morphological similitude explained by their functions. This convergence evinces underlying design constraints or principles. (However, by emphasizing these similarities we are in danger of trivializing significant differences.)

2. *Plants are tremendously tolerant in their capacity to cope with the environment structurally or physiologically.* Provided an organ meets some minimal requirements and maintains a minimal level of performance, its shape, size, and structure can vary widely within these boundary conditions. Hence form and function may be sufficiently relaxed to permit marked variation in form before function becomes seriously impaired.

3. *By contrast, diversity in plant form may reflect extreme fine-tuning.* Subtle differences in the environment may have driven natural selection to produce extremely small, albeit important, morphological and structural differences among species.

4. *The diversity seen in plant morphology and anatomy may reflect the temporal coexistence of phylogenetically distinct groups of plants whose structural and morphological features reflect different phyletic legacies.* (This explanation essentially views plants as evolving out of phase with one another and confuses archaic with primitive features.)

These four hypotheses are not mutually exclusive, nor should they be taken as a comprehensive litany of explication. They simply illustrate that very different opinions can be offered to explain the same observation. Regardless of our philosophical inclination, however, it is obvious that any explanation for the diversity seen in plant form and structure is likely to revolve around the relations between form and function, even if a correlative relation is wholly denied. Accordingly, both the adaptationist, who views life as finely attuned to the environment through natural selection, and the nonadaptationist, who views organisms as functional but not perfectly adapted nor necessarily the products of natural selection, require critical biomechanical analyses.

After laboring through the previous chapters of this book, we are finally in

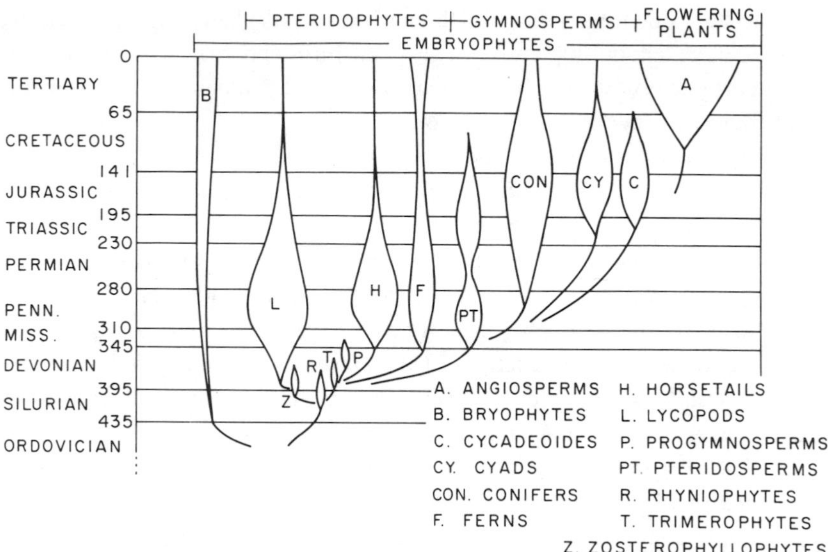

FIGURE 10.1 Evolutionary relationships and first occurrences of major systematic plant groups. Lineages are grouped according to their mode of reproduction (see horizontal bars at top of figure). All lineages are embryophytes (they possess gametophytes with archegonia or, in the case of angiosperms, structures believed to be derived from archegonia). With the exception of the bryophytes (B, see legend), all are vascular plant lineages. For further details, see text.

a position to evaluate the five basic requirements of any plant, whether dead or alive, in a quantitative manner:

1. The interception of sunlight, which is the energy source for photosynthesis.
2. The exchange of gases between the plant body and the atmosphere.
3. The necessity to sustain weight that is elevated above the ground.
4. The absorption and conduction of water and other nutrients from one part of the plant body to another.
5. The requirement to successfully complete the reproductive cycle.

A detailed examination of each of these five requirements is well beyond the scope of this chapter, so here I seek merely to illustrate where, when, and how some of these needs have entered into plant evolution. Perhaps most important is the recognition that each of these five requirements has design specifications that may conflict with one another. The way plants have reconciled these conflicting design specifications is the principal focus of this chapter.

When were the basic solutions to plant survival in the terrestrial environment initially solved? The answer is, During the time between the advent of

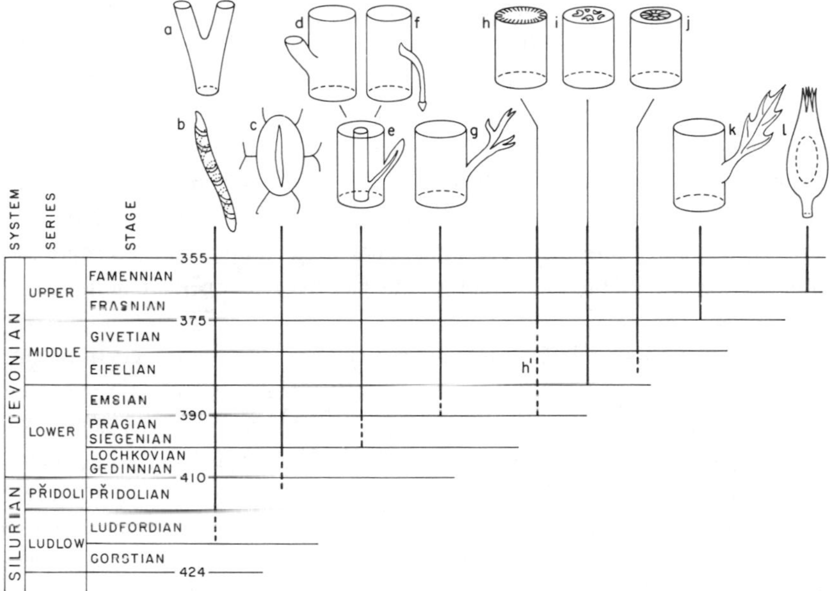

FIGURE 10.2 First occurrences of some morphological and anatomical features in the fossil record of Upper Silurian and Devonian vascular land plants. The Upper Silurian to the Upper Devonian spans an interval of roughly 60 million years, or roughly 15% of the entire history of vascular plant evolution (rough dates given to the right of stage names are in millions of years). The earliest presumed land plant fossils are reported from the Llandoverian (Lower Silurian, not shown). The earliest bona fide vascular plants (such as *Cooksonia,* described from the Přidolian) are characterized by having sporophytes with more or less equally branched axes (a) and a slender vascular strand of tracheids (b). Stomata (c) are reported from Lochkovian (Gedinnian) plants (*Zosterophyllum*). Unequal branching (d) and haplosteles with vascularized leaves (e) are known from Emsian, possibly Pragian (Siegenian) plants (*Psilophyton* and *Drepanophycus,* respectively); adventitious roots (f) are also known from Emsian plants (*Gosslingia, Crenaticaulis*). Planation of lateral branches (g), suggesting a precursor to the megaphyllous leaf, is seen in some Emsian plants (*Pertica*). Wound periderm (h) and a continuous periderm (h) are reported for Emsian and Frasnian plants (*Psilophyton* and *Triloboxylon,* respectively). Dissected steles (i) are reported from Eifelian plants (*Calamophyton*), while secondary xylem (j) is reported for Givetian plants (*Rellimia*). Megaphyllous leaves (planated and webbed) (k) are reported for the Frasnian plant *Archaeopteris.* By the Famennian, seed plants (l) had evolved (*Archaeosperma*). The information presented in this figure is based on Chaloner and Sheerin (1979, text figs. 2–4) and more recent primary literature (see Banks 1981).

terrestrial plant life and the end of the Devonian. It was during this interval that plants evolved the anatomical, morphological, and physiological capacity to deal with an environment radically different from their ancestral aquatic one. It is reasonable, therefore, to assume that in the early radiation of terrestrial plants the physical environment played a dominant role in limiting the scope and progress of plant life on land, and that intra- and interspecific competition was less intense than during subsequent times, owing to an essentially unoccupied terrestrial ecolandscape. With the exception of flowering plants, all the major clades of vascular plants were established by the end of the Devonian, some 355 million years ago (fig. 10.1). By the end of the Devonian, community structures were well diversified, and plant species presumably were experiencing progressively more intra- and interspecific competition. Thus, by the end of the Devonian, plants had evolved operable solutions to life on land.

When land plants made their first appearance has been the subject of recent, intense debate. Evidence for bona fide terrestrial metaphytes is convincing from the Llandoverian of the Lower Silurian. Reasonably convincing evidence for vascular plants comes from Late Ludlowian fossils, while undisputed vascular plants are found in Přidolian strata (fig. 10.2). If we accept these geological dates as rough estimates, then the transition of land plants to vascular plants spans approximately 35 million years (or roughly 8% of the total history of land plant evolution). Similarly, if the Přidolian is accepted as the time of the first appearance of vascular plants, then the time between this occurrence and the end of the Devonian is roughly 60 million years (or roughly 15% of the history of land plants). These dates are all rough estimates, but it becomes evident that considerable evolutionary innovation occurred in a remarkably short time.

The Geometry of Light

First we will consider the geometries of the earliest land plants in terms of their capacity to exchange gases with the external atmosphere and to intercept light. All land plants physiologically exist within a liquid-gas interface created at the surfaces of the wet cell walls of physiologically competent cells and the atmospheric gas phase. Through this interface, water, nutrients, and gases (essential to photosynthesis and respiration) diffuse. As discussed in chapter 1, one of the principal constraints on terrestrial plants is that water vapor is lost as carbon dioxide and oxygen diffuse across this interface. The earliest undisputed vascular plants possessed a cuticle (a complex mixture of cross-linked long-chain esters, fatty acids, and alcohols with a waxy long-chain hydrocarbon layer on its surface) that delineated the size, geometry, and distribution of the liquid-gas interface over that plant body's surface. Water permeability in the cuticles of extant plants is not directly related to the thick-

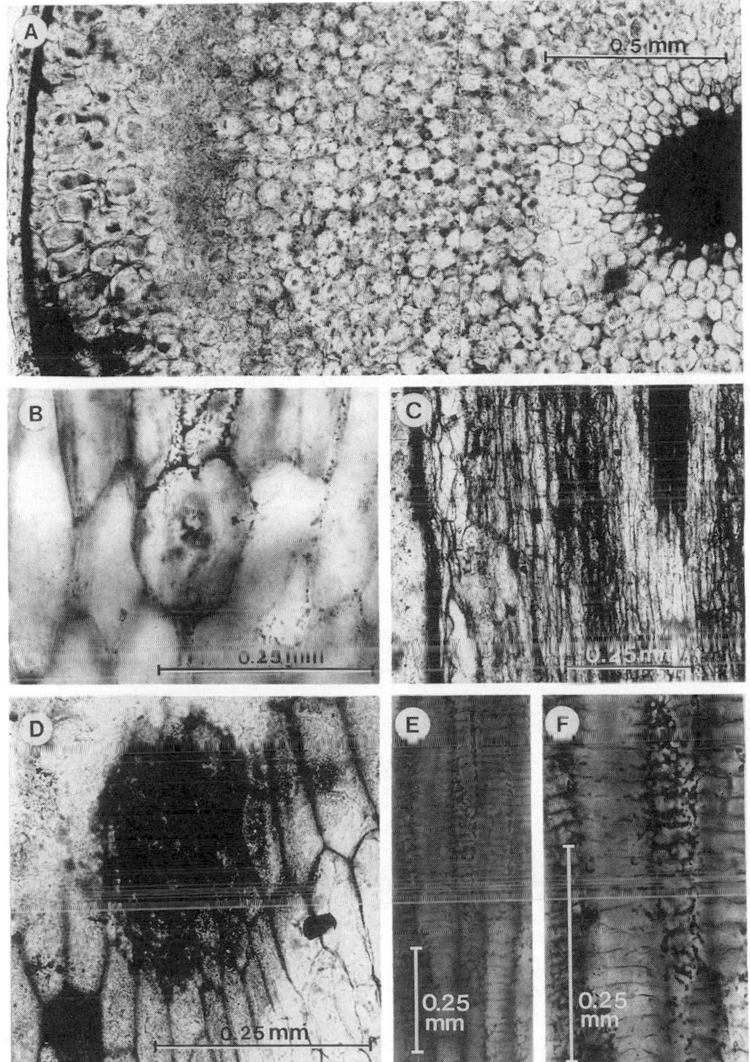

FIGURE 10.3 Anatomical features of *Rhynia*. (A) Cross section of a vertical axis revealing the parenchymatous cortex and a dense cluster of centrally positioned cells that presumably functioned as conducting tissues. (B) Paradermal section showing a stoma with two guard cells. (C) Longitudinal section of a vertical axis (epidermis to left) with an oblique section through the central strand of conducting tissue. (D) Stomalike structure on the epidermis of an axis that may reflect a wound caused by a terrestrial mandibulate invertebrate. (E–F) Longitudinal sections through the (water?) conducting tissue of an axis, revealing darkened bands many believe are secondary wall thickenings comparable to those found in tracheids.

ness of the cuticle but is largely controlled by the chemical composition and thickness of its external waxy layer. (For a given number of long-chain hydrocarbon molecules per unit surface area, water permeability is roughly halved by the addition of each two-carbon unit to the hydrocarbon waxy layer.) Unfortunately, we know very little about the chemistry of the cuticle of the earliest land plants. To maintain growth at some specified level, however, the ratio of the wetted surface area of the plant body (responsible for gas exchange and essential to photosynthesis) to the volume of physiologically active tissues within the plant body must be kept at a reasonably constant value as absolute size increases. In addition, the ratio of total surface area to volume of tissues influences the quantity of photosynthetically active radiation (400–700 nm), called PAR, intercepted by the plant body. The total surface area of terrestrial plants, particularly vascular plants, is difficult to calculate empirically, because a substantial fraction of the wetted surface area may be internalized in the form of linings to intercellular air spaces. This internalization is critical because surface area per unit biomass and respiratory rate per unit biomass tend to change as $-1/3$ and $-1/4$ powers of biomass, respectively (see Peters 1983; Raven 1985, 276). Also, much smaller diffusion gradients can be realized with carbon dioxide than with oxygen, and the loss of water vapor is reduced by the resistance across the openings (pores or stomata) to internal gas chambers.

We can appreciate the importance of the ratio of surface area to the volume of the plant body by noting that the earliest land plants probably lacked a well-defined cuticle and had limited internal air spaces. Indeed, *Rhynia,* one of the best-preserved early land plant fossils, possessed stomata but apparently lacked well-defined intercellular chambers, although the dark areas among neighboring hypodermal cells in the outer cortex may reflect intercellular air spaces that were occluded by minerals during fossilization (fig. 10.3A–C). Thus the external surface of the first terrestrial plants most likely played a critical role in both light interception and gas exchange. We may assume also that most of the earliest land plants had the capacity for indeterminate growth. This assumption is compatible with what is currently known about the earliest land plant fossils and most extant plant species. Yet it imposes a significant constraint in terms of Fick's first law (see chap. 1), since geometric isometry is largely prohibited for any organism that must maintain a constant diffusion rate while at the same time increasing in absolute size. Given the constraint imposed by Fick's first law on any plant with indeterminate growth, a reasonable question is, What geometries are capable of increasing in absolute size while maintaining a constant or near constant ratio of surface area to volume?

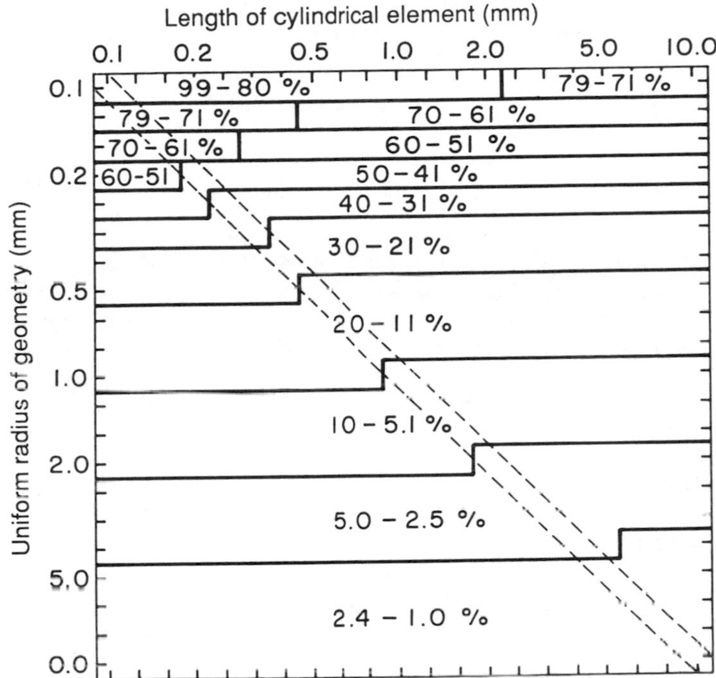

FIGURE 10.4 Changes in the ratio of surface area (*S*) to volume (*V*) as the dimensions of a cylinder capped by two hemispherical ends are varied. For comparison, the diagonal dashed lines represent the values of *S*:*V* of a sphere as its radius increases (from upper left to lower right). All values of *S*:*V* are normalized as percentages against the highest value of *S*:*V* within the domain (the smallest sphere, whose *S*:*V* plots in the upper left corner of the diagram). For any uniform radius, *S*:*V* decreases as the length of the cylinder increases. The smallest reduction in *S*:*V* as length increases occurs for a very slender cylinder (diameter 0.1 mm).

As we will see, one of the answers to this question is a plant very like *Rhynia*, which may be taken as a paradigm for the morphology and possibly the anatomy of the earliest successful land plants.

Some very simple analytical geometry is required to find the shape of plants that can conserve a relatively constant ratio of surface area to volume yet grow in size. One can readily see that spheres (unless they are hollow, like *Volvox*) are essentially useless as geometric solutions, since the ratio of surface area to volume (S:V) of a sphere is proportional to the inverse of the sphere's radius. Thus S:V decreases as the absolute size of the sphere increases. By contrast, spheroids and cylindrical geometries can increase in absolute size yet maintain or minimize the reduction in S:V. Accordingly, if gas diffusion is crudely

TABLE 10.1 Relative Efficiencies of Direct Light Interception for Four Simple Geometries Reminiscent of the Morphologies of Early Land Plants

Spheroids			Cylinder		
b:*a*	Oblate	Prolate	*l*:*d*	Without Hemispherical Ends	With Hemispherical Ends
1.00	44.9 (78.5)	44.9 (100)	1.0	43.3 (100)	40.7 (100)
0.50	50.1 (87.6)	35.7 (79.5)	2.0	40.6 (93.8)	40.2 (96.3)
0.25	53.9 (94.2)	29.0 (64.6)	4.0	38.6 (88.7)	38.1 (93.6)
0.10	54.8 (95.8)	24.3 (54.1)	10	37.4 (86.4)	37.2 (91.4)
0.01	56.9 (99.5)	21.2 (47.2)	100	36.5 (84.3)	36.5 (89.7)
10^{-5}	57.2 (100)	20.8 (46.3)	10^5	36.3 (83.8)	36.3 (89.2)

Source: Data from Niklas and Kerchner (1984, 81, table 1).
Note: Diffuse light interception is neglected. Light interception is calculated at an ambient solar irradiance of 400 watts with geometries positioned at the equator. Values of light interception are in W-hr.; the percentage of the maximum light interception for each geometry is given in parentheses. The aspect ratio for spheroids (oblate and prolate) is the ratio of the semiminor (b) to the semimajor (a) axis of the elliptic cross section (at b:a = 1.0, the spheroid is perfectly spherical). The aspect ratio for cylinders is the ratio of length (l) to diameter (d) (hemispherical ends have a unit radius = $d/2$).

taken to be dependent on S:V, then the morphology of the earliest land plants would have converged on nonspherical shapes. Significantly, light interception also depends on S:V. As it turns out, however, not all geometries can maintain a constant capacity both to intercept sunlight and to maintain S:V, because light interception is a function of the projected surface area of a solid rather than a simple function of the total surface area. Since plants have to intercept light and exchange gases and only a few geometries can do both while increasing in size, we have a method for deducing the morphological domain of early land plants.

Consider three simple shapes (oblate and prolate spheroids and the cylinder) that illustrate this point. (An oblate spheroid is generated by rotating an ellipse around its minor axis, while a prolate spheroid is formed by rotating an ellipse around its major axis. Thus both of these geometries have an elliptical transection that can be measured by the dimensionless ratio of its major axis to minor axis. For those of us that do not instantly visualize an oblate or prolate spheroid, we can liken these two shapes to a flat pancake and a cigar or football, respectively.) An increase from 0.1 to 10 mm in the major axis of an oblate spheroid results in a 50% reduction in S:V, while a comparable geo-

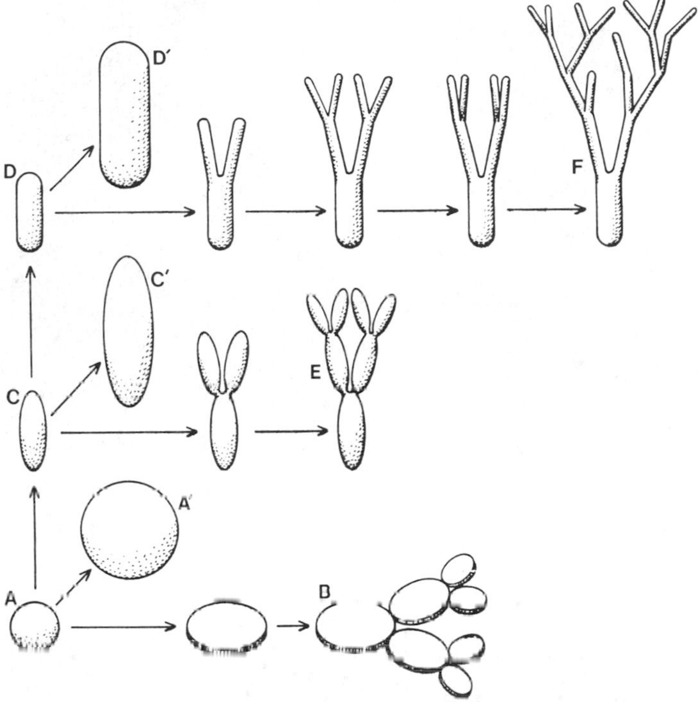

FIGURE 10.5 Hypothetical patterns of early land plant evolution based on the changes in the surface area to volume ratios as various geometries increase in absolute size. Analytical geometry predicts that the shape of the plant body will gravitate toward either a dorsiventral geometry (an oblate spheroidal geometry) (A to B) or a prolate or cylindrical geometry (C and D). The surface area:volume ratio of a spherical morphology (A) decreases as absolute size increases (A to A′). A change in the sphere's aspect ratio into an oblate spheroid (and the reiteration of oblate spheroids) provides for a constant or increasing surface area:volume ratio as absolute size increases (A to B). Other "geometric solutions" to maintaining or increasing the surface area:volume ratio as size increases are to alter the morphology into a prolate spheroid (C) or a cylinder (D). Further increases in the size of these two geometries would require alterations in the aspect ratios (C′ and D′) or reiteration of the prolate or cylindrical geometry to produce "branched" body plans (E and F).

metric change in a prolate spheroid results in less than a 20% reduction in S:V. For a cylinder, elongation reduces S:V, but for any length S:V decreases even more sharply as circumference increases (fig. 10.4). The ability of these geometries to intercept light can be measured by computing the area under a plot of the dimensionless ratio of their projected surface area Ap to total surface area A (or Ap/A) versus the change in the solar angle over a solar cycle. Analysis of this parameter as the absolute sizes of spheroids and cylinders

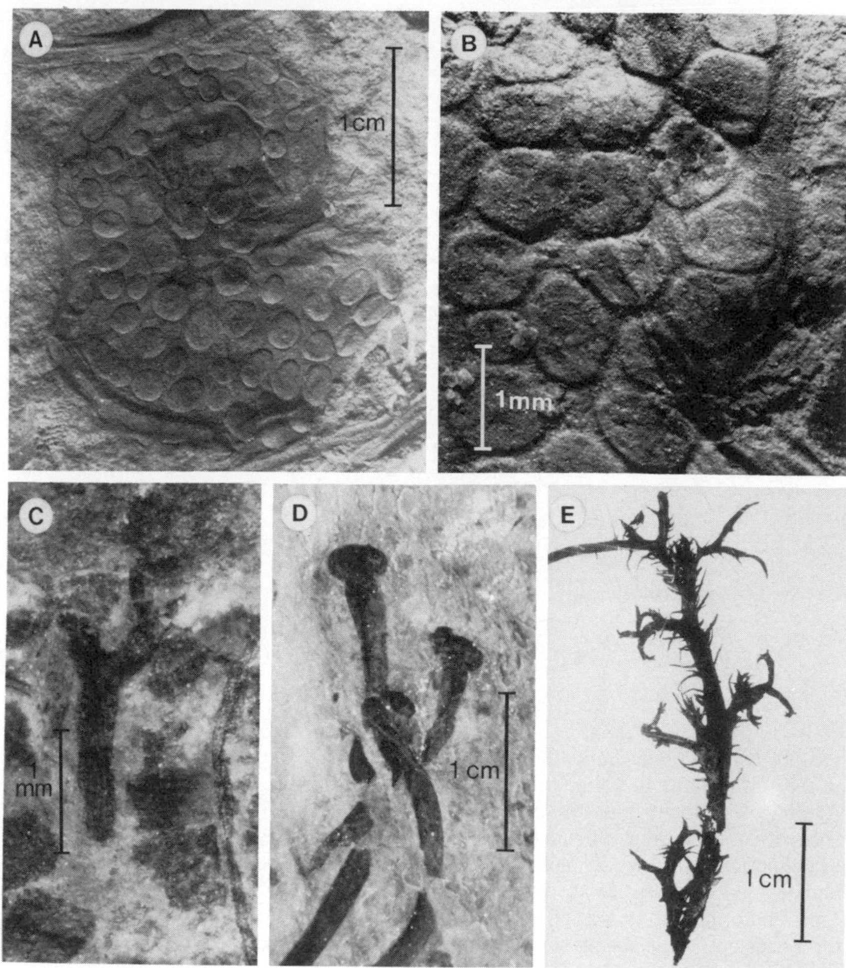

FIGURE 10.6 Dorsiventral (oblate spheroidal morphology) and cylindrical morphologies reflected in some early Paleozoic plants: (A–B) *Parka decipiens,* a thalloid, nonvascular plant with a pseudoparenchymatous tissue construction, similar to that of the algal genus *Coleochaete*. (C–D) Presumed specimens of *Cooksonia* (with terminal sporangia; see D). (E) *Psilophyton,* with numerous enations and lateral branches.

increase indicates that the oblate spheroid optimizes Ap/A as it becomes flatter (as it increases in its major axis:minor axis and becomes more and more like a pancake) (table 10.1). By contrast, prolate spheroids decrease in Ap/A as they get longer (as they increase in their major axis:minor axis and become more and more like a cigar). Cylinders can increase in length by many orders of magnitude without significant reductions in Ap/A (table 10.1).

Depending on their size and aspect ratios, oblate spheroids and cylinders can modulate S:V and Ap/A in compensatory ways. Accordingly, they are highly versatile shapes with which to construct a plant body—gas diffusion and light interception can be maintained at relatively constant values as absolute size increases. Thus our very simpleminded analytical solution to the geometry of early land plants leads to the scenario shown in figure 10.5. If the ancestral aquatic plants giving rise to the first land plants were spherical, spheroidal, or cylindrical (shown to the left of the figure), then either (1) the physical constraints placed on them in a terrestrial environment would have quickly selected for geometries that would have changed the aspect ratio of the simple (unlobed) plant body such that oblate spheroids would become progressively flatter and prolate spheroids or cylinders would become progressively longer as they increased in volume, or (2) the aspect ratios of these geometries would have been retained but the geometries would have been reiterated (by budding or lobing) to effect an increase in the absolute volume of the organism as a whole. This second option is rather intriguing, since most plants show a modular construction involving the reiteration of relatively simple shapes. If we extend our analysis, then we can also demonstrate, by means of analytical geometry, that once surface areas could be internalized (through the developmental invagination of external surface area into the plant body or by schizogony) the geometry of the plant body would be dramatically released from a fundamental constraint. The formation of internal air-filled chambers effectively permits the opportunity to raise the plant body's ratio of surface area to volume without an undue increase in the exposed surface area through which water can be more easily lost. Apparently the internalization of surface areas was not easily achieved, since plants existing well into the Devonian apparently lacked large internal air spaces—for example, *Rhynia*.

It is not surprising, given the scenario offered in figure 10.5, that we find the photosynthetic organs of most extant plants to be constructed on the geometric principle of either the oblate spheroid (flattened leaf) or the cylinder (stem), nor that the evidence available from the fossil record reveals that the earliest land plants had morphologies that were either flattened or cylindrical or both (fig. 10.6). (All of the previous analysis is essentially based on Galileo's principle of similarity, though it is unlikely, given the prevailing social climate of his lifetime, that he would have been permitted to accept credit for these evolutionary deductions.)

THE MECHANICS OF VERTICAL GROWTH

As discussed in chapter 9, the potential dispersal range of airborne spores, pollen, and propagules (seeds and fruits) is increased if these reproductive structures are elevated above the ground. In addition, the elevation of photosynthetic tissues permits plants to avoid being shaded by neighboring objects. Since light is a resource all plants compete for, and since evidence from the fossil record indicates that the earliest land plants dispersed their spores by wind currents, it seems reasonable to speculate that once vertical growth evolved it would have remained in the permanent repertoire of plant development—competition for space, light, and long-distance dispersal would have achieved an added dimension from which there was no turning back. Indeed, compilations of the maximum girth of plant axes from progressively younger rock strata show that stouter (and presumably taller) plants evolved throughout much of the early Paleozoic (see Chaloner and Sheerin 1979, particularly fig. 5).

Returning to the three simple geometries considered previously (oblate and prolate spheroids, and cylinders), we can quickly deduce that only two are adequate to serve as compression members. Prolate spheroids and cylinders are axisymmetric in their transections (the axis defining the direction of continued apical growth), while the oblate spheroid has a reduced second moment of area in one of its two potential planes of bending. Thalloid liverworts, such as *Marchantia* and *Conocephalum,* have gametophytic morphologies very much like an oblate spheroid (crudely mimicked in the bottom drawing in fig. 10.5). That is, they are dorsiventral and typically grow appressed to a substrate. By restricting the direction of light to the rear of the growing tips of their thalli, the apexes of *Marchantia* and *Conocephalum* can be made to recurve upward, presumably as a phototactic response to the unidirectional light source. The vertical support of distally recurved thalli appears limited, however—I have never been able to induce the thalli of these liverworts to continue to grow vertically beyond about 3–5 cm before they recline under the weight added by their growing tips. These observations suggest that a flattened thallus is not a sound geometric design for continued vertical growth. But as we learned in chapter 3, a cylindrical or near-cylindrical geometry, like an attenuated prolate spheroid (which to all intents and purposes is a cylinder with tapered ends) is well suited to a vertical growth posture. Indeed, Stephen Wainwright (1988) devotes much of his book to the mechanical versatility of cylindrical geometries.

What concerns us here is the size (height) limitations on a cylindrical, nonvascularized plant axis. This issue is provoked by evidence from the fossil record that the earliest land plant sporophytes typically adopted a cylindrical geometry that functioned to elevate their sporangia above the ground. Some simple calculations show that these plants could have attained a vertical posture of tens to hundreds of millimeters by means of turgid, thin-walled cells alone. The simplest tissue system we have considered (chap. 6) is parenchyma, and the elastic modulus of this tissue can be used to calculate a theoretical height limit for cylindrical parenchymatous axes. For example, the elastic modulus of fully turgid parenchyma isolated from potato tuber varies from 5.1 to 19.4 $MN \cdot m^{-2}$ depending on the transverse radius of the plug of tissue (0.25 to 0.9 cm; see Niklas 1988a), and the extent of vertical growth can be calculated from the equations from the Elastica (chaps. 3 and 7). For example, with a radius of 0.25 cm and an elastic modulus of 5.1 $MN \cdot m^{-2}$, a cylinder could grow to 13 cm in height while incurring a modest deflection angle of about 10°. With a radius of 0.9 cm (and an elastic modulus of 19.4 $MN \cdot m^{-2}$), the self-loaded cylinder could grow to 48 cm in height. Significantly, the length:radius ratio of both cylinders is roughly 53. By adding the hoop reinforcement that could be provided by an epidermis, calculations indicate that the length:radius ratio of a self-loaded cylinder can be increased by roughly 15%. Thus the maximum theoretical length:radius ratio of a self-loaded plant axis made of parenchymatous core and an epidermal rind is roughly 61. Note that this number would be reduced if the plant required a design factor or safety margin to sustain dynamic loadings or if we considered the added weight of sporangia or appendicular (leaflike) structures. Also, our theoretical plant axis with a length:radius ratio equal to 61 has a slenderness ratio of about 120, which is the cutoff ratio for a safe columnar beam (see chap. 7). Nonetheless, our calculated value of 48 cm for maximum height (with a length:radius ratio of 61) is remarkably close to the maximum heights recorded for extant nonvascular terrestrial plants like *Monoclea forsteri* and *Dawsonia superba,* two mosses known to produce the tallest gametophytes. The thallus of *Monoclea forsteri,* a species native to New Zealand and Patagonia, grows to a maximum height of 20 cm, while *Dawsonia* can grow to 50 cm in height, with a length:radius ratio of about 50 (Parihar 1962, 259). *Dawsonia* gametophytic axes support one another by their interdigitating leaflike phyllids. The plants also grow in stands, thereby creating a boundary layer with a reduced vertical wind profile. So a 50 cm vertical height for a plant without the benefit of vascular tissues seems to be a reasonable limiting case.

Dawsonia superba provides additional insight on the vertical growth of nonvascularized plant axes. This species possesses a central strand of conducting tissues consisting of hydroids (intermixed with steroids) and leptoids. Hydroids and leptoids are functional analogues to tracheids and sieve cells, respectively; steroids are relatively thick-walled cells that contribute to the strengthening of the plant (Hébant 1977, 40). (The conducting tissues of *Rhy-*

nia often appear very like the hydrome and leptome of mosses; see fig. 10.3A, E–F.) Also, peripheral cells seen in transections through the vertical axes of *D. superba* are thick walled and appear as a dense rind girdling relatively thin-walled cells in the cortex. In chapter 7 we examined the core-rind model for vascular plant stems and saw that an outer rind of dense tissue confers significant mechanical benefits in terms of resisting bending moments. Although the conducting strand of *D. superba* may contribute significantly to resisting bending, the outer sterome appears much more mechanically robust when seen in transection (see Hébant 1977, plate 56, fig. 226).

The anatomy of the gametophytic axes of *Dawsonia superba* and other moss genera (particularly in the Polytrichales) raises the issue of the evolution of vascular tissues and lignification. Vertical growth essentially removes portions of a plant from its substrate, which is the source of water and other nutrients. The transport of water through nonspecialized living tissues must keep pace with the rate of transpiration from plant surfaces. Raven (1977) has calculated that xylem consisting of tracheids has a specific conductance of water 10^6 times that of the parenchyma cell pathway. For a plant with a 50% carbon dry weight, roughly 230 g of water must be lost per gram of dry weight gained. Since 0.33 g of carbon is lost as CO_2 in dark respiration, Raven (1977) has calculated that over 300 g of water is required for every gram of dry weight added during growth. Black (1973) has calculated a higher figure for C_3 plants (400 g H_2O of dry weight gained). Thus, growth (as measured by dry weight) is limited by the rate at which water can be supplied, and water transport via a parenchymatous cellular pathway is woefully below the rate at which water can be supplied through a specialized conducting tissue.

From the foregoing it should be apparent that the mechanics of vertical growth cannot be disengaged from the hydraulic requirements of transport and transpiration. The nonvascular strand of plants like *Dawsonia superba* (and possibly *Rhynia*) and the primary vascular tissue system of vascular plants influence how much vertical growth can be achieved by conferring added mechanical strength and the capacity to rapidly supply water as it is lost from photosynthetic plant surfaces. Thus, attempts to discuss the evolution of vascular tissues from the perspective of either conductance or mechanics alone are fruitless, since there is a reciprocity between their effects on plant stature. (This point was raised in chap. 8 when we discussed the work of Rashevsky.)

By the same token, arguments formulated to describe the evolution of lignification can be flawed if the full mechanical effects of lignin are not recognized. For example, it is sometimes stated that lignified secondary cell walls probably evolved as implosion-resistant structures in the water-conducting xylem (Wainwright 1970; Wainwright et al. 1976; 200, 320; Raven 1977). Additionally, we must not ignore the possibility that lignified conducting cells with secondary thickenings may have evolved as a consequence of the need to resist tensile and compressive stresses induced by stem elongation and the expansion of neighboring cells. (See Paolillo and Rubin 1991.) In terms of xylem evolution this may be true. Rapidly moving water within thin-walled tubes exerts a negative pressure that can be great enough to cause walls to collapse inward, blocking fluid transport. Structural reinforcement of walls by secondary thickening (as seen in tracheids and vessel members) increases the implosive stresses walls can sustain before they crimp inward, but the lignification of these cell walls confers added advantages. Lignin lets walls hydrate less and hence stabilizes their elastic modulus; it also provides a bulking agent that helps tissues resist compressive stresses; and finally, it can function as an antiherbivore chemical defense. Therefore lignin can exert a mechanical influence that need not be restricted to conducting tissues. Indeed, lignin has been identified in the secondary cell walls of collenchyma (*Eryngium*), as well as in the walls of some green algae (*Staurastrum*, Gunnison and Alexander 1975; *Coleochaete*, Delwiche, Graham, and Thompson 1989). Clearly, collenchyma and the thalli of green algae, such as *Staurastrum*, are not subjected to implosive stresses owing to a rapid fluid transport. Lignified cell walls are hard to digest and resist microbial hydrolysis, which may account for the evolution of lignification in aquatic, relatively small algae. Indeed, it seems very reasonable to suggest that the mechanical deployment of lignin in the plant bodies of terrestrial plants reflects a transfer from an antimicrobial function in aquatic algae to a mechanical function in terrestrial plants. Significantly, paleobiochemical analyses of fossil plants that significantly predate the earliest vascular plants and that apparently lacked conducting tissues reveal the presence of lignin-like moieties (Niklas 1976). The subsequent evolution of specialized conducting cells (that were subjected to periodically high implosive stresses) could have capitalized on a preexisting metabolic capacity for lignification. In this scenario, lignification can be viewed as an exaptation, sensu Gould and Vrba (1982)—a feature that increases current fitness but was not built by selection for current function.

Hydraulics of Conducting Tissues

The origin and subsequent rapid diversification of vascular plants has been attributed to a suite of evolutionary innovations involving primary xylem tissues. The evolutionary appearance of primary xylem in early land plants permitted photosynthetic organs to remain sufficiently hydrated while elevated well above the ground. In chapter 4 the differences between tracheids and vessels were discussed in terms of water conduction. We saw that the low resistance to longitudinal transport in the xylem is due to the absence of a living protoplast within cell lumens (see Zimmermann 1983; Zimmermann and Potter 1982). However, it was also pointed out that tracheids are suboptimal conducting cells compared with vessel members because water must pass from one tracheid to the next through small openings (pits) separated by a thin membrane, imposing an added resistance to fluid flow. Support for this view comes from direct measurements of water-flow rates in tracheid-dominated versus vessel-dominated xylem (Zimmermann 1983). Needless to say, there are disadvantages to having long, unobstructed vessels, since these can become blocked by air bubbles when water within them freezes, though this is less significant for tracheids, since gas bubbles typically remain trapped in the cell lumen where they form and water can flow around obstructed tracheids.

Water flow through xylem elements (tracheids and vessel members) has been conceptualized in terms of flow through capillary tubes. This was discussed in chapter 4, where we saw that the hydraulic conductance per unit length of capillary tube Lp is given by the Hagen-Poiseuille formula,

$$Lp = \frac{\pi r^4}{8\mu} = \frac{\pi d^4}{128\mu}, \tag{10.1}$$

where r and d are the radius and diameter of the tube and μ is the viscosity of fluid flowing through the tube. For a collection of parallel aligned tubes, Lp can be calculated from the summation of diameters:

$$Lp = \frac{\pi \Sigma d^4}{128\mu}. \tag{10.2}$$

The hydraulic conductance indicates the ability of a tube (or collection of tubes) to permit flow and can be used to predict the volume flowing per unit time Q from the formula

$$Q = Lp\left(\frac{\Delta\psi}{\Delta l}\right) = \frac{\Delta\psi}{R}, \tag{10.3}$$

where $\Delta\psi$ is the water potential difference over the change in length Δl and R is the resistance to water flow. Since Q, Δl, and $\Delta\psi$ can be measured experimentally, empirically determined values of Lp can be compared with those predicted from the Hagen-Poiseuille equation. Studies indicate that tracheid diameter and number apparently control water flow but that estimates of hydraulic conductance predicted by eq. (10.2) are generally twice those actually measured (Gibson, Calkin, and Nobel 1985). In ferns, the data show that up to 70% of the total resistance to water flow through the xylem is due to pit membrane resistances, which are not considered in eq. (10.2) owing to the assumption of capillary tube flow. Gibson, Calkin, and Nobel (1985, 294–96) provide a model for water flow from the middle of one tracheid to the middle of an adjoining tracheid that incorporates pit membrane resistance. The model is based on an electrical circuit analogue to flow and identifies series and parallel resistances. They give the formula for the total resistance R' as

$$R' = R^{tl} + \left[\sum_{i}^{n} \frac{1}{2R_i^{pc} + R_i^{pm}}\right]^{-1}, \quad (10.4)$$

where R^{tl} is the resistance of the tracheid lumen, R_i^{pc} is the resistance of the ith pit canal, and R_i^{pm} is the resistance of the ith pit membrane. Incorporating eq. (10.4) into more traditional calculations of water flow produces a reasonable correspondence between predicted and observed hydraulic conductances in a variety of experimentally examined pteridophytes whose xylem tissues contain tracheids. This is notable, because tracheids are very different from capillary tubes, hence the need for eq. (10.4).

Gibson, Calkin, and Nobel (1985, 301) note that the relatively high values of Lp observed for pteridophytes stands in marked contrast to the general impression that tracheids make poor water-conducting elements. Although this conclusion is not surprising given the long evolutionary history of ferns in particular and pteridophytes in general, analysis of living pteridophytes suggests that appreciable stomatal closure (as the difference in the ratio of leaf to air water vapor concentration increases) may provide the key to the survival of these plants in open and dry habitats (see Nobel, Calkin, and Gibson 1985). Thus the success of early vascular land plants cannot be explained entirely by evolutionary innovations in conducting tissues. This point can be indirectly illustrated by considering the trade-off between gas absorption and water transport. We might assume that the ratio of the absorbed carbon dioxide to the drop in water potential along the length of the organ must play an important physiological role. Recall (from chap. 4) that for water to flow from the base toward the tip of a vascularized organ, there must be a gradient of decreasing water potential from the base of the organ to the tip. Within a leafless plant organ, such as a nearly cylindrical photosynthetic stem, this gradient would be established by the radial diffusion of water from the strand of vascular tissue at or near the center of each cross section and the subsequent loss of water vapor at the exposed surface of the organ. The absorption of carbon dioxide at the surface and its subsequent radial diffusion into photosynthetic tissues beneath the epidermis can occur only when stomata are open, which in turn requires that the epidermis be sufficiently hydrated. It would be reasonable to assume that the ability of such an organ to maximize the ratio of absorbed carbon dioxide to the largest drop in water potential would confer a selective advantage by promoting growth. This ability can be shown to relate to the ratio of the cross-sectional areas of xylem to photosynthetic tissues within the organ. Provided a few assumptions are made, the optimal ratio can easily be computed. Indeed, only four assumptions are required: the evaporative flux of water over the surface of the plant body is uniform; the quantity of carbon dioxide absorbed is directly proportional to the volume of photosynthetic tissue; the geometry of the plant body is somewhat tapered; and the resistance to water flux through the plant body is negligible compared with the resistance to water vapor loss through stomata on the plant body.

We begin by designating the distance z from the base of the organ (with overall length L) at which a cross section is taken, the outer radius of each cross section $R(z)$, and the radius of the conducting tissue $\lambda R(z)$, where λ is some fraction of $R(z)$. Since the organ is conical, $R(z) = R_o - z \tan \alpha$, where R_o is the external radius at the base of the organ, and α is the half-angle subtended between the outer surfaces at the tip of the conical organ. The outer radius of the xylem tissue measured anywhere along z is given by the formula $\lambda R(z) = \lambda[R_o - m(z)]$, where $m = \tan \alpha$. Designating the total carbon dioxide absorbed by the photosynthetic tissue within the organ as Q_{co_2}, we find (from our second assumption) that $Q_{co_2} = k\ (V_T - V_C)$, where k is a proportionality factor, V_T is the total volume of the organ, and V_C is the volume fraction of the conducting tissue. Thus the total CO_2 absorbed by the organ is given by the formula $Q_{co_2} = k(1 - \lambda^2)\pi L R_o[R_o - Lm(L^2m^2/3)]$.

Turning our attention to the diffusion of water through the xylem and ignoring the resistance of water diffusing radially within each cross section, the water flux J in the longitudinal direction measured at a cross section with area A is given by the formula $JA = 2\pi q(1 + m^2)^{1/2} \int_z^L R(z)dz$, where q is the water vapor flux at the surface of the cross section. This formula assumes that the water flux at the very tip of the organ ($z = L$) equals zero. The drop in the water potential per unit distance $d\psi/dx$ equals $[-2\pi q(1 + m^2)^{1/2} \int_z^L R(z)dz]/[\pi\lambda^2 R(z)^2] = [q(1 + m^2)^{1/2}L^2]/\lambda^2 R_o$. Thus, if the plant organ maximizes the ratio of the absorbed carbon dioxide to the largest drop in water potential, then the ratio of xylem tissue area to the total area of a cross section taken anywhere along the length of the organ must equal 0.707; the cross-sectional area of the xylem tissue equals roughly 71% of the total cross-sectional area of the conical organ everywhere along its length. Note that this ratio is relatively indifferent to the geometry of the organ; that is, if the assumption that the organ is conical is relaxed and if we consider other geometries, such as the untapered cylinder or prolate spheroid, then the same optimal ratio is computed.

The fossil record reveals that the "optimal" ratio of 0.707 was not attained until well into the Middle Devonian, long after the first vascular land plants make their evolutionary appearance, suggesting that either the earliest vascular land plants could not maximize the ratio of absorbed carbon dioxide to the vertical drop in water potential or that one or more of the assumptions underlying the previous analysis is in error. Of course, there is no reason to expect the first organisms to evolve in any new lineage to be optimal. The evolutionary acquisition of efficient performance levels requires time and most likely involves "trial and error." It is also likely that one or more of the initial assumptions may be incorrect, such as the assumption that the evaporative flux of water at the surface of the organ was uniform along organ length. Indeed, if stomatal closure figured in the water relations of early land plants, as the Gibson, Calkin, and Nobel (1985) analysis of extant plants suggests, or if a gradient of evaporative flux existed and increased toward the tip of the organ, then significantly lower ratios are calculated; for example, only 10% of each cross section need be devoted to xylem tissue. Lower estimates such as 10% are much more in keeping with the anatomical data for early vascular land plants. One possible consequence of the existence of an evaporative flux gradient is that the volume fraction of photosynthetic tissue within consecutive cross sections would most likely have increased toward the top of the plant organ. One way to test this hypothesis would be to determine the acropetal gradients of air spaces within transverse sections and the frequency distribution of stomata per surface area, since these two features can reflect the extent to which a longitudinal gradient in gas exchange is evident in an organ.

Nonetheless, evidence from the fossil record indicates that hydraulic conductance, *Lp*, may have played a significant role in the early radiation of vascular plants. Zimmermann (1983) and Tomlinson (1983) suggested that even small evolutionary changes in tracheid diameters could have conferred significant hydraulic advantages. Their reasoning was predicated on the Hagen-Poiseuille relationship (eq. 10.1), which as we have seen overestimates *Lp*. However, empirical data suggest that the overestimate of *Lp* may be consistent. Thus, if we are more interested in the *relative* values of *Lp* than in the absolute numbers, then Zimmermann's (and Tomlinson's) prediction is still relevant as a matter of paleobotanical inquiry. Fortunately, the early fossil rec-

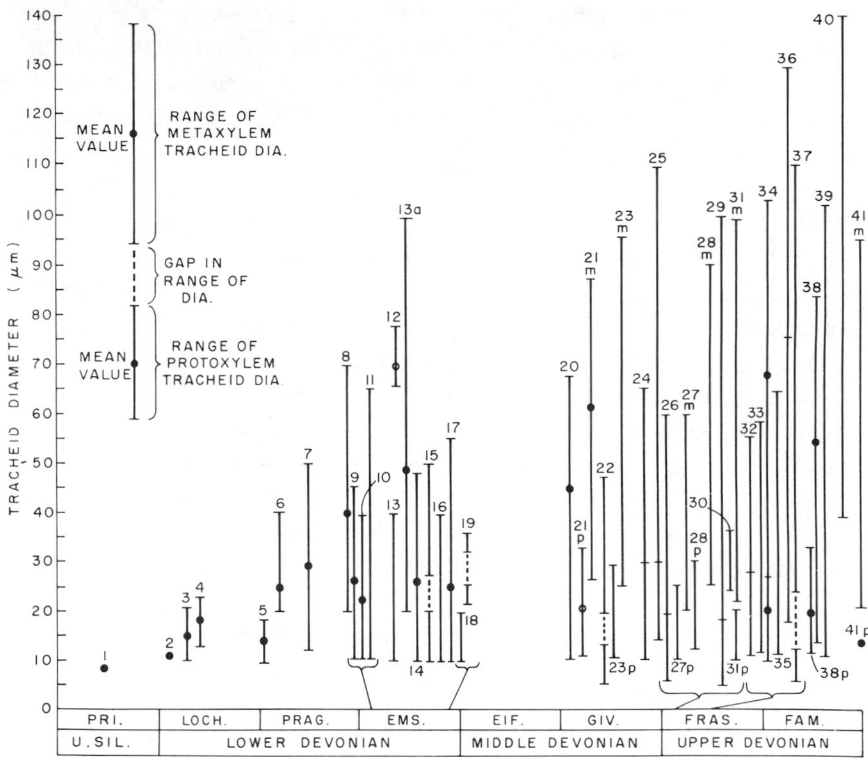

FIGURE 10.7 Tracheid diameters (given in micrometers) of Upper Silurian and Devonian plant fossils plotted against geologic age: p = data for protoxylem; m = data for metaxylem: 1, 3–4 = *Cooksonia;* 2, 18 = *Hostinella;* 5 = *Baragwanathia;* 6 = *Zosterophyllum;* 7 = *Gosslingia;* 8 = *Drepanophycus;* 9–10 = *Rhynia;* 11 = *Asteroxylon;* 12–13a = *Sawdonia;* 14 = *Eogaspesiea;* 15–17 = *Psilophyton;* 19 = *Crenaticaulis;* 20 = *Pseudosporochnus;* 21 = *Leclercqia;* 22 = *Ibyka;* 23 = *Reimannia;* 24, 26 = *Triloboxylon;* 25 = *Arachnoxylon;* 27–28 = *Tetraxylopteris;* 29 = *Rhymokalon;* 30 = *Serrulacaulis;* 31 = *Archaeopteris;* 32 = *Aneurophyton;* 33 = *Sphenoxylon;* 34 = *Proteokalon;* 35 = *Colpodexylon;* 36 = *Stenokoleus;* 37 = *Callixylon* (roots); 38 = *Rhacophyton;* 39 = *Phytokneme;* 40 = *Stenomylon;* 41 = *Laceya*. (From Niklas 1985c.)

ord of vascular plants is good enough to provide a survey of tracheid diameter as a function of geologic time. A survey of the primary literature provides the maximum, minimum, and mean tracheid diameters reported for successively geologically younger taxa (fig. 10.7). The plot reveals that the maximum tracheid diameters of forty-one Silurian and Devonian taxa tend to increase with time almost linearly for all the major (suprageneric) plant groups represented in the data base. Remember that a doubling of the relative diameter of

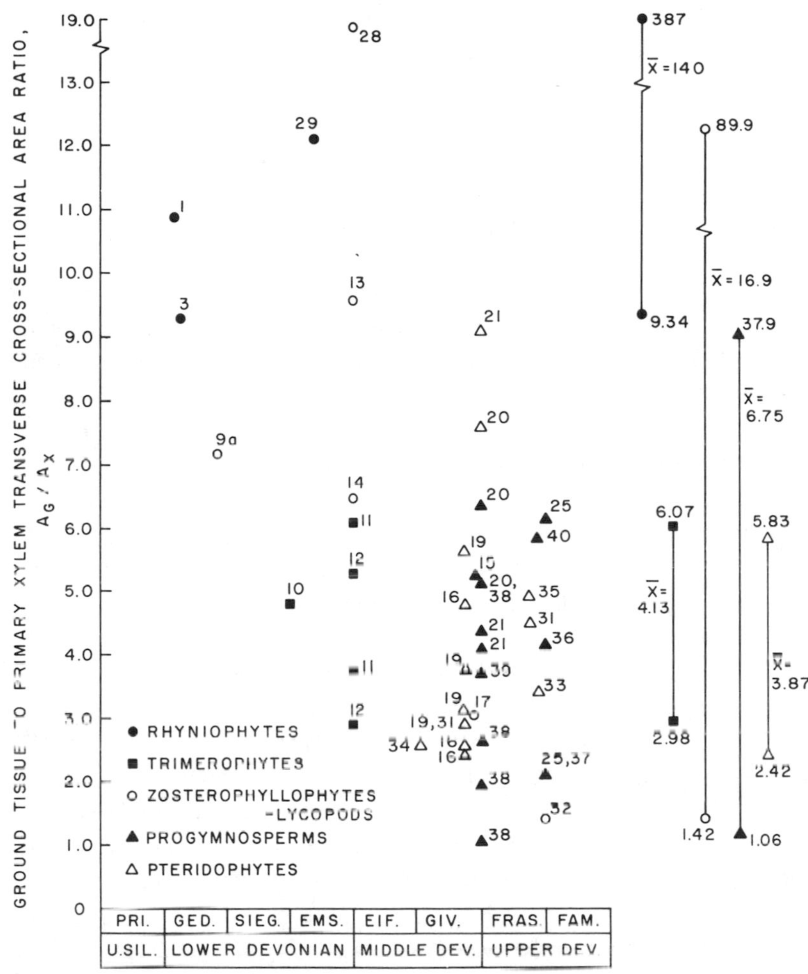

FIGURE 10.8 Ratios of the cross-sectional areas of ground and primary xylem tissue calculated for the axes and stems of various early Paleozoic vascular plants. The three most ancient plant groups (rhyniophytes, zosterophyllophytes, and trimerophytes) typically had high ratios (small amounts of xylem compared with ground tissue) in their axes or stems. More recently evolved plant groups (progymnosperms and pteridophytes) had stems with disproportionately larger amounts of primary xylem. (The numerical key for taxa is provided in the legend to fig. 10.7.) (From Niklas 1984.)

a conducting cell will result in a sixteenfold increase in the relative flow rate of water through the cell. Accordingly, the single character of tracheid diameter shows evidence for a biomechanical improvement in hydraulic conductance.

As we see from eq. (10.2), the hydraulic conductance of an entire xylem strand depends on the number of tracheids. All other factors being equal, a plant with a larger xylem strand would have a significantly higher *Lp* than another plant of the same size but with a more slender strand. Once again, data from the fossil record indicate that the volume fraction of xylem in the axes of Silurian and Devonian plants tends to increase in fossils found in successively younger strata of rock (Chaloner and Sheerin 1979; Niklas 1984). Similarly, the ratio of ground tissue to xylem tissues in these plant axes decreases in younger taxa (Niklas 1984) (fig. 10.8). Let me emphasize, however, that the success of two plant species competing in the same environment is not determined by xylometry alone, particularly in the Silurian-Devonian floras, where water may not have been a limiting resource. Growth rate, stature, and reproductive success (which are dictated by many factors) are important determinants of ecological success (see Tilman 1982).

One other aspect of xylometry is relevant to the topics of hydraulics and early plant evolution—the shape of xylem strands in transverse section. The fossil record of early vascular land plants reveals what appears to be a temporal trend toward increasing vascular complexity. The earliest vascular land plants, such as *Cooksonia* (see fig. 10.3C–D), usually possess a single slender and terete xylem strand. The diversity in the transverse geometry of xylem strands increased by the Middle Devonian to include highly lobed strands, tubular strands (bounding an inner parenchymatous region or pith), and a variety of other complex shapes. One of the first to draw attention to this diversity was the great paleobotanist Frederick O. Bower, who further noted that the geometry of the xylem strand in a single plant axis can sometimes become more complex as the shoot apical meristem assumes its mature configuration and the plant gets larger in girth. Bower (1935) speculated that the ratio of the surface area to volume of the xylem tissue was being maintained at a relatively constant level to ensure adequate lateral transport of water from the xylem to nonvascular, essentially ground tissues. Thus the geometry of the xylem strand was envisioned to undergo allometric changes in shape owing to an absolute increase in the size of the plant axis. (Bower, whether he knew it or not, was using Galileo's principle of similitude.) Bower reasoned that, for small plant axes, a terete strand would be adequate to conduct water laterally through the stem, but as size increased the ratio of surface area to volume of the cylindrical vascular strand would become increasingly less adequate. By inference he concluded that, during plant evolution, allometric adjustments in the shape (surface area:volume ratio) of xylem strands had to occur as plant stature increased. Bower's speculations and observations subsequently became conventional wisdom (Sinnott 1960, 360; Bierhorst 1971, 165–66; Stewart 1983, 206, fig. 18.13), largely because of the work of C. W. Wardlaw.

Unfortunately, a critical examination of the paleobotanical data indicates that Bower's evolutionary hypothesis, based on Galileo's sound logic, is likely not to be correct (Niklas 1984). Indeed, because of their slenderness, the vascular strands of the very first vascular land plants had some of the highest ratios of xylem surface area to volume, while subsequently derived plant groups with complex vascular strands had ratios of surface area to volume lower than those of their phylogenetic progenitors; for example, *Psilophyton* (see fig. 10.6E) has lower S:V than *Rhynia*. Under any conditions, we have seen that the radial resistance of water flow through a cortex is negligible compared with the resistance of the diffusion of water vapor through stomata. The apparent trend toward increasing geometric complexity of xylem is more likely a result of organographic evolution. That is, as the evolution of leaves progressed and shoot morphology became more complex, increasingly greater demands were placed on the hydraulic interconnectedness of the shoot. This in turn required a more complex networking of the vascular system of stem and leaf within a shoot. Hence changes in the S:V of vascular anatomies ensued, but long-term evolutionary trends in S:V are in a sense a biological artifact of evolving leaves and stems—they are more a reflection of organographic specialization and sophistication than the direct result of selection pressures to maintain or aggrandize S:V relations (Niklas 1984). Much the same may be said concerning the change in the xylem anatomy attending the growth of an individual plant. Bower's ontogenetic and evolutionary speculations illustrate a case where selection pressure was incorrectly identified—where an ultimate causality was misapplied to a proximate observation.

Another traditional hypothesis is that the primary vasculature of plants evolved in response to natural selection operating at the functional level of mechanical support. Although this hypothesis is intuitively attractive, more recent work, particularly the application of biomechanical principles to the study of fossil plant anatomy, suggests that it is not entirely correct. Speck and Vogellehner (1988a, b) consider the first occurrence in the fossil record of different xylem geometries in the context of bending effectiveness—the capacity to sustain bending stresses induced by static or dynamic (wind) stresses. They conclude that "the vast majority of the earliest tracheophytes . . . were turgor systems" (Speck and Vogellehner 1988a, 267). Hence the mechanical role of the primary vascular system of the earliest tracheophytes was to supply water to an inflatable core of ground tissue. Speck and Vogellehner (1988a, b) also conclude that mechanical stability was increasingly achieved by the secondary xylem, as well as by an external rind of collenchyma and sclerenchyma, as overall plant stature increased during the Devonian. Their conclusions are entirely compatible with data from extant plants. As we saw (chap. 6), collenchyma and sclerenchyma are mechanically efficacious tissues, while secondary xylem can be accumulated over successive growth cycles to build up a lignified tissue that is extremely strong and textured to resist bending. Further, recall that a primary vascular strand running through the center (centroid axis) of a stem is situated where the intensities of compressive and tensile bending stresses and torsional shear stresses are at their *minimum* values. Thus, a centrally placed, slender vascular strand is not likely to provide much resistance to bending and torsion. Significantly, Speck and Vogellehner's conclusions point toward a diminished role (if any) for the geometry of the primary vascular tissue per se in affording mechanical stability, because as the primary vascular strands of early vascular land plants became more robust so too did hypodermal layers of relatively thick-walled tissues. True, as a central primary vascular strand increases in cross-sectional area with respect to the cross section of the stem, its mechanical role would become more significant. The primary vascular strands of fossil plants, such as *Psilophyton,* were robust (making up as much as 37% of the cross-sectional area of vertical axes) and very likely contributed (to some degree) to mechanical support. Significantly, however, the peripheral tissues in the cross sections of the axes of early vascular plants were often dense and had relatively thick cell walls. From our understanding of core-rind models and the importance of placing stiffer materials as far from the centroid axis as possible, we can see that the outer rind of tissues in the axes of early vascular land plants probably played a more significant mechanical role than even robust centrally placed vascular strands.

The mechanical role of xylem tissue became important with the advent of the developmental capacity for secondary growth. Before this evolutionary innovation, a relatively slender primary vascular tissue system had the capacity to maintain the turgidity of nonvascular, thin-walled ground tissues. A companion tissue in this hydrostatic system was the epidermis, which not only could act as an external tension-bracing system to the inner core of water-inflated ground tissues, but also could maintain the water potential of aerial plant organs in conjunction with appreciable stomatal closure and the presence of an outer cuticle. Nonetheless, this hydrostatic architecture was limited in how long vertical growth could continue before mechanical instability arose. This limitation occurred in many lineages, leading to the evolution of what was apparently an inherent solution (as seen by its polyphyletic appearance)—secondary growth. True, there exist important distinctions among the secondary tissues produced by different lineages, but the functional solution—some type of cambial layer of cells that gives rise to secondary mechanically supportive tissues—is evolutionarily pervasive. Subsequent plant evolution produced a number of plant lineages that innovated the arborescent growth habit (lycopods, horsetails, ferns, gymnosperms, and angiosperms). With few exceptions (principally the ferns), these lineages evolved the capacity to produce large amounts of secondary xylem, whereupon plants achieved truly impressive vertical stature.

TRADE-OFFS AND COMPROMISES

In each of the preceding sections, we have principally focused on the evolutionary background relevant to four of the five requirements for plant survival in a terrestrial environment. I did so intentionally, though recognizing that every plant must perform all five requirements simultaneously and that some of the requirements may have antagonistic design specifications. For example, we have seen that the physiological necessity to exchange carbon dioxide and oxygen with the external atmosphere has unavoidable consequences on water loss through transpiration. By the same token, plants elevate photosynthetic organs above the ground, enhancing their capacity to intercept light, but vertical growth imposes bending and torsional stresses and strains. Other antagonistic design requirements exist as well. The agenda in this section is to examine only a few of the trade-offs and compromises that result from conflicting design requirements and to place these within the context of early tracheophyte evolution. Specifically, we shall focus on the trade-off between light interception and mechanical stability.

Evaluating the trade-off between maximizing light interception and mechanical stability will be considerably easier if for the time being we neglect other design requirements (hydraulics, gas exchange, and reproduction). We can assume that the problems of hydraulics and gas exchange were essentially solved with the evolutionary innovations of conducting tissues and an epidermis with stomata and a cuticle. Additionally, paleoecological data indicate that the earliest land plant floras grew in environments where water was not a limiting resource for vegetative growth. The design requirements for reproduction involved free water for the completion of the life cycle. This aquatic or semiaquatic feature was relevant to the gametophytic generation, however, not to the vascular sporophyte, which is our principal concern here. Nonethe-

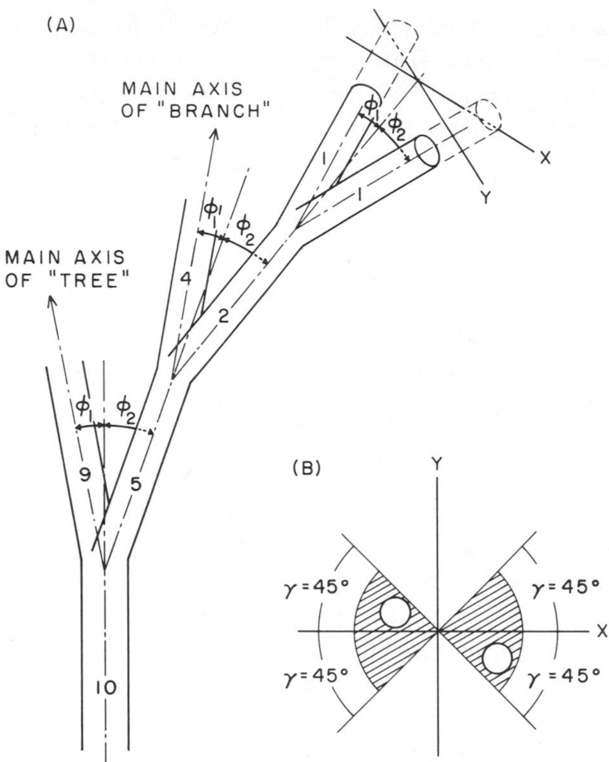

FIGURE 10.9 Geometric parameters sufficient to simulate the branching patterns of some early vascular plants: (A) Branching pattern produced by bifurcation angles (ø). When the two subangles ($ø_1$ and $ø_2$) are equal a symmetrical branching pattern is produced; when the subangles are unequal, an asymmetrical "tree" is produced with a "main axis," bearing "branches." (B) Rotation angle (γ) that defines how far branches rotate with respect to a Cartesian coordinate system in which the *x*-plane is oriented parallel to ground level. Other parameters (not shown) are the probability of branching (*p*) and the length (*l*) of branch elements. Branch elements are numbered in the reverse order generated by a computer program. Each branching pattern can be evaluated in terms of its ability to sustain its own weight and intercept sunlight (see fig. 10.10).

less, the vascular sporophyte does have a reproductive role: it must produce and disperse spores. As discussed in chapter 9, the potential dispersal range of a tracheophyte's spores carried by air currents depends on the elevation of sporangia above the ground. Thus the design requirements for vertical growth and light interception are compatible for the most part with those for the dispersal of spores. Also, the paleobotanical data show that the earliest tracheophytes were organographically simple. That is, the distinctions among leaf, stem, and root were not established, and the primitive vascular sporophyte essentially consisted of naked cylindrical axes. Clearly, the axes of some early tracheophytes produced enations, blebs of tissue protruding from the surfaces of vertical axes (see the enations in fig. 10.6E). By the Devonian, some plant groups had evolved true leaves, often of considerable size and morphological complexity. Nonetheless, if we focus on the early stages in tracheophyte evolution, then the architecture of the vascular sporophyte can be reasonably approximated as a truss of cylindrical photosynthetic axes, much like the sporophyte of *Psilotum nudum* discussed in chapter 8. Later on in this chapter, we will relax this last assumption (leaflessness) and consider the role of broad, flat photosynthetic surfaces in light interception.

Referring to figure 10.9, at least three parameters are required to mathematically construct a branched vascular sporophyte: the bifurcation angle ϕ, which is the angle subtended between two elements in a bifurcation; the rotation angle γ, which is the angle defined by the orientation of a cylindrical element relative to the horizontal plane; and the probability of bifurcation p. The last of these three parameters is the most complex, since it involves the frequency with which the hypothetical apical meristem at the tip of each cylindrical element within the sporophyte branches. Different sporophyte geometries can be simulated by computer by defining the probability of bifurcation at each successively higher level of branching within the simulated plant. A simple linear formula can be used for this purpose:

$$p = \frac{8p_n - (k-1)}{N+k}, \tag{10.5}$$

where the term $[p_n - (k - 1)]$ is the probability of truncating branching at the next generated level of branching, while the term $(N + k)$ designates the previously generated level, where N is the total number of bifurcation events (the number of times an average apical meristem dichotomizes to form two branches). The value of p can be varied from 0 to 0.9. In the scheme used here, when $p = 0$, the highest frequency of bifurcation occurs, and when $p =$ 0.9, the lowest frequency is produced. The parameters ϕ and γ are easily defined and can vary from 0° to 360° (see fig. 10.9).

Clearly, other geometric features need to be considered along with p, ϕ, and γ. We need to specify the tapering of cylindrical elements and the respective lengths of these elements. Also, the bifurcation angle may not be symmetrical with respect to the longitudinal axis of the subtending axis bearing two derived axes. That is, one of the derived axes may be oriented more in

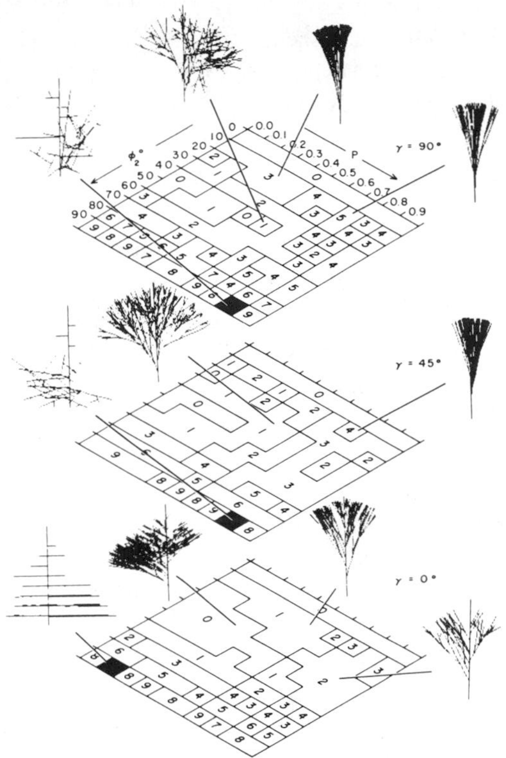

FIGURE 10.10 Computer-generated branching patterns and their abilities to intercept sunlight and sustain their own weight. Three "slices" through a hypothetical universe of mathematically conceivable branching patterns with representative branching geometries are provided. The numbers plotted on each slice are the performance levels of branching patterns relative to the maximum performance level within the entire universe of branching patterns. There are many branching patterns that have the maximum performance level, however; each is shown as an open square. Thus there is no single "optimal" branching pattern in the entire universe of conceivable branching patterns. (From Niklas and Kerchner 1984.)

line with its subtending member than its companion. Therefore two bifurcation angles, ϕ_1 and ϕ_2, can be specified. When $\phi_1 = \phi_2$, symmetrical branching is achieved with respect to the subtending member. When $\phi_1 \neq \phi_2$, asymmetric branching occurs. By the same token, each derived element in a branching event needs to have its length l, taper t, rotation angle γ, and probability of branching p specified. Subscripts 1 and 2 can designate the values of these parameters for each of the two elements produced when an apical meristem bifurcates.

Niklas and Kerchner (1984) employed a computer model like the one just described to generate a universe of mathematically possible branching patterns (see fig. 10.10). In addition, they computed the capacity of each geometry to intercept direct illumination, as well as the total static bending moment at the base of each geometry. Total light interception was computed by integrating the area under the graph of the ratio of the projected surface area with total surface area of each geometry plotted as a function of the ambient direction of light changing over a single diurnal solar cycle. The total bending moment was computed from equations derived from elementary bending theory for cantilevered beams and vertical columns. The bending moment on each element within a single geometry is mathematically dependent on ϕ and γ, since together these two angles define the orientation of the element with respect to the horizontal plane in which gravity acts. Also, the bending moment depends on the length, diameter, and density of each element.

Accordingly, the geometries occupying the universe of sporophyte morphologies can be quantitatively compared with regard to their capacity to garner light and to minimize bending moments. Niklas and Kerchner (1984) found that some geometries were better than others, but that there was no single optimal one. Rather, there existed a number of branching architectures whose geometry reconciled these two biological requirements in nearly equivalent ways. Significantly, when simulations involved isobifurcation—that is, when the branching geometry was symmetric—only small branching architectures were efficient. Large, symmetrically branched sporophytes were too densely branched; they consisted of relatively closely packed branching elements that occluded one another and reduced the capacity for light interception of the geometry as a whole, while at the same time imposing high bending moments. Larger branching patterns with widely spaced elements intercepted much more light but had disproportionately larger bending moments.

When $\phi_1 = 0$, branching structures with a main vertical axis and lateral branches are produced. Comparing these geometries with isobifurcating ones (where $\phi_1 = \phi_2$) revealed that a main vertical trunk substantially increases light interception and reduces the total bending moment at the base of the plant. Additionally, if the lateral branching systems on these plants are planated—if they are caused to flatten horizontally—then light interception can be maximized with respect to the total bending moment. Computer analyses indicate that these geometries are optimal with respect to all others within the hypothetical universe of vascular sporophytes (Niklas and Kerchner 1984; Niklas 1986a, b). Notice the use of the word *these*—there was no single opti-

FIGURE 10.11 Semidiagrammatic representations of some early Paleozoic fossil plants having some morphological correspondence to branching geometries predicted by computer simulations (see plate 4). Plants have been arranged in a sequence (A to J) that corresponds to the sequence of computer-generated branching patterns. (A–C) Upper Silurian and Lower Devonian rhyniophytes: (A) *Cooksonia;* (B) *Rhynia;* (C) *Horniophyton;* (D–F) Late Lower Devonian *Psilophyton* spp.; (G) Late Lower Devonian *Pertica;* (H) single frond of the Upper Devonian *Rhacophyton;* (I) Carboniferous *Calamites;* (J) Carboniferous *Lepidodendron.*

mal geometry, rather there was a suite of shapes of more or less equivalent capacity to maximize light interception and minimize bending moments.

Plate 4 provides a summary of the shifts in geometry that are predicted to progressively maximize the trade-off between light interception and total bending moment. These shifts can be compared with the morphologies of representative fossil tracheophytes ranging from the early experimental period of plant evolution in the Silurian to representatives of well-established lin-

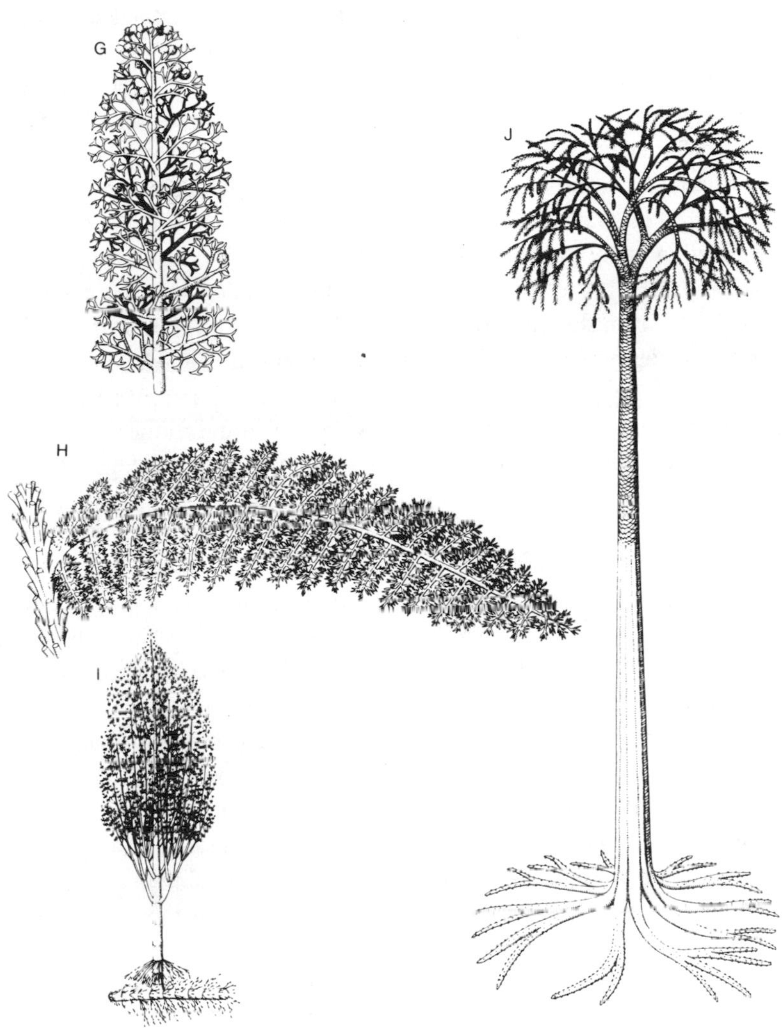

eages found by the end of the Devonian, as well as some plants from the Carboniferous that achieved tree growth habits; for example, *Calamites* and *Lepidodendron* (fig. 10.11). It must be emphasized that the fossil plants illustrated from the Silurian and Devonian reflect some but not all of the morphological variation in early tracheophytes, and the fossil taxa are from more than one lineage of tracheophytes. Nonetheless there are some striking similarities between the geometric changes predicted by the computer simulation and the general morphological patterns evidenced in the fossil record as a whole. The earliest tracheophytes were diminutive and sparsely branched, whereas younger taxa achieved greater stature and were more densely branched. Isobifurcation was replaced by pseudomonopodial branching, in which a main axis is the collective product of each branching event, producing a large and a small axial branch. The end-to-end stacking of the larger axes in successive pairs of branches results in the main axis. Planation in lateral branching systems occurred in some still younger taxa, while arborescence and true leaves had evolved in many tracheophyte lineages by the end of the Devonian.

It is not logical for us to deduce a cause-effect relation between the geometries resulting from the computer model (plate 4) and evolutionary sequence of early tracheophyte morphologies shown in figure 10.11. All one can say of the apparent correspondence between the hypothetical and empirical evolutionary patterns in vascular plant morphology is that the former is consistent with the latter. Thus the assumptions used in the model may be logically valid, but they have not been shown to be *true*. (Logic can show that assumptions are invalid or valid; it can never prove something is true.) Analyses subsequent to those of Niklas and Kerchner (1984) show that other assumptions can yield results equally compatible with the fossil record. For example, Niklas (1986a) constructed computer simulations predicated on plant architectures that maximized their potential for the long-distance dispersal of spores. When different geometries of branching were placed in the same environment and competed for space by means of spore dispersal, pseudomonopodial architectures quickly outcompeted isobifurcating ones. These simulations involved no assumption about light interception, but it is clear that this factor and spore dispersal have mutually compatible design requirements. Accordingly, either or both can be invoked to explain the long-term trends in tracheophyte morphological evolution seen in the early Paleozoic. Perhaps more important, it can easily be demonstrated that all branching geometries are less efficient at gathering light as they continue to branch (Niklas 1986b). That is, with continued growth, a branching pattern can increase self-shading. However, a more vertical posture also helps a plant shade its neighbors. This observation suggests that the increase in plant stature observed during the Silurian-Devonian time period was not necessarily driven by selection pressures favoring more efficient individuals but may have been the result of competition for space and light. A large plant has a greater capacity to shade nearby plants than a smaller one. If this speculation has any element of truth, then increasing plant stature may have been the result of an arms race in which the weapons we call leaves inflicted damage (shade) on other nearby plants, while the weapon we call stems gained more ground (dispersal of spores) in the habitat as a whole. The result was a potentially less efficient individual in terms of light interception but a more efficient species in terms of the potential to shade and outproduce other plant species.

The computer simulations discussed in the context of early land plant evolution can be applied to living plants as well, particularly leafless and much-branched species. This is illustrated by a study of *Salicornia,* an essentially leafless angiosperm (Chenopodiaceae) that often grows in monotypic stands in salt marshes (Ellison and Niklas 1988). Computer simulations based on the morphometrics describing the general mode of growth (branching probability, branching angles, etc.) were used to predict how the general morphology of *Salicornia europaea* would change as a function of the number of individuals in a population. As might be expected from the discussion of early vascular land plants, the simulations predicted that as the number of individuals of *Salicornia* within a population increases, the frequency of branching would decrease and plant height would increase. The geometries of the computer simulations of *Salicornia* were remarkable in how well they mimicked the morphological details of living plants from different successional stages (as populations increased in number of individuals), suggesting that similar approaches could be taken to model the morphology of other extant plants (see Ellison 1989).

LEAVES

Earlier we saw that oblate spheroids and cylinders are geometrically conducive to meeting the biological requirement that surface area : volume and projected surface area:total surface area ratios be manipulated. This characteristic makes these geometries useful in constructing organs that can efficiently intercept light while at the same time permitting the exchange of gases between the plant body and the external environment. Further, we argued that cylindrical geometries are excellent load-bearing members because they are axially symmetrical in their ability to resist bending. Thus, cylinders are an excellent design for both a leaf and a stem. Indeed, many plant species have cylindrical leaves. By contrast, oblate spheroids are ideally suited as light interceptors. Significantly, these conclusions are compatible with the observation that for the most part vascular plants have cylindrical stems and oblate spheroidal leaves. But the way leaves are arranged on stems can vary dramatically even within a single plant genus. For example, within the genus *Plantago,* some species have spirally arranged leaves (*P. major* and *P. lanceolata*), as do most species, whereas *P. indica* has leaves arranged in opposite pairs.

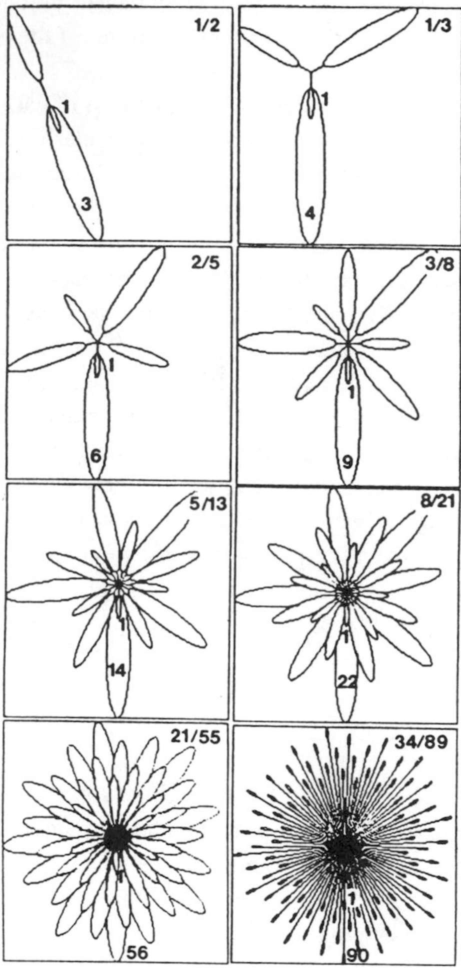

FIGURE 10.12 Computer simulations of different phyllotactic patterns as seen in polar view. The phyllotactic pattern is expressed as a fraction in which the numerator is the number of gyres that must be followed before two superimposed leaves are encountered and the denominator is the number of leaves found within the span of the gyres. See text for further details.

Botanists have long been intrigued by the geometry of leaf arrangement, or phyllotaxy, and have speculated on the adaptive significance of different phyllotactic patterns. It was recognized that the angle swept between successively produced leaves on a shoot frequently conforms to one of a series of angles defined by what is known as the Fibonacci series. In terms of phyllotaxy, the series is in the form of fractions $\frac{1}{2}$, $\frac{1}{3}$, $\frac{2}{5}$, $\frac{3}{8}$, $\frac{5}{13}$, . . . ,$\frac{21}{55}$, $\frac{34}{89}$, $\frac{55}{144}$, and so on, in which each term represents the fraction of the circumference of the stem traversed by the spiral pattern of leaf arrangement. (The numerator represents the number of gyres in the spiral that must be traversed to find two superimposed leaves, while the denominator represents the number of leaves found along this spiral pathway.) Since the circumference of a stem can be represented by 360°, each fraction in the Fibonacci series represents the divergence angle swept between two successive leaves in the spiral phyllotaxy. Thus one-half times 360° equals 180°, which defines the distichous arrangement of leaves; ($\frac{1}{3}$) 360° = 120°, which is the tristichous arrangement, and so forth (fig. 10.12). For plants the Fibonacci series is generated mathematically by adding the two preceding numerators or denominators to give the next numerator and denominator in the series. The higher fractions become more uniform in value and approach a limit decimal fraction of 0.38197, or 137° 30′28″. The relation between the mathematics of the Fibonacci series and the biology of leaf arrangement has long excited the interest of plant morphologists (Schimper 1836; Braun 1831), as well as fascinating the German poet Goethe, and attempts to find a biological meaning for the Fibonacci series are numerous. Yet most plant morphologists are quick to point out that not all patterns of leaf arrangement are based on a spiral geometry. Opposite leaf arrangements are not uncommon. Thus any mechanistic explanation for why leaves are arranged in a Fibonacci series must explain all the other patterns of leaf arrangement.

If we focus not on the developmental way leaves become arranged but rather on the consequences leaf arrangement has on light interception, we can avoid considering processes that are currently not understood but still evaluate the potential functional significance of leaf arrangement. Perhaps one of the first attempts to do so was made by Wright (1873), who mathematically demonstrated that the limit to the Fibonacci series (137° 30′28″) was the angular distance around the stem for which no two leaves would ever precisely overlap when the stem was viewed from above. His solution was based on the recognition that some angles in the Fibonacci sequence yield irrational numbers (like π, which can be calculated in any number right of the decimal place), while others do not. The advantage suggested for this arrangement is that shading of leaves would be minimized regardless of the total number of leaves produced by the stem. Coincidentally, one of the most common phyllotactic patterns seen among plants with spirally arranged leaves is 137°, a value very close to 137° 30′28″.

Computer simulations of different patterns of spiral leaf arrangement are

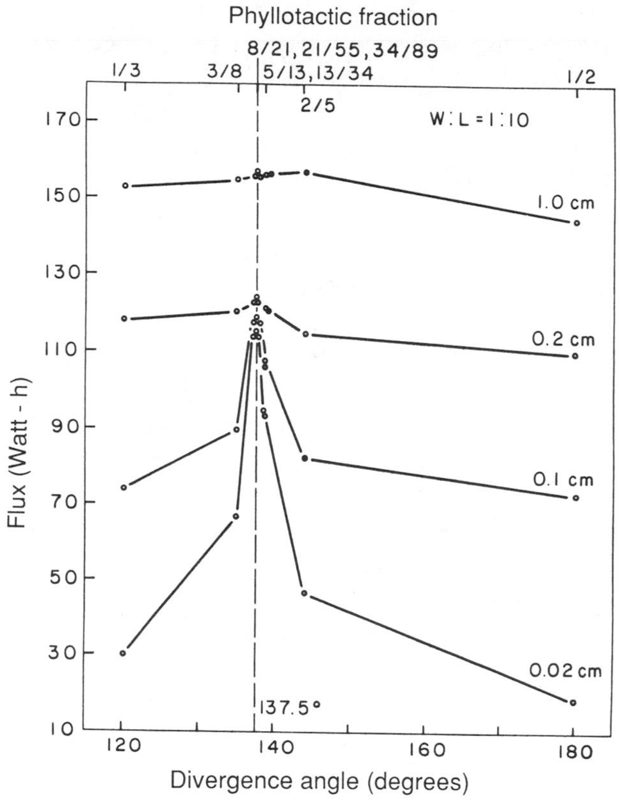

FIGURE 10.13 Computer simulations of the efficiency of light interception (expressed as flux in watt-h) plotted as a function of the phyllotactic divergence angle. All simulations are constructed so that the total leaf area and the ratio of leaf width to length (which equals 1:10) are equivalent. Four graphs are plotted for simulations differing in the internodal distances between successive leaves (0.02, 0.1, 0.2, and 1.0 cm). A maximum flux occurs at some phyllotactic angles (those that converge at 137.5°) when internodal distances are small (0.02–0.2 cm). This maximum gradually disappears as the internodal distance approaches 1.0 cm.

easily constructed (see fig. 10.12), and an evaluation of the capacities of hypothetical plants with different leaf divergence angles is shown in figure 10.13. These simulations show that when leaves are closely packed into a rosette, as in *Plantago,* the capacity to intercept direct sunlight is significantly influenced by the phyllotactic fraction (Niklas 1988c). The shape of leaves, the distance between leaves along the length of a shoot, and the angle at which leaves are held are equally influential, however, and can reduce the significance of the phyllotactic fraction for light interception. Even a modest increase in internodal distance (from 0 to 0.1 cm) can bring the light-gathering capacity of plants with differing spiral leaf arrangements into near parity with one another. Thus the phyllotactic fraction appears not to be an intrinsic constraint on light interception.

If these simulations have any bearing on the issue of plant evolution, it is to suggest that leaf shape, the length of internodes, and the pattern of leaf arrangement act in concert to define the total amount of direct light a shoot can receive. If natural selection operated on the capacity of vertical shoots to intercept light, then selection pressures must have focused on the shoot as a developmentally integrated whole, not on individual parts of the system. Unfortunately, very little is currently known from the fossil record about the evolution of the three-dimensional geometry of leafy shoots. It has been suggested that the leaves of ferns and possibly many seed plants evolved by developmental modifications of meristems that originally gave rise to lateral branches. The planation of lateral branches attached to a main axis or trunk, discussed earlier, may have increased the light-harvesting capacity of primitive tracheophytes. Hence our prior explanation for planation in leafless early vascular plants is entirely compatible with the developmental modifications leading to the evolution of leaves.

The Telome Theory

All of the foregoing discussion concerning design constraints and the resolution of these constraints during early land plant evolution can be placed in context by referring to one of the most comprehensive (and still popular) attempts to explain the major events in the early evolution of the vascular land plants, called the telome theory. The telome theory was originated by Walter Zimmermann, who selected the fossil genus *Rhynia* as the paradigm for the early vascular land plant (Zimmermann 1930, 1965). As we have seen, *Rhynia* has all the basic requirements for survival on land. The sporophyte of *Rhynia* consists of simple (leafless) dichotomizing axes, some of which grew vertically and were presumably photosynthetic, while others grew horizontally and possessed rhizoids to absorb water and other nutrients from the substratum. According to Zimmermann, the basic morphological units of plants like *Rhynia* were the telome and the mesome. A telome, in the very broadest sense, is a single distal axis; a mesome is any intervening axis, comparable in some respects to the internode of a stem. Fertile telomes (those bearing a sporangium) and nonfertile telomes are distinguished.

Zimmermann envisioned five elementary processes that either singly or in various combinations could produce subsequent evolutionary modifications of telomic morphologies like those of *Rhynia* to yield more complex morphologies: (1) overtopping, resulting from unequal branching and producing subordinate lateral telomes and mesomes that were overtopped by more vertical telomes and mesomes; (2) planation, producing telomes and mesomes arranged in a single plane; (3) reduction, resulting in differences in the relative lengths of telomes and mesomes; (4) syngenesis, resulting in the fusion of telomes, mesomes, and even anatomical elements, which in the former two cases involves webbing (the production of nonvascular tissues intervening the spaces among telomes and mesomes); and (5) recurvation, resulting in the bending of telomes and mesomes out of the plane of their branching. More extensive reviews of these processes and of the telome theory in general are available (Stewart 1964; Gifford and Foster 1989, 31–33), and interested readers should review the evidence on which Zimmermann constructed his theory.

It was inevitable that the telome theory would be met with negative as well as positive reaction. Nonetheless, the theory has become a standard part of the paleobotanical repertory (Bierhorst 1971; Stewart 1983), in part because of its apparent comprehensiveness—Zimmermann's five elementary processes can be envisioned to yield virtually any morphology encountered. Or to be more precise, any morphology can be envisioned to have been the result of one or more of these processes. On the negative side, however, the telome theory provides no inferential basis for understanding when and in what circumstances each of the five elementary processes would come into play during plant evolution. The telome theory provides a lexicon for morphological modifications—a vocabulary to describe hypothetical processes engendering complex morphologies seen in the fossil record—but it lacks a functional grammar—a syntax that helps us understand why some of these processes may have been favored over others. In this context the previous biomechanical analyses may provide the grammar for theories such as the telome theory. From our previous discussions about light interception and the mechanics of vertical growth, it is not difficult to imagine that overtopping, planation, and syngenesis are adaptive in some circumstances. Overtopping elevates photosynthetic and reproductive organs above the ground, thereby enhancing the capacity of plants to intercept sunlight and shade neighboring plants and increasing their potential for long-distance dispersal of spores or propagules. Likewise, planation and syngenesis can be shown by computer simulations to minimize self-shading and maximize the ability of subordinate axes to intercept light. Finally, recurvation of some organs can reduce bending moments, lessening the compressive and tensile bending stresses developing within tissues, while reduction can conserve biomass and metabolic energy and can amalgamate a structure, whose whole is greater than the sum of its parts. True, biomechanical analyses do not provide much insight into the developmental mechanisms responsible for these morphological modifications, but they do give us some basis for understanding the adaptive significance of Zimmermann's five elementary processes. Indeed, biomechanical analyses are essential if long-term morphological trends in plant evolution are to be understood at any level of form-function relations.

THE EVOLUTION AND AERODYNAMICS OF THE SEED

So far, the primary focus of this chapter has been on the biomechanical analysis of various vegetative plant organs (stems and leaves), with only a passing nod to the consequences of vegetative innovations for plant reproduction—that is, the advantages vertical growth confers on the long-distance dispersal of spores. From our review of fluid mechanics (chap. 9), however, we could anticipate that the aerodynamic properties of the pollen and ovules of ancient plants played equally significant roles in dictating the relative efficiency of pollination. (The terms seed and ovule will be used in this section without strict regard for the differences between them. For the purist, an ovule becomes a seed when an embryo develops within it.) Indeed, as we shall see, there is good reason to believe that the evolution of the seed habit itself was dictated in part by the physics of pollen capture.

A biomechanical inquiry into the evolutionary origins of the ovule is far from trivial, since the seed habit confers many advantages and its advent profoundly influenced the subsequent course of plant history. The retention of the megagametophyte within parental sporophytic tissue protects and nourishes both the megagametophyte and the embryo within it. The seed habit also released the life cycle of the species from ecological dependence on an external (typically hydrated) microenvironment conducive to the development and survival of the egg-producing generation (the megagametophyte), opening up the possibility for the sporophytic generation of the species to ecologically radiate into and reproduce in drier habitats. Also, the seed habit provided an opportunity to developmentally modify the sporophytic tissues that encapsulate the megagametophyte and its embryo for dispersal. This was seen in chapter 9, where the autogyroscopic seeds of gymnosperms like spruce (*Picea*) were discussed. The wing of the seed is developmentally derived from sporophytic tissue. Finally, the adaptive significance of seeds can be inferred from the simple fact that seed plants rapidly diversified shortly after the seed habit

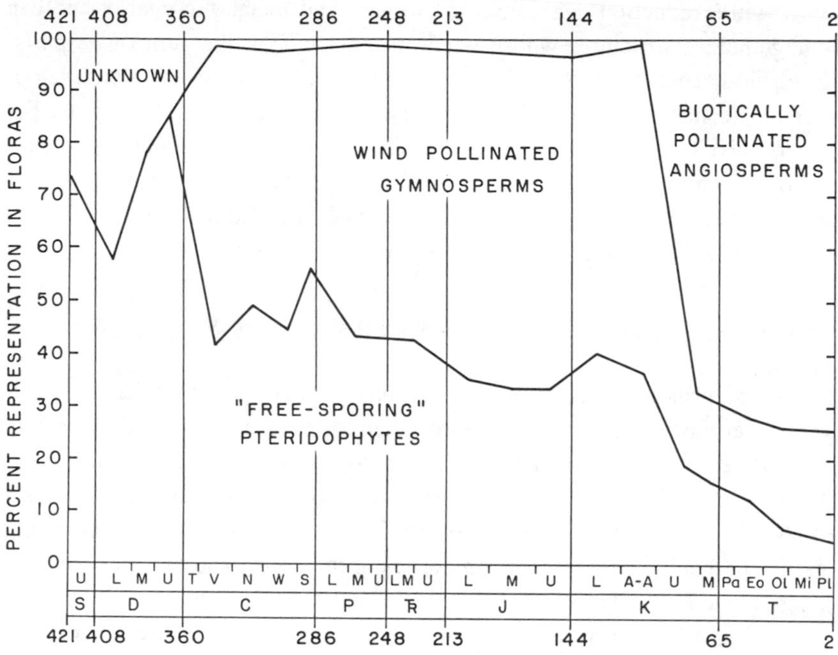

FIGURE 10.14 Patterns of changing modes of sexual reproduction ("free-sporing" pteridophytes; wind-pollinated gymnosperms; and biotically pollinated angiosperms) in terrestrial vascular plants expressed as the percentage of representation in fossil floras plotted against the geologic column.

evolved (fig. 10.14).

It is generally accepted that the evolution from a free-sporing heterosporous ancestral type of plant to a seed plant involved: (1) a reduction in the number of megaspores produced in each megasporangium, concomitant with or followed by (2) the retention of the megaspore within the megasporangium; (3) the modification and elaboration of the tip of the megasporangium to receive airborne microspores (the functional equivalents of pollen); (4) the formation of a sterile layer of tissue, called the integument, surrounding each megasporangium; and (5) the modification of the integument to form the micropyle—a pore formed by thc distal unfused portions of the integument—through which pollen could gain access to the megagametophyte. Among these five evolutionary modifications only one, the formation of the integument with its micropyle, involves an entirely new structure (Pettitt 1970). A transfer of function is envisaged in the evolutionary transition from a megasporangium to a seed; that is, pollen was originally received by the modified tip of the mega-

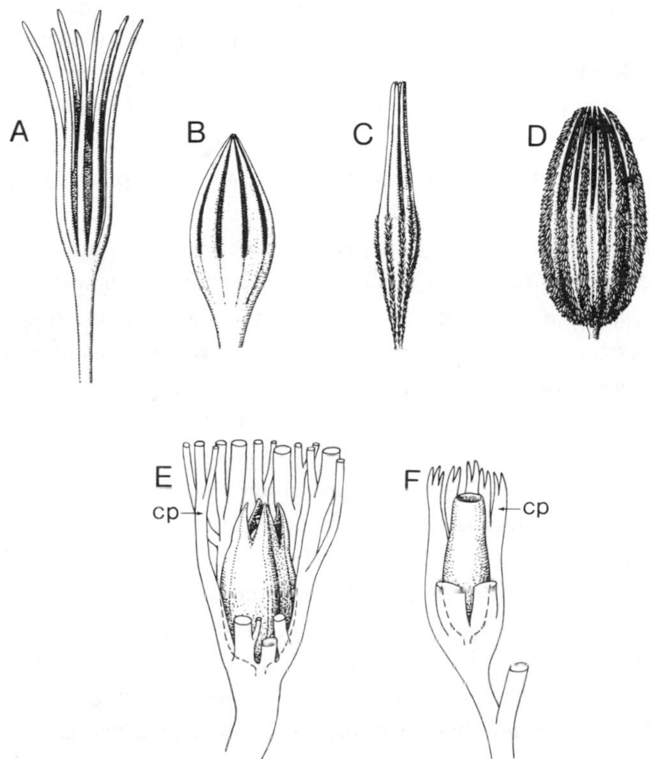

FIGURE 10.15 Reconstructions of early ovulate reproductive structures: (A) *Genomosperma kidstoni;* (B) *Genomosperma latens;* (C) *Salpingostoma dasu;* (D) *Physostoma elegans;* (E) *Eurystoma angulare* (with cupule partially cut away, cp); (F) *Stamnostoma huttonense* (with cupule cut away, cp). See text for further details.

sporangium, but with the evolution of the integument this function was transferred to the micropyle of the ovule.

During the course of seed plant evolution, microspores also evolved and developed features that aided pollination and fertilization. The microspores of some early seed plants had air bladders or sacci (formed by the separation of the outer and inner spore wall layers to yield air-filled chambers). As I mentioned in chapters 1 and 9, flotation devices like sacci promote long-distance dispersal and increase the probability of subsequent capture. Also, during the evolution of the microgametophyte, the pollen tube evolved. The pollen tube is an outgrowth of the microgametophyte that delivers sperm cells to the egg produced by the megagametophyte within the megasporangium. These and other modifications of microspores were important to the evolution of seed plants, but here we will focus on the morphological modifications of the megasporangium leading to the formation of the integument, since these modifications are likely to have influenced the aerodynamics of pollen capture.

The reproductive morphology of the earliest seed plants was diverse, and as might be expected, the reproductive structures of some of the earliest taxa barely qualify as being integumented. Some of these morphologies are shown in figure 10.15. These have been drawn from the detailed reconstructions of fossil specimens provided by a number of paleobotanists, principally A. G. Long (see Long 1966, 1975). These ovulelike megasporangia and ovulate structures have been arranged in a sequence that has been proposed to reflect the evolutionary sequence of morphological modifications leading to the evolutionary appearance of the integumented seed (Andrews 1963; Long 1966). This hypothetical sequence begins with *Genomosperma kidstoni,* which possessed a truss of sterile vascularized lobes surrounding a megasporangium whose tip was extended into a structure called a salpinx, a funnel-like structure through which microspores gained access to the megasporangium. The vascularized lobes or axes on organs of plants such as *Genomosperma kidstoni* are sometimes referred to as "preintegumentary lobes" because they are believed to reflect the condition ancestral to the true integument (Andrews 1963; Long 1966). Since organisms evolve and organs do not, it is more proper to say that the developmental patterns responsible for the preintegumentary lobes are believed to have been evolutionarily modified so that the morphological consequences of development were altered. A number of fossil organs reflect the purported morphological manifestations of these developmental modifications. *Genomosperma latens* was similar in some respects to *G. kidstoni* but possessed a truss of shorter preintegumentary lobes that was partially fused at its base, whereas the preintegumentary truss of *Salpingostoma dasu* was similar to that of *Genomosperma kidstoni. Physostoma elegans, Eurystoma angulare,* and *Stamnostoma huttonense* reflect morphologies that evince integumented megasporangia. The micropyle of *Eurystoma angulare* was flanked by four relatively small lobes, and the integument and micropyle of *Stamnostoma huttonense* were comparable to those seen on the ovules of extant seed plants. Some (possibly all) of these reproductive organs were aggregated within cupules, a cupule being a sterile truss of axes (sometimes fused together, sometimes not) surrounding one or more megasporangia (see fig. 10.15E, F).

The sequence of fossil structures shown in figure 10.15 is misleading in some very meaningful ways, however, particularly since these fossils are more or less geologically contemporaneous, whereas geologically older ovulate structures, known from the Devonian, show morphological features that appear more advanced than their geologically younger counterparts. The oldest seeds currently known are from the late Devonian (Gillespie, Rothwell, and Scheckler 1981). They are cupulate and have preintegumentary lobes that show considerably more fusion than the geologically younger *Genomosperma kidstoni*. Still more recent findings reveal that morphologically "primitive" seedlike structures persist in sediments considerably younger than those containing taxa such as *Stamnostoma* (Galtier and Rowe 1989). Accordingly, the hypothetical sequence starting with *Genomosperma kidstoni* and ending with *Stamnostoma huttonense* is not supported by our current understanding of the stratigraphical relations among early seed plant reproductive organs. Rather, the morphology of these structures may reflect the habitats in which they functioned. We shall return to this important point shortly.

Nonetheless, among most of the known fossil organs, an interesting correlation exists between the presence or absence of a modified megasporangial tip (e.g., salpinx) and how morphologically well defined the integument is. Ovules with preintegumentary lobes typically possess a salpinx, while ovules with well-defined integuments necessarily have well-defined micropyles and a poorly defined salpinx or none at all (Andrews 1963; Long 1966, 1975; Taylor 1982). Accordingly, although the fossil ovules discussed here cannot be arranged into a clearly defined chronological sequence, they can be put in a morphological sequence reflecting the relative importance played by the megasporangial tip versus the micropyle in the reception of airborne microspores.

One potential driving force underlying the evolution of the integument may have been a protective role. This has been the traditional view (see Andrews 1963; Long, 1966, 1975), but it is somewhat at odds with how well some of the hypothetical intermediary morphologies were protected by their preintegumentary lobes. A more attractive hypothesis, and one that is entirely compatible with the first, is that the morphological modifications of preintegumentary lobes to form the integument affected the efficiency of pollen capture. No matter which hypothesis we find most appealing, the morphological variation encountered among the earliest ovules must have had aerodynamic consequences, regardless of the primary adaptive driving force effecting the variation.

Direct experimentation on the physics of pollen capture of extinct species is not possible. But just as an engineer can determine the aeronautical characteristics of a model aircraft by means of a wind tunnel, the paleobiomechanicist can construct scale models of ovules and cupules and test their capacity

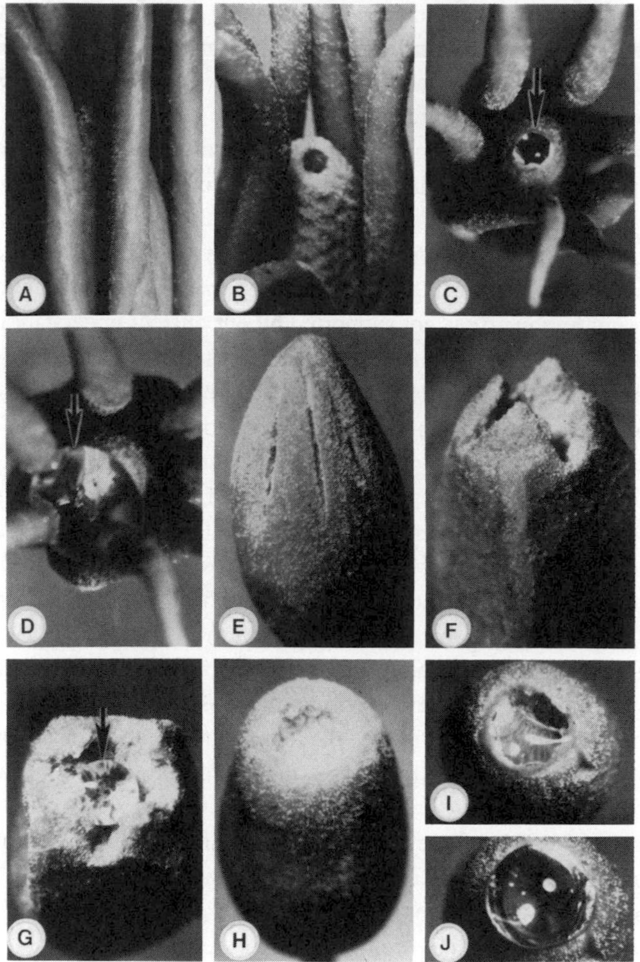

FIGURE 10.16 Life-size scale models of early Paleozoic ovules (see fig. 10.15 for reconstructions) placed in a wind tunnel and subjected to equal numbers of airborne spores carried in the same ambient airflow speed directed perpendicular to the longitudinal axis of each model. Spores adhering to surfaces of models appear as a granular dusting of particulates: (A–B) Side and top view of *Genomosperma kidstoni* model without pollination droplet. (C–D) Top view of *Genomosperma kidstoni* model with small and large pollination droplets (see arrows). (E) Side view of *Genomosperma latens* model. (F) Side view of *Eurystoma angulare* (without pollination droplet). (G) Top view of *Eurystoma angulare* model with small pollination droplet. (H) Oblique view of *Stamnostoma huttonense* model (without pollination droplet). (I–J) Top view of *Stamnostoma huttonense* model with small and large pollination droplets. (From Niklas 1983.)

to trap airborne particulates that mimic pollen grains. From chapter 9 we know that for any given airflow speed, the size, shape, and orientation of a wind-pollinated structure will influence the airflow patterns generated around it and therefore the aerodynamic environment through which airborne particulates must navigate before they can make contact with the structure. Likewise, for any given size, shape, and orientation to airflow, the Reynolds number will increase with the ambient wind speed. Since the order of magnitude of the Reynolds number can change airflow characteristics, the magnitude of the ambient airflow is very important to the physics of pollen capture. Thus, experimental comparison among morphologically diverse ovulate structures requires that they all be built to scale and that different orientations to the direction of ambient airflow, as well as different ambient airflow speeds, be examined. By the same token, we must be mindful of the physical attributes of the airborne microspores, since spores differing in size or density will behave differently within identical airflow patterns.

Life-size models of some of the earliest known ovules were placed in a wind tunnel and variously oriented to the direction of ambient airflow (Niklas 1983). The spores or pollen from living plants were used as surrogates for the pollen of these fossil ovules. Known quantities of spores or pollen were released upstream of the models, and the number and distribution of spores or pollen adhering to the surfaces of the models were quantified. Figure 10.16 shows some of the features of the distribution of spores adhering to the surfaces of the models for some of the ovules purported to illustrate an evolutionary sequence from a nonintegumented megasporangium (*Genomosperma kidstoni*) to an integumented megasporangium (*Stamnostoma huttonense*). (In some instances a drop of water was added to the tip of the megasporangium or the micropyle to mimic a pollination droplet that may have been produced by some of the earliest ovules.) Under the same ambient airflow conditions and with the same quantity of airborne spores, marked differences were seen in the number and distribution of spores adhering to model surfaces. Life-size models with preintegumentary lobes tended to trap airborne spores equally over much of their surfaces, whereas the largest number of spores adhering per unit area was found on or within the micropyle of the model of *Stamnostoma* (fig. 10.16). The presence of a pollination droplet enhanced a model's capacity to collect airborne spores or pollen only slightly. Although the number of spores adhering to model surfaces varied as a function of the ambient airflow speed and direction, for each ambient airflow condition the micropyles of the models of integumented megasporangia were more efficient at trapping pollen than were the tips of nonintegumented megasporangia.

Do these experiments tell us anything about the evolutionary advantages of possessing an integumented megasporangium? The results from these experiments indicate that wind-pollinated integumented megasporangia are more efficient in aerodynamically focusing the trajectories of airborne spores and pollen toward their micropyles than are nonintegumented megasporangia once these particulates become suspended within the airflow patterns they generate. Also, from first principles, we know that the same ovule will vary in its capacity to filter different types of microspores from its immediate airspace. Of course these conclusions are relevant only if the fossil ovules being considered were in fact wind pollinated. The suggestion has been made that some of these plants relied on arthropods to transport pollen (see Taylor and Millay 1979). If so, then an aerodynamic analysis of the ovulate structures of these taxa is largely irrelevant. But if these plants were predominantly (or at the least facultatively) anemophilous, then we must entertain other considerations. For example, the amount of airborne pollen released into the air and the fraction of this quantity that was transported to ovules of the same species most likely depended on community structure and the location of pollen-bearing and ovulate structures within the vertical wind profile of the community. Thus the aerodynamic efficiency of an ovulate morphology could have rendered little advantage if the microhabitat preference of the species was not conducive to long-distance pollen transport. Further, we know comparatively little about the vegetative organs on or near which ovulate organs were produced. The presence of cupules around some of the earliest-known seeds suggests that sterile structures other than preintegumentary lobes and trusses were close to the pollen-receptive sites of ovules. The aerodynamic properties of cupules were similar in many respects to those of the preintegumentary lobes and integuments of some seeds, suggesting that they too influenced pollen-capture efficiency (Niklas 1983). Finally, if the integument evolved as a consequence of its aerodynamic facility to increase the efficiency of pollen capture, then we are left with the puzzling dilemma that its initial hypothetical morphology (the preintegumentary truss) decreased efficiency compared with a relatively streamlined but naked megasporangium. From an aerodynamic point of view the isolated megasporangium of *Genomosperma kidstoni* appears to have been as efficient in capturing airborne pollen as the ovule of *Stamnostoma huttonense*. If so, then the pollen-capture efficiency of some of the intermediate morphologies envisioned for the hypothetical scenario for the evolution of the integument was most likely less than that of their precursors. It is difficult to suggest an adaptationist argument accounting for a dip in the aerodynamic selective advantage of the transition from a nonintegumented to an integumented megasporangium if we maintain a gradualistic scenario for this evolutionary transition.

There is a readily available solution to this dilemma, however, stemming from the fact that many of the ovulate morphologies considered by the scenario were geologically (and possibly ecologically) contemporaneous. Competition among contemporary species differing in their capacities for wind pollination would have selectively removed inefficient species from the game of evolution. The available paleobotanical data suggest that early in the evolution of seed plants there existed a plexus of organisms that encompassed substantial morphological variation in ovulate reproductive structures. Among geologically younger taxa, variations in ovule morphology still persist, but plants bearing ovules with substantial preintegumentary lobes are largely lacking, whereas those bearing ovules with clearly delineated integuments and micropyles are more abundant. Rather than envisioning *Genomosperma kidstoni* as the precursor to *Stamnostoma huttonense* ovules, we should view plants bearing the latter type of ovule as gradually outcompeting the former. Also, we might expect some relatively "primitive" ovulate morphologies to persist ecologically if refugia where their parent plants could survive could be found in geologically younger sediments. Such appears to be the case for the seedlike structures found in early Carboniferous strata in France (see Galtier and Rowe 1989). Under any conditions, the morphological variations in the ovules of early seed plants did not *result* from their selective advantages. Indeed, natural selection, as we currently understand the concept, cannot give rise to anything. Rather, the preintegumentary lobes, trusses, and variously fused envelopes around early seed plant megasporangia are most probably the result of epigenetic pre- or postdevelopmental phenomena attending the evolution of megasporangia.

Nonetheless, aerodynamic analysis of the ovulate organs of the earliest seed plants gives insights into why some taxa survived and could subsequently evolve while others became extinct. To be sure, the survival or extinction of a species depends on a variety of factors. Regardless of their reproductive efficiency, propagules that fail to establish themselves because of vegetative rather than reproductive deficiencies in design provide little continuity for a species. Yet it is fair to assert the truism that the capacity to receive pollen and subsequently produce a seed from an unfertilized ovule was one of the important factors influencing the evolution of seed plants.

HAVE PLANTS EVOLVED ADAPTIVELY?

Evolution is demonstrated by anagenic changes within populations and individual lineages, cladogenic events yielding new lineages, and global changes in the relative frequencies of different organisms through time. That the genetic composition and appearance of individuals within populations and lineages have changed, that new lineages have appeared, and that the relative frequencies of organisms differing in their phylogenetic affiliations have altered the structure of communities over observable and geologic time scales are undeniable facts. But the patterns of evolution, which are the vectors resulting from changes in organisms through time, have been variously interpreted by those who seek global explanations. These interpretations have led to a number of conceptual dichotomies, among which, ironically, two are most relevant to the question addressed here.

The first dichotomy in point of view involves the relative importance of extrinsic and intrinsic forces in evolution. One view sees extrinsic forces as clearly predominant. It stems from the elementary or pure form of the theory of natural selection that advocates a trial-and-error scenario in which organic materials, called species, continuously and blindly reconfigure while the external environment variously accepts or rejects these biological proposals. Accordingly, the patterns of evolution that are so readily apparent, at least to the human mind, may merely reflect the altered remains of an isotropic material that by its nature deforms with equiprobability in all directions and whose plastic strains have magnitude and direction only as a consequence of the magnitude and direction of the externally applied forces. The opposing view, which argues that intrinsic evolutionary forces predominate, sees evolution as the deformation of a very anisotropic material whose prior deformations limit subsequent responses to externally applied forces owing to the often recited pantheon of developmental constraints, pleiotropic effects, and the like. The resulting "stress-strain diagrams" for many lineages share enough features that they collectively engender what we call the "pattern of evolution," with all its elastic and plastic responses, strain hardening, fatigue, and failure. Be that as it may, the merits of viewing evolution as exclusively driven from without or from within are few. The former suggests organisms lack properties that they repeatedly manifest, while the latter borders on orthogenesis. Yet the dichotomy represented by the two extreme perspectives on natural selection is not intellectually sterile, provided we recognize that every set of orthogonal axes defines a plane on whose surface evolutionary events may be plotted. It is this plane, rather than the polarized axes of extrinsic versus intrinsic forces, that provides a realistic interpretation of the extraordinarily complex process we call evolution. Thus evolution involves reciprocity between the properties of the external environment and the properties of things we call organisms.

The second dichotomy involves the nature of the predominant force component operating to deform organic materials. If for now we accept as a truism that external forces play an important role in directing the course of life's history and that the material properties of organisms are anisotropic, then the issue remains whether the magnitude of the biotic force component overwhelms the abiotic force component. Although some argue against Darwin's proposition that biotic interactions engender most evolutionary patterns, it is nonetheless true that biotic competition is typically the most often cited driving force for biomechanical vectors in the history of life. This orthodoxy is expounded more often in the literature treating animals, perhaps, as one may reasonably argue, because the metabolism and neurological complexity of most animals make abiotic factors less important as agents of selection than in plants. In its crudest form, the notion that biotic agents dominate selection is based on two arguments: that abiotic factors have waxed and waned in such an apparently random fashion that their vicissitudes could never provide sufficient directional consistency to engender vectors in organic evolution; and that new organisms can make their appearance within the perpetual crowd of preexisting ones only by wedging out less competitive organisms. But the force of these arguments is radically diminished if we accept that the laws of physics and chemistry, whose consistency appears assured, are abiotic agents and if the "principle of plenitude" is juxtaposed to the reality that unoccupied habitats exist either because they are entirely new (as with a new land mass) or because former life has been extirpated (as with some geologic upheaval). Indeed, before the advent of life on earth, all habitats were unoccupied and the importance of the abiotic component in the environment would have been paramount. As some unoccupied habitats, such as the early terrestrial environment, developed their biological clientele, the overt struggle for continued occupancy may have intensified, shifting importance from the abiotic to the biotic component in terms of natural selection.

Thus the question before us is not whether evolution involves the operation of abiotic or biotic agents for selection or whether the magnitudes of these components differ over time, but rather whether the relative magnitudes of the resulting deformations differ among different types of organisms and whether these deformations reflect long-term adaptive trends. Conventional wisdom argues that because of their meristematic modes of growth, plants are much more deformable materials than animals and that their metabolic life-style makes plants more dependent on, and hence more responsive to, their physical environment. Although these assertions are attractive and possibly true, they contribute little to the interpretation of long-term evolutionary trends of animals or plants, which remains maddeningly elusive. The various morphological and anatomical changes seen in lineages of bryozoans, mollusks, and a host of vertebrates may reflect either escalation in defenses against predators or the refinement of offensive measures to resist abiotic mechanical forces and to acquire the carbon that these organisms cannot synthesize from sunlight and carbon dioxide. Similarly, spines, thorns, prickles, and an arborescent growth habit, with its attending elevation of planated appendages and morphologically complex reproductive organs, may be mechanisms to dissipate heat and offensive strategies to usurp the nutrients and space required by all photoautotrophs, or they may be defenses against heterotrophs that plants blindly shelter and feed. The changes in the morphological, anatomical, and even chemical attributes of plants and animals that paleontologists have compiled as the vectors of biomechanical improvement thus can be variously interpreted as the consequences of biotic competition or of the resulting struggle of organisms to survive in an impartial but often hostile abiotic world. Unfortunately, the resulting ambiguity concerning the nature of the driving evolutionary force or forces necessarily mitigates the vigor with which we assert these trends as evidence for adaptation.

In an effort to seek an approximate solution to these and other conceptual dichotomies, it seems reasonable to argue that the importance of the abiotic component as a driving force in evolution relates to the metabolic and reproductive dependence of a given life form on the abiotic components of its environment. It is a fact that photoautotrophs are intimately dependent on their physical environment for nutrition and, in many cases, reproduction. And even a cursory biomechanical analysis of the fossil record reveals a number of threads holding together the fabric of plant evolution, which strongly suggests that terrestrial plants did not evolve with equiprobability within an isotropic sphere of opportunity, but rather developed in a manner that increased their capacity to acquire nutrients and space and to disperse reproductive structures. Among the evident trends are the specialization of superficial tissues with devices to regulate gas exchange and internal tissues to transport liquids; the specialization of the plant body into organs for attachment, light interception, and mechanical support; the elevation of photosynthetic and reproductive structures above the ground as plants acquired an arborescent growth habit; the advent of secondary growth in its many guises among phyletically distant plant groups; and the appearance of the seed habit in at least two major plant lineages. It is particularly noteworthy that these trends reappear when past biotas have been geologically traumatized and most of their dominant species have been driven to extinction—cladogenesis may be the principal result of mass extinction, but anagenesis within new or surviving lineages has typically led to convergence, shown by biomechanical analyses to increase performance levels.

Evidence for adaptive evolution (based on long-term trends that suggest adaptation) is sometimes dismissed because it is hard to imagine that selection pressures of a given sort can persist over extended geological time. Indeed, the physical environment has changed over the past 400 million years: continents have moved, mountains have formed and eroded, and climatic patterns have changed as oceanic basins expand or contract. The biotic environment has also changed. The world's floras are no longer dominated by pteridophytes or gymnosperms. The great Paleozoic swamps no longer exist, nor do most of the plant genera that occupied them. But throughout all these changes certain biotic and abiotic factors have remained relatively unaltered, and the physical laws governing phenomena like gravity, gas diffusion, convective heat loss, the distribution of stresses and strains in cylindrical stems, and the physics of pollen transport have remained constant. The cast of players in the drama of evolution may have changed, but the play itself has altered little. Thus certain features of plant biology have likely experienced persistent natural selection over millennia, if not hundreds of millions of years, while the laws of physics and chemistry have driven different organisms to mimic their predecessors. Although the same may be said of animals, plants more clearly appear to be influenced by their physical environment. The plant body is a structural solution to its photosynthetic metabolism, and no other type of metabolism is as closely attuned to the physical environment as that of a photoautotroph.

Perhaps the most convincing argument for the adaptive nature of plant evolution is based on the convergence seen among so many phyletically distinct plant groups. Structures like leaves have evolved independently in virtually every algal phylum, the mosses and liverworts, ferns, horsetails, lycopods, and seed plants. Arborescence has evolved in every vascular plant lineage. Even the seed habit, so diagnostic of gymnosperms and angiosperms, had been approached by the monarchs of the Paleozoic, the arborescent lycopods. These observations suggest that there are only so many ways a terrestrial photoautotroph can deal with growth, survival, and reproduction. Evolutionary "experiments" that have diverged too radically from the blueprint of the basic vascular plant have typically met with extinction—the ultimate arbiter of an organism's evolutionary success. True, some radical experiments have met with success as well. But as Bruce Tiffney once said of the history of evolutionary innovation, "Today's revolutionaries quickly become tomorrow's conservatives." This observation appears as relevant to organic evolution as to political or economic affairs. The earliest angiosperms may have been weedy, riparian, and rapidly reproductive species and so capable of quickly invading the strongholds of the more cumbersome gymnosperms, but the flowering plants quickly reinvented the tree habit and use it with great success even today.

Debates in biology, unlike those in the pure physical sciences, are rarely canonically resolved. There is no equation for adaptation, nor are there any currently known formulas to adequately assess the prevalence of adaptive versus nonadaptive evolution in plants. But plants have evolved through a corridor in time and space defined in large part by the physical environment, one that is dominated by the laws of physics and chemistry and interpretable in terms of biomechanics.

Glossary

This glossary defines some of the most important botanical and engineering terms used in this book.

Abaxial. Oriented away from the axis. Opposite of adaxial.

Abiotic environment. The component of the total environment of an organism provided by physical factors, individually or in concert.

Abscission. The shedding of plant parts (leaves, flowers, stems, or any other structure). Typically occurs after the formation of an abscission zone.

Achene. A dry indehiscent fruit produced by a single, uniovulate carpel.

Acropetal. Developing or differentiating in a sequence toward the apex of an organ. Opposite of basipetal.

Actuator disk. A term from the momentum theory of rotors, whereby a rotor is treated as an infinite number of airfoils producing a discontinuous but uniformly distributed pressure rise.

Adaptation. Any characteristic or property of an organic process, or an organ or organism, that contributes to survival. Any process that maintains or improves organic function and survival.

Adaxial. Oriented toward the axis. Opposite of abaxial.

Adventitious. Referring to plant structures developing from unusual sites, such as roots originating on stems or leaves instead of on other roots.

Aerenchyma. Parenchymatous tissue with large intercellular spaces resulting from the breakdown (lysogeny) or shearing and ripping (schizogeny) of cells.

Algae (singular, alga). Term referring collectively to all eukaryotic plants that are not embryophytes, thereby distinguishing a grade of plant organization rather than a single phyletic group of plants.

Allometry (heterogony of Julian Huxley). A constant relative growth often expressed in terms of the formula $y = bx^k$, where y and x are two growth variables (e.g., length and weight or weight and elastic modulus), b is the value of y when x is of some arbitrary magnitude, and k is the ratio of the growth rate of y to that of x.

Allowable stress (safe stress). See Working stress.

Alternation of generations (diplobiontic life cycle). The condition of comprising more than one multicellular type of organism in a complete life cycle. In the embryophytes, the alternation of generations involves a diploid multicellular individual (the sporophyte) that produces meiospores and a haploid multicellular individual (the gametophyte) that produces gametes.

Anemophily. Pollination using wind as the transport vector for pollen.

Angiosperms. Flowering plants; considered by most to be a natural taxon. Plants in which the ovule is borne within a closed structure (carpel).

Anisotropic. Having different material properties (e.g., elastic modulus) when measured along the principal axes of symmetry. Anisotropic materials include axisymmetric materials (those that have two principal axes of symmetry) and orthotropic materials (those that have different mechanical properties along three of their principal axes of symmetry). Opposite of isotropic.

Annual growth layer. A layer of secondary tissue produced in the plant body during one growth season. Annual growth layers are produced by the vascular cambium (secondary wood) and the phellogen ("cork") and are typically referred to as "growth rings" when seen in transverse section.

Annulus. A region of cells with differentially thickened cell walls found on the sporangia of some ferns.

Anther. The pollen-bearing part of the stamen.

Anthesis. The developmental period in seed plants from the appearance of a receptive pollen-receiving structure to fertilization.

Apical meristem. A lenticular cluster of cells found at the tips of roots and stems that initiate vegetative and reproductive organs by cellular division.

Apoplast. The portion of the plant body not composed of the living protoplast (symplast); typically refers to the complex of cell walls that is continuous throughout the plant body.

Apposition. In reference to cell wall growth, the successive deposition of cell wall layers in both the primary and secondary cell wall.

Autorotation. The rotation of an object resulting from its relative motion with respect to the fluid it is immersed in.

Autotroph. An organism that synthesizes organic metabolic requirements from inorganic precursors. The source of energy may be either light via photosynthesis (photoautotroph) or chemical reactions independent of light (chemoautotroph). Opposite of heterotroph.

Axil. The angle between the adaxial surface of an organ and the organ it is attached to, as with a leaf axil or branch axil.

Axillary bud (axillary meristem). Juvenile apical meristem found in the axil of a leaf. May be dormant for a time or may commence development immediately after initiation. Gives rise to either a vegetative shoot (a branch) or a flower (a determinate, reproductive shoot).

Axis. In botany, a cylindrical stemlike organ whose homology with the three principal plant organs (stem, leaf, root) is not specified (as in early land plants whose organographic constructions are unclear). The longitudinal dimension of an elongated plant structure. In engineering or mathematics, one of the three Cartesian coordinates that specify the three-dimensional geometry of an object.

Axisymmetric. Having equivalent material properties (e.g., elastic moduli) when measured along two of the three principal axes of symmetry.

Bark. Nontechnical term applied to all primary and secondary tissues external to the vascular cambium of an organ exhibiting secondary growth. Outer bark refers to nonliving tissues external to the vascular cambium; inner bark refers to living tissues (typically just the phloem) external to the vascular cambium.

Basipetal. Developing or differentiating in a succession toward the base of an organ. Opposite of acropetal.

Bast fiber. Phloem fiber or any fiber not found in the xylem tissue.

Beam. A structural support member placed in bending. See also Column; Shaft.

Bending moment. The product of a bending force and the length of the radius arm, at right angles to the direction of application of the bending force.

Bending stresses. Stresses caused by bending, of which there are three kinds: tensile ($\sigma +$), compressive ($\sigma -$), and shear bending stresses (τ). Bending tensile and compressive stresses increase in intensity toward the perimeter of cross sections; bending shear stresses increase in intensity toward the centroid axis.

Biaxial stresses (tension or compression). A pair of coaxial forces operating orthogonally to one another. See also Coaxial.

Bilateral symmetry (zygomorphy). The condition of having two geometrically complementary sides so that the structure can be symmetrically divided by a single longitudinal plane into two mirror images.

Biomass. The total weight of living matter of an organ, an organism, or a population of organisms.

Biotic (environment). The component of the total environment of an organism that results from another organism or the interrelation among other organisms.

Bluff bodied. Having a shape that lacks streamlining.

Body force. Any force operating within the volume of an object, such as gravity. Opposite of surface force.

Boundary conditions. A time-independent or physical constraint. Boundary conditions are classified as either geometric or kinetic. An example of the former is the way a beam is anchored yet free to vibrate; an example of the latter is the forces or moments applied to a beam.

Boundary layer. The layer of fluid (gas or liquid) immediately surrounding an object, in which viscous forces predominate in dictating fluid dynamic behavior. The dimensions (thickness) of the boundary layer around the same object vary as a function of the ambient velocity (speed and direction of flow) of the fluid. Boundary layers also differ among objects differing in geometry or absolute size but experiencing the same ambient flow conditions.

Bract. A modified leaf that subtends or surrounds a plant organ. Generally provides protection.

Brazier buckling. A shortwave mode of mechanical failure in very long, thin-walled tubes resulting from a localized crimping when a large bending moment is applied. It is assumed that buckling occurs in a plane of symmetry of a cross section; that is, no torsional buckling is involved.

Breaking load. The load that results in the mechanical failure of a structure. The strength of a structure expressed in units of weight.

Breaking stress (breaking strength). The stress level at which a nonductile material breaks. The strength of the material expressed in units of force per area (stress).

Bryophytes. Nonvascular embryophytes encompassing the mosses (Musci), the liverworts (Hepaticae), and the hornworts (Anthocerotae). The gametophyte is the morphologically dominant and free-living generation in the life cycle.

Bud. A short vegetative or reproductive shoot bearing a densely packed series of leaves and intervening stem segments (internodes).

Bulb. A modified perenniating shoot with short internodes and fleshy, scalelike leaves.

Bulk modulus. Symbolized by K: the ratio of the uniformly applied hydrostatic pressure to the relative change in volume. The reciprocal of K is called compressibility.

Cambium. A region of cells configured as a more or less cylindrical layer of embryonic, growing cells typically found in stems and roots. The derivative cells are commonly produced in two directions and are arranged in radial files. The term cambium should be applied only to the two lateral meristems, the vascular cambium and the phellogen (the "cork cambium"). See also Phellogen; Vascular cambium.

Cantilever (cantilevered beam). A nonvertical beam anchored at one end and free to deflect at the other. See also Beam.

Carinal canal. A tubular chamber formed by the breakdown of protoxylem elements in the internodes of horsetails.

Carpel. The leaflike organ in flowering plants (angiosperms) producing one or more ovules.

Cauchy strain (engineering strain or conventional strain). The ratio of the difference between a deformed (l) and original dimension (l_o) to the original dimension: $\varepsilon = (l - l_o)/l_o = (l/l_o) - 1$.

Cavitation. The formation of water vapor bubbles in columns of water subjected to tensile stresses that exceed the tensile strength of water, resulting in the embolism of the conduit conducting water.

Cellular solid. Any fabricated or biological material (whose relative density is equal to or less than 0.3) consisting of a solid phase geometrically arranged in the form of walls (complete partitions, referred to as closed-walled cellular solids) or in the form of strutlike or beamlike interconnected elements (incomplete partitions, referred to as open-walled cellular solids) and a gas or, less commonly, liquid phase. Cellular solids have a complex pattern of mechanical behavior that differs from that of their solid phase and varies as a function of the stress or strain level.

Cellulose. A polysaccharide consisting of long, straight chains of β-D-glucose residues joined by 1, 4 links, in which cellobiose is the repeating polymeric unit. The main component of cell walls in most plants.

Cell wall. The more or less rigid shell secreted by and enveloping the protoplast. The chemical composition of the wall varies from one plant group to another. Typically, however, in vascular and nonvascular land plants the cell wall is composed of cellulose, other carbohydrate polymers, and proteins.

Centrifugal. Developing or produced successively farther from the center of an organ.

Centripetal. Developing or produced successively closer to the center of an organ.

Centroid axis. The longitudinal axis of a structure defined by the center of mass in all successive cross sections. The geometric center of a plane area. The sum of all the elements of area over a plane area multiplied by the distance from any axis through the centroid axis must be zero.

Chemical potential. A measure of the capacity of any substance to do work. When expressed in terms of units of energy per unit volume, the chemical potential of water is referred to as water potential (ψ_w).

Chlorenchyma. Parenchymatous tissue containing numerous chloroplasts, as in the mesophyll of photosynthetic leaves.

Chloroplast. An organelle in which photosynthesis is carried out.

Cladode. A stem modified to look like and function as a photosynthetic leaf; typically found in xerophytes.

Coaxial. Having a common axis, specifically in relation to two opposed, externally applied forces operating along the same axis through a material or structure. Coaxial forces directed inward place a material in uniaxial compression; a pair of coaxial forces directed outward place a material in uniaxial tension. A pair of coaxial forces, in which each set operates orthogonally to the other set, can produce biaxial tension or biaxial compression.

Coenocyte. An organism formed by cytoplasmic growth and nuclear division without the formation of cell wall partitions.

Cohesion theory. A theory explaining the ascent of xylem water within tall vertical vascular plants. Offered in terms of the high cohesive forces developed among neighboring water molecules, the adhesion of water molecules to the inner surfaces of tracheary elements, and transpirational "pull" on columns of water within capillarylike tracheary elements.

Coleoptile. The cylindrical, tubular sheathlike organ enveloping the epicotyl of the embryo of grasses (Poaceae) and other monocots; sometimes considered the first leaf of the epicotyl.

Collenchyma. A supporting tissue composed of more or less elongated living cells with unevenly thickened, nonlignified primary walls. Common in regions of primary growth in stems and leaves.

Column. A structural support member primarily experiencing compressive axial loading. See also Beam; Shaft.

Compliance. Symbolized by *D*, for elastic materials the reciprocal of the Young's modulus. For a viscoelastic material or any material exhibiting creep, compliance is the reciprocal of the tangent modulus. The ratio of strain to stress measured for a viscoelastic material.

Composite material. A material with an infrastructure sufficiently heterogeneous that its material properties cannot be adequately predicted from the properties of any of its constituents. Composite materials may have a periodic infrastructure, in which constituents have a geometric spatial regularity, or they may have a nonperiodic infrastructure, in which constituents have no clearly defined spatial regularity.

Compound middle lamella. The two primary walls and intervening middle lamella of two neighboring cells when the walls and middle lamella are not distinguishable.

Compressibility. The reciprocal of the bulk modulus *K*.

Compression wood. Reaction wood found in conifers, formed on the sides of branches and stems subjected to bending stresses; characterized by dense structure and extensive lignification; capable of expansion.

Compressive (normal) strain. Symbolized by ε, normalized deformation of a material or structure (subjected to compression).

Compressive (normal) stress. Symbolized by σ-, the normal force component divided by the surface area through which the force component operates.

Conductance. The reciprocal of resistance.

Conventional strain. See Cauchy strain.

Convergence (convergent evolution). An evolutionary pattern distinguished by the acquisition of similar characteristics in nonhomologous organs or systematically disparate organisms.

Cork cambium. Nontechnical term for the phellogen.

Cortex. In dicots and many nonflowering plants, the ground tissue found external to the vascular tissue.

Cotyledon. An embryonic leaflike organ produced by angiosperms and gymnosperms; markedly different in form from leaves subsequently produced by the seedling. In dicots, the cotyledon often stores nutrients.

Couple. A pair of forces acting in parallel and separated by a distance. The moment of a couple or torque is the product of one of the two forces and one-half the perpendicular distance between the two forces.

Creep. Mechanical behavior typically exhibited by viscoelastic materials characterized by changes in the magnitude of strain under a constant level of stress.

Critical compressive stress. Symbolized by σ_{cr}, the critical load divided by the cross-sectional area of a column subjected to axial compressive loading.

Critical load. Symbolized by P_{cr}, the compressive load that will result in the elastic buckling of the column. The critical load is calculated from the Euler column formula: if the column is an ideal column anchored at its base and free at its other end, then $P_{cr} = \pi^2 EI/4l^2$.

Cupule. In reference to the ovulate reproductive structures of early Paleozoic seed plants, any sterile, presumably nonfoliar structure surrounding one or more ovules. A cupule may consist of an aggregation of sterile axes or an aggregation of partially or totally fused sterile structures.

Cuticle. A layer of hydrocarbons, fatty acids, and other materials secreted on the outer walls of epidermal cells; serves a variety of functions, one being to reduce the rate of water loss from aerial plant tissues.

Cytoplasm. The portion of the living cell bounded by the plasma membrane and excluding the nucleus and visible vacuoles.

Damping. The absorption of vibrational energy by a structure. Generally, three forms of damping are recognized: material damping, which results when vibration energy is dissipated by the material of a structure; fluid damping, which results when energy is absorbed by the fluid surrounding the structure; and structural damping, which results from the impact and scraping of articulated components making up the structure.

Deformation. The displacement of a structure from its equilibrium position. Any distortion resulting from an externally applied force.

Degree of elasticity. The ratio of the elastic (recovered) deformation to the total deformation when a material is loaded to a given stress level and then unloaded.

Dehiscence. The spontaneous rupture or opening of a plant structure, such as an anther or a fruit, releasing reproductive bodies.

Derivative cell. A cell resulting from the division of a meristematic cell that subsequently differentiates and matures into a nonmeristematic cell within the plant body.

Determinate growth. A type of development in which the number of lateral organs produced by an apical meristem is limited.

Dichotomy (dichotomization). The bifurcation of the apical meristem of a plant axis into two equal-sized apical meristems leading to two axes, neither one developmentally dominant.

Dicot (dicotyledon). Nontechnical term for angiosperms whose embryos have two cotyledons. Dicots are thought to be a natural taxon within the flowering plants. Compare Monocot.

Differentiation. The developmental phase during which physiological and morphological changes lead to the specialization of cell, tissue, organ, or plant body.

Diplobiontic life cycle. See Alternation of generations.

Dipole. A molecule that has two opposite electrical charges somewhere along its atomic structure.

Distal. Farthest from a structure's point of origin or site of attachment to another structure. Opposite of proximal.

Drag. The component of the total aerodynamic force acting on an object that resists motion in the direction of the velocity vector defining the relative motion of the object with respect to a fluid. Drag has three components: normal pressure drag results from the resolved components of pressure forces normal to the surface of an object with motion relative to a fluid; skin-friction drag results from the resolved components of the pressure forces tangential to the surface of an object with motion relative to a fluid; vortex drag (also called induced drag) results from the formation of trailing vortices, that is, leeward vortices that result from the flow of a fluid from a region of high pressure to a region of low pressure.

Drag coefficient. A dimensionless quantity that reflects the fluid resistance associated with the relative motion of an object with respect to a fluid.

Ductile. Term used to describe materials that yield (undergo plastic deformations) once their proportional limits are exceeded. Ductility is the capacity of elastic materials to deform without fracturing, as in metals that can be drawn out into wires.

Dynamic viscosity. Symbolized by μ, the product of the kinematic viscosity (υ) and the density (ρ) of the fluid.

Early wood (spring wood). The secondary xylem (wood) formed at the initiation of a growth layer; characterized by a low tissue density and larger cell diameters than in subsequently formed secondary xylem in the same growth layer (late wood).

Elastic behavior. The ability to instantly restore deformations when the level of stress drops to zero.

Elastic hysteresis. The amount of energy that a material internally dissipates during a loading-unloading cycle. Shown in a stress-strain diagram as an elastic hysteresis loop where the trajectories of the loading and unloading portions of the plot do not coincide.

Elastic limit (proportional limit). The stress level at which elastic behavior is lost. Beyond the elastic (proportional) limit, many elastic materials either yield plastically or break. Thus, either the yield stress or the breaking stress is reached after the elastic limit has been exceeded.

Elastic moduli. The moduli that collectively define the behavior of a material in its elastic range of behavior, that is, E, G, K, and ν.

Elastic modulus (Young's modulus, modulus of elasticity). Symbolized by E, the ratio (material modulus) of normal stress to normal strain measured within the elastic range.

Embryophytes (formerly Embryophyta). Nontechnical term for plants that retain their multicellular embryo within sporophytic tissues. The group of plants encompassing the bryophytes and the tracheophytes (vascular plants).

Enation. Any multicellular outgrowth produced on the axis or stem of an early land plant.

Endarch. A type of primary xylem development in which the metaxylem elements develop outside the protoxylem elements. Same as centrifugal xylem development.

Endodermis. The innermost layer of the cortex, found in the roots and stems of most vascular plants, forming a sheath around the vascularized region of the organ and restricting the flow of water through the symplast.

End-wall effects. The mechanical consequences of placing supports at the ends of a specimen; particularly evident in anisotropic materials stretched in the direction of their greater stiffness. Supports increase the apparent stiffness of a specimen by locally restricting deformations. Since the strains are reduced for each level of stress, the elastic modulus appears higher than its actual value. End-wall effects in uniaxial tensile tests are reduced by ensuring that a specimen has a length:radius ratio ≥ 10 and by measuring stresses and strains along the midspan of the specimen.

Energy. The capacity to do work. Potential energy results from the position of one body with respect to another or relative to the positions of parts of the same body; kinetic energy results from the motion of a body or from the relative motions of parts of a body.

Engineering strain. See Cauchy strain.

Engineering stress. See Nominal stress.

Epicotyl. The shoot axis above the cotylendonary node. Compare hypocotyl.

Epidermis. The outer layer of cells covering the primary plant body, resulting from the protoderm.

Epigeal germination. A type of seed germination in which the cotyledon (or cotyledons) is extended above the surface of the ground as a result of the growth and extension in length of the hypocotyl. Opposite of hypogeal germination.

Epiphyte. A plant that grows on the aerial portions of another plant.

Etiolation. A type of growth characteristic of plants grown under light-limiting conditions; typically refers to plants grown in the dark. Characterized in part by elongated internodes and reduced leaf size.

Eukaryote. An organism whose chromosomes are contained within a nucleus and that possesses membrane-bound organelles. Opposite of prokaryote.

Exarch. A type of xylem development in which the protoxylem elements are farthest from the center of the axis. Same as centripetal xylem development.

Exogenous. Arising in superficial tissue, as with an axillary bud or a leaf primordium.

Extension ratio. The ratio of a deformed (l) to an undeformed (l_o) dimension: l/l_o. Equivalent to the stretch ratio (λ) of rubbery materials.

Fascicular cambium. Vascular cambium originating from procambium within a vascular bundle.

Fiber. An elongated, tapered cell with a lignified or nonlignified secondary wall; may or may not have a living protoplast at maturity.

Fiber angle. The angle subtended between the longitudinal axis of a cell and the longitudinal axis of a cellulosic microfibril within a cell wall layer.

Fibonacci series. Any series of numbers resulting from the successive addition of the last two numbers: 1, 2, 3, . . . , 13, 21, 34. etc. Fibonacci series occur in phyllotactic patterns. Named in honor of Leonardo of Pisa, son (Filius, abbreviated to "Fi") of Bonacci.

Flexural stiffness (flexural rigidity). Symbolized by EI, the product of the elastic modulus and the second moment of area. Measures the ability of a structure to resist bending.

Fluid. A general term encompassing all gases and liquids.

Fluidity. Symbolized by α, the reciprocal of viscosity.

Force. That which changes the state of rest or motion in matter, measured by the rate of change of momentum.

Free energy (Gibbs free energy). The energy within a system that can do work.

Frequency. In uniform circular motion or in any periodic motion, the number of revolutions or cycles completed per unit time.

Gametophyte. The multicellular phase in the diplobiontic plant life cycle that produces gametes.

Gas. A state of matter in which molecules or atoms are practically unrestricted by cohesive forces so that any given quantity has no definite shape or volume.

Gravitational potential. Symbolized by ψ_g, the effect of the force of gravity on the water potential of a system. In reference to xylem water or soil water, the product of the density of water, the acceleration of gravity, and elevation (positive values in the vertical above the ground surface; negative values in the vertical below the ground surface).

Griffith critical crack length. Symbolized by L_G, the limiting length of a crack or flaw within a structure that, when exceeded, will self-propagate a fracture, and when shorter, will remain stable and not propagate a fracture when the structure is subjected to stress.

Ground meristem. One of the three primary meristems derived from an apical meristem, which gives rise to the ground tissue system.

Ground tissue. Any tissue other than the epidermal, peridermal, or vascular tissue.

Ground tissue system. The total ground tissues within a plant body.

Growth layer. See Annual growth layer.

Guard cell. An epidermal cell surrounding a stoma (opening) that regulates the stoma diameter by hydrostatic changes in its size and geometry. The mechanical device whereby CO_2 and O_2 exchange between the plant and the external atmosphere is regulated.

Guttation. Exudation of xylem water, typically occurring on a leaf.

Gymnosperms. A diverse group of seed plants that do not produce ovules within ovaries.

Halophyte. A plant that grows and survives in habitats characterized by high salinity.

Harmonic motion (simple motion). A periodic oscillatory motion in a straight line such that the restoring force is proportional to the magnitude of the displacement.

Heartwood. The innermost growth layers of secondary xylem that have ceased to transport water and that store metabolites. Functions principally for mechanical support and storage.

Hemicelluloses. A chemically heterogeneous group of alkali-soluble polysaccharides (including galactans, glucans, glucomannans, mannans, and xylans) found in the matrix of plant cell walls that are neither pectinaceous nor lignitic fractions of the cell wall.

Henchy strain (true strain). Symbolized by ε_t, the natural logarithm of the extension ratio: $\varepsilon_t = \ln (l/l_o)$.

Heterotroph. An organism that acquires carbon by ingesting organic materials. Opposite of autotroph.

Hookean material. A linearly elastic material. A material that behaves according to Hooke's law.

Hooke's law. The ratio of stress to strain is linearly proportional within the elastic range of behavior: $\sigma = \varepsilon E$, where E (the elastic modulus or Young's modulus) is the proportionality factor.

Horsetails. Nontechnical term for the Sphenopsida, a group of seedless vascular plants characterized by whorls of leaves at nodes, hollow internodes, internodal anatomy with carinal and vallecular canals, and sporangia borne in a conelike structure.

Hydroid. A cell type found in some species of bryophytes that is functionally analogous to the tracheary element. Hydroids transport water.

Hydrophyte. A plant that grows and survives underwater or in water-laden habitats. Opposite of xerophyte.

Hydrostatic pressure. Technically, the product of the distance from the surface of a fluid, the density of the fluid, and the acceleration of gravity; the total force on an area due to hydrostatic pressure.

Hypocotyl. The shoot axis below the cotylendonary node. Compare epicotyl.

Hypodermis. A subepidermal layer of tissue in internodes that is distinct from other cortical tissues found in the stem.

Hypogeal germination. A type of seed germination in which the cotyledon (or cotyledons) is not exerted above the ground by the growth of a hypocotyl. Opposite of epigeal germination.

IAA. Abbreviation for indoleacetic acid, a plant growth hormone (auxin).

Indeterminate growth. A type of development in which the number of lateral organs (branches and leaves) produced by an apical meristem is unrestricted.

Inflorescence. A collection of flowers sharing the same subtending stem (peduncle).

Initial. A meristematic cell that gives rise to two cells, one remaining in the meristem, the other added to the plant body.

Integument. A layer of tissue enveloping the ovule that forms the seed coat during the development of the ovule into a seed.

Intercalary meristem. Meristematic tissue interspersed between two nonmeristematic regions.

Internode. A stem axis between two nodes.

Isotropic. Referring to material whose material properties, as measured in any direction, are equivalent.

Kinematic viscosity. Symbolized by υ, the ratio of dynamic viscosity (μ) to density (ρ).

Laminar flow. Fluid flow characterized by lack of macromolecular mixing among adjacent layers of fluid.

Lateral meristem. Cells of the vascular cambium and the phelloderm (cork cambium); any meristem parallel to the sides of a plant axis that gives rise to a secondary tissue.

Late wood (summer wood). The secondary xylem formed during the latter part of the deposition of an annual growth layer. Late wood is typically denser and composed of smaller cells than early (spring) wood.

Leaf. The principal lateral (appendicular) organ of the plant stem. The leaves of vascular plants are often classified as either microphylls or megaphylls.

Leaf sheath. The basal portion of a strap-shaped leaf that envelops a subtending internode or internodes.

Leptoid. A cell type found in some species of bryophytes that is the functional equivalent of a sieve cell in vascular plants. Leptoids transport cell sap.

Lift. The component of the total aerodynamic forces operating perpendicular to the drag force.

Lift coefficient. A dimensionless quantity that reflects the magnitude of the lift generated on an object with a motion relative to a fluid.

Lignification. The process of impregnating a cell wall with lignin.

Lignin. An organic polymer with a complex three-dimensional configuration of variable composition whose monomeric units include monosaccharides, phenolic acids, aromatic amino acids, and alcohols. Typically found in secondary plant walls, particularly those of the xylem tissue.

Linear elasticity. The property of an elastic material such that stress and strain are proportionally related by a single constant (the elastic modulus E) within the elastic range of behavior: $\sigma = \varepsilon E$. See also Nonlinear elasticity.

Load parameter. The dimensionless ratio of the product of the load and length squared to the flexural stiffness of a beamlike structure: Pl^2/EI.

Lycopods. Nontechnical term for the Lycopsida (Lycophyta), a group of pteridophytes possessing, among other distinguishing features, lateral, reniform sporangia and stems with exarch xylem maturation.

Lysigenous space. Any intercellular space resulting from the breakdown of cells.

Macrofibril. A macrostructural component of the ultrastructure of plant cell walls, composed of a collection of microfibrils and visible at the level of optical resolution achieved by the light microscope.

Mass flow. The bulk transport of materials by means of hydrostatic pressure.

Matric potential. Symbolized by ψ_m, the effects of cell surface areas, colloids, and capillarity on the water potential of a system. Always has a negative value.

Megagametophyte. The gametophyte in a diplobiontic life cycle that produces the gamete functionally equivalent to the egg.

Megaphyll. One of two types of leaves produced by vascular plants. Traditionally characterized as being large, possessing a much-branched vascular trace, and having a vascular connection with the stem that results in a leaf gap (a relatively short discontinuity in the primary vascular system of the stem that marks the point of departure of a leaf's vascular trace). As used in this book, a megaphyll may be defined as any leaf produced by a vascular plant other than a lycopod.

Megasporangium. The multicellular structure in which megaspores are produced.

Megaspore. The meiospore that will give rise to the megagametophyte.

Meristem. A tissue that retains the embryonic capacity to produce new protoplasm and cells by mitotic cell division.

Mesophyll. The photosynthetic parenchymatous tissue of a leaf. It composes the ground tissue found between the upper and lower epidermal layers.

Mesophyte. A plant whose requirements for water are intermediate between those of a hydrophyte and a xerophyte.

Metaphyte. A multicellular plant.

Metaxylem. The primary xylem that differentiates after the protoxylem. Typically possesses secondary wall thickenings that are not annular or spiral and has transverse diameters greater than those of protoxylem elements.

Microfibril. A threadlike, cellulosic component of the cell wall visible only at the level of resolution possible with the electron microscope.

Microgametophyte. The gametophyte in a diplobiontic life cycle that produces the gamete functionally analogous to sperm.

Microphyll. One of two kinds of leaves produced by vascular plants. Traditionally characterized as being relatively small, possessing a single- or little-branched vascular trace, and lacking a leaf gap (see Megaphyll). As used in this book, a microphyll is any leaf produced by a lycopod.

Micropyle. The small tube or pore that remains from the incomplete closure of the integument or integuments of an ovule, through which pollen gains access to the megagametophyte.

Microsporangium. The multicellular structure in which microspores are produced.

Microspore. The meiospore that will give rise to the microgametophyte.

Middle lamella. The intercellular layer of chiefly pectinaceous material found between the primary cell walls of adjoining cells.

Modulus. A mechanical parameter that is the ratio of a stress component to the analogous strain component; for example, elastic modulus, shear modulus.

Modulus of rupture. Symbolized by M_R, the tensile stress at fracture of a material measured in bending.

Molal solution. A solution containing one mole (gram-molecular weight) of a solute per kilogram of solvent.

Molar solution. A solution containing one mole (gram-molecular weight) of the solute per liter of the solution.

Mole. Symbolized by mol, a mass numerically equivalent to the molecular weight of the substance expressed in grams.

Moment arm. The perpendicular distance from the line of action of an externally applied force to the axis about which rotation occurs within a material or structure.

Moment of force. Symbolized by M; the effectiveness of a force to cause rotation about an axis, measured as the product of the applied force and the moment arm (the perpendicular distance from the line of the action of the force to the axis about which rotation occurs).

Moment of torque. Symbolized by T; the effectiveness of a force F to produce rotation about a center at a distance d from the line in which the force acts: $T = Fd$.

Momentum. Quantity of motion measured as the product of mass and velocity.

Monocot (monocotyledon). A nontechnical term for angiosperms whose embryos have a single cotyledon. Monocots are thought to be a natural taxon within the flowering plants. Compare Dicot.

Morphogenesis. Development of the form and structure of an organism or an organ. The latter is more properly referred to as organogenesis.

Mycorrhiza. A symbiotic association between a fungus and the roots of plants.

Natural frequency of vibration. Symbolized by f_i, the frequency at which a linear elastic beam will vibrate when displaced from its equilibrium position and released. The lowest natural frequency of vibration is called the fundamental frequency of vibration. Each frequency is associated with a mode shape of deformation (an eigenvector defined over a structure that describes the relative displacement of any point on the structure as the structure vibrates in a single mode).

Natural group (natural taxon). A group of organisms of any taxonomic rank believed to be descended from a common ancestral group of organisms of equal or lesser taxonomic rank. Equivalent to a monophyletic group.

Natural selection. According to the Darwinian theory of evolution, the principal mechanism responsible for evolution. Competition is believed to result in the death of some individuals and the survival of others whose genetic composition confers a competitive advantage. The survivors pass their genetic advantages on to the next generation, resulting in evolutionary change over sequential generations.

Natural strain. See Henchy strain.

Neutral axis. The axis of zero stress in the cross section of a structure. The neutral and centroid axes precisely coincide for beams composed of isotropic materials, provided the axial load is zero and the beam sustains only a bending load. The neutral and centroid axes also coincide for anisotropic beams experiencing very small deflections resulting from bending loads.

Newtonian fluid. Any fluid for which the viscosity (the ratio of stress to the rate of shear strain) is a constant for a specific temperature and pressure. Additional features of Newtonian fluids exist: for example, they are isotropic materials with Poisson's ratio = 0.5 (contrasted with non-Newtonian fluids, which have high viscosities and for which the ratio of stress to the rate of shear is not constant). Non-Newtonian fluids have a high coefficient of viscous traction.

Nodal diaphragm. A transverse septum of tissue found at the node of a stem possessing hollow internodes.

Node. Botany: The part of a stem where one or more leaves are attached. Dynamic beam theory: A point on a structure that does not deflect during vibration in a given mode.

Nominal stress (engineering stress). Symbolized by σ_n, the stress calculated as the normal force component divided by the original cross-sectional area through which the force component is applied.

Nonlinear elasticity. The condition where the stresses exhibited by an elastic material are not proportionally related to strains by a single constant: $\sigma \neq \varepsilon E$. Rather, the proportionality factor relating a given stress to its corresponding strain is the tangent modulus: $E_T = \Delta\sigma/\Delta\varepsilon$. See also Linear elasticity.

Non-Newtonian fluid. Any fluid for which the viscosity (the ratio of stress to the rate of shear strain) is not constant at a specified temperature and pressure. Non-Newtonian fluids can be extended to form rods or other geometric configurations and will retain their shape for a time when subjected to a stress; they have a high coefficient of viscous traction (a property equivalent to the elastic modulus of solids).

Normal strain components. Symbolized by ε, strain components resulting from normal stresses that operate perpendicular to each of the three principal axes of symmetry (x, y, z) of a material element—ε_x, ε_y, ε_z.

Normal stresses. Symbolized by σ, stresses operating perpendicular to the three principal axes of an element within a material. For materials placed in coaxial tension or compression, only the normal stress components need be considered in terms of evaluating static equilibrium.

Nucellus. The diploid, sporophytic tissue within an ovule, equivalent to the sporangial tissue of non-seed-producing plants. It contains the cells that will undergo meiosis and subsequently develop into megagametophytes.

Ontogeny. The development of an individual organism throughout its lifetime.

Organ. A distinct and visibly differentiated multicellular part of a plant, such as a stem, leaf, or root.

Orthotropic. Referring to any material or structure whose material properties differ when measured in all three of the principal axes of symmetry.

Osmotic adjustment. A metabolic effect that results in the net accumulation of solutes (carbohydrates, organic acids, and inorganic ions) within the protoplasm of water-stressed plants.

Osmotic potential (osmotic pressure). Symbolized by π, the pressure that must be applied to a membrane to prevent the net movement of water molecules across the membrane.

Ovule. The reproductive structure of gymnosperms and angiosperms composed of a megasporangium (nucellus) surrounded by one or more layers of tissues (integument) that develop from the base of the nucellus; it possesses a small tube or pore (the micropyle) resulting from the incomplete closure of the integument.

Ovuliferous scale-bract complex. The basic (repeated) reproductive unit of the coniferous ovulate cone, consisting of a subtending bract (a modified leaf) and an ovuliferous scale (believed to be a highly reduced axillary, ovule-bearing branch). The bract and the ovuliferous scale are fused to varying degrees depending on the species of conifer.

Palisade mesophyll. A layer or layers of mesophyll tissue composed of columnar, prismatic cells possessing a large number of chloroplasts, typically found in the leaves of dicots.

Pappus. A low-density, parachute-like structure developed from the sepals of a single flower. Aids in long-distance dispersal of the subtending fruit.

PAR. Photosynthetically available (active) radiation; light with wavelengths between 400 and 700 nm.

Parenchyma. A tissue composed of more or less isodiametric, thin-walled cells with plasmodesmata found in most of the walls of adjacent cell walls.

Parenchymatous. Characterized by cellular division in each of the three principal axes of a plant structure. Hence any tissue composed of cells that have the potential to share plasmodesmata through adjoining cell walls.

Pectin (pectinaceous substances). A chemically heterogenous group of acidic polysaccharides (typically long-chain, branched or unbranched polymers of arabinose, galactose, galacturonic acid, and methanol) found mainly external to the primary cell wall. Soluble pectins or soluble pectic acids are straight chains of galacturonic acid residues precipitated as calcium and magnesium pectates, forming the middle lamella.

Pedicel. The stem subtending a single flower.

Peduncle. The stem subtending an inflorescence.

Perennial. A plant that has the capacity to grow year after year and frequently can become dormant.

Perforation plate (perforate end). The gap in the primary and secondary cell wall layers found at the ends of adjoining members within a vessel.

Periderm. Secondary tissue derived from the phellogen (cork cambium) that replaces the epidermis. Consists of phellem (cork) and phelloderm.

Petiole. The cylindrical, stalklike axis subtending the lamina of a leaf.

Phellem. The external layer of cells produced by the phellogen (cork cambium).

Phelloderm. A tissue produced internally by the centripetal cell divisions of the phellogen (cork cambium); a component of the periderm.

Phellogen (cork cambium). One of the two lateral meristems that forms the periderm. Produces phellem (cork) centrifugally and phelloderm centripetally by means of tangential cell divisions.

Phloem. The tissue of the vascular plant specialized to conduct cell sap, composed predominantly of sieve elements and various kinds of phloem parenchyma, fibers, and sclereids.

Photosynthesis. The synthesis of organic compounds by the reduction of carbon dioxide in the presence of light energy absorbed by chlorophyll. In green plants, water is the hydrogen donor and the source of oxygen released by the photosynthetic process, represented by the empirical equation $CO_2 + 2H_2O \rightarrow (CH_2O) + H_2O + O_2$.

Phyllotaxy (phyllotaxis). The pattern of the arrangement of leaves on the shoot.

Pit. A porelike cavity in the cell wall where the primary cell wall is not covered by secondary cell wall layers. The adjoining pits of neighboring cells are called pit pairs.

Pith. Ground tissue found toward the center of dicot stems and some dicot and monocot roots.

Pit membrane. An intercellular layer consisting of the middle lamella and the two adjoining primary cell walls that spans a pit cavity externally.

Plasma membrane (plasmalemma). The single membrane enveloping the protoplasm and appressed to the cell wall.

Plasmodesma (plural, plasmodesmata). A symplastic strand passing through the cell walls of two adjoining cells that connects their protoplasts.

Plastic behavior. Irreversible molecular or microstructural deformation within a solid resulting from the application of external forces.

Point load. A point in space having mass but zero moment of inertia for rotation about the center of mass.

Poisson's effect. In uniaxial tension, materials are permitted to contract laterally, whereas in biaxial tension lateral contraction is restrained. Thus materials evince larger strains for a given uniaxial stress level than for the strains measured in biaxial tension.

Poisson's ratio. Symbolized by ν, the ratio of negative lateral strain to the strain measured in the direction of the applied force. For any planar section having two principal axes of symmetry, two Poisson's ratios may be calculated. For isotropic materials, the two Poisson's ratios are equivalent in magnitude.

Polar second moment of area. Symbolized by I_p, the sum of the second moments of area measured in the two orthogonal planes of a cross section: $I_p = I_{xx} + I_{yy}$. For shafts with circular cross sections, I_p equals the torsional constant J.

Pollen (pollen grain). Either the microspore cell wall of the microgametophyte of seed plants or the cell wall and the contained microgametophyte. The microspore-microgametophyte of seed plants.

Pollen tube. A filamentous extension of the microgametophyte of some seed plants, produced after pollination, that provides a transport route for male gametes. The pollen tube of some gymnosperms can be highly branched and serves as a haustorium, absorbing food.

Pollination droplet. A viscous liquid, produced from the disintegration of some nuclear cells, exuded from the micropyle of some gymnosperms, to which pollen grains adhere.

Pressure potential. Symbolized by ψ_p, the effect of any pressure on the water potential of a system. Usually has a positive value, but the pressure potential can be negative in the case of rapidly moving xylem water owing to high transpirational water losses.

Prickle. Typically a woody, ascicular epidermal outgrowth whose principal functions are protection and heat convection.

Primary cell wall. That portion of the cell wall formed mainly while the cell is increasing in size, in which the cellulosic microfibrils are variously oriented from random to more or less parallel to the longitudinal axis of the cell.

Primary growth. That portion of the ontogeny of vegetative and reproductive organs from the time of their meristematic initiation until the completion of their expansion and elongation. In some plants the primary phase of ontogeny is followed by secondary growth.

Primary meristem. Any of the three meristematic tissues derived from an apical meristem: protoderm, ground meristem, and procambium.

Primary tissues. Tissues derived from the apical meristems and the embryo.

Procambium (provascular tissue). The meristematic tissue that gives rise to the primary vascular tissue.

Progymnosperms. An extinct group of Paleozoic plants characterized by having gymnosperm anatomy and a pteridophytic mode of reproduction.

Prokaryote. An organism that does not have its nuclear material enclosed in a nuclear envelope and that lacks other membrane-bound cytoplasmic organelles. Prokaryotic organisms include the cyanobacteria, nonphotosynthetic bacteria, and the mycoplasmas. Opposite of eukaryote.

Propagule. A term encompassing seeds, fruits, and vegetative structures that can propagate an individual plant. Any reproductive structure that is capable of producing a plant either by possessing an embryo or by the vegetative generation of embryonic roots and shoots.

Proportional limit (elastic limit). The stress level below which elastic behavior is maintained and above which a solid material either yields or breaks.

Protoderm. The primary meristem produced by the apical meristem that gives rise to the epidermis.

Protoplasm. The living contents of the cell. Also the inclusive term for the living fraction of an entire organism, the symplast.

Protoxylem. The first-formed tracheary elements. Typically possess annular or spiral secondary wall thickenings and have a smaller girth than the metaxylem elements that differentiate and mature later.

Proximal. Closest to a structure's point of origin or site of attachment to another structure. Opposite of distal.

Pseudoparenchyma. A tissue found in some algae that is constructed from interwoven filaments of cells, giving the appearance of true parenchyma.

Pteridophyte. Nontechnical term for any free-sporing (non-seed-bearing) vascular plant. Encompasses the ferns, horsetails, lycopods, and several extinct seedless plant groups. The pteridophytes were formerly thought to be a natural group and were referred to as the Pteridophyta.

Pteridosperms. A group of now extinct gymnospermous plants that had many of the vegetative features of ferns but reproduced by means of seeds.

Rachis. The part of the axis of a pinnately compound leaf that supports the leaflets including the petiole, which is the basalmost element of the rachis.

Radial. Referring to any plane parallel to a radius through an object or a plane of section through a biological structure.

Radicle. The embryonic root.

Radius of bending. The inverse of the curvature of bending *K*.

Radius of gyration (least radius of gyration). Symbolized by *r*, the square root of the second moment of area (I) divided by the cross-sectional area of a representative transection (A): $r = (I/A)^{1/2}$. For beams with solid circular cross sections, $r = R/2$, where R is the unit radius of the cross section.

Reaction wood. Secondary xylem formed by stems in response to mechanical changes from their original orientations. See also Compression wood; Tension wood.

Relaxation modulus. The ratio of the time-varying stress to the fixed strain.

Relaxation time. Symbolized by T_R, the time required for a given stress to diminish to $1/e$ its original magnitude. Also, the ratio of a material's viscosity to its elastic modulus.

Residual strain. The plastic component of deformation. A perfectly elastic material has no residual strain provided its proportional limit is not exceeded during loading. Most elastic materials are not perfectly elastic, and many exhibit a residual (plastic) strain after they are unloaded.

Resilience. A measure of the elastic energy stored within a body. The capacity to restore original dimensions after an applied load has been removed.

Reuss model (equal stress model). A model assuming that material elements, differing in elastic properties, are aligned normal to the direction of application of an external force so that all elements experience stresses of equivalent magnitude. The reciprocal of the elastic modulus of the composite material is predicted to equal the sum of the quotients of the volume fractions and the elastic moduli of each component material within the composite; that is, $(1/E_{cm}) = (V_1/E_1) + (V_2/E_2) + \ldots + (V_n/E_n)$.

Reynolds number. Symbolized by *Re*, the dimensionless parameter that is the product of a reference dimension (parallel to the direction of ambient fluid flow) and the ambient fluid speed, divided by the kinematic viscosity of the fluid. The Reynolds number expresses the ratio of inertial to viscous forces in the fluid flow in reference to an object obstructing fluid flow. Note that fluid flow does not have a Reynolds number except in reference to an object.

Rhyniophytes. One of the three major groups of vascular Silurian and Devonian plants, characterized by having more or less equal dichotomous branching, terminal sporangia, and a single centrally positioned vascular strand in each of their axes. Thought to be one of the oldest vascular land plant groups.

Root. One of the three primary plant organ types characterized by possessing a rootcap and by exarch xylem differentiation. All but the primary root result from endogenous development.

Rootcap. A thimble-like mass of cells covering the root apical meristem.

Root pressure. A positive pressure exerted on the xylem water by the absorption of water by the root.

Sacci. The two bladderlike extensions of the pollen grains produced by some gymnosperms resulting from the separation of the inner and outer spore walls.

Samara. A fruit possessing an extended bladelike wing or membrane. Aerodynamically, any propagule with a wing that generates lift.

Sapwood. Outermost portion of the secondary xylem of stem or root in which some cells are living and others (tracheary elements) conduct water.

Schizogenous. Producing intercellular spaces by splitting apart neighboring cells along the middle lamellae.

Sclereid. Any sclerenchymatous cell type other than a fiber.

Sclerenchyma. A tissue composed of thick-walled, lignified cells that can provide mechanical support. Also a collective term for sclerenchymatous tissues of an organ or the entire plant body. Includes fibers and sclereids.

Secondary cell wall. That portion of the cell wall deposited in some cells by the protoplast between the plasma membrane and the primary wall after the primary wall ceases to increase in surface.

Secondary growth. Growth characterized by the deposit of secondary tissues, resulting in an increase in the thickness of organs, typically stems and roots. An ontogenetic phase of development seen in some species resulting from the meristematic activity of lateral meristems.

Secondary phloem. The phloem tissue produced by a vascular cambium during secondary growth in a vascular plant.

Secondary xylem. The xylem tissue produced by a vascular cambium during secondary growth in a vascular plant. Colloquially referred to as wood.

Second moment of area. Symbolized by I, the integral of the product of each elemental cross-sectional area and the square of the distance of each elemental cross-sectional area from the centroid axis. A dimensional parameter that quantifies the distribution of mass in each cross section with respect to the center of mass of the cross section.

Section modulus. Symbolized by Z, the ratio of the second moment of area to the characteristic radius of the cross section; for example, for a circular cylinder with radius R, $Z = I/R = \pi R^3/4$.

Seed. A mature ovule containing an embryo or embryos surrounded by a seed coat composed of the mature integument or integuments.

Shaft. A support member subjected to a torque, such as a propeller. See also Beam; Column.

Shear. The result of the tangential application of an external force to the surface of a material or structure.

Shear modulus. Symbolized by G, the ratio of the shear stress (τ) to the shear strain (γ) for a material.

Shear strain. Symbolized by γ, the tangent of the rotation angle (in radians) resulting when a material is subjected to a tangentially applied force.

Shear stress. Symbolized by τ, the shear force component divided by the tangential area over which the shear force component acts.

Shell. A very thin elastic structure whose material is confined to a curved surface. A shell evincing no flexural rigidity is called a membrane.

Siphonaceous. Referring to a coenocyte in which some cellular septa have formed.

S layer. Any secondary cell wall layer. The numerical order of the S layers reflects the growth of the cell wall by apposition; ascending numbers indicate progressively older (inner) cell wall layers.

Slenderness ratio. The length (l) of a column divided by the radius of gyration (r) of the column: l/r. In the case of a column with a circular solid cross section, $l/r = 2\ l/R$, where R is the unit radius of the cross section.

Softwood. The wood of conifers.

Solid. A state of matter characterized by the restricted motion of molecules or atoms such that the shape and volume of any portion are fixed relative to those of any other portion. Opposite of fluid.

Solute. The constituent of a solution that is dissolved in the solvent. The solute is the smaller of the two amounts.

Solute potential. Symbolized by ψ_s, the effects of solutes dissolved in water on the water potential of the system. Always has a negative value.

Somatic. Referring to nonreproductive portions of the plant body or to nonreproductive cells and cell divisions.

Specific conductivity. In reference to the xylem tissue, the volume of water moved per unit time.

Spine. Typically a sharp, sclerotic, highly modified leaf.

Spongy mesophyll. A type of parenchyma containing large intercellular spaces filled with air, found in the leaves of some plants. It aids gas exchange between the plant body and the external atmosphere.

Sporophyte. The multicellular organism in the diplobiontic life cycle of plants that produces meiospores. The dominant organism in the life cycle of vascular plants.

Spring constant. The change in load on a structure (exhibiting linear elasticity) required to produce a unit increment of deflection.

Spring wood. See Early wood.

Stele. The morphologic component of a vascular plant axis comprising the primary tissues produced by the procambium (primary vascular tissues and associated tissues, pericycle) and the endodermis.

Stem. One of the three principal vegetative organs of the vascular plant body. Produced by the shoot apical meristem and bearing appendicular organs (leaves) produced by exogenous growth.

Strain energy. The component of the total energy within an object or a structure that is stored in the form of molecular deformations. Strain energy is a form of potential energy; in elastic materials, within the proportional limits of loading, the strain energy is used to restore the material's original dimensions when the stress is removed. Within the range of elastic behavior, the strain energy is the area measured in the stress-strain diagram.

Streamline. An imaginary line passing through a flow field such that the local velocity vector is tangential to every point on the line. As a consequence, fluid particles flow along but not across streamlines.

Streamlined. Having a body profile that minimizes pressure drag. Opposite of bluff bodied.

Strength. In reference to a structure, the load (breaking load) that will cause the structure to fail. The breaking load will vary from structure to structure. In reference to a material, the stress (breaking stress) that will break the material.

Stress-strain diagram. A plot of stress versus strain for a given material or structure.

Suberin. A fatty hydrophobic substance found in the cell walls of the endodermis and cork cells (phellem).

Summer wood. See Late wood.

Supporting tissue (mechanical tissue). Any tissue composed of thick-walled cells; any primary (collenchyma) or secondary (sclerenchyma) tissue that confers strength on the plant body.

Surface force. Any force operating on the external boundaries of an object, such as hydrostatic pressure. Opposite of body force.

Symplast. The living portion of the plant body; consists of the metabolically functional protoplast of the plant body.

Tangential. Referring to the direction of the tangent (normal to the radius) of a structure. Used in reference to the plane in which a structure is sectioned.

Tangent modulus (instantaneous elastic modulus). Symbolized by E_T, the ratio of the change in the stress to the change in the strain measured at any point along a stress-strain diagram. When measured within the linear portion of the stress-strain diagram of a linearly elastic material, the tangent modulus is called the modulus of elasticity (Young's modulus, elastic modulus).

Telome. Any distal branch of a dichotomized axis, the fundamental morphological unit of an ancient land plant with vascular tissues.

Tension wood. The reaction wood of dicots, formed on the surfaces of branches experiencing stresses; characterized by lack of lignification and often by high content of gelatinous fibers; capable of contraction.

Thigmomorphogenesis. Any morphogenetic response to mechanical perturbation, typically involving some form of growth inhibition.

Thorn. A modified branch, typically serving for protection.

Tissue. Any group of cells organized into a structural or functional unit. The constituent cells may or may not be similar in size and shape.

Tissue system. A tissue (or tissues) in the plant body that is structurally and functionally organized into a unit. The three tissue systems are the dermal, vascular, and fundamental (ground) tissue systems.

Torsion. The mechanical consequence of a torque moment.

Torsional constant. Symbolized by *J*, a parameter describing the geometric contribution of a shaft toward resisting torsion. For shafts with circular cross section, the torsional constant equals the polar second moment of area (I_p). For shafts with nonterete cross sections, the torsional constant is less than the polar second moment of area.

Torsional rigidity. Symbolized by *C*, the product of the shear modulus (*G*) and the torsional constant (*J*). A measure of the ability of a shaft to resist a moment of torque.

Tracheary element. Any water-conducting cell type found in vascular plants; specifically, a collective term for tracheids and vessel members.

Tracheid. A tracheary cell type lacking perforations that transports water. May occur in either the primary or the secondary xylem or in both. Compare vessel member.

Tracheophyte. A vascular plant. Includes ferns, horsetails, lycopods, gymnosperms, and angiosperms; excludes mosses, liverworts, hornworts, and the algae.

Transverse section (transection, cross section). Any plane section taken normal to the longitudinal axis of a structure.

Trimerophyte. One of the three major groups of vascular Devonian plants; characterized by dichotomous growth resulting in lateral branching, terminal spindle-shaped sporangia with longitudinal dehiscence, and vascularized axes possessing a single centrally positioned vascular trace. Believed to be the ancestral plexus of plants that gave rise to all subsequent plant groups except the lycopods.

True strain. See Henchy strain.

True stress. Symbolized by σ_t, the stress calculated from the normal force component divided by the instantaneous cross-sectional area through which it operates. True stress should be calculated when the loading conditions vary over time or when the strains are large ($\geq$ 5%).

Tunica. External layer (or layers) of cells in the apical meristem of a stem that divides to give rise to the protoderm, which in turn gives rise to cells that mature and differentiate into the epidermis.

Turgor pressure. When matric potentials are neglected, the turgor pressure is defined as the difference between the water potential and the solute potential of a cell or tissue.

Uniaxial stresses. A single pair of coaxial forces directed either outward (uniaxial tension) or inward (uniaxial compression) with respect to a material or structure.

Vascular bundle. A threadlike vascular strand consisting of primary xylem and primary phloem.

Vascular cambium. The lateral meristem giving rise to the secondary vascular tissues.

Vascular tissue. The xylem and the phloem; sometimes used to refer to either the xylem or the phloem. Typically the term is used with no distinction between the primary and the secondary xylem or phloem.

Vein. A single vascular bundle in a dorsiventral organ, such as a leaf.

Venation. The vascular architecture of a leaf.

Vessel. A tubelike conduit composed of many vessel members stacked end to end. The members have perforated end walls and offer little resistance to the flow of water through the vessel.

Vessel member (vessel element). A single cellular component of a vessel.

Viscoelastic. Exhibiting the properties of viscosity and elasticity; a viscoelastic material deforms with the application of a force (or forces) and, over time, elastically restores some or all of its deformation when the force is removed.

Viscosity. Symbolized by η, the ability of a fluid to resist shearing deformation. The viscosity of a linear (Newtonian) fluid is the ratio of the shear stress to the resulting velocity gradient.

Viscous modulus. Symbolized by G'', the ratio of the dynamic viscosity (μ) to the frequency (ω) at which strain is varied: $G'' = \mu/\omega$.

Voigt model (equal strain model). A model assuming that material elements, differing in elastic properties, are aligned parallel to the direction of application of an external force so that all elements experience strains of equivalent direction and magnitude. The elastic modulus of the composite material is predicted to equal the sum of the products of the elastic moduli and the volume fractions of each component material within the composite: $E_{cm} = E_1 V_1 + E_2 V_2 + \ldots + E_n V_n$.

Von Karman vortex street. A regular arrangement of vortices in approximately two parallel directions, such that vortices paralleling one another are shed alternately.

Vortex. A rotating body of fluid.

Vortex street. See Von Karman vortex street.

Vorticity. Symbolized by Γ, the rotational motion within a fluid body that at any point in the fluid is defined as twice the angular velocity of a small portion of the fluid surrounding the point.

Wake. A region of disturbed fluid flow behind an object having a motion relative to the fluid.

Water potential. Symbolized by ψ_w, the chemical potential of water or a solution of water and solutes expressed in terms of the units of energy per unit volume of fluid. The water potential of pure water is zero. The water potential of a solution of water and solutes is the ratio of the difference between the chemical potential of the solution and the chemical potential of pure water to the molal volume of water in the solution.

Wood. The secondary xylem within a plant body; any xylem other than the primary xylem.

Working stress (safe stress). Symbolized by σ_w, the stress level for which a structure is designed to operate. Usually well below the yield stress σ_y or the ultimate strength S_U of the materials used in the structure's fabrication; that is, $\sigma_W = \sigma_y/n$ or $\sigma_W = \sigma_U/n_1$, where n and n_1 are factors of safety, which define the magnitude of the working stress. In the case of structural steel, the yield point is used and $n = 2$; in the case of wood, the ultimate strength is usually taken as the basis for determining the working stress.

Xerophyte. Any plant that can grow and survive in a dry habitat. Opposite of hydrophyte.

Xylem. The tissue produced by the procambium, specialized to conduct water and also to store metabolites; also serves as the principal mechanical support tissue in organs exhibiting secondary growth.

Yielding. The initiation of plastic behavior in an elastic, ductile material.

Yield strain. The strain corresponding to the yield stress, providing a measure of the maximum level of deformation that can be elastically recovered when the stress level drops to zero.

Yield stress. The level of stress initiating plastic behavior in an elastic, ductile material. The yield stress represents the limiting stress level for which deformations are recoverable when the stress level drops to zero.

Young's modulus. The elastic modulus E; the proportionality constant relating normal stress to normal strain throughout the linear elastic range of behavior of a material.

Zosterophyllophytes. One of the three major groups of early Silurian and Devonian plants, characterized by possessing lateral, reniform sporangia with lateral dehiscence and exarch steles. Believed to be the ancestral plexus of plants that gave rise to the lycopods.

Zygomorphic. Having bilateral symmetry. Opposite of actinomorphic (having radial symmetry).

References

Ades, C. S. 1957. Bending strength of tubing in the plastic range. *J. Aeronaut. Sci.* 24:605–10.

Albersheim, P. 1976. The primary cell wall. In *Plant biochemistry,* ed. J. Bonner and J. E. Varner, 225–74. New York: Academic Press.

Allmon, W. D., and R. M. Ross. 1990. Specifying causal factors in evolution: The paleontological contribution. In *Causes of evolution: A paleontological perspective,* ed. R. M. Ross and W. D. Allmon, 1–20. Chicago: University of Chicago Press.

Ambronn, H. 1881. Ueber die Entwickelungsgeschichte und die mechanischen Eigenschaften des Collenchyms: Ein Beitrag zur Kenntniss des mechanischen Gewebesystems. *Jahrb. Wiss. Bot.* 12:473–541.

American Institute of Timber Construction. 1974. *Timber construction manual.* 2d ed. New York: John Wiley.

American Society for Testing and Materials. 1984. *ASTM standards.* Vol. 04.08. Philadelphia: American Society for Testing and Materials.

American Society of Civil Engineers Committee on Wood. 1975. *Wood structures: A design guide and commentary.* New York: American Society of Civil Engineers.

Anazado, U. G. N. 1983. Mechanical properties of the corn cob in simple bending. *Amer. Soc. Agr. Eng. Trans.* 26:1229–33.

Andrews, H. N. 1963. Early seed plants. *Science* 142:925–31.

Ang, A. H.-S., and W. H. Tang. 1975. *Probability concepts in engineering planning and design.* New York: John Wiley.

Apfel, R. E. 1972. The tensile strength of liquids. *Sci. Amer.* 227:58–71.

Archer, R. R., and F. E. Byrnes. 1974. On the distribution of tree growth stresses. 1. An anisotropic plane strain theory. *Wood Sci. Tech.* 8:184–96.

Archer, R. R., and B. F. Wilson. 1970. Mechanics of the compression wood response. 1. Preliminary analysis. *Plant Physiol.* 46:550–56.

———. 1973. Mechanics of the compression wood response. 2. On the location, action, and distribution of compression wood formation. *Plant Physiol.* 51:777–82.

Ashby, M. F. 1983. The mechanical properties of cellular solids. *Amer. Inst. Mech. Eng. Metal. Trans.,* ser. A, 14:1755–69.

Atterberg, A. 1911. Die Plastizität der Tone. *Int. Mitt. Bodenk.* 1:10–43.

Auspurger, C. K. 1986. Morphology and dispersal potential of wind-dispersed diaspores of Neotropical trees. *Amer. J. Bot.* 73:353–63.

Bailey, A. J. 1936. Lignin in Douglas fir. *Ind. Eng. Chem.* (Anal. ed.) 8:389–91.

Balashov, V., R. D. Preston, G. W. Ripley, and L. C. Spark. 1957. Structure and mechanical properties of vegetable fibres. 1. The influence of strain on the orientation of cellulose microfibrils in sisal leaf fibre. *Proc. Roy. Soc. London*, ser. B, 146:460–68.

Ball, O. M. 1904. Der Einfluss von Zug auf die Ausbildung von Festigungsgewebe. *Jahrb. Wiss. Bot.* 39:305–41.

Banks, H. P. 1981. Time of appearance of some plant biocharacters during Siluro-Devonian time. *Can. J. Bot.* 59:1292–96.

Becker, G. 1931. Experimentalle Analyse der Genom- und Plasmowirkung bei Moosen. 3. Osmotischer Wert heteroploider Pflanzen. *Zeit. Ind. Abst. Vererb.* 60:17–38.

Bernstein, Z., and I. Lustig. 1985. Hydrostatic methods of measurement of firmness and turgor pressure of grape berries (*Vitis vinifera* L.). *Sci. Hort.* 25:129–36.

Bertram, J. E. A. 1989. Size-dependent differential scaling in branches: The mechanical design of trees revisited. *Trees* 4:241–53.

Bierhorst, D. 1971. *Morphology of vascular plants.* New York: Macmillan.

Black, C. C., Jr. 1973. Photosynthetic carbon fixation in relation to net carbon dioxide uptake. *Ann. Rev. Plant Physiol.* 24:253–86.

Blake, J. R. 1978. On the hydrodynamics of plasmodesmata. *J. Theor. Biol.* 74:33–47.

Bland, D. R. 1960. *Theory of linear viscoelasticity.* London: Pergamon.

Blevins, R. D. 1984. *Formulas for natural frequency and mode shapes.* 2d ed. Malabar, Fla.: Robert E. Krieger.

Bodig, J., and B. A. Jayne. 1982. *Mechanics of wood and wood composites.* New York: Van Nostrand Reinhold.

Bordner, J. S. 1909. The influence of traction on the formation of mechanical tissue in stems. *Bot. Gaz.* 48:251–74.

Bower, F. O. 1935. *Primitive land plants.* London: Macmillan.

Boyd, J. D. 1950. Tree growth stresses. 3. The origin of growth stresses. *Austral. J. Sci. Res.*, ser. B, 3:294–309.

———. 1973. Compression wood force generation and functional mechanics. *New Zeal. J. For. Sci.* 3:240–58.

———. 1974. Compression wood force generation: A rejoinder (letter). *New Zeal. J. For. Sci.* 4:117.

Braam, J., and R. W. Davis. 1990. Rain-, wind-, and touch-induced expression of calmodulin and calmodulin-related genes in *Arabidopsis*. *Cell* 60:357–64.

Braun, A. 1831. Vergleichende Untersuchung über die Ordnung der Schuppen an den Tannenzapfen. *Nova Acta Acad. Car. Leop.* 15:195–401.

Brazier, L. G. 1927. On the flexure of thin cylindrical shells and other "thin" sections. *Proc. Roy. Soc. London*, ser. A, 116:104–14.

Briggs, L. J. 1949. A new method for measuring the limiting negative pressure of liquids. *Science* 109:440.

British Standards Institution. 1975. *Methods of tests for soils for civil engineering purposes.* British Standard 1377. London: British Standards Institution.

Brush, W. D. 1912. The formation of mechanical tissue in the tendrils of *Passiflora caerulea* as influenced by tension and contact. *Bot. Gaz.* 53:453–76.

Buchmann, S. L., M. K. O'Rourke, and K. J. Niklas. 1989. Aerodynamics of *Ephedra trifurca.* III. Selective pollen capture by pollination droplets. *Bot. Gaz.* 150:122–31.

Burrows, F. M. 1987. The aerial motion of seeds, fruits, spores and pollen. In *Seed dispersal,* ed. D. Murray, 1–47. New York: Academic Press.

Burstrom, H. G. 1979. In search of a plant growth paradigm. *Amer. J. Bot.* 66:98–104.

Buss, L. W. 1987. *The evolution of individuality.* Princeton: Princeton University Press.

Bussinger, J. A. 1975. Aerodynamics of vegetated surfaces. In *Heat and mass transer in the biosphere,* ed. D. A. DeVries and N. H. Afgan. New York: John Wiley.

Campbell, G. S. 1977. *An introduction to environmental biophysics.* Heidelberg Science Library. New York: Springer-Verlag.

Carroll, M. M. 1987. Pressure maximum behavior in inflation of incompressible elastic hollow spheres and cylinders. *Quart. Appl. Math.* 45:141–54.

Casada, J. H., L. R. Walton, and L. D. Swetnam. 1980. Wind resistance of burley tobacco as influenced by depth of plants in soil. *Amer. Soc. Agr. Eng. Trans.* 23:1009–11.

Chafe, S. C., and A. D. Wardrop. 1972. Fine structural observations on the epidermis. 1. The epidermal cell wall. *Planta* 107:269–78.

Chaloner, W. G., and A. Sheerin. 1979. Devonian macrofloras. in *The Devonian system,* 145–61. Special Papers in Palaeontology 23. London: Palaeontological Association.

Chappell, T. W., and D. D. Hamann. 1968. Poisson's ratio and Young's modulus for apple flesh under compressive loading. *Amer. Soc. Agr. Eng. Trans.* 11:608–11.

Christensen, R. M. 1971. *Theory of viscoelasticity: An introduction.* New York: Academic Press.

Coleman, B. D., M. D. Gurtin, and I. Herrera. 1967. *Wave propagation in dissipative media.* Berlin: Springer-Verlag.

Cooke, J. R., R. H. Rand, H. A. Mang, and J. G. Debaerdemaeker. 1977. *A nonlinear finite element analysis of stomatal guard cells.* ASAE Paper 77–5511. Saint Joseph, Mich.: American Society of Agricultural Engineers.

Cosgrove, D. 1986. Biophysical control of plant cell growth. *Ann. Rev. Plant Physiol.* 37:377–405.

———. 1989. Characterization of long-term extension of isolated cell walls from growing cucumber hypocotyls. *Planta* 177:121–30.

Cosgrove, D. J., and P. B. Green. 1981. Rapid suppression of growth by blue light: Biophysical mechanism of action. *Plant Physiol.* 68:1447–53.

Cottrell, A. H. 1964. *The mechanical properties of matter.* New York: John Wiley.

Courtet, Y. 1966. Structure anatomique des tiges volubiles de *Mandevilla suaveolens* (Apocynacées). *Ann. Sci. Univ. Besançon* (*Bot.*) 3:24–27.

Cremer, K. W. 1977. Distance of seed dispersal in eucalypts estimated from seed weights. *Austral. For. Res.* 7:225–28.

Crepet, W. L., and E. M. Friis. 1987. The evolution of insect pollination in angiosperms. In *The origin of angiosperms and their biological consequences*, ed. E. M. Friis, W. G. Chaloner, and P. R. Crane, 181–201. Cambridge: Cambridge University Press.

Dean, T. J., and J. N. Long. 1986. Validity of constant-stress and elastic-instability principles of stem formation in *Pinus contorta* and *Trifolium pratense*. *Ann. Bot.* 58:833–40.

Delwiche, C. F., L. E. Graham, and N. Thompson. 1989. Lignin-like compounds and sporopollenin in *Coleochaete*, an algal model for land plant ancestry. *Science* 245:399–401.

Delwiche, M. J. and J. R. Cooke. 1977. An analytical model of the hydraulic aspects of stomatal dynamics. *J. Theor. Biol.* 69:113–41.

Di-Giovanni, F., and P. M. Beckett. 1990. On the mathematical modeling of pollen dispersal and deposition. *J. Appl. Meteorol.* 29:1352–57.

Di-Giovanni, F., P. M. Beckett, and J. R. Flenley. 1989. Modelling of a dispersion and deposition of tree pollen within a forest canopy. *Grana* 28:129–39.

Dinwoodie, J. M. 1968. Failure in timber. Part 1. Microscopic changes in cell wall structure associated with compression failure. *J. Inst. Wood Sci.* 4:37–53.

———. 1974. Failure in timber. Part 2. The angle of shear through the cell wall during longitudinal compression stressing. *Wood Sci. Technol.* 8:56–67.

Easterling, K. E., R. Harrysson, L. J. Gibson, and M. F. Ashby. 1982. On the mechanics of balsa and other woods. *Proc. Roy. Soc. London*, ser. A, 383:31–41.

Ellison, A. M. 1989. Morphological determinants of self-thinning in plant monocultures and a proposal concerning the role of self-thinning in plant evolution. *Oikos* 54:287–93.

Ellison, A. M., and K. J. Niklas. 1988. Branching patterns of *Salicornia europaea* (Chenopodiaceae) at different successional stages: A comparison of theoretical and real plants. *Amer. J. Bot.* 75:501–12.

Esau, K. 1936. Ontogeny and structure of collenchyma and of vascular tissues in celery petioles. *Hilgardia* 10:431–76.

———. 1977. *Anatomy of seed plants*. 2d ed. New York: John Wiley.

Evans, M. L. 1991. Gravitropism: Interaction of sensitivity modulation and effector redistribution. *Plant Physiol.* 95:1–5.

Faegri, K., and L. van der Pijl. 1979. *The principles of pollination ecology*. 3d ed. Oxford: Pergamon.

Falk, S., H. Hertz, and H. Virgin. 1958. On the relation between turgor pressure and tissue rigidity. 1. Experiments on resonance frequency and tissue rigidity. *Physiol. Plant.* 11:802–17.

Finney, E. E., Jr., and C. W. Hall. 1967. Elastic properties of potatoes. *Amer. Soc. Agr. Eng. Trans.* 10:4–8.

Flaskämper, P. 1910. Untersuchungen über die Abhängigkeit der Gefässund Sklerenchymbildung von äusseren Faktoren nebst einigen Bemerkungen über die angebliche Heterorhizie bei Dikotylen. *Flora* 101:181–219.

Ford, C. W., and J. R. Wilson. 1981. Changes in levels of solutes during osmotic adjustment to water stress in four tropical pasture species. *Austral. J. Plant Physiol.* 8:77–91.

Frampton, V. L., M. B. Linn, and E. D. Hansing. 1942. The spread of virus diseases of the yellow type under field conditions. *Phytopathology* 32:799–808.

Fraser, A. I. 1962. The roots and soil as factors in tree stability. *Forestry* 35:117–27.

Fraser, A. I., and J. B. H. Gardiner. 1967. Rooting and stability in Sitka spruce. *For. Comm. Bull.* (U.K.) 40:1–28.

Frey-Wyssling, A. 1954. The fine structure of cellulose microfibrils. *Science* 119:80–82.

Galtier, J., and N. P. Rowe. 1989. A primitive seed-like structure and its implications for early gymnosperm evolution. *Nature* 340:225–27.

Gandar, P. W. 1983. Growth in root apices. 1. The kinematic description of growth. 2. Deformation and the rate of deformation. *Bot. Gaz.* 144:1–19.

Ganong, W. F. 1901. The cardinal principles of morphology. *Bot. Gaz.* 31:426–34.

Gardner, W. R. 1960. Dynamic aspects of water availability to plants. *Soil Sci.* 89:63–73.

———. 1968. Availability and measurement of soil water. In *Water deficits and plant growth,* ed. T. T. Koslowski, 1:107–35. New York: Academic Press.

Gates, D. M. 1965. Radiant energy, its receipt and disposal. *Meteorol. Monogr.* 6:1–26.

Gerard, V. A., and K. H. Mann. 1979. Growth and production of *Laminaria longicruris* (Phaeophyta) populations exposed to different intensities of water movement. *Marine Biol.* 15:33–41.

Gere, J. M., and W. O. Carter. 1963. Critical buckling loads for tapered columns. *Trans. Amer. Soc. Civil Eng.* 128:736–54.

Gertel, E. T., and P. B. Green. 1977. Cell growth pattern and wall microfibrillar arrangement: Experiments with *Nitella. Plant Physiol.* 60:247–54.

Gibson, A. C., H. W. Calkin, and P. S. Nobel. 1985. Hydraulic conductance and xylem structure in tracheid-bearing plants. *IAWA Bull.* n.s., 6:293–302.

Gibson, L. J., and M. F. Ashby. 1982. The mechanics of three-dimensional cellular solids. *Proc. Roy. Soc. London,* ser. A., 382:43–59.

Gibson, L. J., M. F. Ashby, and K. E. Easterling. 1988. Structure and mechanics of the iris leaf. *J. Mater. Sci.* 23:3041–48.

Gifford, E. M., and A. S. Foster. 1989. *Morphology and evolution of vascular plants.* 3d ed. New York: W. H. Freeman.

Gillespie, W. H., G. W. Rothwell, and S. E. Scheckler. 1981. The earliest seeds. *Nature* 293:462–64.

Gillis, P. P. 1973. Theory of growth stresses. *Holzforschung* 26:197–207.

Gillis, P. P., and C. H. Hsu. 1979. An elastic, plastic theory of longitudinal growth stresses. *Wood Sci. Tech.* 13:97–115.

Goebel, K. 1887. *Outlines of classification and special morphology of plants.* (A new edition of Sachs's *Textbook of botany,* book 2.) Oxford: Clarendon Press.

Golenberg, E. M., D. E. Giannasi, M. T. Clegg, C. J. Smiley, M. Durbin, D. Henderson, and G. Zurawaski. 1990. Chloroplast DNA sequence from a Miocene *Magnolia* species. *Nature* 344:656–58.

Gould, S. J. 1990. Foreword. In *Causes of evolution: A paleontological perspec-*

tive, ed. R. M. Ross and W. D. Allmon, vii–xi. Chicago: University of Chicago Press.

Gould, S. J., and E. S. Vrba. 1982. Exaptation: A missing term in the science of form. *Paleobiology* 8:4–15.

Grace, J., and G. Russell. 1977. The effect of wind on grasses. 3. Influence of continuous drought or wind on anatomy and water relations in *Festuca arundinacea. J. Exp. Bot.* 28:268–78.

Green, D. S. 1980. The terminal velocity and dispersal of spinning samaras. *Amer. J. Bot.* 67:1218–24.

Green, P. B. 1960. Multinet growth in the cell wall of *Nitella. J. Biophys. Biochem. Cytol.* 7:289–97.

———. 1980. Organogenesis: A biophysical view. *Ann. Rev. Plant Physiol.* 31:51–82.

Greenberg, A. R., A. Mehling, M. Lee, and J. H. Block. 1989. Tensile behavior of grass. *J. Mater. Sci.* 24:2549–54.

Greenhill, G. 1881. Determination of the greatest height consistent with stability that a vertical pole or mast can be made, and the greatest height to which a tree of given proportions can grow. *Proc. Cambridge Phil. Soc.* 4:65–73.

Greenridge, K. N. H. 1958. Rates and patterns of moisture movement in trees. In *The physiology of forest trees,* ed. K. V. Thimann, 19–41. New York: Ronald.

Gregory, P. H. 1968. Interpreting plant disease dispersal gradients. *Ann. Rev. Phytopathol.* 6:189–212.

Gunnison, D., and M. Alexander. 1975. Basis for the resistance of several algae to microbial decomposition. *Appl. Microbiol.* 29:729–38.

Haberlandt, G. 1909. *Physiologische Pflanzenanatomie.* Vol. 4. Leipzig: W. Engelmann.

Hahn, G. T., and A. R. Rosenfield. 1968. Source of fracture toughness: The relation between K_Ie and the ordinary tensile properties of metals. *ASTM STP* (American Society for Testing and Materials, Philadelphia) 432:5–32.

Haider, K. 1954. Zur Morphologie und Physiologie der Sporangien leptosporangiater Farne. *Planta* 44:370–411.

Hammel, H. T., and P. F. Scholander. 1976. *Osmosis and tensile solvent.* Berlin: Springer-Verlag.

Harper, J. L. 1977. *Population biology of plants.* Orlando, Fla: Academic Press.

Hearle, J. W. S. 1958. The mechanics of twisted yarns: The influence of transverse forces on tensile behavior. *J. Text. Inst.* 49:389–408.

———. 1963. The fine structure of fibers and crystalline polymers. 3. Interpretation of the mechanical properties of fibers. *J. Appl. Polymer Sci.* 7:1207–23.

Hébant, C. 1977. *The conducting tissues of bryophytes.* Vaduz, Ger.: J. Cramer (in der A. R. Gantner Verlag Kommanditgesellschaft).

Hegler, R. 1893. Über den Einfluss des mechanischen Zugs auf das Wachstum der Pflanze. *Beitr. Biol. Pflanz.* 6:383–432.

Henwood, K. 1973. A structural model of forces in buttressed tropical rain forest trees. *Biotropica* 5:83–93.

Hertzberg, R. W. 1983. *Deformation and fracture mechanics of engineering materials.* 2d ed. New York: John Wiley.

Hettiaratchi, D. R. P., and J. R. O'Callaghan. 1978. Structural mechanics of plant cells. *J. Theor. Biol.* 74:237–57.

Hibbeler, R. C. 1983. *Engineering mechanics: Statics.* New York: Macmillan.

Hill, R. 1950. *The mathematical theory of plasticity.* London: Oxford University Press.

Hoerner, S. F. 1965. *Fluid-dynamic drag.* Brick Town, N.J.: S. F. Hoerner.

Hofmeister, W. 1859. Über die Beugungen saftreicher Pflanzenteile nach Erschutterung. *Ber. Verh. Königl.-Sächs. Ges. Wiss. Leipzig* 12:175–204.

Holbrook, N. M., M. W. Denny, and M. A. R. Koehl. 1991. Intertidal "trees": Consequences of aggregation on the mechanical and photosynthetic properties of sea-palms *Postelsia palmaeformis* Ruprecht. *J. Exp. Mar. Biol. Ecol.* 146: 39–67.

Holbrook, N. M., and F. E. Putz. 1989. Influence of neighbors on tree form: Effects of lateral shading and prevention of sway on the allometry of *Liquidambar styraciflua* (sweet gum). *Amer. J. Bot.* 76:1740–49.

Honda, H., and J. B. Fisher. 1978. Tree branch angle: Maximizing effective leaf area. *Science* 199:888–90.

Horn, H. S. 1971. *The adaptive geometry of trees.* Princeton: Princeton University Press.

Huber, B. 1928. Weitere quantitative Untersuchungen über das Wasserleitungssystem der Pflanzen. *Jahrb. Wiss. Bot.* 67:877–959.

———. 1956. Die Gefässleitung. In *Encyclopedia of plant physiology,* ed. W. Ruhland, 3:541–82. Berlin: Springer-Verlag.

Hutte, P. 1968. Experiments on windflow and wind damage in Germany: Site and susceptibility of spruce forests to storm damage. *Forestry,* suppl., 41:20–27.

Huxley, J. S. 1932. *Problems of relative growth.* New York: MacVeagh.

Ingold, C. T. 1953. *Dispersal in fungi.* Oxford: Clarendon Press.

———. 1965. *Spore liberation.* Oxford: Clarendon Press.

Iraki, N. M., R. A. Bressan, P. M. Hasegawa, and N. C. Carpita. 1989. Alteration of the physical and chemical structure of the primary cell wall of growth-limited plant cells adapted to osmotic stress. *Plant Physiol.* 91:39–47.

Jaccard, M., and P. E. Pilet. 1975. Extensibility and rheology of collenchyma. 1. Creep relaxation and viscoelasticity of young and senescent cells. *Plant Cell Physiol.* 16: 1113–20.

Jacobs, M. R. 1938. The fibre tension of woody stems, with special reference to the genus *Eucalyptus. Commonw. For. Bur. Austral. Bull.* 22:1–39.

———. 1945. The growth stresses of woody stems. *Commonw. For. Bur. Austral. Bull.* 28:1–67.

———. 1954. The effect of wind sway on the form and development of *Pinus radiata* Don. *Austral. J. Bot.* 2:35–51.

Jaffe, M. J. 1973. Thigmomorphogenesis: The response of plant growth and development to mechanical stress. *Planta* 114:143–57.

Jaffe, M. J., F. W. Telewski, and P. W. Cooke. Thigmomorphogenesis: On the mechanical properties of mechanically perturbed bean plants. *Physiol. Plant.* 62:73–78.

Jankiewicz, L. S. 1966. Changes in angle width between the main axis and a branch in young pines (*Pinus sylvestris* L.). *Acta Agrobot.* 19:129–42.

Jankiewicz, L. S., and Z. J. Stecki. 1976. Some mechanisms responsible for differences in tree form. In *Tree physiology and yield improvement,* ed. M. G. R. Cannell and F. T. Last, 157–72. New York: Academic Press.

Jennings, J. S., and N. H. MacMillan. 1986. A tough nut to crack. *J. Mater. Sci.* 21:1517–34.

Jensen, R. D., S. A. Taylor, and H. H. Wiebe. 1961. Negative transport and resistance to water flow through plants. *Plant Physiol.* 36:633–38.

Kaplan, D. R. 1987a. The significance of multicellularity in higher plants. 1. The relationship of cellularity to organismal form. *Amer. J. Bot.* 74:617.

———. 1987b. The significance of multicellularity in higher plants. 2. Functional significance. *Amer. J. Bot.* 74:618.

Kaufmann, M. R., and C. A. Troendle. 1981. The evaluation of leaf area and foliage biomass to sapwood conducting area in four subalpine forest tree species. *For. Sci.* 27:477–82.

Kelvin, Lord. 1894. On the homogeneous division of space. *Proc. Roy Soc. London,* ser. B, 55:1–16.

Kennedy, R. W. 1970. An outlook for basic wood anatomy research. *Wood Fiber* 2:182–87.

King, D. A. 1981. Tree dimensions: Maximizing the rate of height growth in dense stands. *Oecologia* 51:351–56.

———. 1987. Load bearing capacity of understory treelets of a tropical wet forest. *Bull. Torrey Bot. Club* 114:419–28.

King, D. A., and O. L. Loucks. 1978. The theory of tree bole and branch form. *Radiat. Env. Biophys.* 15:141–65.

Kirk, J. T. O. 1983. *Light and photosynthesis in aquatic ecosystems.* Cambridge: Cambridge University Press.

Kisser, J. von, and A. Steininger. 1952. Makroskopische und mikroskopische Strukturänderungen bei der Biegebeanspruchung von Holz. *Holz Roh- Werkst.* 11:415–21.

Klauditz, W., A. Marschall, and W. Ginsel. 1947. Zur Technologie verholzter pflanzlicher Zellwande. *Holzforschung* 1:98–103.

Knight, T. A. 1811. On the causes which influence the direction of the growth of roots. *Phil. Trans. Roy. Soc. London* 1811:209–19.

Koehl, M. A. R. 1979. Stiffness or extensibility of intertidal algae: A comparative study of modes of withstanding wave action. *J. Biomech.* 12:634.

———. 1986. Seaweeds in moving water: Form and mechanical function. In *On the economy of plant form and function,* ed. T. J. Givnish, 603–32. Cambridge: Cambridge University Press.

Koehl, M. A. R., and S. A. Wainwright. 1977. Mechanical adaptations of a giant kelp. *Limnol. Oceanogr.* 22:1067–71.

Kokubo, A., S. Kuraishi, and N. Sakurai. 1989. Culm strength of barley. *Plant Physiol.* 91:876–82.

Kramer, P. J. 1983. *Water relations of plants.* Orlando, Fla.: Academic Press.

Kraus, G. 1867. Die Gewebespannung des Stammes und ihre Folgen. *Bot. Zeit.* 25:105–42.

Kraynik, A. M., and M. G. Hansen. 1986. Foam and emulsion rheology: A quasi-static model for large deformations of spatially-periodic cells. *J. Rheol.* 30:409–39.

Kübler, H. 1959. Studies on growth stresses in trees. 2. Longitudinal stresses. *Holz Roh- Werkst.* 17:44–54.

Kuntz, J. E., and A. J. Riker. 1955. The use of radioactive isotopes to ascertain the role of root grafting in the translocation of water, nutrients, and disease-inducing organisms. *Proc. Int. Conf. Peaceful Uses of Atomic Energy* 12:144–48.

Kutschera, U. 1989. Tissue stresses in growing plant organs. *Physiol. Plant.* 77:157–63.

Kutschera, U., R. Bergfeld, and P. Schofer. 1987. Cooperation of epidermis and inner tissues in auxin-mediated growth of maize coleoptiles. *Planta* 170:168–80.

Kutschera, U., and H. Kende. 1988. The biophysical basis of elongation growth in internodes of deep water rice. *Plant Physiol.* 88:361–66.

Lakes, R. 1987. Foam structures with negative Poisson's ratio. *Science* 235: 1038–40.

Lang, J. M., W. R. Eisinger, and P. Green. 1982. Effects of ethylene on the orientation of microtubules and cellulose microfibrils of pea epicotyls with polylamellate walls. *Protoplasma* 110:5–14.

Larson, P. R. 1965. Stem form of young *Larix* as influenced by wind and pruning. *For. Sci.* 11:412–24.

Lawton, R. O. 1982. Wind stress and elfin stature in a montane rain forest tree: An adaptive explanation. *Amer. J. Bot.* 69:1224–30.

Lee, D. R. 1981. Elasticity of phloem tissues. *J. Exp. Bot.* 32:251–60.

Leiser, T. A., and J. D. Kemper. 1973. Analysis of stress distribution in sapling tree trunks. *J. Amer. Soc. Hort. Sci.* 98:164–70.

Levitt, J. 1986. Recovery of turgor by wilted, excised cabbage leaves in the absence of water uptake. *Plant Physiol.* 82:147–53.

Lewis, F. T. 1923. The typical shape of polyhedral cells in vegetable parenchyma and the restoration of that shape following cell division. *Proc. Amer. Acad. Arts Sci.* 58:537–52.

Lin, T.-T., and R. E. Pitt. 1986. Rheology of apple and potato tissue as affected by cell turgor pressure. *J. Text. Stud.* 17:291–313.

Little, C. H. A. 1967. Some aspects of apical dominance in *Pinus strobus*. Ph.D. diss., Yale University.

Littler, M. M., D. S. Littler, and P. R. Taylor. 1983. Evolutionary strategies in a tropical barrier reef system: Functional form groups of marine macroalgae. *J. Phycol.* 19:229–37.

Lockhart, J. A. 1959. A new method for the determination of osmotic pressure. *Amer. J. Bot.* 46:704–8.

Long, A. G. 1966. Some Lower Carboniferous fructifications from Berwickshire, together with a theoretical account of the evolution of ovules, cupules, and carpels. *Trans. Roy. Soc. Edinburgh* 66:345–75.

———. 1975. Further observations on some Lower Carboniferous seeds and cupules. *Trans. Roy. Soc. Edinburgh* 69:267–93.

Lu, R.-F., J. A. Bartsch, and A. Ruina. 1987. *Structural stability of the corn stalk.* ASAE Paper 87–6066. Saint Joseph, Mich.: American Society of Agricultural Engineers.

Lustig, I., and Z. Bernstein. 1985. Determination of the mechanical properties of the grape berry skin by hydraulic measurements. *Sci. Hort.* 25:270–85.

Macagno, E. 1989. *Leonardian fluid mechanics in the manuscript I.* Iowa Institute of Hydraulic Research, no. 11. Iowa City: IIHR.

McCutchen, C. W. 1977. The spinning rotation of ash and tulip tree samaras. *Science* 197:691–92.

McMahon, T. A. 1973. The mechanical design of trees. *Science* 233:92–102.

McMahon, T. A., and R. E. Kronauer. 1976. Trees structures: Deducing the principle of mechanical design. *J. Theor. Biol.* 59:443–66.

Mann, K. H. 1982. *Ecology of coastal waters: A systems approach.* Studies in Ecology, vol. 8. Berkeley: University of California Press.

Margulis, L. 1992. *Symbiosis in cell evolution.* 2d ed. New York: W. H. Freeman.

Mark, R. E. 1967. *Cell wall mechanics of tracheids.* New Haven: Yale University Press.

Marshall, T. J., and J. W. Holmes. 1988. *Soil physics.* 2d ed. Cambridge: Cambridge University Press.

Mason, B. J. 1978. Physics of a raindrop. *Phys. Ed.* 13:414–19.

Masuda, T., and R. Yamamoto. 1972. Control of auxin-induced stem elongation by the epidermis. *Physiol. Plant.* 27:109–115.

Mattheck, C. 1990. Deśign and growth rules for biological structures and their application to engineering. *Fatigue Fract. Eng. Mater. Struct.* 13:535–50.

Matzke, E. B. 1950. In the twinkling of an eye. *Bull. Torrey Bot. Club* 77:222–27.

Maximov, N. A. 1912. Chemische Schutzmittel der Pflanzen gegan Erfrieren. *Ber. Deutsch. Bot. Ges.* 30:52–65, 293–305, 504–16.

Melkonian, J. J., J. Wolfe, and P. L. Steponkus. 1982. Determination of the volumetric modulus of elasticity in wheat leaves by pressure-volume relations and the effect of drought conditioning. *Crop Sci.* 27:116–23.

Mergen, F. 1954. Mechanical aspects of wind-breakage and wind-firmness. *J. For.* 52:119–25.

Metraux, J.-P., P. A. Richmond, and L. Taiz. 1980. Control of cell elongation in *Nitella* by endogenous cell wall pH. *Plant Physiol.* 65:204–10.

Metzger, C. 1893. Der Wind als massgebender Faktor für das Wachsthum der Bäume. *Mundener Forst.* 3:35–86.

Metzger, J., and G. L. Steucek. 1974. Response of barley (*Hordeum vulgare*) seedlings to mechanical stress. *Proc. Pa. Acad. Sci.* 48:114–16.

Milburn, J. A., and R. P. C. Johnson. 1966. The conduction of sap. 2. Detection of vibrations produced by sap cavitation in *Ricinus* xylem. *Planta* 69:43–52.

Millet, B., D. Melin, and P. Badot. 1986. Circumnutation: A model for signal transduction from cell to cell. In *The cell surface in signal transduction,* ed. E. Wagner, H. Greppin, and B. Millet, 137–59. Berlin: Springer-Verlag.

Molè-Bajer, J. 1953. Influence of hydration and dehydration on mitosis. 2. *Acta Soc. Bot. Polon.* 22:33–44.

Monteith, J. L. 1973. *Principles of environmental physics.* New York: American Elsevier.

Münch, E. 1938. Statik und Dynamik des schraubigen Baus der Zellwand, besonders des Druck- und Zugholzes. *Flora* 32:357–424.

Mura, T. 1982. *Micromechanics of defects in solids.* The Hague: Martinus-Nijhoff.

Murray, C. D. 1927. A relationship between circumference and weight in trees and its bearing on branching angles. *J. Gen. Physiol.* 10:725–39.

Nakajima, N., and E. R. Harrell. 1986. Stress relaxation as a method of analyzing stress growth, stress overshoot and steady-state flow of elastomers. *J. Rheol.* 30:383–408.

Nemat-Nasser, S., and T. Iwakuma. 1983. Micromechanically based constitutive relations for polycrystalline solids. In NASA Conference Publication CP 2271, 113–36. Washington, D.C.: NASA.

Nemat-Nasser, S., T. Iwakuma, and M. Hejazi. 1982. On composites with periodic structures. *Mech. Mater.* 1:239–67.

Nemat-Nasser, S., and M. Taya. 1981. On effective moduli of an elastic body containing periodically distributed voids. *Quart. Appl. Math.* 39:43–59.

Neville, A. C. 1985. Molecular and mechanical aspects of helicoid development in plant cell walls. *BioEssays* 3:4–8.

———. 1986. The physics of helicoids: Multidirectional "plywood" structures in biological systems. *Phys. Bull.* 37:74–76.

Newcombe, F. C. 1895. The regulatory formation of mechanical tissue. *Bot. Gaz.* 20:441–48.

Nicholson, J. E., W. E. Hillis, and N. Ditchburne. 1975. Some tree growthwood property relationships of *Eucalyptus*. *Can. J. For. Res.* 5:424–32.

Niklas, K. J. 1976. Chemical examination of some non-vascular Paleozoic plants. *Brittonia* 28:113–37.

———. 1983. The influence of Paleozoic ovule and cupule morphologies on wind pollination. *Evolution* 37:968–86.

———. 1984. Size-related changes in the primary xylem anatomy of some early tracheophytes. *Paleobiology* 10:487–506.

———. 1985a. The aerodynamics of wind pollination. *Bot. Rev.* 51:328–86.

———. 1985b. Wind pollination—A study in controlled chaos. *Amer. Sci.* 73:90–95.

———. 1985c. The evolution of tracheid diameter in early vascular plants and its implications on the hydraulic conductance of the primary xylem strand. *Evolution* 39:1110–22.

———. 1986a. Computer-simulated plant evolution. *Sci. Amer.* 254:78–86.

———. 1986b. Computer simulations of branching patterns and their implications on the evolution of plants. *Lect. Math. Life Sci.* 18:1–50.

———. 1987a. Pollen capture and wind-induced movement of compact and diffuse grass panicles: Implications for pollination efficiency. *Amer. J. Bot.* 74: 74–89.

———. 1987b. Aerodynamics of wind pollination. *Sci. Amer.* 255:90–95.

———. 1988a. Dependency of the tensile modulus on transverse dimensions, water potential, and cell number of pith parenchyma. *Amer. J. Bot.* 75:1286–92.

———. 1988b. Equations for the motion of airborne pollen grains near ovulate organs of wind-pollinated plants. *Amer. J. Bot.* 75:433–44.

———. 1988c. The role of phyllotactic pattern as a "developmental constraint" on the interception of light by leaf surfaces. *Evolution* 42:1–16.

———. 1989a. Nodal septa and the rigidity of aerial shoots of *Equisetum hyemale*. *Amer. J. Bot.* 76:521–31.

———. 1989b. Extracellular freezing in *Equisetum hyemale*. *Amer. J. Bot.* 76:627–31.

———. 1989c. Mechanical behavior of plant tissues as inferred from the theory of pressurized cellular solids. *Amer. J. Bot.* 76:929–37.

———. 1990a. Determinate growth of *Allium sativum* peduncles: Evidence for determinate growth as a design factor for biomechanical safety. *Amer. J. Bot.* 7:762–71.

———. 1990b. The mechanical significance of clasping leaf sheaths in grasses: Evidence from two cultivars of *Avena sativa*. *Ann. Bot.* 65:505–12.

———. 1990c. Biomechanics of *Psilotum nudum* and some early Paleozoic vascular sporophytes. *Amer. J. Bot.* 77:590–606.

Niklas, K. J., and S. L. Buchmann. 1985. Aerodynamics of wind pollination in *Simmondsia chinensis* (Link) Schneider. *Amer. J. Bot.* 72:530–39.

———. 1987. The aerodynamics of pollen capture in two sympatric *Ephedra* species. *Evolution* 41:104–23.

Niklas, K. J., and V. Kerchner. 1984. Mechanical and photosynthetic constraints on the evolution of plant shape. *Paleobiology* 10:79–101.

Niklas, K. J., and F. C. Moon. 1988. Flexural stiffness and modulus of elasticity of flower stalks from flower stalks of *Allium sativum* as measured by multiple resonance frequency spectra. *Amer. J. Bot.* 75:1517–25.

Niklas, K. J., and K. Norstog. 1984. Aerodynamics and pollen grain depositional patterns on cycad megastrobili: Implications on the reproduction of three cycad genera (*Cycas, Dioon, Zamia*). *Bot. Gaz.* 145:92–104.

Niklas, K. J., and T. D. O'Rourke. 1982. Growth patterns of plants that maximize vertical growth and minimize internal stresses. *Amer. J. Bot.* 69:1367–74.

———. 1987. Flexural rigidity of chive and its response to water potential. *Amer. J. Bot.* 74:1033–44.

Niklas, K. J., and D. J. Paolillo, Jr. 1990. Biomechanical and morphometric differences in *Triticum aestivum* seedlings differing in *Rht* gene-dosage. *Ann. Bot.* 65:365–77.

Nilsson, S. B., S. H. Hertz, and S. Falk. 1958. On the relation between turgor pressure and tissue rigidity. 2. Theoretical calculations on model systems. *Physiol. Plant.* 11:818–37.

Nobel, P. S. 1974. Boundary layers of air adjacent to cylinders: Estimation of effective thickness and measurement on plant material. *Plant Physiol.* 54:177–81.

———. 1975. Effective thickness and resistance of the air boundary layer adjacent to spherical plant parts. *J. Exp. Bot.* 26:120–30.

———. 1983. *Biophysical plant physiology and ecology.* New York: W. H. Freeman.

———. 1988. *Environmental biology of agaves and cacti.* New York: Cambridge University Press.

Nobel, P. S., H. W. Calkin, and A. C. Gibson. 1985. Influences of PAR, temperature, and water vapor concentration on gas exchange by ferns. *Physiol. Plant.* 62:527–34.

Norberg, R. A. 1973. Autorotation, self-stability, and structure of single-winged fruits and seeds (samaras) with comparative remarks on animal flight. *Biol. Rev.* 48:561–96.

———. 1988. Theory of growth geometry of plants and self-thinning of plant populations: Geometric similarity, elastic similarity and different growth modes of plants. *Amer. Nat.* 131:220–56.

Okubo, A., and S. A. Levin. 1989. A theoretical framework for data analysis of wind dispersal of seeds and pollen. *Ecology* 70:329–38.

Owens, T. G. 1988. Light-harvesting antenna systems in chlorophyll *a*/*c*-containing algae. In *Light-energy transduction in photosynthesis: Higher plant and bacterial models,* ed. S. E. Stevens, Jr., and D. A. Bryant, 122–36. Rockville, Md.: American Society of Plant Physiologists.

Paleg, L. G., and D. Aspinall, eds. 1981. *The physiology and biochemistry of drought resistance in plants.* New York: Academic Press.

Paolillo, D. J., Jr., and G. Rubin. 1991. Relative elemental rates of elongation and protoxylem-metaxylem transition in hypocotyls of soybean seedings. *Amer. J. Bot.* 78:845–54.

Parihar, N. S. 1962. *An introduction to Embryophyta.* 4th ed. Vol. 1. Allahabad, India: Halcyon Press (Central Book Depot).

Passioura, J. B. 1980. The meaning of matric potential. *J. Exp. Bot.* 31:1161–69.

Pearman, G. I., H. L. Weaver, and C. B. Tanner. 1972. Boundary layer heat transfer coefficients under field conditions. *Agr. Meteorol.* 10:83–92.

Peters, R. H. 1983. *The ecological implications of body size.* Cambridge: Cambridge University Press.

Pettitt, J. 1970. Heterospory and the origin of the seed habit. *Biol. Rev.* 45:401–15.

Plant, R. E. 1983. Analysis of a continuum model for root growth. *J. Math. Biol.* 98:261–68.

Plateau, J. 1873. *Statique expérimentale et théorique des liquides soumis aux seules forces moléculaires.* Paris.

Pohl, F. 1937. Die Pollenerzeugung der Windblutler. *Beih. Bot. Zentralbl.* 56:365–470.

Prandtl, L. 1952. *Fluid dynamics.* London: Blackie Press.

Preston, R. D. 1974. *The physical biology of plant cell walls.* London: Chapman and Hall.

Prince, R. P., and D. W. Bradway. 1969. Shear stress and modulus of selected forages. *Amer. Soc. Agr. Eng. Trans.* 12:426–28.

Putz, F. E., P. D. Coley, K. Lu, A. Montalvo, and A. Aiello. 1983. Uprooting and snapping of trees: Structural determinants and ecological consequences. *Can. J. Bot.* 13:1011–20.

Putz, F. E., and N. M. Holbrook. 1991. Biomechanical studies of vines. In *The biol-*

ogy of vines, ed. F. E. Putz and H. A. Mooney, 65–89. Cambridge: Cambridge University Press.

Rand, R. H. 1978. The dynamics of an evaporating meniscus. *Acta Mech.* 29:135–46.

Rand, R. H., D. W. Storti, S. K. Upadhyaya, and J. R. Cooke. 1982. Dynamics of coupled stomatal oscillators. *J. Math. Biol.* 15:131–39.

Rashevsky, N. 1973. The principle of adequate design. In *Foundations of mathematical biology,* ed. R. Rosen, 3:143–76. New York: Academic Press.

Raupach, M. R., and A. S. Thom. 1981. Turbulence in and above plant canopies. *Ann. Rev. Fluid Mech.* 13:97–129.

Raven, J. A. 1977. The evolution of vascular land plants in relation to supracellular transport processes. *Adv. Bot. Res.* 5:153–219.

———. 1984. Physiological correlates of the morphology of early vascular plants. *Bot. J. Linnean Soc.* 88:105–26.

———. 1985. Comparative physiology of plant and arthropod land adaptation. *Phil. Trans. Roy. Soc. London,* ser. B., 309:273–88.

Ray, P. M. 1967. Radioautographic study of cell wall deposition in growing plant cells. *J. Cell Biol.* 35:659–74.

Rich, P. M. 1986. Mechanical architecture of arborescent rain forest palms. *Principes* 30:117–31.

———. 1987. Mechanical structure of the stem of arborescent palms. *Bot. Gaz.* 148:42–50.

Rich, P. M., K. Helenurm, D. Kearns, S. R. Morse, M. W. Palmer, and L. Short. 1986. Height and stem diameter relationships for dicotyledonous trees and arborescent palms of Costa Rican tropical wet forest. *Bull. Torrey Bot. Club* 113:241–46.

Richards, P. W. 1952. *The tropical rain forest: An ecological study.* Cambridge: Cambridge University Press.

Richmond, P. A. 1983. Patterns of cellulose microfibril deposition and rearrangement in *Nitella:* In vivo analysis by a birefringence index. *J. Appl. Polymer Sci.* 37:107–22.

Richmond, P. A., J. P. Metraux, and L. Taiz. 1980. Cell expansion patterns and directionality of wall mechanical properties in *Nitella. Plant Physiol.* 65:211–17.

Ritman, K. T., and J. A. Milburn. 1988. Acoustic emissions from plants: Ultrasonic and audible compared. *J. Exp. Bot.* 39:1237–48.

Roark, R., and W. C. Young. 1975. *Formulas for stress and strain.* New York: McGraw-Hill.

Robinson, W. C. 1920. The microscopial features of mechanical strains in timber and the bearing of these on the structure of the cell wall in plants. *Phil. Trans. Roy. Soc. London,* ser. B, 210:49–82.

Roelofsen, P. A. 1951. Orientation of cellulose fibrils in the cell wall of growing cotton hairs and its bearing on the physiology of cell wall growth. *Biochem. Biophys. Acta* 7:43–53.

Russell, M. B., and J. T. Woolley. 1961. Transport processes in the soil-plant system. In *Growth in living systems,* ed. M. X. Zarrow, 695–721. New York: Basic Books.

Sachs, J. von. 1865. *Handbuch der Experimentalphysiologie der Pflanzen*. Leipzig: W. Engelmann.

———. 1875. *Textbook of botany.* Oxford: Clarendon Press.

Schimper, C. F. 1836. Geometrische Anordung der um eine Achse peripherische Blattgebilde. *Verh. Schweiz. Ges.* 1836:113–17.

Schneider, E. 1926. Über die Gewebespannung der Vegetationspunkte. *Ber. Deutsch. Bot. Ges.* 44:326–28.

Scholander, P. F., W. E. Love, and J. W. Kanwisher. 1955. The rise of sap in tall grapevines. *Plant Physiol.* 30:93–104.

Schoute, J. C. 1915. Sur la fissure médiane de la gaîne foliaire de quelques palmiers. *Ann. Jard. Bot. Buitenzorg*, ser. 2, 14:57–82.

Schüepp, O. 1917. Über den Nachweis von Gewebespannung in der Sprossspitze. *Ber. Deutsch. Bot. Ges.* 35:703–6.

Schwendener, S. 1874. *Das mechanische Prinzip im anatomischen Bau der Monocotylen.* Leipzig: W. Engelmann.

Seide, P., and V. I. Weingarten. 1961. On the bending of circular cylindrical shells under pure bending. *Amer. Soc. Agr. Eng. Trans.* 28:112–16.

Sellen, D. B. 1983. The response of mechanically anisotropic cylindrical cells to multiaxial stress. *J. Exp. Bot.* 34:681–87.

Shigo, A. L. 1990. Tree branch attachment to trunks and branch pruning. *HortScience* 25:54–59.

Shive, J. B., Jr., and K. W. Brown. 1978. Quaking and gas exchange in leaves of cottonwood (*Populus deltoides* Marsh.). *Plant Physiol.* 61:331–33.

Silk, W. K. 1980. Growth rate patterns which produce curvature and implications for the physiology of the blue light response. In *The blue light syndrome*, ed. H. Senger, 643–55. Heidelberg: Springer-Verlag.

———. 1984. Quantitative descriptions of development. *Ann. Rev. Plant Physiol.* 35:479–518.

———. 1989a. On the curving and twining of stems. *Env. Exp. Bot.* 29:95–109.

———. 1989b. Growth rate patterns which maintain a helical tissue tube. *J. Theor. Biol.* 138:311–27.

Silk, W. K., and R. O. Erickson. 1979. Kinematics of plant growth. *J. Theor. Biol.* 76:481–501.

Silk, W. K., and K. K. Wagner. 1980. Growth-sustaining water potential distributions in the primary corn root. *Plant Physiol.* 66:859–63.

Silk, W. K., L. L. Wang, and R. E. Cleland. 1982. Mechanical properties of the rice panicle. *Plant Physiol.* 70:460–64.

Sinnott, E. W. 1952. Reaction wood and the regulation of tree form. *Amer. J. Bot.* 39:69–78.

———. 1960. *Plant morphogenesis.* New York: McGraw-Hill.

Skalak, R., G. Dasgupta, M. Moss, B. Othen, P. Dullemeijer, and H. Vilann. 1982. Analytical description of growth. *J. Theor. Biol.* 94:555–77.

Smith, A. P. 1972. Buttressing of tropical trees: A descriptive model and new hypotheses. *Amer. Nat.* 106:32–46.

Spark, L. C., G. Darnborough, and R. D. Preston. 1958. Structure and mechanical properties of vegetable fibres. 2. A micro-extensometer for the automatic recording of load-extension curves for single fibrous cells. *J. Text. Inst.* 49: T309–16.

Speck, T. S., and D. Vogellehner. 1988a. Biophysikalische Untersuchungen zur Mechanostabilität verschiedener Stelentypen und zur Art des Festigungssytems früher Gefässlandpflanzen. *Palaeontographica,* ser. B, 210:91–126.

———. 1988b. Biophysical examinations of the bending stability of various stele types and the upright axes of early "vascular" land plants. *Bot. Acta* 101:262–68.

Spielman, L. A. 1977. Particle capture from low-speed laminar flows. *Ann. Rev. Fluid Mech.* 9:297–319.

Sprugel, D. G. 1989. The relationship of evergreenness, crown architecture, and leaf size. *Amer. Nat.* 133:465–79.

Steenbergh, W. F., and C. H. Lowe. 1977. *Ecology of the saguaro,* vol. 2. Scientific Monograph Series 8. Washington, D.C.: National Park Service.

Steponkus, P. L. 1984. Role of the plasma membrane in freezing injury and cold declimation. *Ann. Rev. Plant Physiol.* 35:543–84.

Sterling, C., and B. J. Spit. 1957. Microfibrillar arrangement in developing fibers of *Asparagus*. *Amer. J. Bot.* 44:851–58.

Stewart, W. N. 1964. An upward outlook in plant morphology. *Phytomorphology* 14:120–34.

———. 1983. *Paleobotany and the evolution of plants.* Cambridge: Cambridge University Press.

Strasburger, E. 1891. Über den Bau und die Verrichtungen der Leitungsbahnen in der Pflanzen. *Histol. Beitr.* 3:849–77.

Sutton, R. F. 1969. *Form and development of conifer root systems.* Technical Communication 7. Slough, Eng.: Commonwealth Forestry Bureau.

Taiz, L. 1984. Plant cell expansion: Regulation of cell wall mechanical properties. *Ann. Rev. Plant Physiol.* 35:585–657.

Takeda, K., and H. Shibaoka. 1981a. Changes in microfibril arrangement on the inner surface of the epidermal cell walls of the epicotyl of *Vigna angularis* Ohwi et Ohashi during cell growth. *Planta* 151:385–92.

———. 1981b. Effect of gibberellin and cholchicine on microfibril arrangement in epidermal cell walls of *Vigna angularis* Ohwi et Ohashi epicotyls. *Planta* 151:393–98.

Tandon, G. P., and G. J. Weng. 1986. Stress distribution in and around spheroidal inclusions and voids at finite concentration. *J. Appl. Mech.* 53:511–18.

Tanimoto, E., and Y. Masuda. 1971. Role of the epidermis in auxin-induced elongation of light-grown pea stem segments. *Plant Cell Physiol.* 12:663–73.

Tateno, M., and K. Bae. 1990. Comparison of lodging safety factor of untreated and succinic acid 2,2-dimethylhydrazide-treated shoots of mulberry tree. *Plant Physiol.* 92:12–16.

Taylor, H. M., and L. F. Ratliff. 1969. Root elongation rates of cotton and peanuts as a function of soil strength and soil water content. *Soil Sci.* 108:113–19.

Taylor, T. N. 1982. Reproductive biology in early seed plants. *BioScience* 32:23–28.

Taylor, T. N., and M. A. Millay. 1979. Pollination biology and reproduction in early seed plants. *Rev. Palaeobot. Palynol.* 27:329–55.

Thimann, K. V., and C. L. Schneider. 1938. The role of salts, hydrogen ion concentration and agar in the response of *Avena* coleoptiles to auxins. *Amer. J. Bot.* 25:270–80.

Thompson, D'Arcy W. 1942. *On growth and form.* 2d ed. Cambridge: Cambridge University Press.

Tilman, D. 1982. *Resource competition and community structure.* Princeton: Princeton University Press.

Timoshenko, S. 1976a. *Strength of materials.* Part 1. 3d ed. New York: R. E. Krieger.

———. 1976b. *Strength of materials.* Part 2. 3d ed. New York: R. E. Krieger.

Timoshenko, S., and J. H. Gere. 1961. *Theory of elastic stability.* New York: McGraw-Hill.

Timoshenko, S., and J. N. Goodier. 1970. *Theory of elasticity.* New York: McGraw-Hill.

Tomlinson, P. B. 1962. The leaf base in palms: Its morphology and mechanical biology. *J. Arnold Arboretum* 43:23–50.

———. 1983. Tree architecture. *Amer. Sci.* 71:141–49.

Tritton, D. J. 1977. *Physical fluid dynamics.* Berkshire, Eng.: Van Nostrand Reinhold.

Troll, W. 1937. *Vergleichende Morphologie der höheren Pflanzen.* Vol. 1. *Vegetationsorgane.* Part 3. *Lieferung.* Berlin: Gebrüder Bornträger.

Turgeon, R., and J. A. Webb. 1971. Growth inhibition by mechanical stress. *Science* 174:961–62.

Tyree, M. T., and A. J. Karamanos. 1980. Water stress as an ecological factor. In *Plants and their atmospheric environment,* ed. J. Grace, E. D. Ford, and P. G. Jarvis, 237–61. Oxford: Blackwell.

Upadhyaya, S. K., R. H. Rand, and J. R. Cooke. 1988. Role of stomatal oscillations on transpiration, assimilation, and water use efficiency. *Ecol. Modeling* 47:27–40.

Vautrin, A., and B. Harris. 1987. Acoustic emission characterization of flexural loading damage in wood. *Mater. Sci.* 22:3707–16.

Veen, B. W. 1970. Control of plant cell shape by cell wall structure. *Proc. Koninklijke Ned. Akad. Wet.,* ser. C, 73:118–21.

———. 1971. Cell wall structure and morphogenesis in growing stems of *Pisum sativum* D. Groningen: Rijksuniversiteit te Groningen.

Venkataswamy, M. A., C. K. S. Pillai, V. S. Prasad, and K. G. Satyanarayana. 1987. Effect of weathering on the mechanical properties of midribs of coconut palms. *J. Mater. Sci.* 22:3167–72.

Venning, F. D. 1949. Stimulation by wind motion of collenchyma formation in celery petioles. *Bot. Gaz.* 110:511–14.

Vincent, J. F. V. 1982. The mechanical design of grass. *J. Mater. Sci.* 17:856–60.

———. 1989. Relationship between density and stiffness of apple flesh. *J. Sci. Food Agr.* 47:443–62.

———. 1990a. *Structural biomaterials.* Princeton: Princeton University Press.

———. 1990b. Fracture in plants. *Adv. Bot. Res.* 17:235–82.

Virgin, H. 1955. A new method for determination of the turgor of plant tissues. *Physiol. Plant.* 8:954–63.

Vöchting, H. 1878. *Über Organbildung in Pflanzenreich.* Bonn: Max Cohen.

Vogel, S. 1981. *Life in moving fluids.* Boston: Willard Grant.

Vogel, S., and C. Loudon. 1985. Fluid mechanics of the thallus of an intertidal red alga, *Halosaccion glandiforma*. *Biol. Bull.* 168:161–74.

Wade, M. J., and S. Kalisz. 1990. The causes of natural selection. *Evolution* 44:1947–55.

Wainwright, S. A. 1970. Design in hydraulic organisms. *Naturwissenschaften* 57:321–26.

———. 1988. *Axis and circumference: The cylindrical shape of plants and animals.* Cambridge: Harvard University Press.

Wainwright, S. A., W. D. Biggs, J. D. Currey, and J. M. Gosline. 1976. *Mechanical design in organisms*. New York: John Wiley.

Walker, J. N., and E. H. Cox. 1966. Design of pier foundations for lateral loads. *Amer. Soc. Agr. Eng. Trans.* 9:417–20, 427.

Walker, J. N., and C. T. Haan. 1974. Limiting lateral load on short rigid pier foundations. *Amer. Soc. Agr. Eng. Trans.* 17:516–17.

Walker, K. P., E. H. Jordan, and A. D. Freed. 1989. *Nonlinear mesomechanics of composites with periodic microstructure: First report.* NASA TM-102051. Washington, D.C.: NASA.

Wardrop, A. B. 1964. Reaction anatomy of arborescent angiosperms. In *Formation of wood in forest trees,* ed. M. H. Zimmermann, 405–56. New York: Academic Press.

Ward-Smith, A. J. 1984. *Biophysical aerodynamics and the natural environment*. Chichester, Eng.: John Wiley.

Weibull, W. 1939. *A statistical theory of the strength of materials*. Stockholm: Royal Swedish Institute.

Weisz, P. R., H. C. Randell, and T. R. Sinclair. 1989. Water relations of turgor recovery and restiffening of wilted cabbage leaves in the absence of water uptake. *Plant Physiol.* 91:433–39.

Weng, G. J., and C. R. Chiang. 1984. Self-consistent relation in polycrystalline plasticity with a non-uniform matrix. *Int. J. Solids Struct.* 20:689–98.

Westing, A. H. 1965. Formation and function of compression wood in gymnosperms. *Bot. Rev.* 31:381–480.

Wheeler, W. N., and M. Neushul. 1981. The aquatic environment. In *Physiological plant ecology,* ed. O. L. Lamge, P. S. Nobel, C. B. Osmond, and H. Ziegler, 1:229–47. Encyclopedia of Plant Physiology, n.s., vol. 12A. New York: Springer-Verlag.

White, F. M. 1974. *Viscous fluid flow.* New York: McGraw-Hill.

White, J. 1979. The plant as a metapopulation. *Ann. Rev. Ecol. Syst.* 10:109–45.

Whittaker, R. H., and G. M. Woodwell. 1968. Dimension and production relations of trees and shrubs in the Brookhaven Forest, New York. *J. Ecol.* 56:1–25.

Wilson, B. F., and R. R. Archer. 1977. Reaction wood: Induction and mechanical action. *Ann. Rev. Plant Physiol.* 28:23–43.

Witman, J. D., and T. H. Suchanek. 1984. Mussels in flow: Drag and dislodgement by epizoans. *Marine Ecol. Prog. Ser.* 16:259–68.

Woeste, F. E., S. K. Suddarth, and W. L. Galligan. 1979. Simulation of correlated lumber properties data: A regression approach. *Wood Sci.* 12:73–89.

Wright, C. 1873. On the uses and origin of arrangements of leaves in plants. *Mem. Amer. Acad. Arts Sci.* 9:379–415.

Wright, S. 1932. The roles of mutation, inbreeding, crossbreeding and selection in evolution. *Proc. Sixth Int. Cong. Gen.* 1:356–66.

Zhang, H., J. M. Pleasants, and T. W. Jurik. 1991. Development of leaf orientation in the prairie compass plant, *Silphium laciniatum* L. *Bull. Torrey Bot. Club* 118:33–42.

Zimmermann, M. H. 1983. *Xylem structure and the ascent of sap.* Berlin: Springer-Verlag.

Zimmermann, M. H., and C. L. Brown. 1971. *Trees: Structure and function.* New York: Springer-Verlag.

Zimmermann, M. H., and D. Potter. 1982. Vessel-length distribution in branches, stems, and roots of *Acer rubrum. IAWA Bull,* n.s., 3:103–9.

Zimmermann, M. H., and P. B. Tomlinson. 1972. The vascular system of monocotyledonous stems. *Bot. Gaz.* 133:141–55.

Zimmermann, M. H., A. B. Wardrop, and P. B. Tomlinson. 1968. Tension wood in the aerial roots of *Ficus benjamina* L. *Wood Sci. Tech.* 2:95–104.

Zimmermann, W. 1930. *Die Phylogenie der Pflanzen.* Jena.

———. 1953. Main results of the "telome theory." *Palaeobotanist* 1:456–70.

———. 1965. *Die Telomtheorie.* Stuttgart: Gustav Fischer.

Author Index

Subject Index